DEVELOPMENTS IN SEDIMENTOLOGY 48

Geochemistry of Sedimentary Carbonates

DEVELOPMENTS IN SEDIMENTOLOGY 48

Geochemistry of Sedimentary Carbonates

John W. Morse
Department of Oceanography, Texas A&M University, College Station, TX 77843, U.S.A.

Fred T. Mackenzie
Department of Oceanography, University of Hawaii, Honolulu, HI 96822, U.S.A.

ELSEVIER
Amsterdam – Oxford – New York – Tokyo 1990

ELSEVIER SCIENCE PUBLISHERS B.V.
Sara Burgerhartstraat 25
P.O. Box 211, 1000 AE Amsterdam, The Netherlands

Distributors for the United States and Canada:

ELSEVIER SCIENCE PUBLISHING COMPANY INC.
655, Avenue of the Americas
New York, NY 10010, U.S.A.

ISBN 0-444-88781-4 (Paperback)
ISBN 0-444-87391-0 (Hardbound)

We dedicate this book to Sandy and Judy for their understanding, patience and support, to Angela, Michele and Scott in hope for a peaceful future, and to the memory of our friend and colleague, Bob Garrels.

PREFACE

This book has its roots in the year 1970, when we first met in the carbonates course in Bermuda, Fred as an instructor and John as a student. Over the intervening 20 years, we have been active in research on the geochemistry of sedimentary carbonates in both the field and laboratory and from a theoretical standpoint. During this period, an immense amount of information has been acquired in carbonate geochemistry resulting in a voluminous journal literature and many books on specialized topics. However, with the exception of the now classic book by Bathurst (1974) on carbonate sediments and their diagenesis, no work has been produced that presents an overview of sedimentary carbonate geochemistry. In attempting to teach this topic and through many conversations with colleagues active in research on sedimentary carbonates, we have perceived a need for a book that presents an overview of the field of sedimentary carbonate geochemistry and is reasonably comprehensible to the nonspecialist.

Carbonate minerals are common and important components of sediments and sedimentary rocks. They are of intermediate chemical reactivity among sedimentary minerals, less reactive than evaporite minerals, but more reactive than most silicate minerals. Their chemical characteristics exercise major influences on their accumulation and distribution in sediments, diagenesis, and preservation in sedimentary rocks. The chemistry of the atmosphere and oceans is controlled in part by reactions of these minerals with natural waters. These interactions are important in regulating climate and must be considered in such environmentally important issues as the fate of fossil fuel CO_2. The facts that at least 60% of the world's known petroleum reserves occur in carbonate reservoirs (e.g., Roehl and Choquette, 1985) and that carbonate cements exert a major influence on the porosity and permeability of petroleum reservoirs in other lithic types mean that an understanding of the geochemistry of sedimentary carbonates is of considerable economic importance. In addition, much of our insight into paleoclimatology and paleooceanography is based on geochemical interpretations of the composition of the carbonate minerals.

In these days following the plate tectonic revolution in natural science, there has been an increased propensity for specialization among scientists. This trend is apparent in the field of study of the geochemistry of sedimentary carbonates. Chemical oceanographers deal with the chemistry of the carbonic acid system in seawater. Some marine geologists and geochemists concern themselves with the relationship between factors controlling the lysocline and carbonate compensation depths in the ocean and the distribution of calcareous oozes; others deal with the burial history of these sediments in the deep-sea realm. Sedimentary petrologists concern themselves with the composition and fabric of carbonates to interpret their origin and, particularly if they are associated with the petroleum field, the occurrence and timing of porosity development and hydrocarbon maturation. Sedimentary geochemists appear to be divided into two camps: those who investigate early diagenetic processes in the shallowly-buried carbonate sands and muds of the modern marine environment, and those who are interested in the diagenesis of carbonate rocks. This is not to say that there are no "generalists" in the field, or that cross-fertilization among sub-disciplines is totally lacking. However, the fact that the study of sedimentary carbonates and the carbonic acid system covers areas of research in fields as diverse as oceanography, sedimentary petrology, climatology, soil science, and petroleum geology makes it difficult for scientists to keep abreast of the vast amount of literature generated in these fields on the subject of carbonate geochemistry. In a recent computer search, we found about 1500 publications per year dealing with carbonates and the carbonic acid system. Even for those of us whose research efforts are largely devoted to this field, it is difficult to keep abreast of all the rapid developments occurring in carbonate science. Hence, the trends have been toward specialization in education and in the scope of papers and books published in the field.

Therefore we addressed ourselves to the question of what should be the present-day content of a book designed to educate an informed reader about the field of carbonate geochemistry. What are the major topics that should be included? What should the modern carbonate geochemist-petrologist know abut the field, and what do our students need to know? We chose to cover the range from electrolyte chemistry of carbon-containing waters to the global cycles of carbon and related materials. The goal of this book is to present the broad spectrum of information contained in the field of sedimentary carbonate geochemistry. We have been selective in topics chosen and the depth to which they are pursued, because the field now has grown to the point where no single book can hope to be totally comprehensive. In making these selections, we have tried to present sufficient background material to enable the nonspecialist to understand the basic geochemistry involved, stressing, in our opinion, what are the major accomplishments, failures and unresolved problems in the study of the geochemistry of sedimentary carbonates. We anticipate that students and professionals from fields of oceanography, geochemistry, petrology, environmental science and petroleum geology will find some interest in this book.

This book is divided into four major areas. The first three chapters deal with the more "basic" chemical aspects of carbonate minerals and their interactions with aqueous solutions. The next two chapters are a discussion of modern marine carbonate formation and sediments in the ocean. These discussions are followed by three chapters on carbonate diagenesis, early marine, meteoric and burial. The last major section is two chapters on the global cycle of carbon and human intervention, and the role of sedimentary carbonates as indicators of stability and changes in Earth's surface environment. A final epilogue is given in which we present our views of where we have been, currently are and possibly should be headed in the study of the geochemistry of sedimentary carbonates.

After this book was written and finally sent to the publisher, we discovered how little our own contributions were and how very important was that of our colleagues, their and our students and all the individuals engaged in typing and drafting the manuscript. It is virtually impossible to express our gratitude to all the people involved directly or indirectly in our effort. We mention below some of the individuals who participated in active exchange of ideas and/or who read critically portions of the text.

Our colleagues through the years provided stimulus and discussion. The late Robert M. Garrels led the way in development of the field of geochemistry and in the education of both of us in the tricks that the carbonate system can play on you. Bob Berner most dramatically taught us that simple experiments, like putting a fish in mud, can tell you much about natural processes. The faculty at Northwestern University in the late 60's and 70's honed FTM's approach to the study of sedimentary rocks, and in the presence of H. C. Helgeson, his quantitative knowledge of geochemical processes. K. E. Chave, in his days at Lehigh University, was responsible for FTM's first inklings of the role of geochemistry in the study of sedimentary carbonates. S. V. Smith, J. S. Tribble and F. J. Sansone of the University of Hawaii tried to keep FTM "honest". John Hower was responsible for JWM's entering the field and Don Graf and Eugene Perry, Jr. stimulated his early interest in carbonate geochemistry. Bob Berner has served as JWM's mentor both in graduate school and throughout his career. Frank Millero's patient tutoring of JWM in the finer points of electrolyte chemistry has had a major influence on much of his work.

Our students in both introductory and advanced courses at the University of Miami, Northwestern University, Texas A&M University and the University of Hawaii reacted critically to our ideas and kept our thoughts focused. We are particularly indebted in this work to those involved in carbonate studies: C. R. Agegian, S. S. Anthony, T. Arakaki, R. Arnseth, K. Badiozamoni, L. D. Bernstein, M. Bertram, W. D. Bischoff, N. Cook, J. deKanel, C. H. Dodge, M. L. Franklin, S. He, G. M. Lafon, W. A. Kornicker, A. Mucci, K. L. Nagy, K. J. Roy, J. D. Pigott,

L. N. Plummer, C. S. Romanek, C. L. Sabine, S. V. Smith, D. C. Thorstenson, J. S. Tribble, S. B. Upchurch, H. L. Vacher, L. M. Walter, A. F. White, and J. J. Zullig.

We would like to express our gratitude to those colleagues who read critically initial, rough (very rough at times!) drafts of the text before finalization: W. D. Bischoff of Wichita State University, K. E. Chave, S. V. Smith and J. S. Tribble of the University of Hawaii, R. Wollast and L. Chou of the Université Libre de Bruxelles, S. L. Dorobek of Texas A&M University, R. N. Ginsburg of the University of Miami, C. B. Gregor of Wright State University, J. S. Hanor of Louisiana State University, L. S. Land of the University of Texas, and L. M. Walter of the University of Michigan. Also, we would like to thank our students in recent classes in sedimentary geochemistry for their critical and editorial comments.

We thank our editor, Martin Tanke, of Elsevier for his patience and encouragement from the inception of this book to its completion. There are times we imagine when he thought he would be "old and grey" before seeing a finished product.

Our sincere thanks go to our typists and in particular D. Sakamoto who labored through endless transcriptions of FTM's writing and dictation, and D. Paul, L. Childress, S. Nelson and C. Smith who helped JWM prepare the camera ready copy. The hundreds of fine figures were prepared by L. Chou, L. Cortiavs, L. Deleon, D. Meier and D. Partain, and A.R. Masterson helped immensely with preparation of the index.

In attempting to synthesize such a field as the sedimentary geochemistry of carbonates, we have undoubtedly missed a number of references and slighted a part of the field of study, and we apologize for the pertinent work not cited and for the topical "holes" in our text.

Some of the material in this book has been drawn from research supported by the National Science Foundation and the Petroleum Research Fund administered by the American Chemical Society.

John W. Morse   Fred T. Mackenzie

College Station, Texas   Honolulu, Hawaii

May, 1990

TABLE OF CONTENTS

Chapter 1

THE CO$_2$-CARBONIC ACID SYSTEM AND SOLUTION CHEMISTRY

Basic concepts

One of the most interesting and important aspects of the geochemistry of sedimentary carbonates is that the anion CO_3^{2-} is part of a complex chemical system in natural waters. This is the carbonic acid system, which includes the chemical species of carbonate and bicarbonate (CO_3^{2-} and HCO_3^-) ions, undissociated carbonic acid (H_2CO_3), dissolved carbon dioxide ($CO_{2,aq}$), and in many instances exchange with gaseous carbon dioxide ($CO_{2,g}$). The relationships among these different chemical species of the carbonic acid system can be represented by a sequence of reactions:

$$CO_{2(g)} \leftrightarrow CO_{2(aq)} \tag{1.1}$$

$$CO_{2(aq)} + H_2O_{(l)} \leftrightarrow H_2CO_{3(aq)} \tag{1.2}$$

$$H_2CO_{3(aq)} \leftrightarrow HCO_3^-{}_{(aq)} + H^+{}_{(aq)} \tag{1.3}$$

$$HCO_3^-{}_{(aq)} \leftrightarrow CO_3^{2-}{}_{(aq)} + H^+{}_{(aq)} \tag{1.4}$$

The fact that the hydrogen ion is an important chemical species in these reactions is indicative of the major role that carbonic acid plays in influencing the pH and buffer capacity of natural waters. Furthermore, the activity of the carbonate anion in part determines the degree of saturation of natural waters with respect to carbonate minerals. Determination of the activity or concentration of CO_3^{2-} is not an easy task; nevertheless, it is necessary to the interpretation of a myriad of processes, including carbonate mineral and cement precipitation-dissolution and recrystallization reactions.

The relative proportions of the different carbonic acid system species can be calculated using equilibrium constants. If thermodynamic constants are used, activities must be employed instead of concentrations. The activity of the ith dissolved species (a_i) is related to its concentration (m_i) by an activity coefficient

(γ_i) so that:

$$a_i = \gamma_i \, m_i \qquad (1.5)$$

Approaches for calculating activity coefficients will be discussed later in this chapter. Three important concepts are introduced here. The first is that in dilute solutions the activity coefficient approaches 1 as the concentration of all electrolytes approaches zero. The second is that the activity coefficient must be calculated on the same scale (e.g., molality, molarity, etc.) as that used to express concentration. The third is that activity in the gas phase is expressed as fugacity. Because the fugacity coefficient for CO_2 is greater than 0.999 under all but the most extreme conditions for sediment geochemistry (e.g., deep subsurface) , the partial pressure of CO_2 (P_{CO2}) may be reasonably used in the place of fugacity.

Based on these considerations, it is possible to write the expressions for the carbonic acid system thermodynamic equilibrium constants (K_i).

$$K_H = \frac{a_{CO_2(aq)}}{P_{CO_2}} \qquad (1.6)$$

$$K_0 = \frac{a_{H_2CO_3(aq)}}{a_{CO_2(aq)} \, a_{H_2O(l)}} \qquad (1.7)$$

$$K_1 = \frac{a_{HCO_3^-(aq)} \, a_{H^+(aq)}}{a_{H_2CO_3(aq)}} \qquad (1.8)$$

$$K_2 = \frac{a_{CO_3^{2-}(aq)} \, a_{H^+(aq)}}{a_{HCO_3^-(aq)}} \qquad (1.9)$$

Values for these constants are a function of temperature and pressure.

An often confusing convention generally used in describing carbonic acid system equilibria is to combine reactions 1.1 and 1.2, thus representing an equilibrium between gaseous CO_2 and carbonic acid [$H_2CO_3(aq)$]. Although this is a convenient means of handling the equilibria, it is also rather misleading because in most solutions greater than 99 percent of $[CO_{2(aq)} + H_2CO_3]$ is $CO_{2(aq)}$ not H_2CO_3. The combination of the two components is sometimes represented as $H_2CO_3^*$, according to the reaction:

$$CO_{2(g)} + H_2O_{(l)} \leftrightarrow H_2CO_{3(aq)}^* \tag{1.10}$$

This commonly used approach will not be followed here. Instead the conventions of Plummer and Busenberg (1982) will be used. They represent the equilibria more reasonably using a reaction notation in which equations 1.2 and 1.3 are combined and the Henry's Law constant (K_H) is retained. We, therefore, eliminate K_0, and redefine K_1 according to the reaction:

$$CO_{2(aq)} + H_2O_{(l)} \leftrightarrow HCO_3^-{}_{(aq)} + H^+{}_{(aq)} \tag{1.11}$$

$$K_1 = \frac{a_{HCO_3^-(aq)}\ a_{H^+(aq)}}{a_{CO_2(aq)}\ a_{H_2O(l)}} \tag{1.12}$$

The values computed at 25°C for K_H, K_1, and K_2 from Plummer and Busenberg (1982) are, respectively, 0.03405 ($10^{-1.468}$), 4.448 x 10^{-7} ($10^{-6.352}$), and 4.690 x 10^{-11} ($10^{-10.329}$).

It is now possible to write equations for the activities of aqueous carbon dioxide, bicarbonate, and carbonate in terms of carbon dioxide partial pressure and hydrogen ion activity, by rearranging equations 1.6, 1.9, and 1.12.

$$a_{CO_2(aq)} = K_H\ P_{CO_2} \tag{1.13}$$

$$a_{HCO_3^-(aq)} = K_H\ K_1 \left(\frac{P_{CO_2}\ a_{H_2O}}{a_{H^+}}\right) \tag{1.14}$$

$$a_{CO_{3(aq)}^{2-}} = K_H K_1 K_2 \left(\frac{P_{CO_2} \, a_{H_2O}}{a_{H^+}^2} \right)$$

(1.15)

Use of equation 1.5 for the relation between activity and concentration makes it possible to express these activities as concentrations. Except for high ionic strength solutions such as brines (ionic strength > 1), it is usually sufficient to approximate the activity of water as unity, thus eliminating the a_{H_2O} term. The resulting equations for concentrations are:

$$m_{CO_{2(aq)}} = K_H \left(\frac{1}{\gamma_{CO_{2(aq)}}} \right) P_{CO_2}$$

(1.16)

$$m_{HCO_{3(aq)}^-} = K_H K_1 \left(\frac{1}{\gamma_{HCO_{3(aq)}^-}} \right) \left(\frac{P_{CO_2}}{a_{H^+}} \right)$$

(1.17)

$$m_{CO_{3(aq)}^{2-}} = K_H K_1 K_2 \left(\frac{1}{\gamma_{CO_{3(aq)}^{2-}}} \right) \left(\frac{P_{CO_2}}{a_{H^+}^2} \right)$$

(1.18)

In these equations m_{H^+} is not substituted for a_{H^+}, because a_{H^+} is usually obtained from a pH measurement. The complexities of the relation between a_{H^+} and measured pH values are discussed later in this chapter.

We now introduce two other important relationships in the carbonic acid system, total CO$_2$ (ΣCO_2) and alkalinity (A). These quantities are important because they are usually determined by chemical analysis. If any two of the four analytic parameters pH, alkalinity, P_{CO_2} and total CO$_2$ are known, the other two parameters can be calculated as well as the various components of the CO$_2$ - carbonic acid system.

Total CO$_2$ is also commonly referred to as total inorganic carbon and represents the amount of carbon dioxide that can be obtained from a solution if it is acidified and the resulting CO_2 removed by gas stripping. It is defined as:

$$\Sigma CO_2 = m_{CO_{2(aq)}} + m_{H_2CO_3} + m_{HCO_3^-} + m_{CO_3^{2-}} \tag{1.19}$$

Because there is only a small proportion of H_2CO_3 present, equation 1.19 is well approximated by:

$$\Sigma CO_2 = m_{CO_{2(aq)}} + m_{HCO_3^-} + m_{CO_3^{2-}} \tag{1.20}$$

ΣCO_2 can be computed from a_{H^+} and P_{CO_2} using equations 1.5, 1.6, 1.9, and 1.12.

$$\Sigma CO_2 = \left(\frac{K_H \, P_{CO_2}}{\gamma_{CO_{2\,(aq)}}}\right) + \left(\frac{K_H \, K_1 \, P_{CO_2}}{\gamma_{HCO_3^-} \, a_{H^+}}\right) + \left(\frac{K_H \, K_1 \, K_2 \, P_{CO_2}}{\gamma_{CO_3^{2-}} \, a_{H^+}^2}\right) \tag{1.21}$$

For the ideal case of a dilute solution in which all activity coefficients are equal to unity, equation 1.19 simplifies to:

$$\Sigma CO_2 = K_H \, P_{CO_2} \left(1 + \frac{K_1}{a_{H^+}} + \frac{K_1 \, K_2}{a_{H^+}^2}\right) \tag{1.22}$$

Alkalinity refers to the acid neutralization capacity of the solution in moles of protons, generally referred to as "equivalents", per unit volume. For our purposes, it can be divided into the alkalinity associated with the carbonic acid system, carbonate alkalinity (A_c), and the alkalinity associated with all other solution components.

$$A_t = A_c + \Sigma A_i \tag{1.23}$$

Where A_t is total alkalinity and A_i are other alkalinity contributing components. In many natural waters, numerous components such as boric acid, hydrogen sulfide, phosphate, and organic compounds contribute to alkalinity, making precise determination of carbonate alkalinity difficult. It is convenient to envision components such as H^+ as contributing to alkalinity in a negative manner.

Carbonate alkalinity is defined as:

$$A_c = m_{HCO_3^-} + 2m_{CO_3^{2-}} \tag{1.24}$$

It can also be computed as ΣCO_2 was from a_{H^+} and P_{CO2}, by using the carbonic acid system relationships:

$$A_c = \left(\frac{K_H \, K_1 \, P_{CO2}}{\gamma_{HCO_3^-} \, a_{H^+}}\right) + 2\left(\frac{K_H \, K_1 \, K_2 \, P_{CO2}}{\gamma_{CO_3^{2-}} \, a_{H^+}^2}\right) \tag{1.25}$$

As before, for the ideal case of a dilute solution as described for ΣCO_2, equation 1.25 simplifies to:

$$A_c = K_H \, K_1 \, P_{CO2}\left(\frac{1}{a_{H^+}} + \frac{2K_2}{a_{H^+}^2}\right) \tag{1.26}$$

Changing the P_{CO2} of a solution will change its carbonate alkalinity, but the total alkalinity will not be changed. This is easily demonstrated for a dilute solution in which the only alkalinity contributing components are carbonic acid and water. In such a solution, total alkalinity is described by:

$$A_t = A_c + A_w = m_{HCO_3^-} + 2\,m_{CO_3^{2-}} + m_{OH^-} - m_{H^+} \qquad (1.27)$$

where $A_w = m_{OH^-} - m_{H^+}$. Some investigators have used this definition of A_t for A_c. We do not, because it can result in a situation where there are no CO_2 species in a solution yet it would have an A_c. Combining equations 1.4 and 1.11 illustrates that in increasing the $CO_{2(aq)}$ of a system for each HCO_3^- produced a H^+ results, and for each CO_3^{2-} produced, $2H^+$ are produced.

$$CO_{2(g)} + H_2O \leftrightarrow HCO_3^- + H^+ \leftrightarrow CO_3^{2-} + 2H^+ \qquad (1.28)$$

Consequently, according to equation 1.27 the net change in alkalinity is zero as P_{CO2} increases, forcing reaction 1.28 to the right. In such a solution the carbonate alkalinity and ΣCO_2 would increase but pH decreases.

With the above equations and relationships in mind, we can now consider a pictorial representation of the carbonic acid system. The relative distribution of carbonic acid system chemical species is commonly presented as a function of pH relative to ΣCO_2 in a plot known as a Bjerrum diagram (Figure 1.1). It can be seen from this diagram that the point of equal concentration between $CO_{2(aq)}$ and HCO_3^- is at a $pH = pK_1$ ($pK_i = -\log K_i$) and for equal concentrations of HCO_3^- and CO_3^{2-} at a $pH = pK_2$. At the pH where $CO_{2(aq)}$ and CO_3^{2-} are at the same concentration, addition of HCO_3^- to the solution does not result in a pH change. It is also obvious that over the pH range of most natural waters, HCO_3^- is the most abundant dissolved inorganic carbon species. It should be kept in mind, however, that activity coefficients can severely alter effective concentrations in complex solutions (e.g., $\gamma_{CO_3^{2-}}$ in seawater is only about 0.035).

Because one of our primary goals is to understand the behavior of carbonate minerals in natural waters, which depends on $a_{CO_3^{2-}}$, it is desirable to be able to calculate carbonate ion activity using any pair of the four quantities pH (gives a_{H^+}), P_{CO2}, A_c, and ΣCO_2. The following equations can be derived from the fundamental relationships previously described but are summarized here to enable the reader to make calculations. Equation 1.15 gives $a_{CO_3^{2-}}$ from a_{H^+} and P_{CO2}. The other appropriate equations for $a_{H^+} - A_c$, $a_{H^+} - \Sigma CO_2$, $A_c - P_{CO2}$, $P_{CO2} - \Sigma CO_2$ and $A_c - \Sigma CO_2$ are, respectively:

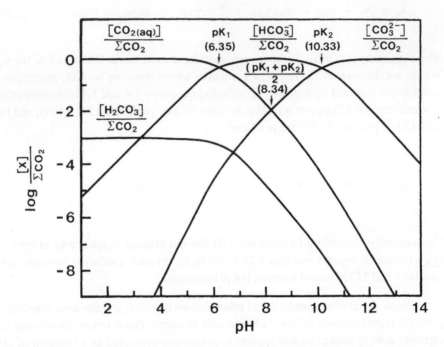

Figure 1.1. A Bjerrum diagram for the relative proportions of carbonic acid system chemical species as a function of pH, for the case where all activity coefficients are equal to 1. pK's are values at 25°C.

a_{H^+} - A_c

$$a_{CO_3}{}^{2-} = \frac{A_c \; \gamma_{HCO_3^-}}{2\left(\frac{\gamma_{HCO_3^-}}{\gamma_{CO_3}{}^{2-}}\right) + \left(\frac{a_{H^+}}{K_2}\right)}$$

(1.29)

a_{H^+} - ΣCO_2

$$a_{CO_3}{}^{2-} = \Sigma CO_2 \; \gamma_{CO_3}{}^{2-} \; (1 - x - y)$$

(1.30)

where:

$$x = \frac{a_{H^+}^2}{a_{H^+}^2 + \left(\dfrac{\gamma_{CO_2}}{\gamma_{HCO_3^-}}\right) K_1\, a_{H^+} + \left(\dfrac{\gamma_{CO_2}}{\gamma_{CO_3^{2-}}}\right) K_1\, K_2}$$

$$y = \frac{K_1\, a_{H^+}}{\left(\dfrac{\gamma_{HCO_3^-}}{\gamma_{CO_2}}\right) a_{H^+}^2 + K_1\, a_{H^+} + \left(\dfrac{\gamma_{HCO_3^-}}{\gamma_{CO_3^{2-}}}\right) K_1\, K_2}$$

A_c-P_{CO2}

The easiest approach using these variables is to calculate a_{H^+} from P_{CO2} and A_c, then use a_{H^+} and P_{CO2} in equation 1.15 to determine $a_{CO_3^{2-}}$. Let:

$$\beta = \frac{A_c}{K_H\, K_1\, P_{CO_2}} \quad \text{then}$$

$$a_{H^+} = \frac{1}{2\beta}\left[\frac{1}{\gamma_{HCO_3^-}} + \left(\frac{1}{\gamma_{HCO_3^-}^2} + \frac{4\beta K_2}{\gamma_{CO_3^{2-}}}\right)^{\frac{1}{2}}\right]$$

$$(1.31)$$

P_{CO2} - $\Sigma CO2$

A good approach is to calculate a_{H^+} from P_{CO2} and ΣCO_2, then use a_{H^+} and P_{CO2} in equation 1.15 to determine $a_{CO_3^{2-}}$. Let:

$$\alpha = \frac{\Sigma CO_2}{K_H\, K_1\, P_{CO_2}} - \frac{1}{\gamma_{CO_2}\, K_1} \quad \text{then:}$$

$$a_{H^+} = \frac{1}{2\alpha}\left[\frac{1}{\gamma_{HCO_3^-}} + \left(\frac{1}{\gamma_{HCO_3^-}^2} + \frac{4\alpha K_2}{\gamma_{CO_3^{2-}}}\right)^{\frac{1}{2}}\right]$$

(1.32)

A_c - ΣCO_2

The same general approach can again be used here in solving for a_{H^+}, but then using it with A_c in equation 1.29. Let:

$$a = \left(\frac{K_2}{\gamma_{CO_3^{2-}}}\right)\left(\frac{2\Sigma CO_2}{A_c} - 1\right), \quad b = \left(\frac{1}{\gamma_{HCO_3^-}}\right)\left(\frac{\Sigma CO_2}{A_c} - 1\right), \quad \text{and} \quad c = \left(\frac{-1}{\gamma_{CO_2}}\right)$$

then:

$$a_{H^+} = \frac{2a}{\left[-b + \left(b^2 - 4ac\right)^{\frac{1}{2}}\right]}$$

(1.33)

Activity coefficients in solutions

Activity coefficients are of major importance to the study of carbonate minerals and solutions. These coefficients provide a bridge between the observable composition of a solution and the thermodynamic theory necessary to interpret this composition in terms of such factors as the saturation state of the solution with respect to a particular carbonate mineral. Evaluation of activity coefficients in natural solutions, to the generally required accuracy, has proven difficult. There are many competing theories, computational methods and values for parameters in the literature, making this aspect of carbonate geochemistry among the most difficult to comprehend for anyone not deeply involved in research in the area of electrolyte solution chemistry. To a large extent as a result of these difficulties, geochemists working in particularly important solutions of relatively constant composition such as seawater have commonly avoided using activity coefficients and have taken the lengthy approach of measuring "apparent" or stoichiometric constants that involve only observable quantities. Several books and numerous review articles are available which deal specifically with the theories and details of how to calculate activity coefficients (e.g., Pytkowicz, 1979), and no attempt will be made here to provide a similar in-depth treatment of this topic. Instead, the major methods and concepts will be briefly summarized so that later in this book when a particular type of activity coefficient evaluation is referenced or used, the reader will have a basic idea of what is involved.

A fundamental concept in all theories for determining activity coefficients is that ionic interactions are involved. These interactions cause a deviation in the free energy associated with the ions from what it would be if they did not occur. Consequently, at the limit of an infinitely dilute solution, activity coefficients go to 1 because there are no ionic interactions. This basic consideration also leads to the idea that as the concentration of ions increases, their extent of interaction must also increase. Ionic strength is a measure of the overall concentration of ions in a solution and the fact that more highly charged ions exert a greater influence on ionic interactions. It is calculated as:

$$I = \frac{1}{2} \Sigma \; m_i \; z_i^{\;2} \tag{1.34}$$

where I is ionic strength, and m_i and z_i are, respectively, the concentration and charge of ion i. A term for ionic strength occupies an important position in all models for the calculation of activity coefficients. Most models, or in the case of some of the newer models the number of terms needed, are presented as being valid over a given range of solution ionic strength. A rough generality is that solutions with an ionic strength of less than 0.1 are considered "dilute," a range of ionic strength of 0.1 to 1 is considered "intermediate," and solutions with an ionic strength greater than 1 are "concentrated.". Examples of natural solutions falling into these three classes are river water, seawater, and brines.

Before proceeding to look at some of the more commonly used methods of calculating activity coefficients, it is important to point out some basic differences in the major types of models used. The approach that has been most commonly employed in the past is based on the idea that two types of interactions influence the activity of an ion in solution. The first influence is purely electrostatic and, hence, is not concerned with what other ions may be present in solution, but only their contribution to the overall solution ionic strength. The second influence is based on the idea that particular ions may have especially strong interactions that result in the formation of "ion pairs." When ions react to form these new species, they are effectively removed from the solution as an individual entity. The formation of ion pairs can change the overall ionic strength of the solution and the concentration of the unpaired or "free" ions. The influences of pairing and ionic strength must be combined to calculate a "total" ion activity coefficient. Thus the total ion activity coefficient is given by the relation:

$$a = \gamma_f \, m_f = \gamma_t \, m_t \qquad (1.35)$$

where the subscripts t and f refer to the total and free ion species. When a significant portion of the ions is paired, a process of circular refinement must be used to calculate a new ionic strength, then new values for the free ion activity coefficients and ion pairs, and finally new total activity coefficients are recalculated. This process is repeated to reach the desired degree of precision for the total ion activity coefficient values. This is the general method that has been used in such classic works as the Garrels and Thompson (1962) model for seawater.

In dilute solutions two equations for simple ion activity coefficients as a function of ionic strength have been commonly used. The Davies equation is the simpler of the two and contains no adjustable parameters for different ions. It is:

$$\log \gamma_i = -0.5 \, z_i^2 \, f(I) \qquad (1.36)$$

where:

$$f(I) = \left(\frac{I^{\frac{1}{2}}}{1 + I^{\frac{1}{2}}} - 0.2I \right) \left(\frac{298}{t + 273} \right) \qquad (1.37)$$

and t is temperature in °C (good over the range of about 0 to 50°C). The second is the Debye-Hückel equation.

$$\log \gamma_i = \frac{-A \, z_i^2 \, I^{\frac{1}{2}}}{1 + å \, B \, I^{\frac{1}{2}}} \qquad (1.38)$$

where A and B are constants that are a function of both temperature and the type of solution, and å is the approximate radius of the hydrated ion under consideration. Values for A and B in water are presented in Table 1.1, and values for å are given in Table 1.2. In Figure 1.2 the free ion activity coefficients calculated up to an ionic strength of 0.1 are presented for univalent and divalent ions using the Davies equation, and for HCO_3^- and Ca^{2+} using the Debye-Hückel equation. The difference at an ionic strength of 0.1 between the -1 ion and HCO_3^- activity coefficients is only about 1%, and for a +2 ion and Ca^{2+} about 10%. This indicates that for many applications the simpler Davies equation may be adequate.

Table 1.1. Debye-Hückel equation A and B constants.

Temperature (^{o}C)	A	B ($\times 10^{-8}$)
0	0.4918	0.3248
10	0.4989	0.3264
20	0.5070	0.3282
25	0.5115	0.3291
30	0.5161	0.3301
35	0.5212	0.3312
40	0.5262	0.3323
50	0.5373	0.3346
60	0.5494	0.3371
80	0.5767	0.3426
100	0.6086	0.3488

Based on Bockris and Reddy (1977).

Table 1.2. Ion size parameters for the Debye-Hückel equation.

å x 10^8	Ion
2.5	NH_4^+
3.0	K^+, Cl^-, NO_3^-
3.5	OH^-, HS^-, MnO_4^-, F^-
4.0	SO_4^{2-}, PO_4^{3-}, HPO_4^{2-}
4.0-4.5	Na^+, HCO_3^-, $H_2PO_4^-$, HSO_3^-
4.5	CO_3^{2-}, SO_3^{2-}
5	Sr^{2+}, Ba^{2+}, S^{2-}
6	Ca^{2+}, Fe^{2+}, Mn^{2+}
8	Mg^{2+}
9	H^+, Al^{3+}, Fe^{3+}

From Berner (1971) as adapted from Klotz (1950).

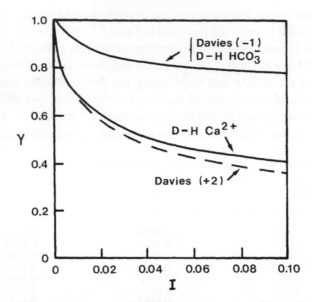

Figure 1.2. The change in the value of the ion activity coefficient as a function of ionic strength for -1 and +2 ions calculated from the Davies equation, and HCO$_3^-$ and Ca^{2+} calculated using the Debye-Hückel method.

In Table 1.3 values are presented for the thermodynamic constants at 25°C used to calculate ion pairs for some of the commonly encountered ions in natural waters. It is important to note that some of the ion pairs are charged (e.g., CaHCO$_3^+$) and must have a free ion activity coefficient assigned. In addition, it is also noteworthy that neutral ions such as uncharged ion pairs and dissolved gases can have activity coefficients which deviate significantly from 1. The methods and supporting theories for evaluating these neutral ion activity coefficients are considerably more tenuous than for charged ions and it is frequently necessary to fall back on direct experimental results (for discussion see Whitfield, 1979).

Table 1.3. Values of ion association constants for common major ions (as log K$_a$).

| Cation | Anion | | |
	SO$_4^{2-}$	HCO$_3^-$	CO$_3^{2-}$
Na$^+$	0.95 ± 0.05	-0.19 ± 0.1	1.02 ± 0.2
K$^+$	0.88 ± 0.02	-	-
Mg^{2+}	2.21 ± 0.02	0.95 ± 0.19	2.9 ± 0.1
Ca^{2+}	2.28 ± 0.02	1.8 ± 0.1	3.2 ± 0.05

Taken from selected values of Millero and Schreiber (1982).

These methods of determining total ion activity coefficients for dilute solutions by calculating free ion activity coefficients using the Debye-Hückel equation and correcting for ion pairing have been applied to mean river water. Results are presented in Table 1.4. Even in this dilute solution ($I \approx 2 \times 10^{-3}$), ion pairing is important, with about 10% of the Ca^{2+} and SO_4^{2-} being present as ion pairs. Plummer and Busenberg (1982) and Sass et al. (1983) have stressed the importance of making ion pairing corrections for calcite and aragonite solubility determinations in dilute solutions. The conclusion is that it is important not to ignore these interactions even if the solution is dilute.

Table 1.4. Composition and activity coefficients in mean river water ($I = 2.08 \times 10^{-3}$, pH = 7.0).

Component	Concentration (mmole kg^{-1} H$_2$O)	γ_f (Debye-Hückel)	f	γ_t	a (x 10^6)
Na$^+$	0.274	0.95	1.00	0.95	260
K$^+$	0.059	0.95	1.00	0.95	56
Mg^{2+}	0.169	0.83	0.97	0.81	136
Ca^{2+}	0.374	0.82	0.91	0.75	282
Cl$^-$	0.220	0.95	1.00	0.95	209
SO$_4^{2-}$	0.117	0.82	0.90	0.74	87
HCO$_3^-$	0.958	0.95	0.97	0.92	881

Composition of mean river water based on Livingstone (1963). f= fraction of total ions which are unpaired.

The fact that simple approaches to calculating free single ion activity coefficients at ionic strengths greater than about 0.1, such as the use of the Debye-Hückel equation, have proven generally unsatisfactory has led to the development of more complex methods for higher ionic strengths. The method frequently used in the past for intermediate ionic strength solutions, in spite of both theoretical and practical weaknesses, was the mean salt method. This approach starts out by choosing a 1:1 reference salt such as KCl (MacInnes convention) or NaCl and assigning equal values to the activity coefficient for the cation and anion (e.g., γ_{K^+} = γ_{Cl^-}). Then the dubious assumption is made that the value of the free ion activity coefficient for the reference cation or anion is the same in any solution of equivalent ionic strength. The final assumption is that the free ion activity coefficient of any other ion in any solution is the same as it would be for that ion in a solution of equivalent ionic strength with the appropriate reference salt cation or anion. These free ion activity coefficients are then used in conjunction with ion pairs to calculate total ion activity coefficients as previously described.

During the last decade another approach has been steadily gaining wider acceptance. This approach uses the "specific interaction" concept, in which continuous equations are employed to represent all the ionic interactions. The most widely used model employs the Pitzer equation (e.g., Pitzer, 1979). In this approach concentrations of specific ion pairs are not calculated, but rather "higher order" terms in the equations are used to represent increasingly complex ionic interactions. This approach to calculating single ion activity coefficients, while mathematically more complex and providing less of a direct picture of interactions than simple ion pairs, will probably become the method of choice in the immediate future. It is proving especially useful in calculating activity coefficients and mineral solubilities in high ionic strength solutions such as brines (e.g., Harvey and Weare, 1980). The current major limitations are that all the needed interaction terms have not been experimentally determined as yet and little has been done in providing data on the influence of temperature or pressure on interaction parameters.

There has been some controversy over the theoretical basis for the "Pitzer equation", the physical meaning of some of its terms, and its ability to deal with very strong ion-ion interactions and neutral species. It should be remembered that most values for the parameters used in the equation are primarily determined through fitting data, rather than by purely theoretical means. Consequently, the equation may best be regarded as empirical. This does not diminish its utility, and its theoretical weaknesses are probably no worse than those of other available means for calculating activity coefficients in moderate to high ionic strength waters. The generalized form of Pitzer's equation (Pitzer, 1979) for a single ion activity coefficient in a mixture of electrolytes is:

$$\ln \gamma_M = z_M^2 \, f^\gamma + 2\sum_a m_a \left[B_{ma} + (\sum_a m_a \, z_a) \, C_{ma} \right] + \sum\sum_{c \; a} m_c \, m_a \, (z_m^2 \, B_{ca}' + z_m \, C_{ca})$$

$$+ 2\sum_c m_c \, (\Theta_{MC} + {}^E\Theta_{Mc} + \sum_a m_a \, \Psi_{Mca}) + \sum\sum_{a \; a'} m_a \, m_{a'} \, (z_M^2 \, {}^E\Theta'_{aa'} + \Psi_{Maa'})$$

$$+ \sum\sum_{c \; c'} m_c \, m_{c'} \, z_M^2 \, {}^E\Theta'_{cc'} \tag{1.39}$$

Subscripts M and c represent cations and a represents anions. For the activity coefficient of a negative ion, the subscripts are interchanged. The double sums are taken over all pairs of different ions of the same polarity. Concentration is in molal units (m) and z is ionic charge.

f^{γ} is a Debye-Hückel term that accounts for long-range electrostatic forces between ions. The B, B', and C terms account for short-range interactions between cations and anions. These three terms are products of an ionic-strength-dependent coefficient (B, B', or C) and ionic concentrations. The other terms, Θ, ψ, $^{E}\Theta$, and $^{E}\Theta'$, are independent of ionic strength and small in value. They become increasingly important, however, with higher ionic strength because they are multiplied by ionic concentrations. The short-range interaction terms, Θ and ψ, represent ion pair repulsions and triplet ion interactions, respectively. The $^{E}\Theta$ and $^{E}\Theta'$ terms account for long-range electrostatic forces between pairs of unsymmetrical ions of the same charge sign. All terms except $^{E}\Theta$ and $^{E}\Theta'$ include one or more parameters that are determined from experimental data on single or mixtures of electrolytes. $^{E}\Theta$ and $^{E}\Theta'$ terms are calculable. It should also be noted that many of these terms have equations from which they are calculated. These secondary equations in some cases again contain terms from which they are calculated. As previously discussed, most of these terms are ultimately empirical in nature and determined by fitting these equations to experimental data.

For natural dilute solutions, such as river water, single ion activity coefficients can be predicted using the simplest form of Pitzer's equation. The simple version includes only the f^{γ}, B, B', and C terms and is adequate to ionic strengths of about 2 molal (Whitfield, 1975). An intermediate version that includes Θ terms provides only slightly better predictions. For higher ionic strength solutions (to > 20 molal), such as evaporitic brines, the complete version of the equation is required (Harvie and Weare, 1980). Millero (1982a) indicated that the complete version is also necessary at the ionic strength of seawater (0.7). Table 1.5 gives examples of single ion activity coefficients for mean river water, seawater, and Dead Sea brine using the three versions of Pitzer's equations.

The values of some total ion activity coefficients calculated from different models can be checked when a solid composed of the same ions is equilibrated with a solution without substantially changing its composition. Under these circumstances the activity coefficient product can be determined by dividing the thermodynamic solubility product by the observed equilibrium concentration product.

$$\gamma_{AB} = \frac{K_{sp,AB}}{m_A m_B}$$

(1.40)

Table 1.5. Pitzer equation activity coefficients (based on Nagy, 1988).

Component	Molality	γ_1	γ_2 w/Θ and ψ	γ_3 w/Θ, and ψ, $^E\Theta$ and $^E\Theta'$
Mean River Water[1]			Ionic strength = 0.002076	
Na	0.00027	0.9503	0.9503	0.9486
K	0.00006	0.9502	0.9501	0.9484
Mg	0.00017	0.8114	0.8114	0.8107
Ca	0.00037	0.8080	0.8080	0.8073
Cl	0.00022	0.9519	0.9520	0.9517
HCO$_3$	0.00096	0.9500	0.9501	0.9497
SO$_4$	0.00012	0.7882	0.7882	0.7852
Seawater, S = 35			Ionic strength = 0.7226	
Na	0.48610	0.6429	0.6428	0.6378
K	0.01058	0.6059	0.5961	0.5886
Mg	0.05474	0.2198	0.2199	0.2029
Ca	0.01066	0.2065	0.2065	0.1910
Sr	0.00009	0.1685	0.1685	0.1458
Cl	0.56577	0.6912	0.6905	0.6887
HCO$_3$	0.00193	0.5691	0.5903	0.5878
Br	0.00087	0.7181	0.7181	0.7150
B(OH)$_4$	0.00009	0.4421	0.4421	0.4402
P	0.00007	0.2174	0.2174	0.2164
SO$_4$	0.02927	0.1233	0.1206	0.1064
CO$_3$	0.00020	0.0463	0.0418	0.0370
Dead Sea Brine[2]			Ionic strength = 8.9603	
Na	1.841	0.8968	0.8900	0.8411
K	0.204	0.5373	0.4629	0.2975
Mg	1.840	1.6265	1.6411	1.4772
Ca	0.46	0.8018	0.8021	0.7146
Cl	6.580	2.8408	2.8357	2.6874
HCO$_3$	0.00397	1.69	2.2794	2.3076
Br	0.070	4.3527	4.3491	4.4064
SO$_4$	0.0054	0.0644	0.0471	0.0344

See text for description of different methods of calculation.
[1]Data from Whitfield (1979).
[2]Data from Neev and Emery (1967) as recalculated by Krumgatz and Millero (1982).

While this does not yield single ion coefficients, the calculated product can be compared with that obtained by multiplying the appropriate individual total ion activity coefficients together. Often one of these is much more uncertain than the other, so the approach can be a reasonable check on the more questionable values.

This procedure is demonstrated (see Table 1.6) for $\gamma_{Ca^{2+}} \gamma_{CO_3^{2-}}$ using calcite and aragonite values from Plummer and Busenberg (1982) and Sass et al. (1983), and equilibrium calcium carbonate concentration products of calcite and aragonite in seawater of S=35 of Morse et al. (1980). The values for K_{sp} calcite are the same, whereas those for aragonite differ by about 9%. The values of $\gamma_{Ca^{2+}} \gamma_{CO_3^{2-}}$ calculated from calcite solubility and Sass et al. aragonite solubility are in excellent (better than 1%) agreement. Although the value of $\gamma_{Ca^{2+}}$ in seawater is also a matter of some debate, most researchers would agree that it is less uncertain than that for $\gamma_{CO_3^{2-}}$, and probably close to 0.20. If this value for $\gamma_{Ca^{2+}}$ is accepted, then the resulting best guess for $\gamma_{CO_3^{2-}}$ in seawater of S=35 is 0.035.

Table 1.6. $\gamma_{Ca^{2+}} \gamma_{CO_3^{2-}}$ calculated from solubility measurement or solution models at 25°C in seawater.

$\gamma_{Ca^{2+}} \gamma_{CO_3^{2-}}$	Method of Calculation
7.59 x 10^{-3}	Calcite -- Plummer and Busenberg (1982) or Sass et al. (1983) and Morse et al. (1980)
6.87 x 10^{-3}	Aragonite -- Plummer and Busenberg (1982) and Morse et al. (1980)
7.53 x 10^{-3}	Aragonite -- Sass et al. (1983) and Morse et al. (1980)
7.07 x 10^{-3}	Complete Pitzer equation (Nagy, 1988)
6.61 x 10^{-3}	Pitzer and pairing (Millero and Schrieber, 1982)

Influences of temperature and pressure

Temperature and pressure variations in natural systems exert major influences on carbonate mineral solubility and the distribution of carbonic acid chemical species. For example, the solubility of calcite decreases with increasing temperature, as does the solubility of CO$_2$ gas in water. These two effects on solubilities can lead to precipitation of calcite as a cement in a marine sediment-pore water system that undergoes moderate burial.

The thermodynamic principles for evaluating the influences of T and P on equilibrium are well established and will not be dealt with in detail here. One basic concept is that the combined influences of temperature and pressure are path independent functions, which simply means that the order in which they are calculated is not important. In practice, however, the availability of data may determine the sequence of calculation. It is generally necessary to calculate the temperature effect first for most systems of practical concern to the sedimentary carbonate chemist.

The equation representing the influence of temperature on an equilibrium constant is:

$$\frac{d \ln K}{dT} = \frac{\Delta H^o_r}{RT^2} \tag{1.41}$$

where ΔH^o_r is the enthalpy of the reaction. ΔH^o_r may also vary with temperature. This variation is relatively small over a moderate temperature range and, consequently, ΔH^o_r is often considered a constant. When ΔH^o_r is treated as a constant, empirical equations are generally used of the form:

$$\ln K = A + \frac{B}{T} + C \ln T + DT + ET^2 + \frac{F}{T^2} \tag{1.42}$$

where A, B, C, D, E, and F are experimentally determined constants and T is absolute temperature. The variations of K_H, K_1, and K_2 for the carbonic acid system with temperature are presented in Figure 1.3. Values for the constants in

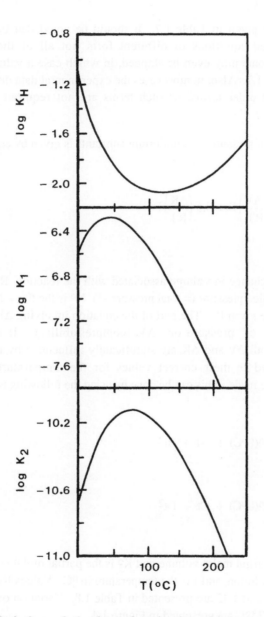

Figure 1.3. Variation of the log of carbonic acid system constants with temperature. (After Plummer and Busenberg, 1982.)

equation 1.42 are given in Table 1.7. It should be noted that because different authors have used equations of different form not all of the constants are determined and some may even be skipped, in which case a value of 0 has been assigned in Table 1.7. Also, in most cases the experimental data do not warrant the use of all higher order terms, or such terms are not required for the needed precision.

The effect of pressure on equilibrium constants is given by equation 1.43:

$$\ln\left(\frac{K^P}{K^0}\right) = \left(\frac{-\Delta V}{RT}\right)P + 0.5\left(\frac{\Delta K}{RT}\right)P^2$$

(1.43)

where ΔV is the change in volume associated with the reaction, R is the ideal gas constant, P is applied pressure (= total pressure -1), K^0 is the the value of K at 1 atm and K^P is K at the given P. The part of the equation involving ΔK is a correction for the influence of pressure on ΔV (compressibility). It is important to remember that both ΔV and ΔK are significantly influenced by temperature and must be modified to their correct values for the temperature at which the calculation is to be made. This can be done by using the following equations:

$$V^0(t) = V^0(0^oC) + At + Bt^2$$

(1.44)

$$K^0(t) = K^0(0^oC) + Ct + Dt^2$$

(1.45)

where V^0 is the partial molal volume and K^0 is the partial molal compressibility in water at infinite dilution, and t is the temperature in ^oC. Values for the parameters in equations 1.44 and 1.45 are presented in Table 1.8. Variation of pK values with pressure at 0 and 25^oC are presented in Figure 1.4.

Table 1.7. Influence of temperature on selected equilibrium constants.

Equilibrium Constant	Equation 1.42 Parameters					T°C Range	pK Values			Ref.
	A	B	C	D	F		0°C	25°C	50°C	
$K_{(B(OH)_3)}$	148.0248	-8966.9	-24.4344	0	0	0-50	9.50	9.24	9.08	1
K_H	249.5708	-15932.9	-40.45154	0.04570836	1541279	0-250	1.11	1.47	1.71	2
K_1	290.9097	-14554.21	-45.0575	0	0	0-50	6.58	6.35	6.29	1
K_1	-820.438	50275.82	126.8339	-0.1402736	-3879685	0-250	6.58	6.35	6.29	2
K_2	207.6548	-11843.79	-33.6485	0	0	0-50	10.63	10.33	10.17	1
K_2	-248.4208	11862.51	38.92561	-0.0749001	-1298007	0-250	10.63	10.33	10.18	2
$K_{(calcite)}$	303.1308	-13348.09	-48.7537	0	0	0-50	8.36	8.43	8.63	1
$K_{(calcite)}$	-395.8319	6537.816	71.595	-0.17959	0	0-90	8.38	8.48	8.66	2
$K_{(calcite)}$	-395.8293	6537.773	71.595	-0.17959	0	0-90	8.38	8.48	8.66	3
$K_{(aragonite)}$	303.5363	-13348.09	-48.7537	0.17959	0	0-50	8.18	8.26	8.48	1
$K_{(aragonite)}$	-395.9949	6685.122	71.595	-0.17959	0	0-90	8.22	8.34	8.54	2
$K_{(aragonite)}$	-395.9180	6685.079	71.595	-0.17959	0	0-90	8.18	8.30	8.50	3
$K_{(vaterite)}$	-396.3454	7079.777	71.595	-0.17959	0	0-90	7.74	7.92	8.16	2

1- Millero (1979). 2-Plummer and Busenberg (1982). 3-Mucci (1983). All values of parameter E = 0.

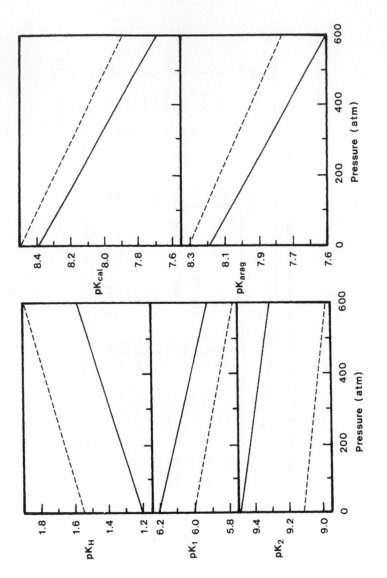

Figure 1.4. Influence of pressure on pK values at 25°C (dashed line) and 0°C (solid line).

Table 1.8. Values of V^o and K^o at 0^oC, and parameters for equations 1.44 and 1.45.

Ion	$V^o(0^oC)$	$-K^o(0^oC, x10^3)$	A	$-B(x10^3)$	$C(x10^3)$	$-D(x10^6)$
H^+	0	0	0	0	0	0
OH^-	-7.58	6.41	0.2324	3.536	0.1310	1.160
H_2O	18.02	0	0.0043	0	0	0
Ca^{2+}	-19.61	7.43	0.0970	1.091	-0.0284	-0.844
HCO_3^-	21.07	5.30	0.1850	2.248	0.1310	1.160
CO_3^{2-}	- 8.74	11.04	0.3000	4.064	0.2240	3.056
$B(OH)_4^-$	15.85	10.32	0.3368	3.643	0.2930	4.400
$B(OH)_3$	36.56	0	0.0043	0	0	0

Selected from Millero (1982b, 1983), except for H_2O where values are approximate based on data in handbook of Chemistry and Physics (Weast, 1985), and $B(OH)_3$ data from Ward and Millero (1974).

The influence of temperature on the volume of a solid is expressed as expansibility. This is reasonably constant over a moderate temperature range. The volume of the solid (V_s) is given by:

$$V_s = V_s(0^oC) + \alpha t \qquad (1.46)$$

where α is usually 20 to 100 x 10^{-5} $^oC^{-1}$ and t is ^oC. For a change in temperature of about 50^oC, the volume change is usually small enough (~ 0.04 cm^3 mol^{-1}) to be neglected. Pressure influences are given by:

$$K_s = \frac{-\partial V_s}{\partial P} + V_s \beta_s \qquad (1.47)$$

where β_s is the isothermal compressibility. The change in V_s up to pressures of about 1 kbar can also usually be neglected.

The same general equation forms and considerations presented for equilibrium constants also apply to activity coefficients; that is:

$$\frac{\partial \ln \gamma_i}{\partial T} = \frac{H - H^o}{RT^2}$$

(1.48)

$$\frac{\partial \ln \gamma_i}{\partial P} = \frac{V - V^o}{RT}$$

(1.49)

where the H or V refer to partial molal quantities and H^o or V^o to the reference state (pure water) quantity. For many solutions of interest, however, the data do not exist for precise calculations of activity coefficients over extended temperature and pressure ranges.

The carbonic acid system in seawater

The chemistry of the carbonic acid system in seawater has been one of the more intensely studied areas of carbonate geochemistry. This is because a very precise and detailed knowledge is of this system necessary if carbon dioxide cycling and deposition of carbonate sediments are to be understood in the marine environment. In this section the effects of salinity, temperature, and pressure on carbonate equilibria in seawater will be examined. The factors influencing carbonate mineral solubility in seawater will be discussed in Chapter 2, although general solubility equations are included in this section where appropriate. Processes influencing the distribution of dissolved inorganic carbon within the ocean and the deposition of calcium carbonate in pelagic sediments are topics included in Chapter 4. In Chapter 5 shoal-water oceanic carbonate systems are examined.

One major concept applicable to problems dealing with the behavior of carbonic acid and carbonate minerals in seawater is the idea of a "constant ionic media". This concept is based on the general observation that the salt in seawater is close to constant in composition, i.e., the ratios of the major ions are the same from place to place in the ocean. Seawater in evaporative lagoons, pores of marine sediments, and near river mouths can be exceptions to this constancy. Consequently, the major ion composition of seawater can generally be determined from salinity. It has been possible, therefore, to develop equations in which the influences of seawater compositional changes on carbonate equilibria can be described simply in terms of salinity.

In theory, it should be possible to deal with all carbonate geochemistry in seawater simply by knowing what the appropriate activity coefficients are and how salinity, temperature, and pressure affect them. In practice, we are only now beginning to approach the treatment of activity coefficients under this varying set of conditions with sufficient accuracy to be useful for most problems of interest. That is why "apparent" and stoichiometric equilibrium constants, which do not involve the use of activity coefficients, have been in widespread use in the study of marine carbonate chemistry for over 20 years. The stoichiometric constants, usually designated as K^*, involve only the use of concentrations, whereas expressions for apparent equilibrium constants contain both concentrations and a_{H^+} derived from "apparent pH". These constants are usually designated as K'. Examples of these different types of constants are:

Stoichiometric constant. $\quad K^*_{calcite} = m_{Ca^{2+}}\ m_{CO_3^{2-}}$ \hfill (1.50)

Apparent constant.

$$K'_2 = \frac{m_{CO_3^{2-}}\ a_{H^+}}{m_{HCO_3^-}}$$

\hfill (1.51)

It should be noted that in seawater the molinity concentration scale (moles kg^{-1} of seawater) is often used, and care must be taken to make certain that stoichiometric and apparent constants are on the same concentration scale as the measured values. The ratios of thermodynamic constants to their apparent or stoichiometric constant are activity coefficients, for example:

$$\frac{K_2}{K'_2} = \frac{\gamma_{CO_3^{2-}}\ f(H^+)}{\gamma_{HCO_3^-}}$$

\hfill (1.52)

where $f(H^+)$ is the ratio of the "apparent" hydrogen ion activity, derived from apparent pH, to the true hydrogen ion activity.

It is appropriate at this point to discuss the "apparent" pH, which results from the sad fact that electrodes do not truly measure hydrogen ion activity. Influences such as the surface chemistry of the glass electrode and liquid junction potential between the reference electrode filling solution and seawater contribute to this complexity (see for example Bates, 1973). Also, commonly used NBS buffer standards have a much lower ionic strength than seawater, which further complicates the problem. One way in which this last problem has been attacked is to make up buffered artificial seawater solutions and very carefully determine the relation between measurements and actual hydrogen ion activities or concentrations. The most widely accepted approach is based on the work of Hansson (1973). pH values measured in seawater on his scale are generally close to 0.15 pH units lower than those based on NBS standards. These two different pH scales also demand their own sets of apparent constants. It is now clear that for very precise work in seawater the Hansson approach is best.

As a practical matter, pH values of seawater are generally measured at 25°C and 1 atmosphere total pressure. To apply these pH measurements to *in situ* conditions it is necessary to correct for pressure and temperature changes. For salinities between 30 and 40, temperatures of 0 to 40°C, and pH=7.6 to 8.2, the temperature correction on the NBS scale is (Millero, 1979):

$$pH_t = pH_{25} + A(t - 25) + B(t - 25)^2 \qquad (1.53)$$

where:

$$10^3 A = -9.702 - 2.378(pH_{25} - 8) + 3.885(pH_{25} - 8)^2 \qquad (1.54)$$

$$10^4 B = 1.123 - 0.003(pH_{25} - 8) + 0.993(pH_{25} - 8)^2 \qquad (1.55)$$

and on the Hansson (1973) scale:

$$pH_t = pH_{25} + C(t - 25) + D(t - 25)^2 \qquad (1.56)$$

where:

$$10^3C = -9.296 + 32.505(pH_{25} - 8) + 63.806(pH_{25} - 8)^2 \qquad (1.57)$$

$$10^4D = 3.916 + 23.000(pH_{25} - 8) + 41.637(pH_{25} - 8)^2 \qquad (1.58)$$

pH_t refers to the pH at a given t.

The influence of pressure on pH for S=32 to 38 and t=0 to 25°C (Millero, 1979) is:

$$pH_{t,P} = pH_t - EP \qquad (1.59)$$

where:

$$-10^3E = 0.424 - 0.0048(S - 35) - 0.00282T - 0.816(pH_t - 8) \qquad (1.60)$$

and $pH_{t,P}$ refers to the pH under the given t and P (in bars) conditions.

Another practical consideration when dealing with the seawater carbonic acid system is that a number of components can contribute to total alkalinity in addition to carbonate alkalinity, H^+ and OH^-. The seawater constituent that is generally most important is boric acid. It contributes about 0.1 meq L^{-1} alkalinity under most conditions, and it is usually taken into consideration when making calculations. Nutrient compounds that are highly variable in concentration in seawater, such as ammonia, phosphate, and silica, can also influence alkalinity. They must be taken into account for very precise work. In anoxic pore waters a number of compounds, such as hydrogen sulfide and organic matter can be significant contributors to alkalinity (e.g., see Berner et al., 1970). In addition, for Gran (1952) type titrations truly accurate results can be obtained only when interactions between H^+ and sulfate and fluoride ions are included.

To further complicate matters in interpreting and using alkalinity measurements, apparent constants can only be used with data in which pH was measured and alkalinity calculated in exactly the same manner as that done for the samples. A classic example of the difficulties that can arise when these factors are not carefully dealt with is the GEOSECS carbonate data set (see Bradshaw et al., 1981, for discussion). Fortunately, for many open ocean water column studies of direct concern to carbonate geochemistry, only the boric acid correction needs to be made to alkalinity, and the treatment of the carbonic acid system here will be restricted to this correction.

The general equation determined by Millero (1979) for the influence of temperature and salinity on apparent or stoichiometric constants in seawater is:

$$\ln K'_i = \ln K^o_i + (a_0 + a_1/T + a_2 \ln T) S^{1/2} + b_0 S \qquad (1.61)$$

where a_0, a_1, a_2, and b_0 are constants given in Table 1.9a. Mucci (1983) has used a similar equation (1.62) to describe the temperature-salinity influence on calcite and aragonite solubility.

$$\ln K^*_{sp} = \ln K^o_{sp} + (b_0 + b_1 T + b_2/T) S^{1/2} + c_0 S + d_0 S^{3/2} \qquad (1.62)$$

Values for the parameters are presented in Table 1.9b. Variations of the pK' or pK* value of the constants at 0, 25, and 50°C with salinity are presented in Figure 1.5.

The influence of pressure on the apparent or stoichiometric constants can be determined by using equations 1.43, 1.44, and 1.45 with the data in Table 1.10. It should be noted that the currently available information for V' and K' is restricted to S=35. A reasonable approximation of pressure effects at other salinities, however, can be made assuming a linear variation of V' and K' between pure water and S=35 seawater as a function of the square root of ionic strength. The variations of pK values with pressure at 0 and 25°C for S=35 seawater are presented in Figure 1.6.

Weiss (1974), based on literature data and his own experiments, arrived at equation 1.63 for K'_H (K_0 in his notation) as a function of temperature and salinity:

$$\ln K'_H = A_1 + A_2 \left(\frac{100}{T}\right) + A_3 \ln \left(\frac{100}{T}\right) + S \left[B_1 + B_2 \left(\frac{100}{T}\right) + B_3 \left(\frac{100}{T}\right)^2 \right]$$

$$(1.63)$$

Table 1.9a. Values of parameters for equation 1.61 from Millero (1979).

Constant	pH Scale	a_0	a_1	b_0 ($\times 10^2$)	pK' (25°C)	Original Reference
K'_B	NBS	0.0473	49.10	-----	8.692	1
	Hansson	0.5998	-75.25	-1.767	8.612	2
K'_1	NBS	0.0221	34.02	-----	6.002	3
	Hansson	0.5709	-84.25	-1.632	5.859	2
K'_2	NBS	0.9805	-92.65	-3.294	9.112	3
	Hansson	1.4853	-192.69	-5.058	8.943	2

1-Lyman (1957); 2-Hansson (1972); 3-Mehrbach et al. (1973). All values of $a_2 = 0$.

Table 1.9b. Values of parameters for equation 1.62 (based on Mucci, 1983).

Constant	b_0	$b_1 (\times 10^3)$	b_2	c_0	$d_0 (\times 10^3)$	pK^* (25°C)
$K^*_{calcite}$	-1.78938	6.5453	410.64	-0.17755	9.4979	6.370
$K^*_{aragonite}$	-0.157481	3.9780	202.938	-0.23067	13.6808	6.189

Table 1.10. Selected values and parameters for ion partial molal volumes and compressibilities in seawater (from Millero, 1982b and 1983).

Ion	$V'(0°C)$	$-K'(0°C, \times 10^3)$	A	$B(\times 10^3)$	$C(\times 10^3)$
H^+	0	0	0	0	0
OH^-	- 2.01	4.24	0.1110	1.304	0.0767
H_2O	18.01	-0.78	0.0020	0	-0.0028
Ca^{2+}	-15.93	5.82	-0.0196	-0.668	-0.4280
HCO_3^-	24.72	3.01	0.1967	3.006	0.773
CO_3^{2-}	7.40	7.88	0.3725	5.653	0.1673
$B(OH)_4^-$	28.55	4.06	0.1746	0.216	0.0847
$B(OH)_3$	39.10	0	0.0544	0	0

V' in cm^3 $mole^{-1}$; K' in cm^3 $mole^{-1}$ bar. Determined only for $S = 35$.

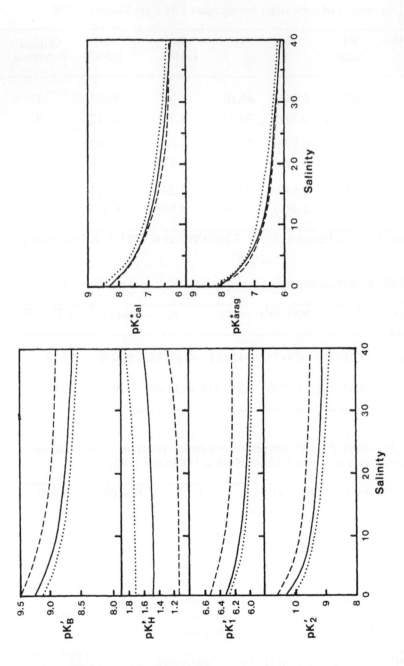

Figure 1.5. Influence of salinity on pK' and pK* values at 0°C (dotted line), 25°C (solid line) and 50°C (dashed line) for NBS based constants.

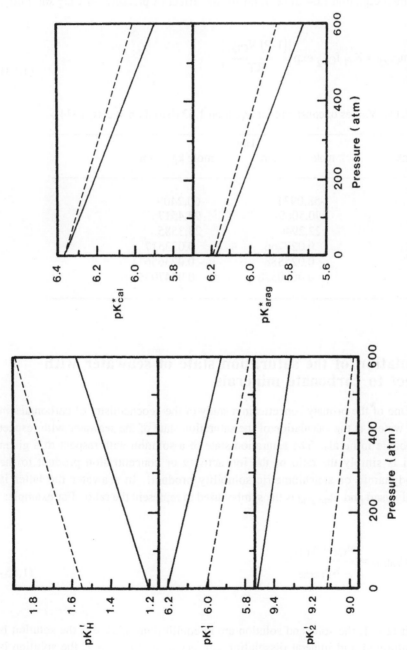

Figure 1.6. Influence of pressure on pK' and pK* values at 0°C (solid line) and 25°C (dashed line) in seawater of S=35.

Values for constants A_1, A_2, A_3, B_1, B_2 and B_3 are given in Table 1.11. He also determined equation 1.64 to be valid for the effect of pressure on CO_2 solubility.

$$m_{CO_2} = K'_H \, f_{CO_2} \, \exp\left[\frac{(1\text{-}P)\,V_{CO_2}}{RT}\right]$$

(1.64)

Table 1.11. Values of constants for equation 1.63 (based on Weiss, 1974).

Constant	Units	
	mole L^{-1} atm	mole kg^{-1} atm
A_1	-58.0931	-60.2409
A_2	90.5069	93.4517
A_3	22.2940	23.3585
B_1	0.027766	0.023517
B_2	- 0.025888	- 0.023656
B_3	0.0050578	0.0047036

Calculation of the saturation state of seawater with respect to carbonate minerals

One of the primary concerns in a study of the geochemistry of carbonates in marine waters is the calculation of the saturation state of the seawater with respect to carbonate minerals. The saturation state of a solution with respect to a given mineral is simply the ratio of the ion activity or concentration product to the thermodynamic or stoichiometric solubility product. In seawater the latter is generally used and $\Omega_{mineral}$ is the symbol used to represent the ratio. For example:

$$\Omega_{calcite} = \frac{m_{Ca^{2+}} \, m_{CO_3^{2-}}}{K^*_{calcite}}$$

(1.65)

If $\Omega = 1$, the solid and solution are in equilibrium; if $\Omega < 1$, the solution is undersaturated and mineral dissolution can occur; and if $\Omega > 1$, the solution is supersaturated and precipitation should occur.

For normal seawater the calcium concentration can be calculated from salinity for most purposes (accuracy is almost always better than 1%).

$$m_{Ca2+} = 10.28 \text{ (mmole kg}^{-1} \text{ seawater) } (S/35) \qquad (1.66)$$

Care must be taken, however, in pore waters and special areas such as coastal waters, carbonate banks, and lagoons where significant deviations from normal seawater concentration ratios can occur. As an aside, it should also be noted that the apparent constants are not accurate where seawater composition is altered. This composition change can be especially important in anoxic environments where extensive sulfate concentration changes occur.

The calculation of $m_{CO_3^{2-}}$ in seawater from the possible analytic parameters pH, P_{CO_2}, ΣCO_2, and A_t is similar to that described for $a_{CO_3^{2-}}$ (e.g., Butler, 1982). When A_t is involved, however, corrections must be made, at the very least, for the contribution of boric acid to alkalinity (A_B). For normal seawater where the salinity is known and pH is known or calculated as shown later:

$$A_B = \frac{K'_B\, m_B}{K'_B + a_{H^+}} \qquad (1.67)$$

where :

$$m_B = \frac{\left(1.2 \times 10^{-5}\right) S}{1 - (S/1000)} \qquad (1.68)$$

The equations and methods given in this chapter can be used to calculate the distribution of carbonic acid system components and the saturation state of a solution with respect to a carbonate mineral under varying temperature, pressure, and composition. To illustrate the type of changes that occur, a calculation has been done for seawater, and the results summarized for nine different cases in Table 1.12. Case 1 is used as a reference typical of surface, subtropical, Atlantic seawater in equilibrium with the atmosphere. In all other cases the salinity and total

Table 1.12. Variation of carbonic acid components and calcite saturation in seawater under different conditions.

Case	T (°C)	P (applied atm)	pH (NBS)	P_{CO_2} (µatm)	A_c (meq kg⁻¹)	ΣCO_2 (mmole kg⁻¹)	$CO_{2(aq)}$ (µmole kg⁻¹)	HCO_3^- (µmole kg⁻¹)	CO_3^{2-} (µmole kg⁻¹)	Ω_C	R_Ω
1	27	0	8.28	330	2.30	2.03	9	1729	293	6.70	1.00
2	2.2	0	8.59	116	2.28	2.03	7	1765	258	6.10	0.91
3	2.2	0	8.21	330	2.35	2.24	19	2096	128	3.03	0.45
4	27	500	8.12	599	2.29	2.03	8	1755	266	2.99	0.45
5	2.2	500	8.40	215	2.26	2.03	6	1785	238	2.26	0.33
6	2.2	500	8.15	430	2.32	2.11	12	2013	151	1.44	0.21
7	2.2	500	8.03	600	2.34	2.24	17	2102	118	1.12	0.17
8	2.2	500	7.92	800	2.36	2.28	22	2167	94	0.90	0.13
9	2.2	500	7.83	1000	2.37	2.32	28	2210	78	0.75	0.11

For S=36.5 seawater with A_t = 2.43 meq kg⁻¹. The underlined parameter is the value which was used with A_t to calculate the other parameters under the given conditions. R_Ω is the ratio of Ω values between the case under consideration and Case 1.

alkalinity were kept constant, and temperature, pressure, or the partial pressure of CO_2 varied in the calculations. Note that the surface seawater in case 1 is almost seven times supersaturated with respect to calcite and that in all cases bicarbonate is by far the most abundant dissolved inorganic carbon species.

The first change considered is lowering the temperature from 27 to 2.2°C (close to bottom seawater temperature). If this were done in a sealed bottle of water where no CO_2 exchange with the atmosphere could take place (ΣCO_2 content constant), case 2 would result. In this case, A_t and ΣCO_2 are known. The lowering of temperature by close to 25°C changes the saturation state only by about 10%, but the P_{CO2} is dropped to almost one third its original value for equilibrium with the atmosphere and pH increases by about 0.3 units. If the bottle is then opened, exchange of CO_2 with the atmosphere will occur until equilibrium is reached. For this case (3), A_t and P_{CO2} are known. The infusion of CO_2 from the atmosphere increases ΣCO_2 by about 10% and the pH is lowered to close to its original value. The most dramatic effect is cutting the supersaturation with respect to calcite about in half.

Next, the influence of pressure is considered. We take a sample of the original water (case 1) and again seal it in a bottle, but with a small piston to equalize pressure, and insulate it so no heat will be exchanged. Case 4 is an approximate result if the bottle were lowered to 5000 m where the pressure is about 500 atm. The total alkalinity and ΣCO_2 remain constant, but the P_{CO2} almost doubles whereas the saturation state drops to less than half its original value. Case 5 results if the insulation is removed, and the sample is cooled to 2.2°C. A_t and ΣCO_2 remain constant again, but the P_{CO2} drops to about a third its value in case 4, and the saturation state drops by another quarter to about one third its original value. An important point to be noted for this case is that, if no chemical exchange takes place, the seawater remains over two times supersaturated with respect to aragonite.

It is observed (see Chapter 4) that seawater at this depth of 5000 m in the Atlantic and Pacific Oceans is substantially undersaturated with respect to both calcite and aragonite. This undersaturation can only come about through compositional changes. The primary one is addition of CO_2 prior to deep water formation and its addition in the deep sea through the oxidation of organic matter. Cases 6 through 9 illustrate the influence of increasing P_{CO2} in lowering dramatically the saturation state of the water under deep sea conditions.

Concluding remarks

In this chapter, we introduced the reader to some basic principles of solution chemistry with emphasis on the CO$_2$-carbonate acid system. An array of equations necessary for making calculations in this system was developed, which emphasized the relationships between concentrations and activity and the bridging concept of activity coefficients. Because most carbonate sediments and rocks are initially deposited in the marine environment and are bathed by seawater or modified seawater solutions for some or much of their history, the carbonic acid system in seawater was discussed in more detail. An example calculation for seawater saturation state was provided to illustrate how such calculations are made, and to prepare the reader, in particular, for material in Chapter 4. We now investigate the relationships between solutions and sedimentary carbonate minerals in Chapters 2 and 3.

Chapter 2

INTERACTIONS BETWEEN CARBONATE MINERALS AND SOLUTIONS

Sedimentary carbonate minerals

Basic Concepts

Sediments and sedimentary rocks contain a variety of carbonate minerals. Calcite and dolomite are by far the most important carbonate minerals in ancient sedimentary rocks. Aragonite is rare, and other carbonate minerals are largely confined to very special deposits such as evaporites or iron formations or present as trace minerals. The fact that calcite and dolomite are the dominant carbonate minerals in sedimentary rocks is consistent with equilibrium thermodynamic considerations for the environmental conditions under which most sedimentary rocks are formed and persist. Calcitic and dolomitic carbonate rocks constitute about 10-15% of the mass of sedimentary rocks and make up approximately 20% of the Phanerozoic geologic column based on the measurement of stratigraphic sections. Paleozoic carbonates contain a great deal of dolomite, whereas Mesozoic and Cenozoic age carbonates contain progressively less dolomite. This trend seems to be a worldwide phenomenon. In general, the proportion of dolomite relative to calcite appears to increase with increasing geologic age into the late Precambrian; the cause of this trend and indeed its actual presence are hotly debated topics today (see Chapter 10).

It is appropriate to divide carbonate minerals in marine sediments into those found in relatively shallow and deep water environments. The factors controlling the source, mineralogy, and diagenesis of carbonates in these environments are very different (see Chapters 4 and 5). The vast majority of deep sea, carbonate-rich sediments today are composed of calcite low in magnesium (>99% $CaCO_3$). This material is primarily derived from pelagic skeletal organisms, with coccolithophores usually being the most important quantitatively, followed by foraminifera, in abundance. In sediments overlain by water of intermediate depth, aragonite derived from pteropods and heteropods can be found. Calcite cements containing abundant magnesium can also occur in these deep water sediments, but they are relatively rare.

Shallow water, carbonate-rich sediments are largely confined today to the subtropic and tropic climatic zones, but are found even at high latitudes. Their occurrence is strongly influenced by the factors of water temperature and terrigenous input. These sediments are generally dominated by aragonite, followed by calcites rich in magnesium. Low-magnesium calcite is usually a minor component and dolomite occurs only in special environments, and even then it is generally not a major portion of the sediment. Shallow water carbonates are primarily produced by the disintegration of the skeletons of benthic organisms, such as corals, echinoids, molluscs, forams, and algae. In some environments, inorganic precipitates such as cements and oöids are also abundant.

In the following section we briefly discuss some characteristics of sedimentary carbonate minerals. This discussion provides necessary initial information expanded on in later chapters.

Characteristics of Sedimentary Carbonate Minerals

CaCO3 minerals Three polymorphs of $CaCO_3$ are known to occur in sediments and organisms. The rhombohedral mineral, calcite, is the most abundant and is thermodynamically stable (see Table 2.1 for thermodynamic and solubility data on carbonate minerals). Aragonite, the orthorhombic form, is also abundant and about 1.5 times more soluble than calcite. Although aragonite has the larger cation sites, it is the more dense phase and, consequently, the high pressure phase. The pressure-temperature phase relations between calcite and aragonite are shown in Figure 2.1.

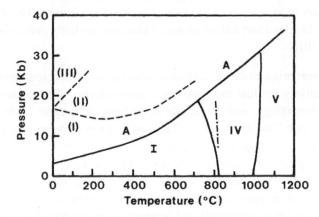

Figure 2.1. Generalized phase relations in the unary system $CaCO_3$. "**A**" - aragonite; "**I**" through "**V**" - calcite polymorphs with metastable fields indicated by (). Dash-dot line at 800°C represents transition encountered on cooling runs; solid line at lower T represents transition encountered on heating runs. (After Carlson, 1980.)

Table 2.1. Thermodynamic properties of carbonate minerals.

Mineral	Formula	Formula Wt (g)	Density (g cm^{-3})	Crystal System	$\Delta G_{f_{298}}$ (J mole^{-1})	- log K_{sp} (calc)	- log K_{sp} (ref 2)	- log K_{sp} (other)
Calcite	CaCO$_3$	100.09	2.71	Trig.	-1128842[3]	8.30	8.35	8.48[6,8], 8.46[9]
Aragonite	CaCO$_3$	100.09	2.93	Ortho.	-1127793[3]	8.12	8.22	8.34[6], 8.30[8]
Vaterite	CaCO$_3$	100.09	2.54	Hex.	-1125540[3]	7.73	-	7.91[6]
Monohydrocalcite	CaCO$_3$.H$_2$0	118.10	2.43	Hex.	-1361600[3]	7.54	-	7.60[17]
Ikaite	CaCO3.6H$_2$0	208.18	1.77	Mono.	-	-	-	7.12[5]
Magnesite	MgCO$_3$	84.32	2.96	Trig.	-1723746[3]	8.20	7.46	5.10[7], 8.10[10]
Nesquehonite	MgCO$_3$.3H$_2$0	138.36	1.83	Trig.	-1723746[3]	5.19	4.67	
Artinite	Mg$_2$CO$_3$(OH)$_2$.3H$_2$0	196.68	2.04	Mono.	-2568346[3]	18.36	-	
Hydromagnesite	Mg$_4$(CO$_3$)$_3$(OH)$_2$.3H$_2$0	359.27	-	Mono.	-4637127[4]	36.47	-	30.6[7]
Dolomite	CaMg(CO$_3$)$_2$	184.40	2.87	Trig.	-2161672[3]	17.09	-	
Huntite	CaMg$_3$(CO$_3$)$_4$	353.03	2.88	Trig.	-4203425[3]	30.46	-	
Strontianite	SrCO$_3$	147.63	3.70	Ortho.	-1137645[3]	8.81	9.03	9.13[9], 9.27[11]
Witherite	BaCO$_3$	197.35	4.43	Ortho.	-1132210[3]	7.63	8.30	8.56[9,18]
Barytocalcite	CaBa(CO$_3$)$_2$	297.44	-	Trig.	-2271494[4]	17.68	-	
Rhodochrosite	MnCO$_3$	114.95	3.13	Trig.	-816047[3]	10.54	9.30	10.59[14]
Kutnahorite	CaMn(CO$_3$)$_2$	215.04	-	Trig.	-195058[1]	55.79	-	
Siderite	FeCO$_3$	115.85	3.80	Trig.	-666698[3]	10.50	10.68	10.91[12]
Cobaltocalcite	CoCO$_3$	118.94	4.13	Trig.	-650026[1]	11.87	9.68	
-	CuCO$_3$	123.56	-	Trig.	-	-	9.63	11.51[12]
Gaspeite	NiCO$_3$	118.72	-	Trig.	-613793[1]	7.06	6.87	
Smithsonite	ZnCO$_3$	125.39	4.40	Trig.	-731480[3]	9.87	10.00	
Otavite	CdCO$_3$	172.41	4.26	Trig.	-669440[3]	11.21	13.74	
Cerussite	PbCO$_3$	267.20	6.60	Ortho.	-625337[3]	12.80	13.13	12.12.15[15]
Malachite	Cu$_2$CO$_3$(OH)$_2$	221.11	4.00	Mono.	-	-	33.78	33.46[13]
Azurite	Cu$_3$(CO$_3$)$_2$(OH)$_2$	344.65	3.88	Mono.	-	-	45.96	
Ankerite	CaFe(CO$_3$)$_2$	215.95	-	Trig.	-1815200	19.92[19]	-	
Natronite	Na$_2$CO$_3$.1OH$_2$O	285.99	-	Ortho.	-3428997[4]	1.03	-	
Thermonatrite	Na$_2$CO$_3$.H$_2$O	124.00	-	Mono.	-1286538[4]	-0.403	-	-0.54[16]
Trona	NaHCO$_3$.Na$_2$CO$_3$.2H$_2$0	229.00	2.25	Mono.	-2386554[4]	2.07	-	1.00[16]
Nahcolite	NaHCO$_3$	84.01	2.16	Mono.	-851862[4]	0.545	-	0.39[16]
Natron	NaHCO$_3$.H$_2$O	102.03	-	-	-	-	-	0.80[16]

1. Weast, 1985
2. Smith and Martel, 1976
3. Robie et al.,1979
4. Garrels and Christ, 1965
5. Suess,1982 at 1.6°C
6. Plummer and Busenberg, 1982
7. Langmuir, 1973
8. Sass et al ., 1983
9. Millero et al ., 1984
10. Christ and Hostetler, 1970
11. Busenberg et al .,1984
12. Reiterer et al ., 1981
13. Symes and Kester, 1984
14. Johnson, 1982
15. Bilinski and Schindler, 1982
16. Monnin and Schott, 1984
17. Hull and Turnhull, 1973
18. Busenberg and Plummer, 1986a
19. Woods, 1988

Most other divalent cation carbonate minerals are classified as being in the "calcite-like" or "aragonite-like" groups based on structural similarities. Carbonate minerals of divalent cation size less than calcium are in the "calcite-like" group, whereas those with divalent cations larger than calcium are in the "aragonite-like" group (see Table 2.2).   The transition metal carbonates, of which siderite ($FeCO_3$) and rhodochrosite ($MnCO_3$) are the most important phases found in sediments, have "calcite-like" structures. Strontianite ($SrCO_3$) and witherite ($BaCO_3$) have structures similar to aragonite.

Table 2.2. Anhydrous rhombohedral and orthorhombic carbonates. (After Speer, 1983.)

Cation	Ionic Radius*	Mineral	Unit Cell Volume, Å^3
Rhombohedral carbonates			
Ni	0.69	gaspeite	269.6
Mg	0.72	magnesite	279.6
Co	0.745	spherocobaltite	281.1
Zn	0.74	smithsonite	281.7
Fe	0.78	siderite	292.7
Cu	0.73	---	308.4
Mn	0.83	rhodochrosite	309.6
Cd	0.95	otavite	341.7
Ca	1.00	calcite	367.9
Orthorhombic carbonates			
Ca	1.18	aragonite	226.8
Yb	--	---	237.7
Eu	1.30	---	259.1
Sr	1.31	strontianite	259.1
Sm	--	---	260.7
Pb	1.35	cerussite	269.6
Ba	1.47	witherite	304.2
Ra	--	---	335.3

*Data from Shannon, 1976.

Vaterite is the third anhydrous $CaCO_3$ phase and has a hexagonal structure. It is approximately 3.7 times more soluble than calcite and 2.5 times more soluble than aragonite. It rarely has been observed in natural systems and biogenic material, and always under conditions where it is metastable relative to calcite and aragonite (e.g., see Albright, 1971, for summary). However, it can be produced in the laboratory under experimental conditions that give rise to high precipitation rates of carbonate phases. Although there has been disagreement (e.g., Albright, 1971; Turnbull, 1973) over metastable phase relations, it is clear that vaterite is always a metastable phase under the environmental conditions found in sediments and sedimentary rocks.

Hydrated $CaCO_3$ minerals also are found occasionally in sediments. The occurrence of monoclinic $CaCO_3 \cdot 6H_2O$ (ikaite) in Antarctic sediments has been reported from the Bransfield Straight (Suess et al., 1982). Its formation is favored by sub-zero temperatures in interstitial waters where active microbial decomposition of organic matter is taking place.

Monohydrocalcite has been observed rarely in lake sediments and biologic material (Hull and Turnbull, 1973). It generally is found as spherulitic aggregates which may slowly convert to aragonite or calcite. Like ikaite, its formation is favored by low temperatures and high hydrostatic pressures.

An important subgroup of the $CaCO_3$ minerals is the magnesian calcites. They are common components of shallow water, marine sediments, where they are derived from the skeletons of organisms and by direct precipitation of marine cements. Their solubility is strongly influenced by their magnesium content. In seawater, biogenic magnesian calcite containing ~11 mole % $MgCO_3$ may have about the same solubility as aragonite. The characteristics and complex properties of this variety of calcite are extensively discussed in Chapter 3.

Dolomite   Dolomite is one of the most abundant sedimentary carbonate minerals, yet after years of intense study its mode of formation remains controversial, and its properties under Earth surface conditions are less well known than for most other carbonate minerals. The primary reason for this seems to be that its formation is kinetically hindered by its complex and well-ordered structure. Another problem in understanding dolomite may be as Lynton Land says "there are dolomites and dolomites and dolomites". The topic of dolomite genesis will be dealt with in several later sections of this book.

The phase relations between dolomite and calcite are summarized in Figure 2.2. It should be noted that, over a wide range of temperatures from 500 to nearly 1100°C and up to a bulk composition of about 50 mole % $MgCO_3$, a two-phase assemblage calcite plus dolomite is stable. The stable calcite becomes progressively enriched in magnesium and the stable dolomite becomes progressively enriched in calcium with increasing temperature. Projection of the calcite limb of the calcite-dolomite solvus to lower temperatures would imply that at surface temperatures, a calcite containing virtually no magnesium would be the stable phase coexisting with a nearly 50-50 mole % $CaCO_3$-$MgCO_3$ dolomite. At bulk compositions greater than 50 mole% $MgCO_3$, and at higher temperature, the stable mineral assemblage would be dolomite plus a nearly pure $MgCO_3$ phase--magnesite. In marine and many other natural waters, dolomite is the stable carbonate mineral phase, and these waters generally are the most supersaturated with respect to this carbonate mineral.

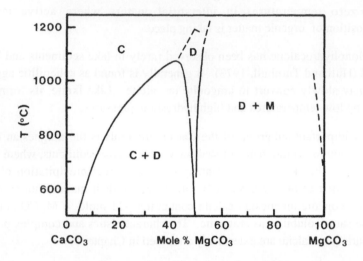

Figure 2.2.   Phase diagram for the system $CaCO_3$-$MgCO_3$. **C** - calcitic phase, **D** - dolomitic phase, **M** - magnesite-rich phase. The region in which dolomite is observed as an ordered compound is limited by a dashed line (data from Goldsmith and Heard, 1961). (After Goldsmith, 1983.)

Other carbonate phases Magnesium carbonates present many problems in sedimentary geochemistry because of their tendency to form a variety of hydrates. Magnesite (Figure 2.2) cannot form directly from aqueous solutions under near Earth surface conditions. Instead, a variety of magnesium carbonate hydrates and hydroxyhydrates form from solution. The primary reason for this observation is believed to be the high energy of hydration of the magnesium ion. Lippmann (1973) has presented an extensive discussion of the chemistry of these minerals and the difficulties in understanding their phase relations. They are minor minerals found in sedimentary deposits, and generally reflect evaporitic conditions.

Substitutions of iron and manganese in the calcite structure produce a range of carbonate mineral compositions and structure. Figure 2.3 illustrates phase diagrams for the $CaCO_3$-$MnCO_3$ and $CaCO_3$-$FeCO_3$ systems. Siderite and manganoan-rich carbonate minerals are important phases found in iron formations. The compositional complexity of the carbonate minerals found associated with these important rock types are illustrated in Figure 2.4. Recent studies of carbonate phase relationships have led to new insight into the formation of the iron formations (Woods, 1988). Some of this work is discussed in Chapter 10.

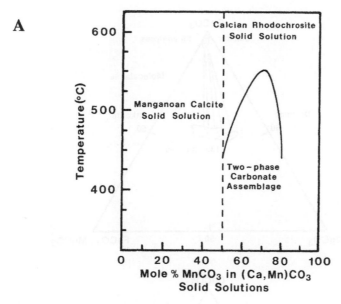

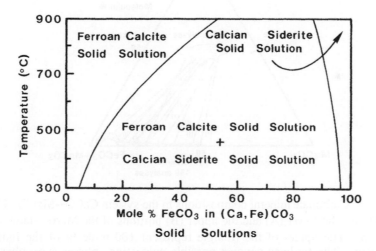

Figure 2.3. Experimental phase diagrams of the system (A) $CaCO_3$-$MnCO_3$ from 325-625°C, and (B) $CaCO_3$-$FeCO_3$ from 300-900°C. The regions under the curves are those in which a two-carbonate assemblage was observed. Elsewhere homogeneous solid solutions were obtained. (After Goldsmith, 1983, and Goldsmith et al., 1962.)

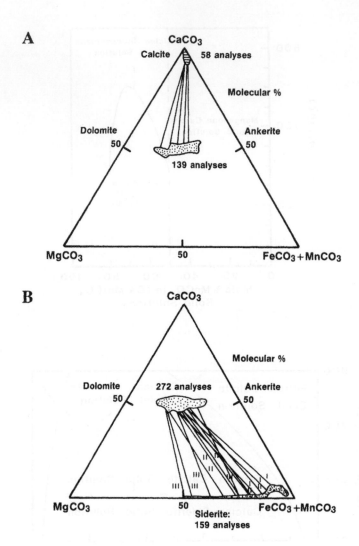

Figure 2.4. Carbonate mineral compositions in the system $CaCO_3$-$MgCO_3$-$FeCO_3$ - $MnCO_3$ for the lower (A) and the upper (B) portions of the Marra Mamba Iron-formation. The apices of the triangle represent 100 mole % of the indicated component. The tie-lines connect possible coexisting phases (i.e., physically touching pairs of carbonate minerals). The numbers I-III on the tie lines represent three different possible configurations. Data from Klein and Gole (1981), as represented in Woods (1988).

Alkali (monovalent) carbonate minerals also can occur in sediments. These phases have been studied in association with lake sediments. Their formation in these lacustrine environments has been used to deduce the conditions under which they formed based on their phase relations (see Figure 2.5). These phases usually precipitate during the late stages of evaporation of saline waters.

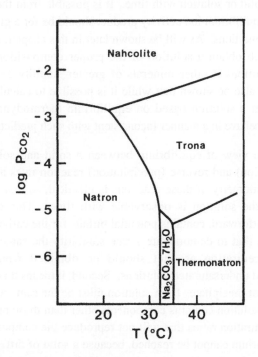

Figure 2.5. Stability fields of sodium carbonates as a function of P_{CO_2} and temperature. (After Monnin and Schott, 1984.)

Solubility behavior of carbonate minerals

General Considerations

A central concept important in studies of the geochemistry of carbonate systems is that of carbonate mineral solubility in natural waters. It is the touchstone against which many of the most important processes are described. In the previous chapter, methods for the calculation of the saturation state of a solution relative to a given carbonate mineral were presented. In addition, equations were given for determining the value of calcite and aragonite solubility products under different conditions. Here we will examine in more detail the concept of carbonate mineral solubility and the methods employed to determine the composition of a solution in equilibrium with respect to a given carbonate mineral.

A basic premise of solubility considerations is that a solution in contact with a solid can be in an equilibrium state with that solid so that no change occurs in the composition of solid or solution with time. It is possible from thermodynamics to predict what an equilibrium ion activity product should be for a given mineral for a set of specified conditions. As will be shown later in this chapter, however, it is not always possible to obtain a solution of the proper composition to produce the equilibrium conditions if other minerals of greater stability can form from the solution. It shall also be shown that while it is possible to calculate what mineral should form from a solution based on equilibrium thermodynamics, carbonate minerals usually behave in a manner inconsistent with such predictions.

The kinetic view of equilibrium between a solid and solution is that the forward (dissolution) and reverse (precipitation) reaction rates are the same, and that the net stoichiometry of these reactions is identical, so that no change in the composition of the solution is observable with time. This definition, while seemingly straightforward, contains potential pitfalls for the carbonate geochemist. First, it demands that to demonstrate a true solubility the same equilibrium ion activity or concentration product should be obtained from both initially supersaturated and undersaturated solutions. Second, it means that the composition of the material that precipitates from solution must be the same as that of the bulk solid. When the solution contains components other than those present in the bulk solid or in concentration ratios that will not reproduce the composition of the bulk solid, true equilibrium cannot be reached, because a solid of different composition may precipitate. A related problem is that if the ratio of components in the precipitating solid is different from that of the solution and significant precipitation takes place, on a solid to solution ratio basis, a heterogeneous solid will result. Under such conditions only the very outermost portion of the solid can be in short-term equilibrium with the solution and the composition of both phases should change with time.

The solid carbonate can also present several potential difficulties in solubility studies. These can be broken down into two major areas: heterogeneity in composition, and excess free energy associated with lattice strain or defects and surface free energy. The problem of solid heterogeneity presents itself in most sedimentary carbonates, and is especially important in biogenic carbonates such as magnesian calcites. The problem of lattice strain and high defect density is most commonly encountered as a result of crushing and grinding material for experimental use, and probably in nature because of mechanical and bio-erosional processes. Solution of this problem may not be amenable to thermodynamic treatment. It is extremely hard to quantify, but it can usually be overcome experimentally through annealing the solid.

The influence of surface free energy on solubility can be determined from straightforward thermodynamic considerations. The solubility of a solid can be calculated from the standard Gibbs free energy of reaction according to the relation:

$$\ln K_{sp} = \frac{-\Delta G_r^0}{RT} \tag{2.1}$$

The free energy change can be altered by addition of excess surface free energy.

$$\Delta G_T = \Delta G^0_r - G_s \tag{2.2}$$

where ΔG_T is the total Gibbs free energy change, and ΔG^0_r and G_s are, respectively, the standard free energy of the reaction and surface free energy on a molar basis. Note that G_s is subtracted from ΔG^0_r because it contributes to the energy of the solid that is associated with the reactant. G_s is the product of the surface free energy per unit area (σ) times the surface area of the solid per mole (A).

$$G_s = \sigma A \tag{2.3}$$

The total equation is then:

$$\ln K_{sp} = \frac{-(\Delta G_r^0 - G_s)}{RT} = \frac{-\Delta G_r^0}{RT} + \frac{\sigma A}{RT} \tag{2.4}$$

If we let $\ln K_b = -\Delta G^0_r / RT$, $E_s = \sigma/RT$ and take the exponential, the resulting equation is:

$$K_{sp} = K_b \, e^{E_s A} \tag{2.5}$$

It is clear from this equation that the solubility of a solid will increase exponentially with increasing surface area. For most solids, including carbonates, the influence of excess surface free energy is not large for particles unless they are under 0.1 microns in diameter (see Figure 2.6). It is important, however, to remember that

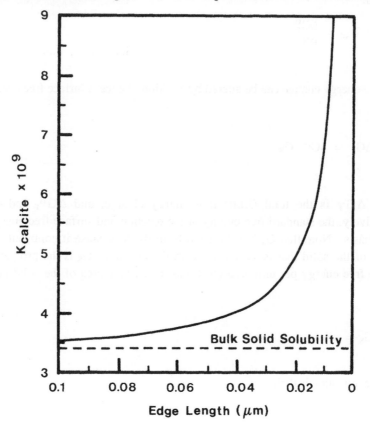

Figure 2.6. Change in calcite solubility with particle size, assuming cubic shape and $\sigma = 85$ ergs cm^{-2}.

only a few such particles, in a powder dominantly composed of much larger particles, can significantly influence solubility measurements. These theoretical conclusions will be discussed further when we consider carbonate solubility studies later in this chapter. Also, natural biogenic carbonate particles can be fine-grained, compositionally heterogeneous, structurally disordered, and complex in micro-architecture; all these factors can contribute to a carbonate mineral "solubility".

Before proceeding to the examination of recent experimental studies, one other factor concerning solubility must be briefly mentioned. This is the matter of equilibration time. As previously noted, one of the major criteria for determining if a solution is in equilibrium with a solid is the constancy of the solution composition with time. Many investigators have failed to carry out experimental solubility studies long enough to establish clearly this constancy in a satisfactory manner. All too often no observable change over a period of only an hour to a day has been used as a satisfactory time period. While it is generally not possible to carry out equilibrium experiments over years, considerable caution needs to be used in defining steady-state solution compositions in experimental work.

Calcite and Aragonite Solubility

During the past decade there have been four major studies of calcite solubility in dilute solutions, three of which also included aragonite (Jacobsen and Langmuir, 1974; Berner, 1976; Plummer and Busenberg, 1982; Sass et al., 1983). These studies were aimed at more accurate and precise determinations of the thermodynamic solubility product for these minerals. Earlier work on this topic was well summarized by Jacobsen and Langmuir (1974).

Jacobsen and Langmuir (1974) determined a value for pK_{sp} (25°C) for calcite of 8.42 ± 0.01, whereas Berner's (1976) value was 8.45 ± 0.01. Berner also determined the pK_{sp} for aragonite at 25°C to be 8.28 ± 0.03. An aspect of particular interest of both studies was that to obtain internal consistency for the carbonic acid system or constant values for the solubility products over the range of conditions studied, it was necessary to neglect ion pair formation. The potentially important ion pairs that could have formed in the experimental solutions are $CaHCO_3^+$ and $CaCO_3^0$. The former is by far the most important species, and a vast body of previous literature supported its existence (see Plummer and Busenberg, 1982, for summary).

Largely as a result of this unexpected need to neglect ion pairing to get consistent solubility results, Plummer and Busenberg (1982) and Sass et al. (1983) initiated further studies aimed at resolving the problem. It was found in both studies that ion pairing needed to be evaluated. An example of the difference in pK_{sp} results obtained by including or ignoring ion pairing for the study of Sass et al. (1983) is presented in Figure 2.7. The values obtained at 25°C for pK_{sp} calcite and aragonite, respectively, were 8.48 ± 0.02 and 8.34 ± 0.02 by Plummer and Busenberg (1982), and 8.48 ± 0.02 and 8.30 ± 0.02 by Sass et al. (1983). Thus, the results of these two studies are the same for calcite solubility, and close to being in agreement for aragonite solubility. Table 2.3 shows that when the appropriate ionic models are applied to earlier solubility studies for calcite and aragonite, they also are in good agreement with the more recent data.

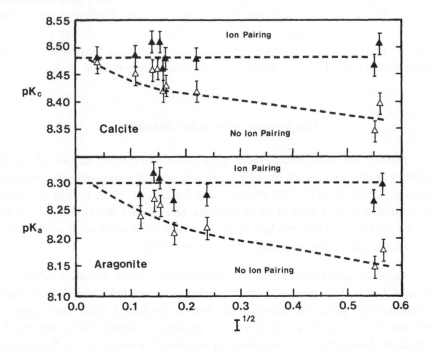

Figure 2.7. The influence of including or not including ion pairing in the calculation of calcite and aragonite solubility (pK_c or pK_a) as a function of ionic strength. (After Sass et al., 1983.)

The question then becomes one of why in both the studies of Jacobsen and Langmuir (1974) and Berner (1976), they found it necessary to neglect ion pairing. One of the reasons is that their studies were done in very dilute solutions over a rather narrow compositional range. If a wider range of solution compositions had been used, the need to include ion pairing would have become clear. Sass et al.

(1983) hypothesized that in the very dilute experimental solutions only a very small amount of calcium carbonate could dissolve, so that if any fine particles were present they could cause an apparent enhancement of solubility. They substantiated this hypothesis with an experiment in which both normal reagent grade calcite powder and the powder after acid leaching to remove fines were equilibrated with a very dilute solution (I = 0.003). The pK_{sp} value for the unleached powder was 8.42 ± 0.01 and for the leached calcite 8.48 ± 0.01.

Table 2.3. Summary of solubility measurements at 25°C.

Author	Solid	pK_{sp} Open System	Closed System
Sass et al. (1983)	Calcite	8.48 ± 0.02	8.49 ± 0.03
Jacobsen & Langmuir (1974)	"	8.48 ± 0.01	-
Berner (1976)	"	8.48 ± 0.02	-
Nakayama (1968)	"	8.50 + 0.03	-
Plummer & Busenberg (1982)	"	8.48 ± 0.02	-
Sass et al. (1983)	Aragonite	8.28 ± 0.02	8.31 ± 0.01
Berner (1976)	"	8.31 ± 0.04	-
Plummer & Busenberg (1982)	"	8.34 ± 0.02	-

Based on Sass et al., 1983.

In the previous chapter, the fact that stoichiometric and apparent constants have been widely used in seawater systems was discussed. Berner (1976) reviewed the problems of measuring calcite solubility in seawater, and it is these problems, in part, that have led to the use of apparent constants for calcite and aragonite. The most difficult problem is that while the solubility of pure calcite is sought in experimental seawater solutions, extensive magnesium coprecipitation can occur producing a magnesian calcite. The magnesian calcite should have a solubility different from that of pure calcite. Thus, it is not possible to measure pure calcite solubility directly in seawater.

The central question was, how major is the influence of coprecipitated magnesium? Berner (1976) used a clever approach to find out. His reasoning was that whereas calcite has a major coprecipitate from seawater (Mg) capable of altering its solubility, aragonite that can be precipitated from seawater is greater than 99% pure, and its solubility in seawater should be measurable. He also pointed out that if the solubility ratio of calcite to aragonite was precisely known, it would be possible to predict what the solubility of pure calcite in seawater should be. While Berner's line of reasoning was correct, the solubility constants that he used to test his hypothesis have subsequently proven questionable.

Let us examine this problem using more recently published constants. The first question is what value should be used for the aragonite to calcite solubility ratio? The value obtained from the study of Plummer and Busenberg (1982) is 1.38, whereas Sass et al. (1983) found a ratio of 1.51. A value of 1.45 ± 0.07 is consistent with most measurements and their uncertainty.

The next problem is that of aragonite solubility in seawater. Morse et al. (1980) made a study of the solubility of calcite and aragonite in seawater. Their findings for calcite solubility were in good agreement with earlier work (e.g., Ingle, 1975). They found a much lower solubility, however, for aragonite than that observed by previous investigators. They demonstrated that the primary reason for this difference was that other investigators had not waited a sufficiently long period of time to reach equilibrium. The value obtained by Morse et al. (1980) for aragonite was in good agreement with the value predicted from thermodynamic solubility products and ion activity coefficients in seawater. In addition, their value for the solubility ratio of aragonite to calcite in seawater was 1.51, in good agreement with the dilute solution studies. From these considerations, it appears that the coprecipitation of magnesium (~8 mole % $MgCO_3$) in calcite forming slowly from seawater does not change its solubility beyond our current uncertainty in the measurements, which probably is less than 10 percent. The topic of magnesian calcite solubility will be dealt with in detail in the next chapter.

Before continuing, it is important to note that the solubility behavior of aragonite in seawater has proven rather strange, in that for equilibration periods of less than about a month, different solubility values can be obtained in an apparently reversible manner. This result points to some sort of short-term, solubility controlling surface phase and may have implications for processes such as oöid formation which is discussed in Chapter 5.

Methods for the Calculation of Equilibrium Solution Composition Under Different Conditions

Whereas the basic principles for calculating the composition of a solution in equilibrium with a carbonate mineral are relatively straightforward, the application of these types of calculations to real world situations is commonly less obvious and fraught with difficulties. Consequently, we will present a series of calculations applied to natural geochemical systems, inspired by those originally given by Garrels and Christ (1965). A major difference in our presentation will be a more precise approach to calculated solution compositions in which terms will not be dropped or other major approximations made. Our approach necessitates at least a programmable calculator or preferably a minicomputer, because in many cases iterative solutions of lengthy equations are involved.

The different cases are presented in terms of a story about the passage of a raindrop through a series of environments with which it is required to come to equilibrium. These situations are summarized in Figure 2.8. The raindrop 1) equilibrates with the atmosphere, 2) falls on a limestone rock and equilibrates with calcite, 3) passes through a calcium carbonate-free soil where carbon dioxide is added, 4) enters a limestone aquifer and equilibrates with calcite, and 5) emerges from a spring and precipitates calcite as it equilibrates with the atmosphere. These situations have direct application to problems dealing with freshwater reaction pathways in the natural environment, and provide the reader with some idea of how calculations involving carbonate mineral-water reactions are done.

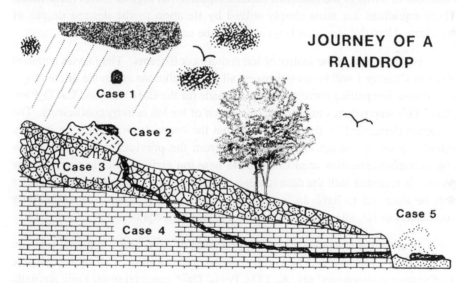

Figure 2.8. Journey of the raindrop (heavy black) from sky (1) to calcite rock (2) to soil (3) to limestone aquifer (4) and out of a spring (5).

Before proceeding, a few general comments will be helpful. In most cases at least one of the four carbonate system analytical quantities (total alkalinity, total CO_2, carbon dioxide partial pressure, and pH) will be known. The equilibrium equations relating these quantities, along with those for water and calcite, will be used frequently in calculations. Also, in cases where calcite dissolution and precipitation occur, the charge balance equation demands, in these fresh water systems, that:

$$\Delta m_{Ca^{2+}} = 0.5 \, \Delta A_t \qquad\qquad (2.6)$$

In systems that are closed to the outside exchange environment, the mass balance relation:

$$\Delta m_{Ca^{2+}} = \Delta \Sigma CO_2 \qquad\qquad (2.7)$$

can also be used. The general objective will be to find, through suitable algebraic manipulation, one of the unknowns, which in our examples will generally be pH (as m_{H^+} initially). It is not always possible to write a direct solution for the final equations in terms of the unknown because equations as high as fourth order result. These equations are most simply solved by iteration to the desired degree of precision. Here this precision is set at 1% for the unknown.

Finally, there is the matter of ion activity coefficients. The Davies equation given in Chapter 1 will be used, because all of the solutions are in the dilute range. In addition, ion pairing corrections will be made for the $CaHCO_3^+$ and $CaCO_3^0$ ion pairs. This step requires iteration in calculation of the ion activity coefficients. The sequence demanded by the problems is that the concentrations must be initially calculated using ion activity coefficients from the previous case. These new concentrations are then used to calculate new ion activity coefficients, and the process is repeated until the desired degree of precision is reached. Neutral species will be assumed to have an ion activity coefficient of 1, and the ion activity coefficients of H^+, OH^-, and $CaHCO_3^+$ will be assumed equal.

In all of the examples given, a temperature of 25°C and 1 atm total pressure are assumed. The task will be to calculate the concentration of all of the carbonic acid system components, pH, A_t, CO_2, P_{CO_2}, Ca^{2+} concentration, ionic strength, and the appropriate ion activity coefficients. The values of the thermodynamic constants used in these calculations were calculated from appropriate equations in Chapter 1. Concentrations are on the molal scale, and results for each case are summarized in Table 2.4.

Table 2.4. Composition of the raindrop for different cases.

Case	P_{CO_2}	pH	ΣCO_2	A_t	A_c	$m_{Ca^{2+}}$
1	3.30×10^{-4}	5.65	1.34×10^{-5}	0	2.22×10^{-6}	0
2	3.30×10^{-4}	8.25	1.01×10^{-3}	1.02×10^{-3}	1.02×10^{-3}	5.17×10^{-4}
3	1.19×10^{-1}	5.71	5.08×10^{-3}	1.02×10^{-3}	1.02×10^{-3}	5.17×10^{-4}
4	4.32×10^{-2}	6.86	7.66×10^{-3}	6.20×10^{-3}	6.20×10^{-3}	3.08×10^{-3}
5	3.30×10^{-4}	8.69	2.79×10^{-3}	2.86×10^{-3}	2.85×10^{-3}	7.72×10^{-5}

Case	m_{CO_2}	$m_{HCO_3^-}$	$m_{CO_3^{2-}}$	$\gamma_{Ca^{2+}}$	$\gamma_{HCO_3^-}$	$\gamma_{CO_3^{2-}}$	γ_{H^+}
1	1.12×10^{-5}	2.23×10^{-6}	4.68×10^{-11}	0.99	1.00	0.99	1.00
2	1.12×10^{-5}	9.86×10^{-4}	1.47×10^{-5}	0.80	0.94	0.54	0.96
3	4.06×10^{-3}	1.02×10^{-3}	4.36×10^{-8}	0.80	0.93	0.54	0.96
4	1.46×10^{-3}	6.19×10^{-3}	6.66×10^{-6}	0.56	0.83	0.28	0.91
5	1.12×10^{-5}	2.70×10^{-3}	2.94×10^{-5}	0.69	0.95	0.78	0.96

Case 1. A raindrop of pure water equilibrates with atmospheric CO_2. Let a raindrop of pure water with a pH = 7 ($A_t = 0$) form and come to equilibrium with CO_2 in the atmosphere at $P_{CO_2} = 330$ μatm. We assume that there is no Ca in the initial water. Also, A_t cannot change because a neutral species (CO_2) is being added. The first thing that must be done is to establish clearly the conditions before the CO_2 enters the raindrop. From the pH being equal to 7, it follows that $a_{H^+} = 10^{-7}$ and $a_{OH^-} = (K_w / a_{H^+}) = 10^{-7}$. Because H^+ and OH^- are not involved in ion pairing reactions, from the Davies equation their ion activity coefficients are equal. Therefore, their concentrations will also be equal.

Because A_t is known (0) and cannot change upon addition of only CO_2, the equation for A_t can be readily written for the addition of CO_2. Using the equations for the equilibrium relations in the carbonic acid system, it is possible to write an equation for alkalinity in terms of P_{CO_2} (known) and m_{H^+} (our unknown).

$$A_t = \left(\frac{K_H K_1 P_{CO_2}}{\gamma_{HCO_3^-} \gamma_{H^+} m_{H^+}}\right) + 2\left(\frac{K_H K_1 K_2 P_{CO_2}}{\gamma_{CO_3^{2-}} \gamma_{H^+}^2 m_{H^+}^2}\right) + \left(\frac{K_W}{\gamma_{H^+}^2 m_{H^+}}\right) - m_{H^+} = 0$$

(2.8)

This equation can be rearranged to a form that is more convenient for solving for the value of m_{H^+} by iteration.

$$m_{H^+}^3 - m_{H^+}\left[\left(\frac{K_H K_1 P_{CO_2}}{\gamma_{HCO_3^-} \gamma_{H^+}}\right) + \left(\frac{K_W}{\gamma_{H^+}^2}\right)\right] - 2\left(\frac{K_H K_1 K_2 P_{CO_2}}{\gamma_{CO_3^{2-}} \gamma_{H^+}^2}\right) = 0$$

(2.9)

Because it is known from the reactions involved that the net direction of pH change will be to decrease, it is possible to make an initial estimate of the final pH of our raindrop. The estimate is based on the idea that both OH^- and CO_3^{2-} concentrations will be relatively small under acidic conditions. The pH then should be near the square root of the HCO_3^- term (pH = 5.7).

Once m_{H^+} is known, the other carbonic acid system components can be calculated from m_{H^+} and P_{CO_2} using the relations given in Chapter 1. It is then possible to calculate the ionic strength and activity coefficients. These activity coefficients can be substituted back into the equation for m_{H^+} using the previously calculated value for the initial guess. This process is repeated until m_{H^+} is constant to within 1%. Then the other parameters are calculated as before.

Case 2. The raindrop falls on a limestone rock and comes to equilibrium with calcite, while remaining in equilibrium with the P_{CO_2} of the atmosphere. This process results in dissolution of the rock, and addition of calcium and alkalinity to the raindrop. The knowns are P_{CO_2}, the initial composition of the solution, and that it must be in equilibrium with calcite. In this case, as in all others, electroneutrality must apply, but mass balance will not because of CO_2 exchange with the atmosphere. First the equations will be written for calcium.

$$m_{Ca^{2+}} = \left(\frac{K_C}{\gamma_{Ca^{2+}} \gamma_{CO_3^{2-}} m_{CO_3^{2-}}}\right) = \frac{1}{2}\Delta A_t$$

(2.10)

Because the initial alkalinity is 0, the change in alkalinity is simply the final value ($\Delta A_t = A_t$). Using these relations and the alkalinity definition we can write:

$$\left(\frac{2K_C}{\gamma_{Ca^{2+}}\ \gamma_{CO_3^{2-}}\ m_{CO_3^{2-}}}\right) = \left(m_{HCO_3^-} + 2m_{CO_3^{2-}} + m_{OH^-} - m_{H^+}\right)$$
$$- \left[m_{HCO_3^-}^i + 2m_{CO_3^{2-}}^i + m_{OH^-}^i - m_{H^+}^i\right] \tag{2.11}$$

where the superscript i indicates initial values. The quantities in brackets are known from the first case; let them equal α. Because there is no initial alkalinity in this case, $\alpha = 0$. Substituting the equilibrium definitions in terms of m_{H^+} and P_{CO_2}, and rearranging yields:

$$m_{H^+}^4\left(\frac{2\gamma_{H^+}^2\ K_C}{\gamma_{Ca^{2+}}\ K_H K_1 K_2 P_{CO_2}}\right) + m_{H^+}^3 + \alpha\ m_{H^+}^2 - m_{H^+}\left[\left(\frac{K_H K_1 P_{CO_2}}{\gamma_{HCO_3^-}\ \gamma_{H^+}}\right) + \left(\frac{K_W}{\gamma_{H^+}^2}\right)\right]$$
$$- 2\left(\frac{K_H K_1 K_2 P_{CO_2}}{\gamma_{CO_3^{2-}}\ \gamma_{H^+}^2}\right) = 0 \tag{2.12}$$

The value of m_{H^+} is determined by the previously described technique from this equation along with the other carbonic acid system components. The calcium concentration is calculated from the relation given in equation 2.10.

Case 3. The raindrop now runs off the calcite rock and flows into a calcium-carbonate-free soil, where carbon dioxide is added from the decomposition of organic matter until the total CO_2 is five times its original value. For simplicity, it will be assumed that no other components are added and no other reactions take place between the soil and the raindrop. In this case calcium concentration and total alkalinity are constant, and the total CO_2 and initial concentrations are known. From the definition of terms and equilibrium relations we can write:

$$\Sigma CO_2 - A_t = m_{CO_2} - m_{CO_3^{2-}} - m_{OH^-} + m_{H^+}$$
$$= \left(\frac{K_H P_{CO_2}}{\gamma_{CO_2}}\right) - \left(\frac{K_H K_1 K_2 P_{CO_2}}{\gamma_{CO_3^{2-}}\ \gamma_{H^+}^2\ m_{H^+}^2}\right) - \left(\frac{K_W}{\gamma_{H^+}^2\ m_{H^+}}\right) + m_{H^+} \tag{2.13}$$

The equation contains two unknowns: P_{CO_2} and m_{H^+}. The alkalinity definition can be used to get P_{CO_2} in terms of A_t and m_{H^+}. The resulting equation is:

$$P_{CO_2} = \frac{A_t + m_{H^+} - \left(\dfrac{K_W}{\gamma_{H^+}^2\, m_{H^+}}\right)}{\left(\dfrac{K_H K_1}{\gamma_{HCO_3^-}\, \gamma_{H^+}\, m_{H^+}}\right) + \left(\dfrac{2 K_H K_1 K_2}{\gamma_{CO_3^{2-}}\, \gamma_{H^+}^2\, m_{H^+}^2}\right)}$$

(2.14)

If the numerator of the above equation is represented by X and the denominator by Y, the following equation which has only m_{H^+} as an unknown results. Again, m_{H^+} and the other quantities are calculated as before:

$$\left(\frac{X}{Y}\right)\left[\left(\frac{K_H}{\gamma_{CO_2}}\right) + \left(\frac{K_H K_1 K_2}{\gamma_{CO_3^{2-}}\, \gamma_{H^+}^2\, m_{H^+}^2}\right)\right] - \left(\frac{K_W}{\gamma_{H^+}^2\, m_{H^+}}\right) - m_{H^+} + A_t - \Sigma CO_2 = 0$$

(2.15)

   Case 4. The raindrop continues on its journey, now flowing into a limestone (calcite) aquifer. Here it again comes to equilibrium with calcite, but under closed system conditions. The only knowns are the initial composition of the solution and that it must react to equilibrium with calcite. This is a particularly difficult case for an exact solution. Both mass balance and electroneutrality equations must be used. Let us write these out again.

Mass Balance

$$\Delta m_{Ca^{2+}} = \Delta m_{CO_2} + \Delta m_{HCO_3^-} + \Delta m_{CO_3^{2-}}$$

(2.16)

Charge Balance

$$2\Delta m_{Ca^{2+}} = \Delta m_{HCO_3^-} + 2\Delta m_{CO_3^{2-}} + \Delta m_{OH^-} - \Delta m_{H^+}$$

(2.17)

If we double the mass balance equation and subtract the charge balance equation, the result is:

$$2\Delta m_{CO_2} + \Delta m_{HCO_3^-} + \Delta m_{H^+} - \Delta m_{OH^-} = 0 \tag{2.18}$$

Garrels and Christ (1965) in a somewhat similar problem were able to drop the CO_2 and H^+ terms, but because of the high initial concentrations of the components in this case it is not possible to do so. To make the equations easier to follow, the knowns from initial concentrations can be collected and the above equation rewritten in terms of m_{H^+} and P_{CO_2}. It is then possible to solve for P_{CO_2} in terms of m_{H^+}. Let:

$$S = 2\, m^i_{CO_2} + m^i_{HCO_3^-} + m^i_{H^+} - m^i_{OH^-}$$

then:

$$2K_H P_{CO_2} + \left(\frac{K_H K_1 P_{CO_2}}{\gamma_{HCO_3^-}\, \gamma_{H^+}\, m_{H^+}} \right) + m_{H^+} - \left(\frac{K_W}{\gamma^2_{H^+}\, m_{H^+}} \right) - S = 0 \tag{2.19}$$

$$P_{CO_2} = \frac{S - m_{H^+} + \left(\dfrac{K_W}{\gamma^2_{H^+}\, m_{H^+}} \right)}{2K_H + \left(\dfrac{K_H K_1}{\gamma_{HCO_3^-}\, \gamma_{H^+}\, m_{H^+}} \right)} \tag{2.20}$$

The next step to remember is that the solution is in equilibrium with calcite. This equilibrium condition can be expressed as:

$$m_{Ca^{2+}} = \frac{\gamma^2_{H^+}\, m^2_{H^+}\, K_C}{\gamma_{Ca^{2+}}\, K_H K_1 K_2 P_{CO_2}} \tag{2.21}$$

The mass balance equation can be used next with the following algebraic manipulation to yield an equation where P_{CO_2} and m_{H^+} are the only unknowns.

$$\Delta m_{Ca^{2+}} = \left(\frac{\gamma_{H^+}^2 \, m_{H^+}^2 \, K_C}{\gamma_{Ca^{2+}} \, K_H K_1 K_2 P_{CO_2}} \right) - m_{Ca^{2+}}^i = \Sigma CO_2 - \Sigma CO_2^i$$

(2.22)

and

$$\left(\frac{\gamma_{H^+}^2 \, m_{H^+}^2 \, K_C}{\gamma_{Ca^{2+}} \left(K_H K_1 K_2 P_{CO_2} \right)} \right) = K_H P_{CO_2} + \left(\frac{K_H K_1 P_{CO_2}}{\gamma_{HCO_3^-} \, \gamma_{H^+} \, m_{H^+}} \right)$$

$$+ \left(\frac{K_H K_1 K_2 P_{CO_2}}{\gamma_{CO_3^{2-}} \, \gamma_{H^+}^2 \, m_{H^+}^2} \right) - \Sigma CO_2^i + m_{Ca^{2+}}^i$$

(2.23)

Then:

$$P_{CO_2} \left[K_H + \left(\frac{K_H K_1}{\gamma_{HCO_3^-} \, \gamma_{H^+} \, m_{H^+}} \right) + \left(\frac{K_H K_1 K_2}{\gamma_{CO_3^{2-}} \, \gamma_{H^+}^2 \, m_{H^+}^2} \right) \right]$$

$$- \left(\frac{1}{P_{CO_2}} \right) \left(\frac{\gamma_{H^+}^2 \, m_{H^+}^2 \, K_C}{\gamma_{Ca^{2+}} \left(K_H K_1 K_2 \right)} \right) + m_{Ca^{2+}}^i - \Sigma CO_2^i = 0$$

(2.24)

From our two equations (2.20 and 2.24) with two unknowns it is possible to write an equation where m_{H^+} is the only unknown. It and the other parameters are then evaluated as before:

$$\left[\frac{S - m_{H^+} + \left(\dfrac{K_W}{\gamma_{H^+}^2 \, m_{H^+}}\right)}{2K_H + \left(\dfrac{K_H K_1}{\gamma_{HCO_3^-} \, \gamma_{H^+} \, m_{H^+}}\right)}\right]\left[K_H + \left(\frac{K_H K_1}{\gamma_{HCO_3^-} \, \gamma_{H^+} \, m_{H^+}}\right) + \left(\frac{K_H K_1 K_2}{\gamma_{CO_3^{2-}} \, \gamma_{H^+}^2 \, m_{H^+}^2}\right)\right]$$

$$- \left[\frac{2K_H + \left(\dfrac{K_H K_1}{\gamma_{HCO_3^-} \, \gamma_{H^+} \, m_{H^+}}\right)}{S - m_{H^+} + \left(\dfrac{K_W}{\gamma_{H^+}^2 \, m_{H^+}}\right)}\right]\left(\frac{\gamma_{H^+}^2 \, m_{H^+}^2 \, K_C}{\gamma_{Ca^{2+}} \left(K_H K_1 K_2\right)}\right) + m_{Ca^{2+}}^i - \Sigma CO_2^i = 0$$

$$(2.25)$$

There is one rather nasty twist in the ion pair evaluation where a negative value for free carbonate ion concentration results, because the initial value which must be used for the carbonate ion activity coefficient is much larger than it is under the new conditions. This demands some maneuvers with both the calcite solubility constant and the calcium carbonate ion pair association constant. These will not be gone into here.

Case 5. The raindrop now emerges from the aquifer in a spring and can again come into equilibrium with the atmosphere. The release of CO_2 to the atmosphere causes the pH to rise and results in the solution precipitating calcite until it is again in equilibrium with both calcite and the atmosphere. The knowns are the initial composition and the two equilibrium constraints. Charge balance applies, but mass balance does not. The solution to this problem can be found by using equation 2.12 and substituting $\beta = \alpha + m_{Ca^{2+}}^i$ for α.

Unfortunately(?), the sun came out and evaporated our raindrop before it could continue its journey and have further adventures. So it goes (Vonnegut, 1967)!

Surface chemistry of carbonate minerals

Basic Principles

Chemical interactions occurring at or near the solid-solution interface exert major influences on the behavior of carbonate minerals in the natural environment. The control of calcite and aragonite solubility, retention of the supersaturated state of near surface seawater with respect to calcite and aragonite, the inhibition of calcite and aragonite dissolution and precipitation, control of crystal morphology, and the control of the rate and pathways of carbonate mineral diagenesis are among the most significant of these influences.

The interface of a solid in contact with a solution represents a discontinuity in the system that has unique characteristics. In fact, this boundary between phases has usually been modeled as a separate phase with its own special thermodynamic properties. The surface of a solid or liquid is a region where structure is disrupted and bonding requirements are not satisfied in a manner typical of the bulk phase. These features give rise to macroscopic manifestations such as surface tension and charge. The surface is an area of excess free energy that can significantly alter the free energy of the system. Major changes in the solubility of the phase (see previous section) can result from this excess free energy.

Because of the excess free energy and disrupted bonding occurring in the surficial region, it is a particularly likely area for interactions to obtain. These are usually generalized as adsorption reactions when dealing with solid interactions with gas or dissolved fluid phase constituents. Two major classes of interaction are generally identified. They are physical adsorption in which purely physical processes predominate (e.g., helium adsorption on platinum metal), and chemisorption in which chemical bonding occurs (e.g., phosphate adsorption on calcite). Actually, neither form of adsorption usually occurs by itself. Physical adsorption has proven much more amenable to modeling. For solid-solution interactions, classical models for the interfacial region based on physical interactions include the Helmholtz, Stearn, and Gouy-Chapman models. Discussions of these and more sophisticated recent models are available in many books and articles (e.g., Adamson, 1982, Davis and Hayes, 1986; Stumm, 1987). Because physical adsorption is not nearly as important for carbonates interacting with solutions as chemical adsorption, these models will not be discussed in detail.

An important concept in treating the adsorption of a component from one phase onto another is that of the adsorption isotherm. An adsorption isotherm is simply a means of representing the equilibrium distribution of a component between its dissolved and adsorbed forms at a constant temperature. This relationship is almost always expressed as a plot of the surface concentration versus the dissolved concentration. There are, however, some subtle points to be remembered. First, the idea of equilibrium implies that the reaction should be readily reversible. In many instances where chemisorption is involved, this reversibility is not the case. Under these circumstances the boundary between adsorption and coprecipitation reactions becomes unclear.

Second, there are problems in using concentrations, because the true driving force for the adsorption reaction is chemical potential, which is related to activity. For the dissolved component this means that adsorption isotherms measured in one solution do not necessarily apply to other solutions; a point commonly overlooked by geochemists, particularly with regard to organic compounds. Also, variation in solution composition can result in the introduction of other ions or compounds that have an affinity for the surface. These ions may severely alter the adsorption of the component of interest.

The situation is even more complex when dealing with the adsorbed component. There are no generally applicable methods for estimating activity coefficients for adsorbed compounds. Because the surfaces of most real solids are heterogeneous, the energy of interaction between the solid and adsorbate may cover a wide spectrum making simple theoretical treatments almost impossible. Even the concept of surface concentration can become very complex. If solids with complex microporosity such as biogenic carbonates are involved, it is usually not possible to determine what fraction of the total surface area of the solid is interacting with the solution. Also, complex interactions such as the formation of two-dimensional micelles on surfaces can lead to a patchy type of adsorption for which average surface concentration values are misleading.

A final area of difficulty is in the application of data analysis to specific models of adsorption isotherms. This difficulty results from the fact that different models for adsorption isotherms generate plots of surface versus dissolved concentration that have characteristic shapes. If a plot of observational data results in a curve with a shape similar to that generated by a model, this result is often taken as proof that the particular model applies. Unfortunately, this assumption has been made for situations where many of the basic requirements of the model are violated in the system under study. The Langmuir adsorption isotherm model has suffered considerable abuse by geochemists in this regard. It should be remembered that "shapes" of adsorption isotherms are far from proof that a specific model applies.

Two other closely related concepts are connected with the idea of adsorption isotherms. These are the distribution coefficient (usually called K_D) which is the ratio between the proportion of a component that is dissolved relative to that associated with solids, and surface affinity, which is the tendency of a dissolved component to adsorb on a specific substrate. K_D's are commonly rather poorly defined in terms of the chemistry of the solution in which the adsorption is occurring and the particular substrate upon which adsorption takes place. These coefficients are generally obtained from a site-specific data base, but commonly are used in a general manner. Surface affinity is a more precise parameter in that it is related to the slope of the linear portion of an adsorption isotherm for a particular substrate.

The concept of chemisorption is of central importance in understanding sorption reactions between carbonate minerals and dissolved components of natural waters. The type of bonding involved can range from relatively weak hydrogen bonding with organic compounds to ionic or covalent bonds that are difficult to break once formed. The fact that considerable activation energy usually is needed to break such bonds means that from a practical standpoint the adsorption reaction can be largely irreversible. In addition, surface heterogeneity can play a major role in the type and extent of bonding that occurs. The BCF (Burton-Cabrerra-Frank) model has proven useful in visualizing some of the types of surface sites which can exist (all of which have different energies) and how an ion may interact with them. An illustration of the different major types of surface sites is presented in Figure 2.9. This type of model has been used by Berner and Morse (1974) in describing the influence of phosphate sorption on carbonate mineral reaction kinetics. Surface sites with different energies can react at different rates with sorbates. This can result in long periods of adsorption during which the rate of uptake exponentially decreases with time (e.g., see deKanel and Morse, 1978).

Chemisorption raises basic questions for the carbonate geochemist about the boundary between sorption and coprecipitation. If the adsorption reaction takes place in a solution that is also supersaturated with respect to the carbonate mineral substrate, then the adsorbed ions can be buried in the growing layers of the mineral and become coprecipitates. This mechanism can result in distribution coefficients that are dependent on growth rates. Also, when chemisorption is involved, an entirely new phase or a coprecipitate can form in the near-surface region of the carbonate (e.g., see Morse, 1986; Davis et al., 1987). A classic example is apatite formation on calcite in dilute solutions (e.g., Stumm and Leckie, 1970).

The basic properties of calcium carbonate mineral surfaces have not been extensively studied. The work of Somasundaran and Agar (1967) on the zero point of charge (ZPC) on calcite certainly has at least qualitative application to other

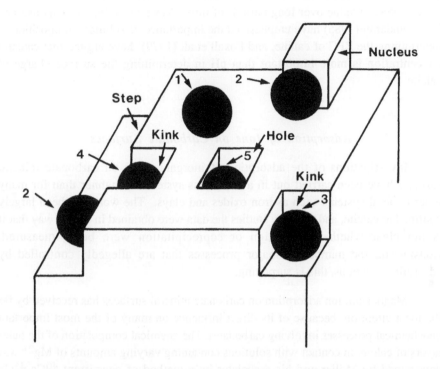

Figure 2.9. BCF model for surface with adsorbed ions (black spheres). The terms step, kink, hole and nucleus refer to different types of surface sites, and the numbers refer to the number of chemical bonds likely.

carbonate minerals, and is important in identifying many of the complex processes occurring on carbonate mineral surfaces. They investigated the calcite ZPC by measuring streaming potential, solution equilibrium, and flotation response as a function of pH. It was found that the ZPC could range from a pH of 8 to 9.5. This means that in most natural solutions, if no other ions were to interact with the calcite surface, the surface would have a positive to close to neutral charge. Thermodynamic calculations indicated that Ca^{2+}, HCO_3^-, CO_3^{2-}, H^+, and OH^- could all participate as potential determining ions. Somasundaran and Agar (1967) hypothesized that preferential hydrolysis of surface ions or the adsorption of complexes formed in solution could all play an important role in establishing surface charge. They observed that when calcite was added to solutions with an initial pH near the isoelectric point, the zeta potential and pH initially changed in one direction and then later in the opposite direction. This result indicated that a series of fast and slow reactions were responsible for establishing the charge development at the surface. They also found that the surface charge could exhibit a slow change in value over long periods of time. More recently, Amankonah and Somasundaran (1985) have emphasized the importance of solution composition in determining the ZPC of calcite, and Foxall et al. (1979) have argued that calcium concentration is more important than pH in determining the surface charge of calcite.

Adsorption of Ions on Carbonate Surfaces

Investigations of the adsorption of inorganic ions on carbonate mineral surfaces have been carried out in a much less systematic manner than for many other mineral systems such as iron oxides and clays. The work has been largely confined to calcite, and in many studies the data were obtained in such a way that it is not clear whether adsorption or coprecipitation were being measured. Considering the number of major processes that are allegedly controlled by adsorption reactions, this is surprising.

Magnesium ion adsorption on carbonate mineral surfaces has received by far the most attention, because of its direct influence on many of the most important geochemical processes involving carbonates. The chemical composition of the outer layers of calcite in contact with solutions containing varying amounts of Mg^{2+} was determined by Möller and his associates by a method of concurrent ^{40}Ca-^{45}Ca isotope and Ca^{2+}-Mg^{2+} ion exchange. Measurements were done on both calcite powders and single crystals. It was established that for a single calcite crystal, not more than a monolayer participates in the isotope exchange process (Möller and Sastri, 1974). These studies showed that for a solution Mg^{2+} to Ca^{2+} molar ratio of between 2 and 10, the Mg to Ca molar ratio on the surface of calcite is about 1.

The applicability of scanning Auger spectroscopy to the analysis of carbonate mineral surface reactions was demonstrated by Mucci and Morse (1985), who carried out an investigation of Mg^{2+} adsorption on calcite, aragonite, magnesite, and dolomite surfaces from synthetic seawater at two saturation states. Results are summarized in Table 2.5.

Aragonite is the only one of the four carbonate minerals examined that does not have a calcite-type rhombohedral crystal structure. For all the minerals examined, with the exception of aragonite, the two solution saturation states studied represent supersaturated conditions, because at a saturation state of 1.2 with respect to calcite, the seawater solution is undersaturated (0.8) with respect to aragonite.

Table 2.5. The Mg to Ca concentration ratio on the surface of four carbonate mineral crystals after extended exposure to synthetic seawater at two different saturation states. (After Mucci and Morse, 1985.)

| Mineral | $([Mg^{2+}]/[Ca^{2+}])_{surf}$ | |
	$\Omega_c = 1.2$	$\Omega_c = 8$
Aragonite	0.63	0.03
Calcite	0.83	1.34
Dolomite	3.05	2.67
Magnesite	120.00	8.00

At the higher saturation state, the seawater solution is more than 5 times supersaturated with respect to aragonite so that aragonite would be expected to precipitate on the aragonite seed crystal. Results indicated that Mg^{2+} is adsorbed between 25 to 40 times less on aragonite than on calcite from solutions supersaturated with respect to both minerals.

The results from surface seawater solutions indicated that there was an overgrowth on the magnesite crystal. This overgrowth had a composition corresponding to a 1.5 mole % calcian magnesite. It is likely that this overgrowth was composed of a mixture of minerals, possibly of hydrated magnesium carbonate and calcite, formed through heterogeneous precipitation. The depth profiling Auger analysis did not detect the presence of any overgrowth on the magnesite crystals exposed to the less supersaturated solution. A Mg to Ca surface ratio of 120 for magnesite was estimated.

The Mg to Ca surface ratios for dolomite in both seawater solutions were statistically identical. A Mg to Ca surface ratio of 3 is predicted from a simple site-specific adsorption model, and is in reasonable agreement with observed values.

The Mg to Ca surface ratios for calcite in both supersaturated seawater solutions were nearly identical. The lower Mg to Ca surface ratio obtained in the less supersaturated solution may be the result of incomplete coverage of the pure calcite crystal by the magnesian calcite overgrowth. The Mg to Ca surface ratio on calcite exposed to both saturation state solutions is in close agreement with the value of 1 obtained in a solution with a Mg^{2+} to Ca^{2+} ratio of 5 by Möller and his associates.

A number of adsorption studies have been conducted for the adsorption of transition and heavy metals on carbonates. Mn^{2+} adsorption has been the most intensely studied mechanism. This is possibly because of its importance to carbonate mineral behavior in marine sediments (see Chapter 4). Studies of Mn^{2+} interaction with the surface of calcite in dilute aqueous solutions indicated that Mn^{2+} uptake is approximately balanced by Ca^{2+} release (McBride, 1979). It was suggested that the calcium release may not be the result of direct displacement, but rather results from the following reactions:

$$Mn^{2+} + HCO_3^- \rightarrow MnCO_3 + H^+ \tag{2.26}$$

$$H^+ + CaCO_3 \rightarrow Ca^{2+} + HCO_3^- \tag{2.27}$$

Electron spin resonance (ESR) studies indicated that no discrete new phase formed at low Mn^{2+} surface concentrations, but that at high concentrations of Mn^{2+} on the calcite surface $MnCO_3$ was nucleated. The pattern of uptake showed that there is an initial rapid adsorption, followed by a period of little uptake, during which $MnCO_3$ nucleation occurs, and finally a slow but steady growth of $MnCO_3$. By considering the calcite surface as a two dimensional solid (a monolayer of unit calcite cell dimensions), McBride was able to calculate a surface distribution coefficient of 10 to 25, a range of values in good agreement with those determined for the solid. This surface distribution coefficient decreased as surface Mn^{2+} concentration increased. These studies were extended by Franklin and Morse (1983) into other solutions with the primary objective being the determination of Mg^{2+} influence on Mn^{2+} adsorption. The interpretation for the uptake of Mn^{2+} from dilute solutions onto calcite was that initial rapid uptake represents an adsorption reaction. This reaction is followed by a period of relatively little uptake, during which time $MnCO_3$ nucleation takes place. In seawater only a two-stage process was found in which initial rapid uptake was followed by steady Mn^{2+} removal, with no intervening quiescent period. The rate of uptake following the initial fast adsorption phase was found to be first order with respect to dissolved Mn^{2+} concentration. The first-order rate constant measured in seawater is approximately 4000 times less than the first-order rate constant found in dilute solutions. Because the activity coefficient for Mn^{2+} in the two solutions differs by less than an order of magnitude, the difference in observed rate constants cannot be explained by differences in dissolved Mn^{2+} activity and must represent a change in processes occurring on the calcite surface.

It is interesting to note that the uptake patterns found for Mn^{2+} on the surface of calcite are similar to those observed for orthophosphate uptake on calcite. Stumm and Leckie (1970) observed a three-stage process for phosphate uptake on calcite surfaces from dilute solutions, representing adsorption, nucleation, and apatite growth. However, deKanel and Morse (1978) found a two-stage pattern of phosphate uptake on calcite from seawater, with an initial rapid adsorption phase followed by slow but steady uptake.

The interactions of many other metal ions with carbonate mineral surfaces have been studied. These were recently reviewed by Morse (1986) and will not be covered here, because, while they may be important to the geochemistry of these metals, they are generally of only secondary importance to an understanding of carbonate geochemistry. A recent paper by Davis et al. (1987) on Cd^{2+} adsorption on calcite, however, has offered an alternative interpretation to earlier work. They hypothesize that coprecipitation, as well as adsorption, takes place in the near-surface region, and that the influence of Mg^{2+} on adsorption may be more the result of its inhibition of reaction kinetics than site competition. This explanation is possibly true during long-term experiments, but it is less likely during short-term (~1 hour) experiments.

The importance of the interaction of organic compounds with calcium carbonate surfaces has long been recognized. It has been demonstrated that aragonite precipitation is inhibited by uncharacterized dissolved organic matter (Chave and Suess, 1967), and that humic and fulvic acids, and certain aromatic carboxylic acids, inhibit seeded aragonite precipitation from seawater (Berner et al., 1978). The selective adsorption of amino acids on carbonate substrates has received considerable attention. A preferential adsorption of aspartic acid has been shown from humic and fulvic acids and proteinaceous matter (Carter and Mitterer, 1978; Carter, 1978; Mitterer, 1971).

Some of the most interesting studies of calcium carbonate surface chemistry have been concerned with the interaction of fatty acids with calcite and aragonite. Adsorption of several members of the straight chain fatty acid homologous series on both calcite and aragonite from solutions ranging from pure water to seawater was studied by Zullig and Morse (1988). An important finding was that great care must be taken to avoid bacterial degradation of the fatty acids lest spurious data be obtained. They found that the surface affinity of fatty acids is highly dependent on carbon chain length, with little or no adsorption occurring for acids with less than 12 carbons and almost complete adsorption occurring for stearic (C_{18}) acid. The pattern is roughly the inverse of the solubility of the fatty acids. This pattern reflects the hydrophobic nature of the acids and indicates that attachment comes about through interaction between the carbon chain and the carbonate surface rather than by the carboxylic group interacting with surface calcium.

Carbonate dissolution and precipitation kinetics

Basic Principles

Carbonate minerals are among the most chemically reactive common minerals under Earth surface conditions. Many important features of carbonate mineral behavior in sediments and during diagenesis are a result of their unique kinetics of dissolution and precipitation. Although the reaction kinetics of several carbonate minerals have been investigated, the vast majority of studies have focused on calcite and aragonite. Before examining data and models for calcium carbonate dissolution and precipitation reactions in aqueous solutions, a brief summary of the major concepts involved will be presented. Here we will not deal with the details of proposed reaction mechanisms and the associated complex rate equations. These have been examined in extensive review articles (e.g., Plummer et al., 1979; Morse, 1983) and where appropriate will be developed in later chapters.

The degree of disequilibrium (super or undersaturation) is one of the primary factors controlling the rate of reaction of carbonate minerals in aqueous solutions. The general observation is that the rate of reaction increases with increasing disequilibrium. In fact, the common observation is that the ratio of the dissolution and precipitation reaction rates change with increasing disequilibrium, and that these rates are the same at equilibrium. In addition, different chemical and physical processes are usually involved in any chemical reaction. For a given set of conditions one of these processes will be slower than the others. This process is the rate controlling step.

Reaction Kinetics in Simple Solutions

The primary focus of research on calcium carbonate reaction kinetics in simple solutions (low ionic strength without multiple components outside the carbonate system) has been on calcite dissolution kinetics. One major objective of these investigations has been to determine the relative importance of transport and surface control of reaction rates.   There is general agreement that in highly undersaturated solutions the rate of calcite dissolution is controlled by transport processes between the mineral surface and the bulk solution. This process is referred to as diffusion controlled dissolution because the rate limiting transport process is through a stagnant boundary layer extending out from the mineral surface. The thickness of this stagnant layer is generally controlled by the hydrodynamics of the solution. Purely transport controlled reaction kinetics will not be discussed here because it generally occurs well outside of the range of saturation states observed to occur in sedimentary systems.

A number of factors have been investigated that influence calcite dissolution in relatively simple systems. These include the important variables of temperature, pH, and the partial pressure of CO_2. The equations describing the net dissolution rate of carbonate minerals in simple experimental solutions differ somewhat, depending on experimental conditions and the interpretations placed on the experimental results by various investigators (cf., for example, Plummer et al., 1978; Chou et al., 1989).

Recently, Chou et al. (1989) studied the dissolution kinetics of various carbonate minerals in aqueous solution. Figure 2.10 illustrates the experimental results for aragonite, calcite, witherite, dolomite, and magnesite. These data can be fit by rate equations, an example of which is shown in equation 2.28 for calcite.

$$\text{Rate (mole cm}^{-2}\text{sec}^{-1}) = k_1\, a_{H^+} + k_3 - k_6\, m_{Ca^{2+}}\, m_{CO_3^{2-}} \qquad (2.28)$$

Where k_1, k_3, and k_6 are first-order rate constants that are temperature dependent. As can be seen in the figure over a range of pH, aragonite and calcite have about the same rate of dissolution whereas magnesite and dolomite dissolve at a slower rate. These results are what we might anticipate from our knowledge of the relative reactivities of calcium carbonate, dolomite and magnesite. Initial experiments on magnesian calcite dissolution rates (Bertram, 1989) bear similarities to those on calcite, but the overall dissolution rate of calcites containing Mg is higher than that of pure calcite. Experiments of this nature are necessary to an understanding of the rate of dissolution of the various carbonate phases in natural environments. However, this type of research has just begun, and it will be necessary to understand the variation in rates with pressure, temperature, and other solution components before these rate studies can be applied to natural situations, such as the burial diagenesis environment. Presently it appears that a decrease in pH or an increase in P_{CO_2} or temperature will increase the rate of dissolution of carbonate minerals.

The kinetics of calcium carbonate precipitation in simple solutions have received less attention than those of dissolution reactions. This perhaps reflects the fact that most sedimentary carbonates are initially formed biogenically and that the primary interest in carbonate precipitation reactions has been directed at reaction

inhibitors and coprecipitation reactions. Scant data exist for aragonite reaction
kinetics in simple solutions.

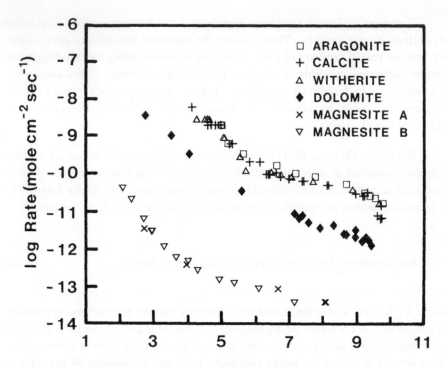

Figure 2.10. Dissolution rates as a function of pH for aragonite, calcite, witherite,
dolomite, and magnesite. (After Chou et al., 1989.)

Reaction Kinetics in Complex Solutions

The reaction kinetics of calcite and aragonite in solutions containing
dissolved constituents in addition to those of the carbonic acid system, calcium, and
sodium or potassium chloride have been of intense interest, both in basic and
applied sciences. Among the many important processes in which the influence of
other dissolved components may play an important role are: nucleation of calcium
carbonate from seawater, aragonite to calcite diagenetic transformations,
coprecipitation of heavy metals with calcium carbonate, retention of supersaturated
states in cooling waters, buffering of fresh waters, accumulation of carbonates in
deep sea sediments, and formation and destruction of carbonate cements.
Magnesium has received by far the most attention because of its common

occurrence in natural waters and large influence on calcite solubility and kinetics. Phosphate and phosphorus compounds have also received a good deal of attention because of their potential as reaction inhibitors for both calcite and aragonite even at very low concentrations. Heavy metals and organic compounds have also been studied. Reactions in seawater have been given particular emphasis because this solution is a major "constant ionic medium."

Before proceeding to specific systems, it is worth considering some basic aspects of the influence of "foreign" ions on reaction kinetics. There are two primary ways in which ions can influence reaction rates. The first is in the solution phase, where they can form complexes with the reactive ions. This interaction can alter both activity coefficients, and hence the saturation state of the solution, and the rate at which transformation reactions occur. The second is that the "foreign" ions can adsorb on the surface of the reacting solid. The adsorption of "foreign" ions can influence reaction rates in two basic ways. The additional "foreign" ions, if cations, can cause a concurrent increase in surface carbonate concentration, or, if anions, an increase in surface calcium concentrations. Most recent models of surfaces point to the fact that they are heterogeneous and that adsorption is preferred at higher energy sites. These sites are also favored for crystal growth and dissolution.

Mg^{2+} influences calcite dissolution rates the same way, but not to the same extent as Ca^{2+}. The inhibition effects of Mg^{2+} can be described in terms of a Langmuir adsorption isotherm. Sjoberg (1978) found he could model results for the combined influences of Ca^{2+} and Mg^{2+} in terms of site competition consistent with ion exchange equilibrium. The inhibition effects of Mg^{2+} in calcite powder runs increase with increasing Mg^{2+} concentration and as equilibrium is approached.

A major portion of the studies on calcium carbonate reaction kinetics has been done in seawater because of the many significant geochemical problems related to this system. Morse and Berner (1979) summarized the work on carbonate dissolution kinetics in seawater and their application to the oceanic carbonate system. The only major seawater component in addition to Mg^{2+} that has been identified as a dissolution inhibitor is SO_4^{2-} (Sjoberg, 1978; Mucci et al., 1989). Sjoberg's studies of other major and minor components (Sr^{2+}, H_3BO_3, F^-) showed no measurable influence on dissolution rates. Morse and Berner (1979) and Sjoberg (1978) found that for near-equilibrium dissolution in phosphate-free seawater, the dissolution rate could be described as:

$$R = k \ (1-\Omega)^n \tag{2.29}$$

where n is approximately 3. Sjoberg (1978) found that the kinetics could also be described by:

$$R = k \ A \ (1-\Omega^{1/2})^n \tag{2.30}$$

where n is 2 in phosphate-free seawater. Similar findings were reported by Morse et al. (1979) for synthetic aragonite dissolution in seawater. Increasing pressure increases aragonite dissolution rates in seawater (Acker et al., 1987). It is important to emphasize that all data indicate that calcite and aragonite dissolution rates are not simply proportional to the extent of undersaturation, as commonly modeled in ocean systems (e.g., Broecker and Peng, 1982), but are exponential functions of disequilibrium.

No area of carbonate reaction kinetics has been more controversial than the precipitation of calcite from solutions containing magnesium. This reaction is central to many major sedimentary and oceanic processes, including retention of the supersaturation of seawater, formation of carbonate cements, control of the carbonate phase precipitated, and inhibition of the aragonite to calcite transition. The fact that major mole fractions of $MgCO_3$ can coprecipitate with the calcite means that it is not possible to deal simply with calcite of constant solubility or surface properties. The solubility of magnesian calcite has been a constant source of controversy for almost 20 years (see Chapter 3). This problem has made interpretation of kinetic results difficult. In fact, the dissolution kinetics of magnesian calcites has been at the center of much of the solubility controversy (e.g., Walter and Morse, 1984a).

After the initial studies of Weyl (1965), which provided generally qualitative information on the inhibition of calcite precipitation by magnesium, interest shifted to the inhibition of calcite nucleation from solutions containing Mg^{2+}. This work was primarily concerned with either the retention of the supersaturated state of seawater (e.g., Pytkowicz, 1965), or the aragonite to calcite transformation. The central idea was that when calcite nuclei attempted to form, magnesium adsorbed on their surfaces. This process prevents the growth of nuclei beyond subcritical size by either raising the surface free energy or causing a magnesian calcite to form of substantially higher solubility.

Berner (1975) made a detailed study of the inhibition of calcite and aragonite precipitation by Mg^{2+} in seawater solutions with different Mg^{2+} concentrations. Results for aragonite and calcite precipitation runs are presented in Figure 2.11. No Mg^{2+} influence on aragonite precipitation kinetics was observed. A major retardation of the precipitation of calcite was found in solutions containing Mg^{2+}. Berner (1975) found that at low levels of Mg^{2+}, no retardation of growth rate occurred. He interpreted this result to indicate that Mg^{2+} did not act primarily as a poison blocking reactive sites. By conducting precipitation runs for extended periods of time, he was able to form enough overgrowth for analysis. The overgrowth from seawater contained 7 to 10 mol % $MgCO_3$. Berner's calculations indicated that the magnesian calcite should be substantially more soluble, and that this enhanced solubility resulted in the solutions being less supersaturated with respect to the solid that formed. This effect, he concluded, was the major reason for the lower precipitation rates in solutions containing Mg^{2+}. Direct reversible equilibrium measurement of the solubility of calcite in seawater solutions of different Mg concentration (Mucci and Morse, 1984), however, indicates that at the seawater Mg^{2+} to Ca^{2+} ratio the change in calcite solubility owing to Mg^{2+} incorporation may be less than 10%.

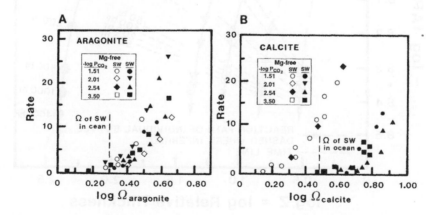

Figure 2.11. (A) Rate of seeded precipitation (in arbitrary units) versus degree of supersaturation with respect to aragonite, $\Omega_{aragonite}$. SW = artificial seawater, Mg-free SW = artificial seawater of same ionic strength, but made up without magnesium. S = 35, T = 25°C. Seed is synthetic aragonite. All rates are initial. (B) Rate of seeded precipitation (in arbitrary units) versus degree of supersaturation with respect to calcite, $\Omega_{calcite}$. Symbols and conditions as in Figure A. Seed is Fisher reagent grade calcite. (After Berner, 1975.)

More recent studies have generally concluded that the inhibiting influence of Mg^{2+} results from difficulties in rapid dehydration of the Mg^{2+} ion, or from crystal poisoning by adsorption of Mg^{2+} at reactive sites. Mucci and Morse (1983) found that the log of the rate constant was a linear function of the solution Mg^{2+} to Ca^{2+} ratio and that the empirical reaction order increased from 3.07 to 3.70 as the Mg^{2+} to Ca^{2+} increased from 1 to 10.3.

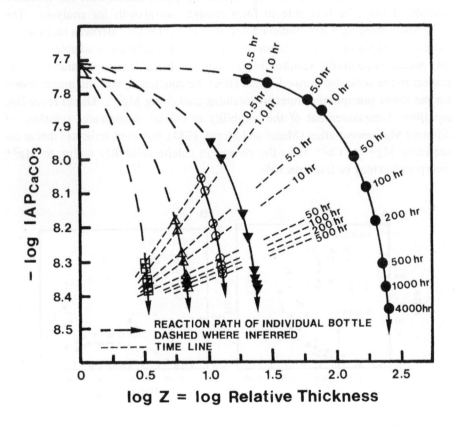

Figure 2.12. The relationship of solubility ($-\log IAP_{CaCO_3}$) to the log of relative coating thickness (log Z), and the variation of that relationship with time for natural seawater systems. These experiments were performed by suspending different amounts of calcite in seawater in a closed system, and monitoring pH and total alkalinity with time. g calcite cm^{-3} seawater = 0.001 (●), 0.010 (▼), 0.020 (○), 0.040 (△), 0.080 (□). (After Schoonmaker, 1981.)

An added complication to experimental studies investigating the rate of precipitation of carbonate on calcite seeds suspended in seawater is the possibility that the initial thin surficial precipitate may continuously recrystallize (Wollast et al., 1980; Schoonmaker, 1981). Figure 2.12 demonstrates the relationship between the inferred solubility of a carbonate coating precipitated on calcite seeds in seawater and the thickness of the coating. It can be seen that as the coatings grow in thickness for various suspension concentrations, the solubility of the coating approaches that of pure calcite with a pK $\cong$ 8.48; a value like that of Plummer and Busenberg (1982). Schoonmaker (1981) interpreted this continuous change in coating solubility, expressed as -log IAP_{CaCO_3}, to reflect continuous recrystallization of an initially high magnesian calcite coating to one near pure calcite in composition. It was found that the recrystallization rate for both natural and artificial seawater systems was a function of the saturation state of the solution with respect to calcite and the thickness of the coating (Figure 2.13). The application of these results to natural systems is discussed in later chapters. Here we simply emphasize that the order of dependence of recrystallization rate of these coatings on supersaturation in natural and artificial seawater solutions is 4.3 and 4.1, respectively, values not too unlike those obtained by Mucci and Morse (1983) for the calcite crystallization rate as a function of solution Mg^{2+}/Ca^{2+} ratio.

The role of temperature and saturation state on the relative rates of aragonite and magnesian calcite precipitation from seawater was investigated by Burton and Walter (1987). They found that at 5°C precipitation rates of both phases are nearly equivalent, but that as temperature increases, the rate of aragonite precipitation becomes significantly faster than that of magnesian calcite. They also observed that the Mg content of the magnesian calcite increases with increasing temperature. It is possible that the differences in rate observed with increasing temperature could be the result of the formation of a more soluble magnesian calcite. The solution would then be less supersaturated with respect to this phase and the reaction rate might, therefore, be expected to slow.

Phosphate has been found to be an extremely strong inhibitor of carbonate reaction kinetics, even at micromolar concentrations. This constituent has been of considerable interest in seawater because of its variability in concentration. It has been observed that phosphate changes the critical undersaturation necessary for the onset of rapid calcite dissolution (e.g., Berner and Morse, 1974), and alters the empirical reaction order by approximately a factor of 6 in going from 0 to 10 mM orthophosphate solutions. Less influence was found on the log of the rate constant. Walter and Burton (1986) observed a smaller influence of phosphate on calcite

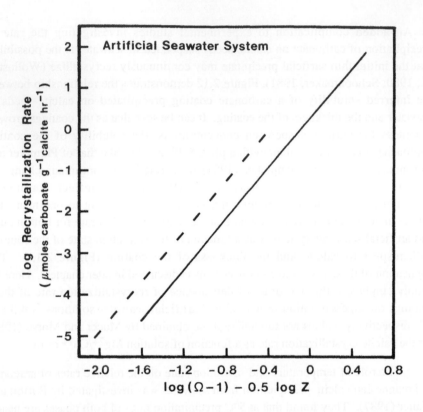

Figure 2.13. Log of the rate of recrystallization plotted as a function of log $(\Omega - 1)$ - 1/2 log Z for artificial seawater systems in which calcite seeds are suspended in seawater, and the saturation state and coating thickness (Z) calculated from measurements of pH and total alkalinity. The solid line represents the data for the majority of experiments, whereas the dashed line is for a system containing a great deal of calcite in which the pH was measured in the sediment rather than in the supernatant seawater. (After Schoonmaker, 1981.)

dissolution than reported by other investigators, and Mucci (1986) determined that phosphate inhibits the precipitation rate of magnesian calcite, but does not influence the coprecipitation of Mg. Phosphate also inhibits aragonite dissolution (Walter and Burton, 1986) and precipitation (Berner et al., 1978).

Studies of phosphate inhibition of calcite dissolution both in the presence and absence of Ca^{2+} and Mg^{2+} have shown that the inhibiting effect of phosphate increases with increasing Ca^{2+} concentration. Results in solutions with Mg^{2+} were less reproducible, indicating the same trend, but not as strong an inhibition relation (Sjoberg, 1978), and a decrease in apparent equilibrium with increasing phosphate, similar to the change in "critical undersaturation" found by Berner and Morse (1974). Similar results were found in artificial seawater, but the influence of phosphate was less, probably because of site competition with sulfate. The surface blocking of dissolution by adsorption of phosphate selectively at high energy sites is consistent with the findings of deKanel and Morse (1978).

The influence of heavy metals on calcium carbonate reaction rates has not been extensively studied. Experiments have shown that many metals exhibit inhibitory effects on calcite dissolution. Ions tested by Terjesen et al. (1961), in decreasing order of effectiveness, were Pb^{2+}, La^{3+}, Y^{3+}, Sc^{3+}, Cd^{2+}, Cu^{2+}, Au^{3+}, Zn^{2+}, Ge^{4+}, and Mn^{2+}, and those found to be about equal were Ni^{2+}, Ba^{2+}, Mg^{2+}, and Co^{2+}. The general trend follows the solubility of the metal carbonate minerals, with the exception of Zn^{2+} and the "about equal" group whose solubilities are all greater than calcite.

When all else fails to explain carbonate kinetic behavior, organic matter is generally given the credit (e.g., Mitterer, 1971). It is tempting to ascribe major kinetic influences to organic compounds because they are associated with most biogenic and sedimentary carbonates. In evaluating the role of organics on reaction kinetics of natural materials, one must take into account that the organic matter may not be chemically interactive with the carbonate mineral surface, but be present simply as a physical coating such as a bacterial or algal slime, or adhering dead organic matter. The observations of Morse (1974) and Sjoberg (1978) indicate that calcite dissolution is unaffected by a wide range of organic compounds. Morse's (1974) finding that, even after soaking carbonates in organic-rich pore waters, no inhibition of dissolution occurs supports the idea that most natural inhibition of carbonate dissolution is due to physical covering rather than to chemical adsorption processes.

Considerable work has been done on the influences of organics on carbonate precipitation. A general trend observed is that the influence of organic matter is roughly proportional to the association constant of the organic compound with Ca^{2+} (Kitano and Hood, 1965). Three classes of organic inhibitors were determined as strong (citrate, malate, pyruvate, glycylglycerine, glycogen), moderate (arginine, glutamate, glycine, glycoprotein, succinate, taurine, chondroitin sulfate), and weak (galactose, dextrose, acetate). In experimental solutions containing organic compounds, strong organic inhibitors, because they slow carbonate precipitation rates, lead to calcite precipitation rather than aragonite. Kitano and Kanamori (1966) also found that sodium citrate and sodium malate favor the formation of magnesian calcites. Again, Ca^{2+} complexation and rate reduction were invoked as the mechanism.

The most complete study of the inhibition of calcium carbonate precipitation by organic matter was carried out by Berner et al. (1978), where primary concern was the lack of carbonate precipitation from supersaturated seawater. Both synthetic organic compounds and organic-rich pore waters from Long Island Sound were used to measure the inhibition of aragonite precipitation. Natural marine humic substances and certain aromatic acids were found to be the strongest inhibitors. The rate of precipitation in pore waters was also found to be strongly inhibited.

The dissolution kinetics of biogenic carbonates has been chosen for special consideration, because these phases represent the major source of sedimentary carbonates and their metastability and complex microstructure result in unique behavior. These carbonates are produced by metabolic processes, which in many cases result in the formation of carbonates quite different from those that would be produced by direct precipitation from the surrounding water. Once exposed to the surrounding water, the biogenic carbonates are usually metastable (e.g., aragonite), or unstable (e.g., high magnesian calcite). The factors influencing their persistence in sediments and transformation during diagenesis have been of considerable interest and a primary area of application of carbonate reaction kinetics.

During much of the past decade, primary interest focused on the behavior of carbonates associated with deep sea sediments. The carbonate in these sediments is dominantly low magnesian calcite (usually < 1 mole % $MgCO_3$) derived from foraminifera and coccoliths. Because these biogenic materials are nearly mineralogically homogeneous, the dominant controls on dissolution which have been investigated are "effective" surface area and inhibitors. Typical results of dissolution experiments using different grain size fractions of deep sea sediments from different ocean basins are presented in Figure 2.14. Morse (1978) found that the > 62 μm fraction dissolved at a rate only about 10% of that predicted from its surface area, whereas the < 62 μm fraction dissolved at 2.5 times the predicted rate. Walter and Morse (1984b) conducted a detailed investigation of the influence of grain size and microstructure on biogenic carbonate reactivity. Table 2.6 presents a comparison between the measured surface area and that calculated for smooth particles of the given sizes. For rhombic calcite, the agreement is good, but for biogenic carbonates the predicted value can be in error by over a factor of 500. Also, the trend in the surface area with size does not follow that predicted by a geometric model. To determine what fraction of the biogenic carbonate surface area was available for reaction, dissolution rates were determined for different biogenic carbonates as a function of grain size. Typical results are presented in Figure 2.15.

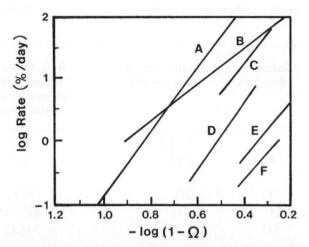

Figure 2.14. The log of dissolution rate in percent per day versus the log of $(1-\Omega)$. A = whole Indian Ocean sediment dissolved in deep-sea sediment pore water; B = whole Pacific Ocean sediment dissolved in Atlantic Ocean deep seawater; C = whole Atlantic Ocean sediment dissolved in Long Island Sound seawater (Morse and Berner, 1972); D = > 62 μm size fraction of the Indian Ocean sediment dissolved in Atlantic Ocean deep seawater; E = the 125 to 500 μm size fraction of Pacific Ocean sediment dissolved in Atlantic Ocean deep seawater; F = 150 to 500 μm Foraminifera dissolved in the Pacific Ocean water column. (After Morse, 1978.)

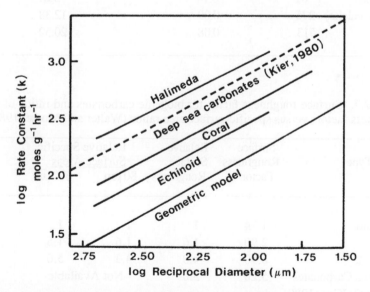

Figure 2.15. Log rate constant, k, as a function of log reciprocal diameter for dissolution of various biogenic carbonates: a comparison with the geometric model. (After Walter and Morse, 1984b.)

Table 2.6. Specific surface area as a function of grain size (Walter and Morse, 1984b).

Grain Type	Median Grain Size (μm)	Specific Surface Area ($m^2\ g^{-1}$)	BET-Measured Surface Area / Geometric Predicted Surface Area
Rhombic	5	$0.45 \pm .005$	1.13
calcite	81	0.03	1.20
Halimeda	81	2.04	82.82
(aragonite)	215	2.10	225.75
	513	2.11	541.22
Coral	51	0.23	5.87
(aragonite)	81	0.22	8.91
	275	0.17	23.38
	513	0.12	30.78
Echinoid	81	0.14	5.67
(Mg-calcite)	275	0.09	12.38
	513	0.08	20.52

Table 2.7. Surface roughness factor for biogenic carbonates and ratios of surface roughness factors versus specific surface area ratios (Walter and Morse, 1984b).

Grain Type	Surface Roughness Factor	Relative Surface Roughness	Relative Specific Surface Areas 80 μm	500 μm
Echinoid	1.74	1	1	1
Coral	2.90	1.6	1.6	1.5
Halimeda	6.90	4.0	4.3	5.0
Deep-sea Carbonates (average) Kier (1980)	5.20	3.0	--Not Available--	

The dissolution kinetics of different grain size fractions of each biogenic carbonate had the same reaction order but had rate constants that were an inverse function of grain radius. BET surface areas show only a slight decrease at larger grain sizes, so the dissolution rate constant cannot be modeled as a direct function of specific surface area. A roughness constant modifies the inverse radius relationship for each biogenic grain type. The constant had a value of one for rhombic calcites, and a maximum of 7 for *Halimeda*, the most intricate grain surface texture investigated. Deep sea carbonates have roughness constants (Keir, 1980) similar to shallow water carbonates (Table 2.7). When the solubility values of Walter and Morse (1984b) were used to calculate saturation states in dissolution runs, empirical reaction orders similar to those obtained for synthetic calcite and aragonite were found.

Walter and Morse (1984b) were able to document the relative importance of microstructure and mineralogy for the dissolution of biogenic carbonates. If the solution is supersaturated with respect to one carbonate but undersaturated with respect to another, then mineralogy is the control on relative dissolution. However, if the solution is undersaturated with respect to all the carbonates present, then microstructure may play a more important role in determining relative rates of dissolution than simply the degree of disequilibrium. These results are indicative of the difficulties likely to be encountered when attempting to make predictions about biogenic carbonate reactivity from anything but direct measurements. Studies such as those by Wollast and Reinhard-Derie (1977), indicating that at least some biogenic magnesian calcites are heterogeneous, and Keir and Hurd (1983), demonstrating the importance of "internal sediments" (fine-grained material in chambers of tests), further complicate the interpretation of dissolution and kinetic solubility data for biogenic carbonates.

It will be shown later that biogenic magnesian calcites are structurally disordered and chemically heterogeneous; aragonite also shows some evidence of structural disorder. Both these factors play a role in determining the "solubilities" of these phases and in their reactivity in natural systems.

Concluding remarks

In natural systems, carbonates react with a variety of solutions at different pressures and temperatures. The processes involved in these reactions are complex, but depend significantly on the solubilities of the carbonate minerals, their surface chemistries, and dissolution and precipitation kinetics. In this chapter, we have reviewed these attributes of carbonate minerals. One is left with the feeling that much research work has been done, some of which is controversial and conflicting, that much work remains to be done, and that the applications of some of these studies to natural systems may be difficult. In the next chapter, we consider another complexity of carbonate solids -- they are generally not pure phases but contain trace to even major amounts of cations and anions other than components of calcium, carbon, and oxygen in solution, resulting from coprecipitation reactions involving compositionally complex natural waters.

Chapter 3

COPRECIPITATION REACTIONS AND SOLID SOLUTIONS OF CARBONATE MINERALS

General concepts

Background Information

Natural carbonate minerals do not form from pure solutions where the only components are water, calcium, and the carbonic acid system species. Because of the general phenomenon known as coprecipitation, at least trace amounts of all components present in the solution from which a carbonate mineral forms can be incorporated into the solid. Natural carbonates contain such coprecipitates in concentrations ranging from trace (e.g., heavy metals), to minor (e.g., Sr), to major (e.g., Mg). When the concentration of the coprecipitate reaches major (>1%) concentrations, it can significantly alter the chemical properties of the carbonate mineral, such as its solubility. The most important example of this mineral property in marine sediments is the magnesian calcites, which commonly contain in excess of 12 mole % Mg. The fact that natural carbonate minerals contain coprecipitates whose concentrations reflect the composition of the solution and conditions, such as temperature, under which their formation took place, means that there is potentially a large amount of information which can be obtained from the study of carbonate mineral composition. This type of information allied with stable isotope ratio data, which are influenced by many of the same environmental factors, has become a major area of study in carbonate geochemistry.

As our understanding of the chemistry of these processes has progressed, many complicating factors have been discovered. Their significance has usually been overlooked by investigators attempting to interpret observations of natural carbonate mineral compositions. Also, in some cases, predictions based on laboratory studies fail to agree well with what is observed in the field, indicating that our understanding of the chemistry or the actual processes at work under natural conditions is still not adequate.

In this chapter we will examine the basic chemical concepts of coprecipitation and solid solutions, and the partition coefficients of different elements and compounds in major sedimentary carbonate minerals will be presented. A brief summary of information on oxygen and carbon isotope fractionation in carbonate minerals will also be presented. A major portion of this chapter is devoted to magnesian calcites. This is because of their importance in shallow water marine sediments, the level of sophistication that has been reached in their study relative to other coprecipitation reactions, and our own recent interest in the subject. Application of this information to natural systems will be largely reserved until later chapters where natural processes are examined.

Basic Chemical Considerations

Solid solution theory   The chemical theories of primary importance to understanding factors controlling carbonate mineral compositions in natural systems are associated with solid solutions. Carbonate minerals of less than pure composition can be viewed as mixtures of component minerals (e.g., $SrCO_3$ and $CaSO_4$ in $CaCO_3$). If the mixtures are of a simple mechanical type then the free energy of formation of the resulting solid will be directly proportional to the composition of the aggregate. Thus, for a two component , a and b, mixture:

$$G_{soln-mech} = x_a G_a^o + x_b G_b^o \qquad (3.1)$$

where x_i and and G_i^o are, respectively, the mole fraction and Gibbs free energy of formation of component i. In this system $x_a + x_b = 1$.

In true chemical solutions mixing on the molecular level leads to substantial increases in the entropy of the system and, consequently, negative deviations from the linear relationship between composition and the free energy of the solution (Raoult's law). If this is the only deviation from the Henry's law behavior of equation 3.1, then the solution is referred to as being ideal.

$$G_{soln-ideal} = G_{soln-mech} + G_{mix} \qquad (3.2)$$

Equation 3.2 can be formally written as:

$$\overbrace{G_{soln-mech}}\qquad\overbrace{G_{mix}}$$
$$G_{soln-ideal} = x_a\mu_a^o + x_b\mu_b^o + RT\,[x_a \ln(x_a) + x_b \ln(x_b)] \qquad (3.3)$$

where μ_i^o is the standard chemical potential of component i.

The basis of the ideal solution model is that the thermodynamic activities of the components are the same as their mole fractions. Implicit in this assumption is the idea that the activity coefficients are equal to unity. This is at best an approximation and has been found to be invalid in most cases. Solutions in which activity coefficients are taken into account are referred to as "real" solutions and are described by equation 3.4.

$$G_{\text{soln-real}} = G_{\text{soln-ideal}} + RT\,[x_a\,\ln\,(\lambda_a) + x_b\,\ln\,(\lambda_b)] \qquad (3.4)$$

λ_i is the solid state activity coefficient of component i. This coefficient usually, but not necessarily, is greater than 1 and, consequently, leads to positive deviations from Henry's and Raoult's laws. The term $\{RT\,[x_a\,\ln\,(\lambda_a) + x_b\,\ln\,(\lambda_b)]\}$ is commonly written as $\{W_a x_a x_b^2 + W_b x_b x_a^2\}$, where W_a and W_b are empirical coefficients that vary with pressure and temperature but not with composition. Solutions following this rule are called regular solutions. If $W_a = W_b$, then the solutions are symmetrical regular solutions. If W_a and W_b are not equal, the solution is said to be an asymmetrical regular solution. The other major possibility is that W_a and/or W_b are not independent of composition. In this case an "irregular" solution results.

Partition coefficients The definition of a partition coefficient which generally appears in the Earth science literature is based on the non-thermodynamic Henderson-Kracek (1927) relation:

$$D = \frac{(c_i/c_j)_{\text{solid}}}{(c_i/c_j)_{\text{soln}}} \qquad (3.5)$$

where c_i and c_j are the concentrations of components i and j, and D is the partition coefficient. It is important to stress that this relation is not a thermodynamic definition, because thermodynamic activities of the components are not used. This type of partition coefficient is usually referred to as a homogeneous distribution coefficient, because it applies to compositionally homogeneous solids.

The thermodynamic basis for distribution coefficients rests in the previously discussed area of solid solution theory. The equilibrium requirement is that the chemical potentials of the components must be the same in both the solid and solution phases. The concepts that the solid is homogeneous and that it is the stable phase under the specified P and T conditions are also implicit in defining true equilibrium. If this is not so, the best that one can achieve is the description of a metastable equilibrium. The thermodynamic definition of a distribution coefficient is:

$$K = \frac{(a_i/a_j)_{solid}}{(a_i/a_j)_{soln}}$$

(3.6)

where a_i and a_j are the thermodynamic activities of components i and j, and K is the thermodynamic distribution coefficient. Ideally, the two types of coefficients are related by the expression:

$$D = K \, (\lambda_i /\lambda_j) \, / \, (\gamma_i/ \gamma_j)$$

(3.7)

where λ and γ are, respectively, solid and solution phase activity coefficients.

The non-thermodynamic partition and thermodynamic distribution coefficients can be equal only when the above activity coefficient ratio is unity. Also, their relative values will change differentially if factors such as pressure, temperature, and composition change the activity coefficient ratios in the solid and solution phases. This fact is a potential source of disagreement between partition and distribution coefficients measured in different media (e.g., dilute solutions and seawater). If a partition coefficient measured in one solution is to be applied to another solution, at the very least a correction should be made for the change in the ratio of solution phase activity coefficients, a practice rarely followed. This conclusion suggests that we should reevaluate many of the partition coefficients measured in solutions that have undergone major changes in composition during precipitation. These types of measurements include those in which spontaneous nucleation is used to initiate precipitation, and the solution is then allowed to "free drift" toward equilibrium (e.g., Kitano, 1962; Kinsman and Holland, 1969; Raiswell and Brimblecombe, 1977; Pingitore and Eastman, 1984). Under such experimental conditions the solution phase activity coefficient ratio can change substantially.

Examples of the magnitude of the problem for Mg^{2+}, Sr^{2+} and Cu^{2+} have been calculated by Morse and Bender (1989) using starting and final solution compositions during typical spontaneous precipitation experiments from Kitano (1962). In the experimental solutions, the variation in Mg^{2+}/Ca^{2+}, Sr^{2+}/Ca^{2+} and Cu^{2+}/Ca^{2+} ion activity coefficient ratios between starting and ending solution composition were, respectively, -28%, -26% and -42%. If partition coefficients determined in these solutions were applied to seawater the errors in D_{Mg}, D_{Sr} and D_{Cu} would be, respectively, -12%, -32% and -89%. Thus, ignoring changes in solution phase activity coefficient ratios can lead to significant difficulties in applying partition coefficients to natural waters. They also call into question the validity of many published values for partition coefficients for calcite that were determined under conditions where substantial changes in solution composition took place.

The behavior of solid phase activity coefficients is a more complex issue. We can view this behavior in two ways. The first is where only a trace mole fraction of component i ("trace component") is coprecipitated in a relatively pure component j ("carrier component"). The activity of the carrier component is well approximated as being equal to unity. When this is true, the activity coefficient of the trace component in the solid is given by the following relation:

$$\lambda_t = (\gamma_t / \gamma_c)(D/K) \tag{3.8}$$

where the subscripts t and c refer to the trace and carrier components. At very low trace component concentrations, λ_t is usually nearly constant over wide concentration ranges, but care should be exercised in using this assumption. It should also be noted, that even at low concentrations, the solid state activity coefficient can be orders of magnitude greater than unity. The second case is one in which the composition of the solid undergoes substantial compositional changes. This case is typified in carbonate minerals by the magnesian calcite solid solution. Under these conditions the "major" component solid phase activity coefficient undergoes significant changes, and all other solid phase activity coefficients change as well. A final point to remember is that both distribution and partition coefficients are dependent on temperature and pressure. Although some work has been done on their behavior as a function of temperature, little direct data has been obtained on pressure influences which are generally expected to be less important.

In the vast majority of experiments carried out to determine carbonate partition coefficients, the trace component concentration has been allowed to vary substantially during the precipitation process. This experimental approach results in constantly changing solution and solid compositions. The approach that has usually been used to compute partition coefficients from the resulting experimental data is that of Doerner and Hoskins (1925), in which the following relation holds:

$$\frac{dB}{dA} = D_{D\text{-}H} \frac{[B]}{[A]}$$

(3.9)

and upon integration is:

$$D_{D\text{-}H} = \frac{\ln\left([B_i] / [B_f]\right)}{\ln\left([A_i] / [A_f]\right)}$$

(3.10)

where $D_{D\text{-}H}$ is the partition coefficient for a solid that is heterogeneous in composition, dB and dA are increments of B and A deposited, and [ ] are solution concentrations. The equation is valid only if no recrystallization takes place. This assumption represents an ideal situation that can only be approximated under actual experimental conditions. Factors such as temperature and reaction rate can strongly influence how well this condition is approximated.

One of the most controversial topics in the recent literature, with regard to partition coefficients in carbonates, has been the effect of precipitation rates on values of the partition coefficients. The fact that partition coefficients can be substantially influenced by crystal growth rates has been well established for years in the chemical literature, and interesting models have been produced to explain experimental observations (e.g., for a simple summary see Ohara and Reid, 1973). The two basic modes of control postulated involve mass transport properties and surface reaction kinetics. Without getting into detailed theory, it is perhaps sufficient to point out that kinetic influences can cause both increases and decreases in partition coefficients. At high rates of precipitation, there is even a chance for the physical process of occlusion of adsorbates to occur. In summary, there is no reason to expect that partition coefficients in calcite should not be precipitation rate dependent. Two major questions are: (1) how sensitive to reaction rate are the partition coefficients of interest? and (2) will this variation of partition coefficients with rate be of significance to important natural processes? Unless the first question is acceptably answered, it will obviously be difficult to deal with the second question.

An additional complication has been demonstrated by Reeder and Grams (1987). They found that coprecipitation of Mg and Mn in calcites is different for different crystal faces. This results in the phenomenon of sector zoning.

Consequently, it is probable that most reported partition coefficients represent average values. Their results imply that partition coefficients may be dependent on factors controlling crystal morphology.

Coprecipitation of "foreign" ions in carbonate minerals

Examples of Coprecipitation Reactions

Numerous studies have been made of the coprecipitation of a wide variety of ions and compounds with calcite and aragonite. In the past decade, results of these studies have clearly demonstrated the complexity of the processes involved in coprecipitation and shown that equilibrium thermodynamics may not be appropriate for dealing with many low temperature reactions. These considerations have raised serious questions about the application of partition coefficients to natural systems, where a detailed knowledge of the mechanisms and rates involved in carbonate mineral formation and diagenesis are not available. The idea that an unique value of a partition coefficient can be used, even at constant temperature, for paleoenvironmental interpretation now seems generally inadequate. These observations and additional factors to be discussed later in this section have led Morse and Bender (1989) to stress that most experimentally determined and naturally observed partition coefficients are not thermodynamic quantities, but rather phenomenologic measurements generally applicable to only a narrow range of conditions.

Mucci and Morse (1989) have reviewed much of the research on coprecipitation reactions with calcite and aragonite, and the interested reader is referred to their paper for a detailed discussion of this literature. Here we present examples of the complex coprecipitation behavior of some of the most important ions in natural systems with carbonate minerals. The ions that we have selected are Mg^{2+}, Sr^{2+}, Na^+, Mn^{2+}, and SO_4^{2-}.

Magnesium The coprecipitation of Mg^{2+} with calcite has been studied far more extensively than any other coprecipitation reaction involving carbonate minerals. One of the major reasons for this is that both biogenic calcites and, presumably, non-biogenic calcite cements are commonly found containing variable amounts of Mg, with up to 30 mole % $MgCO_3$ in the case of some biogenic calcites. This Mg-enriched group of calcites is generally referred to as magnesian calcite or magnesium calcite. In shallow water, carbonate-rich sediments, these phases are usually the dominant type of calcite and typically make up from 5 to 20 weight % of these sediments (Chapter 5). The addition of a major $MgCO_3$ component to calcite substantially alters its solubility. The solubility behavior of magnesian calcites will be discussed later in this chapter.

The study of the coprecipitation of Mg^{2+} in calcite has been an active area of research, frequently marked by controversy over experimental results and their applicability to natural systems. The literature on this topic is prolific, and we will not attempt to review all of it (for a general review see Mackenzie et al., 1983). The literature is divided into studies where direct measurement of the solids formed from solutions have been made, studies where properties of the solids have been inferred from their interactions with solutions (e.g., Schoonmaker et al., 1982), and papers where authors have estimated values of the partition coefficient by deduction (e.g., Lahann and Seibert, 1982; Given and Wilkinson, 1985a). Here the discussion will be confined to the experimental studies where the compositions of the solids have been directly determined.

There are two relatively recent summaries of the direct measurements of Mg partition coefficients in calcite; those of Berner (1978) and Mackenzie et al. (1983), which serve as good starting points for this topic. Berner's summary of data with his comments are presented in Table 3.1. This table and the observation of Berner that much of the variability in results could be ascribed to differences in reaction rate, have been the cornerstones for those investigators supporting the argument that precipitation rate controls the Mg content of natural calcites.   Unfortunately, many of the experiments were done in a manner where conditions changed dramatically during precipitation, including variation of Mg to Ca ratios in solution. Perhaps the most relevant observation that can be made from this rather messy set of data is that the only cases in which magnesian calcites of greater than 10 mole % $MgCO_3$ were produced were those in which the starting calcium carbonate ion activity product exceeded 120 times that of the calcite solubility product. For comparison it should be noted that surface seawater is about 7 times supersaturated with respect to calcite, and that the porewaters of calcium carbonate-rich shallow water sediments are less supersaturated than the overlying seawater (e.g., Berner, 1966; Morse et al., 1985).

The set of data in Table 3.1 indicates that high magnesian calcites could only be produced, under the experimental conditions used, at extremely high rates of precipitation. A chemical factor missing in Table 3.1 is the Mg to Ca ratio of the solutions in which the reactions were carried out. The influence of Mg^{2+}/Ca^{2+} on calcite composition was addressed by Mackenzie et al. (1983), who noted the "impressive agreement" of the large amount of diverse data (Figure 3.1) indicating a strong correspondence between the solution Mg^{2+}/Ca^{2+} ratio and magnesian calcite composition.

Table 3.1. Laboratory precipitation of magnesian calcite from seawater (Berner, 1978).

Method Reference	Nuclei	Time to Precipitate	Ω_c	% $MgCO_3$	
pH-stat at constant Ω (4 runs)	reagent calcite	10-50 hr	5-7	7-10	Berner (1975)
$HCO_3^-+CO_3^{2-}$-added (3 runs)	glass dust et cetera	~10 min	120-180	17-18	Berner (1978)
$HCO_3^-+CO_3^{2-}$+water soluble extract of marine grass added (2 runs)	"	overnight	170	16-17	Berner (1978)
SW evaporated to S=70 salinity; $HCO_3^-+CO_3^{2-}$ added	" "	~10 min	~150	22	Berner (1978)
$HCO_3^-+CO_3^{2-}$ added drop by drop (0°C) (9 runs)	"	days	~50	0-1	Kinsman & Holland (1969)
$HCO_3^-+CO_3^{2-}$ added drop by drop (15-17°C) (8 runs)	"	days	~50	8-10	Kinsman & Holland (1969)
$HCO_3^-+CO_3^{2-}$ added (0°C)	"	minutes	>200	14	Glover & Sippel (1967)
Florida Bay sediment + CO_2; then degas by air bubbling (2 runs)	low-Mg calcite, aragonite,quartz, organic matter	~1 day	???	8	Winland (1969)
Florida Bay sediment + CO_2; then degas by diffusion (3 runs)	"	~3 days	???	8-10	Berner (1978)
R.G. Calcite + CO_2; then degas by diffusion	reagent	~3 days	???	9-10	Berner (1978)

"Time to precipitate", except in pH stat runs, refers to first appearance of white precipitate and not total run length. In almost all runs, magnesian calcite was accompanied by aragonite, hydrocalcite, or vaterite. Unless otherwise indicated, runs were conducted at room temperature.

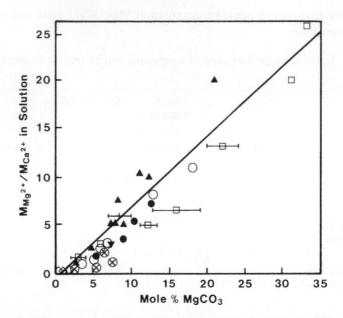

Figure 3.1. $MgCO_3$ content of precipitated magnesian calcites as a function of the Mg to Ca molar ratio in solution. (+) Berner (1975), pH-stat experiments; (▲) Mucci and Morse (1983), pH-stat; (●) Kitano and Kanamori (1966), free drift; (□)Glover and Sippel (1967), free drift; (◇) Katz (1973), free drift; (⊗) McCauley and Roy (1974), free drift; (○) Devery and Ehlmann (1981), free drift; (▼)Last (1982), composition of magnesian calcites precipitating in water column, Lake Manitoba, Canada. (After Mackenzie et al., 1983.)

Mucci and Morse (1983) found no statistically significant dependence of the partition coefficient for Mg in calcite precipitated from seawater on reaction rate over a range of seawater supersaturations from about one half (Ω =3) to close to three times (Ω=17) that of typical surface seawater. However, their experiments had durations of from only a few hours to days. The compositions of magnesian calcites grown very near equilibrium (Ω=1.2) over periods of several months were determined by Mucci et al. (1985). They found excellent agreement with the results of Mucci and Morse (1983) even though different experimental techniques were used. The rate independence of the partition coefficient of Mg in calcite, therefore, has been found to be independent of reaction rate over about 12 orders of magnitude in seawater.

Several studies have been conducted on the influence of temperature on the partition coefficient of Mg in calcite (see Figure 3.2 for summary). The recent studies of Burton and Walter (1987), Mucci (1987) and Oomori et al. (1987) are all in good agreement, whereas the results of earlier studies scatter significantly. The influence of temperature is large. For example, calcite precipitated from seawater at 25°C has about 8 mole % $MgCO_3$, whereas at 5°C it has 5 mole % $MgCO_3$, and at 40°C about 13 mole % $MgCO_3$. These results are corroborated in a general manner by the observation that natural calcite cements formed in cold seawater usually contain less Mg than those formed in warm seawater (e.g., Videtich, 1985).

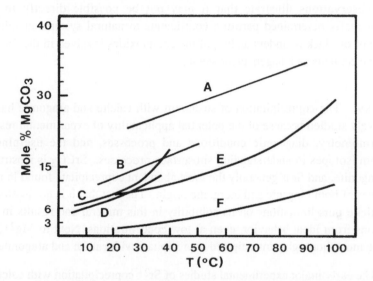

Figure 3.2. The influence of temperature on Mg coprecipitation with calcite. A-Katz (1973), B-Füchbauer and Hardie (1976), C-Oomori et al. (1987-calculated), D-Mucci (1987), and Burton and Walter (1987), E and F-Chilingar (1962).

An important aspect of the coprecipitation of Mg^{2+} with calcite clearly demonstrated by the studies of Mucci and Morse (1983) is the strong dependence of the partition coefficient on the mole % $MgCO_3$ in the calcite. The mole % $MgCO_3$ has also been observed to influence strongly the coprecipitation of Sr^{2+} and other coprecipitates.

A major problem in the application of Mg^{2+} partition coefficients in calcite to natural carbonates has been the failure to predict commonly observed compositions of carbonate cements. The Mg content of shallow water Holocene marine cements is typically 1.5 to 3 times that predicted using experimentally measured partition coefficients, and assuming that the solution from which the cements formed had the same Mg to Ca ratio as normal seawater. This problem has been the basis for arguments against the use of experimentally determined partition coefficients (e.g., Given and Wilkinson, 1985a,b). Other observations of natural marine cements indicating much lower, not higher, values for the partition coefficient of Mg in calcite have been made by Baker (1981) and Baker et al. (1982) on Deep Sea Drilling Project samples. They hypothesized that the values they observed for the Mg^{2+} partition coefficient for calcite represented true thermodynamic equilibrium, whereas the higher values observed in Holocene cements and laboratory experiments were the result of metastable equilibrium. These observations illustrate that it may not be possible directly to apply experimentally determined partition coefficients to natural systems at this time, because of our lack of understanding of the complexities involved in the formation of calcitic cements and diagenetic processes.

Strontium   The coprecipitation of strontium with calcite and aragonite has been extensively studied because of the potential applicability of experimental results to geothermometry, diagenetic conditions and processes, and the usefulness of strontium isotopes in understanding subsurface processes. $SrCO_3$ is isostructural with aragonite, and Sr is generally the most abundant coprecipitate found in marine aragonites of both biogenic and inorganic origin. The fact that it is not isostructural with calcite puts limitations on its solubility in this mineral and results in major deviations from ideal behavior, even at low concentrations. Next to Mg^{2+}, it has been the most thoroughly investigated coprecipitate with calcite and aragonite.

The early major experimental studies of Sr^{2+} coprecipitation with calcite and aragonite were undertaken by Holland et al. (1963, 1964), who obtained a ratio of the partition coefficient of strontium in aragonite to that in calcite of 7.8 ($\pm$1) for temperatures between 90 and 100°C. This large difference in the strontium partition coefficient between calcite and aragonite has traditionally been ascribed to the fact that ions smaller than calcium are favored in calcite, whereas those larger than calcium are favored in aragonite. It is also probable that the previously mentioned crystallographic structural factors play a role in influencing partition coefficients.

A detailed investigation of the influence of temperature on the partition coefficient of Sr in aragonite was made by Kinsman and Holland (1969). They obtained a value for the partition coefficient in aragonite at 25°C of 1.13. The validity of their results was backed up by a wide range of analyses performed on marine samples by Kinsman (1969), who noted higher than expected concentrations of Sr in oölites. He suggested that this observation may be the result of organic influences.

The results of Lorens (1981) for the partition coefficient of Sr in calcite, in 0.7 M NaCl solution, and at different precipitation rates, are presented in Figure

3.3. Along with Lorens' controlled precipitation rate results, data of other workers for the Sr partition coefficient in calcite are also plotted. These other data are restricted to experiments for solutions with 0.5 or 0.7 M NaCl, and free of Mg. The results of Katz et al. (1972) and Baker et al. (1982), and perhaps Pingitore and Eastman (1986, "normal precipitation rates"), reflect the partition coefficient at the very slow precipitation rates associated with $CaCO_3$ recrystallization, whereas the spontaneous precipitation rate experiment of Lorens and the partition coefficient typical of biogenic low Mg calcite reflect precipitation at very rapid rates. Clearly the partition coefficient systematically increases from a value of about 0.045 at low rates to about 0.12 at high rates. Many of these complexities in the partition coefficient of strontium in calcite are summarized in Figure 3.4.

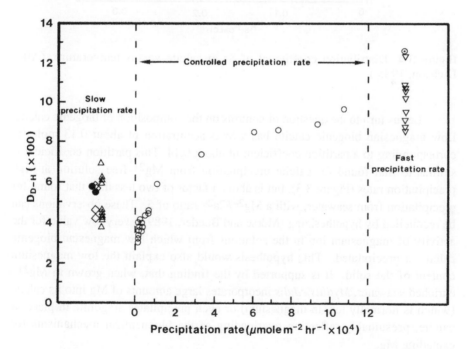

Figure 3.3. Influence of precipitation rate on the distribution of Sr in calcite. ○ Lorens (1978) controlled precipitation rate; ⊙ Lorens (1978) spontaneous precipitation rate; ◇ Baker (1981) 60 to 80°C; ● Katz (1973) 40 to 98°C; △ Pingitore and Eastman (1986) "normal" precipitation rate; ▽ Pingitore and Eastman (1986) "very fast" precipitation rate. (After Morse and Bender, 1989.)

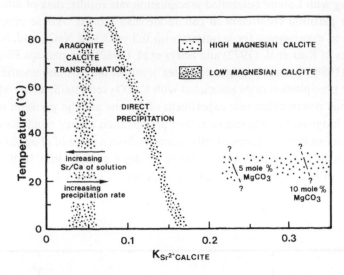

Figure 3.4. Distribution coefficient of Sr in calcite versus temperature. (After Dickson, 1985.)

Let us turn to the question of controls on the composition of biogenic calcite. Low magnesian biogenic calcite has a Sr concentration of about 0.13 mole %, corresponding to a partition coefficient of about 0.14. This partition coefficient is similar to that found for calcite precipitation from Mg^{2+}-free solution at high precipitation rates (Figure 3.3), but is about a factor of two less than that found for precipitation from seawater, with a Mg^{2+}/Ca^{2+} ratio of 5. These observations can be reconciled by hypothesizing (Morse and Bender, 1989) a very low value for the activity of magnesium ion in the solutions from which low magnesian biogenic calcite is precipitated. This hypothesis would also explain the low magnesium content of the solid. It is supported by the finding that, when grown in Mg^{2+}-enriched seawater, *Mytilus edulis* incorporates large amounts of Mg into its calcite (which is normally low in magnesium) or even precipitates aragonite in place of calcite, presumably reflecting overloading of the biochemical mechanisms for excluding Mg.

Many organisms secrete shells high in Mg, rather than low magnesian calcite (see Milliman, 1974 for a summary and Chapter 5). The Sr concentration of these shells varies considerably, but is generally about a factor or two higher than that of low magnesian biogenic calcite. As Ohde and Kitano (1984) have shown, this difference is readily explained by the observation of Mucci and Morse (1983) that increasing amounts of Sr are partitioned into calcite as the Mg^{2+} to Ca^{2+} ratio of solution and solid rises.

The $SrCO_3$-$CaCO_3$ (aragonite) solid solution up to 10 mole % Sr in aragonite was investigated by Plummer and Busenberg (1987). A major influence of composition on solubility was found. They present a detailed examination of the thermodynamics of the system indicating that the solid solution is not regular but unsymmetrical.

Sodium Sodium is present in recent marine carbonate minerals greatly in excess of equimolar chloride concentrations, with sodium concentrations generally exceeding 1000 ppm. Land and Hoops (1973) observed that the sodium concentration of carbonate minerals generally declined with increasing age of the sampled carbonate and suggested that the study of sodium concentrations in carbonate rocks might help in our understanding of such processes as dolomite formation and give indications of paleosalinities (also see White, 1977).

Experimental studies of the partitioning of sodium into aragonite, calcite, and dolomite have been carried out by White (1977, 1978), who concluded that the mode of substitution was as Na_2CO_3 rather than as Na bicarbonate or hydroxide. The behavior of Na^+ does not follow that normally described by a simple partition coefficient, but rather coprecipitation is better described by a Freundlich-type adsorption isotherm equation, implying that adsorption is important.

The behavior of sodium in dolomite and calcite was shown by White (1978) to be complex. His results for calcite are portrayed in Figure 3.5. At low Na^+ concentrations the behavior follows the general relation found for aragonite. At intermediate concentrations, coprecipitated Na deviates from this relation by becoming pH dependent, and progressively more insensitive to additional increases in the aqueous Na^+ to Ca^{2+} ratio. White concluded that these changes could represent a change from Na_2CO_3 precipitation to $NaHCO_3$ precipitation. He also noted that Mg apparently had no influence on Na substitution. Based on the data presented by White in these papers, we estimate the partition coefficients in seawater for sodium in aragonite and in calcite to be, respectively, 2.1×10^{-3} and $\sim 10^{-5}$.

Ishikawa and Ichikuni (1984) also conducted a study of the coprecipitation of sodium with calcite. They suggested that Na^+ is located at interstitial sites in the calcite lattice, and good agreement was observed between their predicted Na concentrations and those in natural calcites for waters up to about a salinity of 10. At higher salinities the natural samples generally had higher Na concentrations than predicted, and a wide scatter in Na concentration was observed. A number of possible explanations were speculated upon, but none were demonstrated.

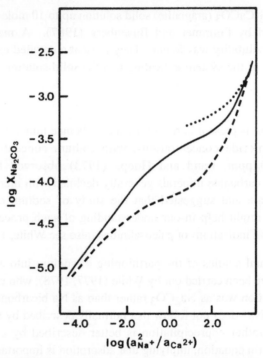

Figure 3.5. Na distribution in calcite expressed as the logarithmic mole fraction of Na_2CO_3 plotted against the logarithmic aqueous Na^+/Ca^{2+} activity ratio at pH 6.8 (solid line), 7.4 (dashed line), 8.4 (dotted line), and at 25°C. (After White, 1978.)

Busenberg and Plummer (1985) also studied the coprecipitation of Na^+ with calcite. Their conclusions agreed with previous concepts that normal solid solution coprecipitation was not occurring, and that the sodium was present at crystal defects. A major contribution of their work was to demonstrate the rate dependence of Na^+ coprecipitation, and in doing so to resolve the differences in Na^+ behavior observed in earlier studies. They also found that increasing concentrations of coprecipitated sulfate increased Na^+ coprecipitation, probably by increasing lattice defects.

Manganese   Numerous investigations have been made of the coprecipitation of transition and heavy metals with calcite and aragonite. The two elements that have received the most attention are Mn^{2+} and Zn^{2+}. This attention probably stems from their potential importance for studying diagenetic processes (e.g., Pingitore, 1978).

Unfortunately, there is generally a large scatter in the values obtained for these partition coefficients. A possible reason for this scatter, as shown by the study of Lorens (1981), is probably the major effect of precipitation rate on the values of the partition coefficients. It is interesting to note that the partition coefficients for the transition and heavy metals in calcite, studied by Lorens (Cd^{2+}, Mn^{2+}, Co^{2+}), have a negative linear log partition coefficient- log precipitation rate relation, whereas Sr^{2+} has a positive relation. This behavior may be explained by the fact that the transition metal carbonates are isostructural with calcite, whereas strontium carbonate is isostructural with aragonite. Also, as precipitation rates increase, partition coefficients tend towards unity.

Manganoan calcite overgrowths were precipitated from artificial seawater of various Mn^{2+} concentrations at 25°C using a constant disequilibrium technique by Mucci (1988). He confirmed Lorens' observation that the amount of manganese coprecipitated with calcite decreased with increasing precipitation rate (also see Pingitore et al., 1988), but he found that the partition coefficient varied with Mn^{2+}/Ca^{2+} of the parent solution, probably, as pointed out by Holland (1966), as a result of the change in solid state activity coefficients. Mucci proposed that the amount of manganese in the carbonate solid solution was determined by the relative precipitation rate of the magnesian calcite and kutnahorite end-member components. A partition coefficient of about 11 can be derived from his results for a precipitation rate equivalent to the log R = 0 of Lorens (1981). At least 50% of the difference between the results of Lorens (1981) and Mucci (1988) can be accounted for by considering variations in solution phase activity coefficients.

There are two related studies on Mn of interest. Using electron paramagnetic resonance (E.P.R.), Wildeman (1969) found that Mn in carbonate minerals exists in highly ionic lattice sites that are isolated in the carbonate minerals. He also established that Mn partitioning in Mg or Ca sites in dolomites is different, with the Mg site being preferred. Reeder and Prosky (1986) have also found compositional zoning in dolomites for a number of metals including Mn, indicating that the partition coefficient may be different on nonequivalent crystal faces.

Sulfate The coprecipitation of relatively few anions with carbonate minerals has been studied and, with the exception of sulfate, these studies have generally not been as detailed as many of those with cations. However, coprecipitation reactions can be important for the removal of ions such as fluoride, borate, and phosphate from seawater (e.g., Morse and Cook, 1978; Okumura et al., 1983). It is also probable that anions will eventually gain a greater stature in the study of diagenesis as their complementary nature with respect to cations becomes more fully appreciated.

Sulfate has been recognized as a significant component of biogenic marine carbonates in ancient rocks. A detailed study of sulfate coprecipitation with calcite was conducted by Busenberg and Plummer (1985), who investigated a variety of parameters that can significantly influence the partition coefficient. The partitioning of sulfate into calcite was observed to depend on the saturation state of the solution with respect to both calcite and gypsum, and covaried with sodium content. The coprecipitation reaction was also found to be log linear with precipitation rate, but not in a manner that followed the distribution law, because the partition coefficient, with respect to concentration, was not constant even at constant supersaturation.

General Models for Partition Coefficients in Carbonates

Relatively few attempts have been made to establish general relationships or models for partition coefficients in carbonate minerals. General statements can be made, such as cations larger than calcium are favored in aragonite, whereas cations smaller than calcium are favored in calcite, but these statements are not quantitative and do not predict relations among cations falling into any of the subcategories. Attempts to show relations within these subgroups, based on simple factors such as differences between the trace cation and calcium ion free energies of formation (e.g., Sverjensky, 1984) have shown poor correlations with observational data (see Figure 3.6). This result is perhaps not surprising considering that more recent and detailed experimental studies have shown that partition coefficients can be precipitation rate dependent, and, therefore, not representative of truly equilibrium conditions. They are also dependent on trace component concentration. These observations both nullify the generally used Doerner-Hoskins (1925) relation and the use of an unique value for a partition coefficient. Also, no attempts have been made to construct predictive models for anions or cations such as Na^+ and K^+, which do not follow relatively standard behavior and appear to be intimately associated with surface-solution interactions and lattice defect densities.

For models to be useful in studies of natural carbonate minerals, they must also be able to deal with the problem of multiple coprecipitates. The importance of this statement in terms of Sr^{2+}, Na^+ and SO_4^{2-} concentrations has previously been discussed in this chapter. This problem is made particularly formidable by the observations of Angus et al. (1979). They measured the partition coefficients of several ions in calcite and aragonite using electron spin spectroscopy (ESR),

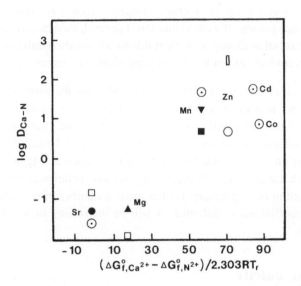

Figure 3.6. Experimental determination of distribution coefficients for different cations in calcite at 25°C compared with the difference in free energies between Ca^{2+} and N^{2+}, where N is the cation. ● Katz et al. (1972);   ▲ Katz (1973); ▼ Bodine et al. (1965); ■ Michard (1968); ○ Crocket and Winchester (1966) and Dardenne (1967);   ⌀ Tsuesue and Holland (1966);   ⊙ Lorens (1981); □ Mucci and Morse (1983). (After Sverjensky, 1984.)

Table 3.2. Comparison of calculated and observed partition coefficients.

		Calculated		Observed		
	r (Å)	D^{Do}	$D^{Cc/Dol}$	$D^{Cc/Dol}$ (mean)	s	n
Mg	0.80	3.67	0.43	0.28	0.08	12
Fe	0.86	1.54	0.79	0.85	0.16	11
Mn	0.91					
Ca	1.08					
Sr	1.33	0.0	2.0	2.3	0.30	8
Ba	1.50	0.0	2.0	1.8	0.58	13

r = ionic radius. D^{Do} is the within-dolomite distribution coefficient - atomic fraction on Mg site ÷ atomic fraction on Ca site. $D^{Cc/Dol}$ calcite/dolomite distribution coefficient - atomic fraction in calcite ÷ atomic fraction in dolomite. s = standard deviation. n= number of data. (After Kretz, 1982.)

thermal analysis and X-ray diffraction. Impurity ions were observed to group together into trace phases. If such a situation is general, then clearly it is impossible to apply solid solution theory to such solids or assign thermodynamically based distribution coefficients to results of coprecipitation experiments.

On a brighter side, the model of Kretz (1982) for the partitioning of cations between calcite and dolomite has shown impressive correlations between observational data and predicted partition coefficients (see Table 3.2). He proposes, based on the earlier work of Jacobsen and Usdowski (1976), that different partition coefficients exist for a trace cation at Mg and Ca sites in dolomite, which are principally influenced by the size differences between the trace cation and calcium or magnesium. He has made a number of predictions of metal ion partition coefficients in dolomite. It will be interesting to see if future work bears them out.

Magnesian calcite

General Considerations

It is doubtful that the solubility of any mineral in low temperature aqueous solutions has been more fully investigated or debated than that of magnesian calcites. Contrary to studies of phase relationships in other mineral systems, in which inorganic phases form the basis of interpretation of mineral behavior and stability, magnesian calcites of biogenic origin have been the principal material used to interpret the behavior of these phases at low temperatures and pressures. The rationale for the use of biogenic phases has been their ready availability, and presumably, their dominance in natural systems. The argument goes: why study well-ordered, well-crystallized compositionally homogeneous products of an inorganic nature, if the magnesian calcites found in most natural systems are "sloppy"? This argument has merit, but the approach is a major source of concern in interpretation of the behavior of these phases. This point will become evident in the discussions that follow.

Values of the equilibrium solubility product for magnesian calcites have been reported in numerous studies (e.g., see summary in Mackenzie et al., 1983). The majority of experimental investigations involved dissolution of biogenic magnesian calcites in aqueous solutions including distilled water in equilibrium with a fixed P_{CO2}, synthetic Ca^{2+}-Mg^{2+}-HCO_3^- solutions, and seawater. Besides dissolution, other experimental approaches have included use of a flow-through apparatus, Ca-Mg exchange studies, and precipitation studies.

As a starting point for the following discussion, a compilation of solubility determinations by numerous investigators is given in Figure 3.7. In this diagram, solubility is expressed as the stoichiometric solubility product, although the use of this expression for solubility data has been debated (see discussion below). Curves have been drawn through three sets of data for future reference. The large amount of scatter in Figure 3.7 undoubtedly reflects many factors including differences in materials, solutions, and experimental and analytical techniques. Another source of divergence between the results of different authors is the use of different models for the aqueous carbonate system. In Figure 3.7, the experimental data have been used along with the recent aqueous model of Plummer and Busenberg (1982) and the solubility product for pure magnesite from Christ and Hostetler (1970) to calculate IAP.

In addition to problems resulting from differences in experimental procedures, theoretical interpretation of the experimental data has been a matter of debate. Dissolution experiments, in particular, are fraught with difficulties, because the incongruent nature of the reaction prevents attainment of equilibrium. We will not go into the detailed arguments regarding the theoretical interpretation of experimental data on magnesian calcite solubility, as this subject has been extensively discussed by Mackenzie et al. (1983), but the following section highlights some problems.

The Fundamental Problems

Throughout the ensuing discussion, dealing in part with dissolution experiments designed to determine the stability (free energies of formation) of the magnesian calcites, the reader should be kept aware of two fundamental problems. The first is that in most experiments to calculate stabilities the magnesian calcites have been treated as solids of fixed composition of one component, whereas they are actually a series of at least two-component compounds forming a partial solid solution series. The second is that these magnesian calcite phases during reaction with aqueous solution dissolve incongruently, leading to formation of a phase different in composition from the original reactant solid (e.g., Plummer and Mackenzie, 1974; Pytkowicz and Cole, 1980). To overcome the inherent incongruency problem, experimentalists have extrapolated their data for the early congruent solution reaction leg to infinite time. These extrapolated data are used to calculate ion activity products (IAP), usually expressed as $IAP = a_{Ca2+}^{1-x} \, a_{Mg2+}^{x} \, a_{CO32-}$ where a is activity of the subscripted species and x is the mole fraction of $MgCO_3$, and subsequently free energies of formation.

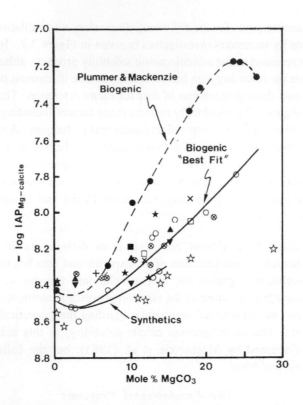

Figure 3.7. Solubilities of the magnesian calcites as a function of $MgCO_3$ content, expressed in terms of -log $IAP_{Mg-calcite}$. The solid curve represents the general trend of dissolution experiment results.

○ Chave et al. (1962) a variety of biogenic materials; distilled water, P_{CO2} = 1 atm, 25ºC;

▲ Schmalz (1967) *Pinna* (2.5 mole % $MgCO_3$), *Lytechinus* spines (10 mole % $MgCO_3$), and *Amphiroa* (15.4 mole % $MgCO_3$); distilled water, P_{CO2} = $10^{-0.01}$ atm, 25ºC; ▼ Schmalz (1965), same as previous, distilled water, P_{CO2} = $10^{-3.5}$ atm, 25ºC; ★ Weyl (1967) a variety of biogenic materials, Ca^{2+} - Mg^{2+} -HCO_3^- solution, flow-through apparatus, 30ºC; ☆ Land (1967) a variety of biogenic materials, distilled water, P_{CO2}=$10^{-0.01}$ atm, 25ºC; ● Plummer and Mackenzie (1974) (revised by Thorstenson and Plummer, 1977), a variety of biogenic materials, distilled water, P_{CO2}= 1 atm, 25ºC; ■ Garrels, Wollast, and König (unpub. data, 1977) *Sphaerechinus granularis* (11 mole % $MgCO_3$); distilled water, pure and spiked with Ca^{2+} or Mg^{2+}, P_{CO2}= 1 atm, 25ºC; □ Walter and Hanor (1979) *Tripneustes* (12 mole % $MgCO_3$) and *Neogoniolithon* (18 mole % $MgCO_3$); distilled water, pure and spiked with Ca^{2+} or Mg^{2+}, P_{CO2} = 1 atm, 25ºC; + Wollast and Marijns (1980) *Sphaerechinus granularis* spines (5 mole % $MgCO_3$); distilled water and solutions spiked with $CaCl_2$, $MgCl_2$ and $NaHCO_3$, P_{CO2}=1 atm, 25ºC; × Walter and Morse (1984a) *Clypeaster* (12 mole % $MgCO_3$) and *Goniolithon* (18 mole % $MgCO_3$); distilled water spiked with $CaCl_2$ and $MgCl_2$, P_{CO2}= $10^{-2.5}$ atm, 25ºC; ⊗ Bischoff et al. (1987) a variety of biogenic phases in distilled water, P_{CO2}= $10^{-2.5}$ atm, 25ºC. The curve fitting the data of Plummer and Mackenzie (1974) is dashed, and the curve for synthetic magnesian calcites is based on the work described in the text.

The former problem was nicely discussed by Garrels and Wollast (1978) for the magnesian calcites, and more recently by May et al. (1986) for smectites, whose dissolution behavior in aqueous solution bears some similarity to that of magnesian calcites. Figure 3.8 is an idealized diagram adapted from Garrels and Wollast illustrating relations involving the stabilities of magnesian calcites, solid solutions, and magnesium-rich carbonates of fixed composition in the system $CaCO_3$-$MgCO_3$-H_2O-CO_2 with a P_{CO_2} maintained at one atmosphere. The saturation curve for solution compositions in equilibrium with various solid solution phases is shown by curve SS'. Point Z on this curve is the solution composition in equilibrium with a magnesian calcite solid solution of composition $Ca_{0.85}$ $Mg_{0.15}$ CO_3. If this phase were of fixed composition, its equilibrium constant could be written $K = a_{Ca^{2+}}{}^{0.85}$ $a_{Mg^{2+}}{}^{0.15}$ $a_{CO_3^{2-}}$, and its solubility curve would be defined by VV' in Figure 3.8.

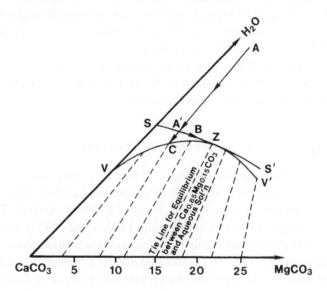

Figure 3.8. Schematic relations for equilibrium between magnesian calcite of fixed composition or a magnesian calcite solid solution series and aqueous solution. SS' is a saturation curve and VV' is a solubility curve. Tie lines are hypothetical. See text for discussion.

Now consider the dissolution of mineral $Ca_{0.85}$ $Mg_{0.15}$ CO_3 in pure water with one atmosphere CO_2. If this mineral acted like a solid solution phase, it would dissolve congruently along reaction path AA' until it reached at A' saturation with a calcium-rich magnesian calcite less soluble than itself. We will see later in this chapter that this initial solution behavior is followed by magnesian calcites dissolving in CO_2-H_2O aqueous solutions. After reaching A', the solution will follow the pathway ABZ as the solid recrystallizes towards a phase of composition $Ca_{0.85} Mg_{0.15} CO_3$. The extent of recrystallization will depend on time (reaction kinetics) and solid:solution ratio (bulk system composition). Conceivably the solution would not reach equilibrium with the solid at Z, because of the above constraints, and remain at rest at a point along SS', perhaps B.

To overcome the above difficulties, experimentalists have extrapolated the aqueous parameters (e.g., pH) of the congruent portion of the reaction path AA' to infinite time, with the implicit assumption that this extrapolation leads, in our example, to point C, the solution composition in equilibrium with a solid of fixed composition $Ca_{0.85} Mg_{0.15} CO_3$. C is interpreted as a solution composition lying on the locus of solution compositional points forming the solubility curve VV' for the mineral $Ca_{0.85} Mg_{0.15} CO_3$ treated as a phase of fixed composition. Calculation of $IAP = a_{Ca2+}^{0.85} a_{Mg2+}^{0.15} a_{CO3^{2-}}$ at point C for our example magnesian calcite gives the value of the constant, representing equilibrium between aqueous solution and the solid solution mineral $Ca_{0.85} Mg_{0.15} CO_3$ at point Z. Herein lies the interpretational problem that has led to considerable debate (see Am. J. Sci., 1978), and in our opinion, still is unresolved. In the following discussion, the reader should keep these fundamental problems in mind.

Experimental Observations

Dissolution experiments As shown in Figure 3.7, many investigators have attempted to determine the solubilities of magnesian calcites. The three most comprehensive studies, prior to investigations of synthetic minerals, in terms of compositional range studied (Chave et al., 1962; Land, 1967; Plummer and Mackenzie, 1974) all involved dissolution of biogenic magnesian calcites in distilled water saturated with 0.97 atm CO_2. Thus, the results of these three studies may be compared directly. We will not consider here the numerous other studies in which values of the equilibrium solubility product for some single magnesian calcites were obtained generally under different experimental conditions (e.g., in seawater).

Chave et al. (1962) limited their measurements to changes of pH during the dissolution reaction, and estimated the pH at infinite time by extrapolation of plots of pH versus the reciprocal of the square root of time. These plots were chosen empirically, because they usually yield linear plots as infinite time is approached when calcite, aragonite, or other simple carbonate minerals are used (Garrels et al., 1960). Chave et al. claimed that in the case of magnesian calcites, however, only relative solubilities are obtained because the dissolution reactions are irreversible, and therefore the material can have no true equilibrium solubility.

Solubilities of annealed skeletal carbonates were estimated by Land (1967) by allowing the system to reach a steady state pH and by determining dissolved Ca^{2+} and Mg^{2+} concentrations. Dissolution of the magnesian calcites in Land's experiments appeared to be congruent in the sense that the steady state solution had a molar Ca^{2+}/Mg^{2+} ratio similar to that of the solid.

At the suggestion of Owen Bricker, Plummer and Mackenzie (1974) studied the behavior of magnesian calcites during dissolution in greater detail. The pH and concentrations of Mg^{2+} and Ca^{2+} were monitored throughout the experiments. They were able to distinguish up to three stages in the dissolution reactions (Figure 3.9). The first stage was congruent dissolution, presumably of the most soluble

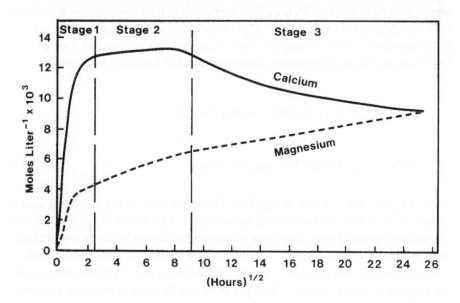

Figure 3.9. Evolution of Ca^{2+} and Mg^{2+} during the dissolution of *Amphiroa rigida* in distilled water equilibrated with a P_{CO_2} of 1 atmosphere. Stage 1 represents congruent dissolution; the ratio of Ca^{2+} to Mg^{2+} released to solution is fixed and equal to the ratio in the solid phase. At the onset of Stage 2, the dissolution reaction becomes incongruent. The solution is supersaturated with a less Mg-rich magnesian calcite and precipitation occurs, although dissolution of *Amphiroa r.* continues. The net effect is to lower the ratio of Ca^{2+} to Mg^{2+} in solution. During Stage 3, bulk precipitation of a Ca-rich phase from solution takes place, resulting in a decrease of Ca^{2+} in solution. (After Plummer and Mackenzie, 1974.)

phase present when the biogenic magnesian calcite sample was not an initially homogeneous phase. The congruent portion of dissolution was followed by a stage in which dissolution continued, but a low magnesium calcite was precipitated on grain surfaces. Solution chemistry indicated more than a two-fold supersaturation with respect to pure calcite during stage 2. Finally, these authors attributed stage 3 to the precipitation of a Ca-rich phase from the bulk solution. It is interesting to note that the congruency of the initial step of dissolution is still maintained when the aqueous solution initially contains dissolved Ca^{2+} or Mg^{2+} (Wollast and Marijns, 1980).

As mentioned earlier, the distinction between congruent and incongruent dissolution is important in the extrapolation of experimental data. Plummer and Mackenzie extrapolated their pH data to infinite time on plots of pH versus the reciprocal of the square root of time, as did Chave et al. (1962), but used only the congruent portions of the reactions. They assumed that the extrapolated pH values represented equilibrium with the initial magnesian calcite composition determined by X-ray diffraction, and the Ca^{2+}/Mg^{2+} ratio of the experimental solution during the congruent reaction. These pHs were then used with P_{CO_2} and the Ca^{2+} and Mg^{2+} concentrations, to calculate solubilities from the following expression:

$$Ca_{1-x} Mg_x CO_3 \rightarrow (1-x)Ca^{2+} + xMg^{2+} + CO_3^{2-}$$

$$IAP_{Mg\text{-}calcite} = a_{Ca^{2+}}{}^{(1-x)} \ a_{Mg^{2+}}{}^x \ a_{CO_3 2-} \qquad (3.11)$$

where x is the mole fraction of $MgCO_3$. This expression (3.11) is the one used in Figure 3.7 to express magnesian calcite solubility. As a matter of interest, the final pH values obtained during dissolution of magnesian calcites in distilled water in the three studies mentioned above generally increase with increase in the $MgCO_3$ content of the reacting magnesian calcite (Mackenzie et al., 1983), the more soluble the phase, the higher the pH. Thus, pH alone can be used as an index of relative solubility of the magnesian calcites.

Bertram (1989) showed, in dissolution studies of magnesian calcites, that their solubilities decrease with increasing temperature. This trend and its relationship to solubility vs. temperature trends for calcite and aragonite are discussed in Chapter 7.

There are experimental problems inherent to dissolution studies, of which the reader should be aware. Chave and Schmalz (1966) noted the effect of grain size and grinding-induced strain on magnesian calcite stabilities. The influence of sample preparation and of other experimental conditions such as the ratio of solid to solution was examined in detail by Walter (1983) and Walter and Morse (1984a) for two magnesian calcite compositions (12 and 18 mole % $MgCO_3$). Walter and Morse demonstrated that the high solubilities obtained by Plummer and Mackenzie could only be duplicated using water-rinsed material at the highest solid to solution ratio. After ultrasonically cleaning and annealing the material, the extrapolated values of pH to infinite time gave consistent results for the two phases, independent of the experimental conditions (solid:solution ratio, initial pH, or solution Mg:Ca ratio). They estimated that the solubilities of these two phases were three times smaller than the values previously determined by Plummer and Mackenzie (see Figure 3.7). It should be noted that Walter and Morse used the Mg content of the bulk solid in calculating and reporting their results. Their results may also explain in part why the results of Land (1967) obtained with annealed samples gave final pH values generally lower than those of Chave et al. (1962) and Plummer and Mackenzie (1974).

Submicron-size particles and strained crystal surfaces affect not only the kinetic data used in extrapolating to an equilibrium or steady state, but these factors are also known to decrease substantially the apparent stability of solid phases. It is presently difficult to evaluate how much these factors affect the solubility of the magnesian calcites occurring naturally. Thus, dissolution experiments using either synthetic magnesian calcites or carefully annealed natural samples are needed to evaluate the fundamental parameters governing the solubility of magnesian calcites. Grain size distribution, surface area, crystallographic defects, and compositional heterogeneities of natural samples should be studied more thoroughly to understand their departure from perfect crystals and their behavior in natural systems. The next section deals with some current research of this nature.

Stability of synthetic magnesian calcites in aqueous solutions   Recently Mackenzie and co-workers (Bischoff et al., 1983, 1985, 1987; Bischoff, 1985) took another approach to the determination of the stability of magnesian calcites. This approach is discussed in some detail, because it may have application to other sedimentary mineral groups. These authors argued that one reason for the discrepancies in IAP versus composition (Figure 3.7) for calcites was that biogenic samples were used in most determinations of stability (solubility). These phases can be distinctly different in their chemical and physical properties from well-crystallized, chemically-homogeneous, synthetic samples of similar *bulk* composition. Therefore, they chose an approach of investigating the chemistry and mineralogy of magnesian calcites by comparison of the properties of synthetic and natural materials using a variety of analytical techniques including X-ray diffraction, atomic adsorption, electron microprobe, scanning and transmission electron microscopy, and infrared and Raman spectroscopy. These crystal chemical investigations formed the basis for interpretation of dissolution experiments performed to determine magnesian calcite stabilities of well-characterized natural and synthetic samples. For details of this work, the reader is referred to the aforementioned publications. We will highlight some of the conclusions in this section.

In Figure 3.10 the unit cell volume and cell edge c/a ratio of biogenic magnesian calcites are plotted as a function of bulk $MgCO_3$ content. The relationship between unit cell volumes and the c/a ratio versus composition for

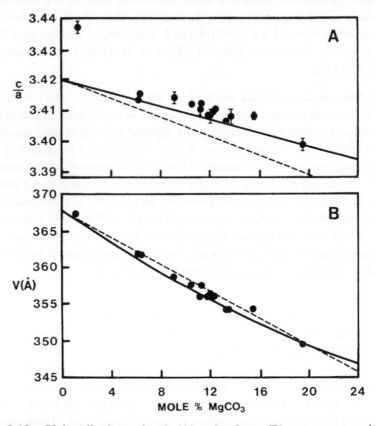

Figure 3.10.   Unit cell edge ratio c/a (A) and volume (B) versus composition for biogenic magnesian calcites. Solid line is quadratic least-squares fit to data for synthetic calcites, and dashed line represents straight line between calcite and disordered dolomite. (After Bischoff et al., 1985.)

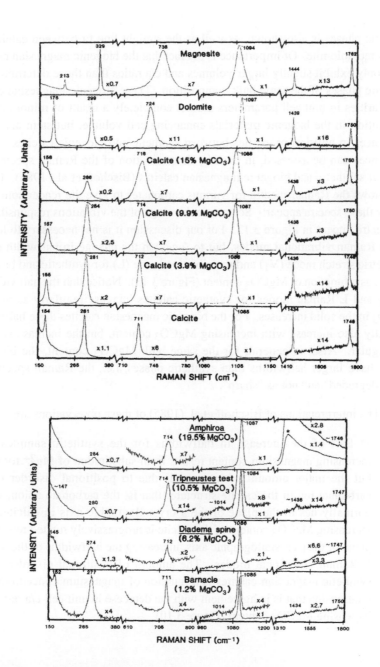

Figure 3.11.   Raman spectra of synthetic and biogenic calcites, dolomite and magnesite. (After Bischoff et al., 1985.)

synthetic phases is also shown, as well as the straight line fit between calcite and disordered dolomite. Of importance is the fact that the biogenic magnesian calcites commonly exhibit slightly larger volumes and c/a ratios than those determined for synthetic phases of the same bulk composition. These biogenic magnesian calcite irregularities in unit cell parameters are not completely a result of minor element substitution in the biogenic materials enhancing cell volume, but stem also from differences in physical properties between synthetic and biogenic calcites. These differences can be assessed, in part, by determination of the Raman spectra of a series of synthetic and biogenic magnesian calcites (Bischoff et al., 1985). Figure 3.11 shows the Raman spectra for various carbonates to give the reader some idea of how these spectra appear. Schematic drawings of the vibrations responsible for the spectra appear in Figure 3.12. For our discussion it is not necessary to look at all the Raman vibrational modes, but to focus on the change in halfwidth of the symmetric stretch mode (V_1) and the vibration mode (L) for synthetic and biogenic calcites as a function of $MgCO_3$ content (Figure 3.13). Notice that the halfwidths of the V_1 and L Raman modes for synthetic calcites increase regularly as mole % $MgCO_3$ in the solid increases. For the biogenic magnesian calcites these halfwidths generally also increase with increasing $MgCO_3$ content, but the increase is much less regular. Also, in general, for the same bulk $MgCO_3$ content, the biogenic phases have larger halfwidths than synthetic phases, i.e., the Raman spectra are more "degraded" and not as "sharp".

The interpretations of Bischoff et al. (1985) of these observations are:

1) Part of the increase in halfwidths for the synthetic samples with increasing magnesium content results from substitution of Mg^{2+} for Ca^{2+}, but the major amount of increase is due to positional disorder of the carbonate ion in the calcite structure; that is, the carbonate anion, which normally lies in the plane of the a crystallographic axis in calcite, with increasing $MgCO_3$ content of the calcite is progressively rotated toward the plane of the c crystallographic axis, increasing the halfwidths of the V_1 and L Raman vibrational modes. The amount of positional disorder in the synthetic magnesian calcites is a function of magnesium concentration, a relationship that is in agreement with the decrease in unit cell c/a ratio with increasing $MgCO_3$ content (Figure 3.10).

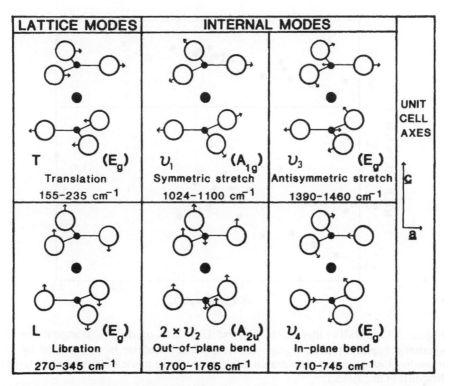

Figure 3.12. Schematic drawings of Raman modes of vibration in rhombohedral carbonates. Notation and symmetry of the vibration and the range of Raman shift for all rhombohedral carbonates are included for each mode. Hexagonal unit cell axes are shown for orientation. (After White, 1974.)

2) The larger halfwidths of the Raman V_1 and L vibrational modes of biogenic magnesian calcites relative to synthetic calcites of similar composition reflect greater positional disorder of the carbonate anion in the biogenic phases, as well as magnesium heterogeneities (Mg-rich areas in the phase) and substitution of trace elements. The larger cell volumes and c/a ratios of biogenic phases of bulk composition similar to synthetic phases suggest that positional disorder is the major factor involved.

3) The physical and chemical heterogeneities of biogenic materials *a priori* disqualify them as starting phases for dissolution experiments aimed at understanding the thermodynamics of the $CaCO_3$-$MgCO_3$ solid solution at low temperature. Comparison of the results of dissolution experiments on

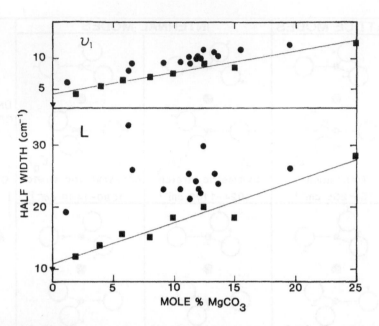

Figure 3.13. Halfwidths of the V_1 and L Raman modes vs. composition for synthetic (squares) and biogenic (circles) phases. Calcite is represented by the triangle. Least squares line drawn through synthetic phases for reference. (After Bischoff et al., 1985.)

synthetic and natural magnesian calcites, however, can provide invaluable information concerning the reactivity of natural magnesian calcites in the sedimentary environment. The results of Bischoff et al's. (1987) calculations of -log IAP versus $MgCO_3$ composition for synthetic and cleaned, annealed, and size-fractionated biogenic materials dissolved in CO_2-H_2O solutions are shown in Figure 3.14. These results and their significance are discussed in the next section of this chapter. Before doing so let us consider results from precipitation experiments.

Precipitation experiments   Mucci and Morse (1984) reported solubility data for several magnesian calcites produced from precipitation experiments (see Figures 3.7 and 3.14). In their experiments overgrowths of magnesian calcite were precipitated on calcite seeds in a pH-stat which maintained constant solution composition. The overgrowth compositions were determined by atomic absorption spectrophotometry and X-ray diffraction. The Mg contents of the overgrowths varied only with the Mg:Ca ratio in solution, and the overgrowths were shown to be at exchange equilibrium with the solution. Mucci and Morse (1984) then precipitated magnesian calcite overgrowths on calcite seeds in closed, free-drift

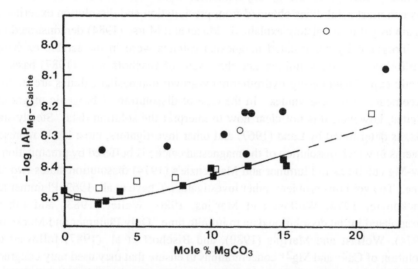

Figure 3.14. Stabilities of calcite, and synthetic (closed squares) and biogenic (closed circles) magnesian calcites as a function of composition. Stabilities are expressed as -log IAP$_{Mg-Calcite}$. The curve is a hand-drawn "best" fit to the synthetic data. Also plotted are the results of precipitation experiments by Mucci and Morse (1984, open squares) and biogenic dissolution experiments by Walter and Morse (1984a, open circles). (After Bischoff et al., 1987.)

experiments using the same solutions as in their pH-stat experiments (Mucci and Morse, 1983). The magnesium content of the overgrowths was inferred from the solution composition and the overgrowth compositions from the pH-stat runs. Mucci and Morse suggested that the steady-state solution compositions from the free-drift experiments may be interpreted to represent thermodynamic equilibrium, and on this basis used them to calculate IAP$_{Mg-calcite}$ values. These data define a much less dramatic increase in solubility with Mg content than most of the data from dissolution experiments (see Figure 3.7), but they are in close agreement with the dissolution data on synthetic calcites (Figure 3.14). Bischoff et al. (1987) contend that this similarity may indicate that reversibility for the magnesian calcite IAP values has been demonstrated, and that these values should be used to model the thermodynamic behavior of the magnesian calcite solid solution. It is also interesting to note that the biogenic dissolution data in closest agreement with the Mucci and Morse data are those of Land (1967) which, as mentioned previously, may represent systems buffered by precipitation of a low-magnesian calcite.

Summary interpretation of experimental data The extreme discrepancies between most solubilities obtained from precipitation and dissolution experiments remain as yet incompletely explained. Mucci and Morse (1984) demonstrated that the inorganically-precipitated magnesian calcites were in metastable exchange equilibrium with the solutions, and the results of Bischoff et al. (1987) based on kinetic experiments using hydrothermally-grown magnesian calcites are in good agreement with these values. In the case of dissolution of biogenic magnesian calcites, however, it is not clear how to interpret the solution data. Steady-state pHs, as determined by Land (1967) and other investigators, most likely represent systems in which dissolution of the magnesian calcite is buffered by precipitation of low-Mg calcite as in Plummer and Mackenzie's (1974) dissolution states two and three. To avoid this problem, other investigators (Chave et al., 1962; Plummer and Mackenzie, 1974; Wollast and Marijns, 1980; Walter, 1983; and others) extrapolated initial dissolution data to infinite time. Only Plummer and Mackenzie (1974), Wollast and Marijns (1980), and Bischoff et al. (1987) followed the evolution of Ca^{2+} and Mg^{2+} concentrations to ensure that they used only congruent leg data in their extrapolations. The meaning of extrapolated dissolution data has been a matter of debate. As mentioned previously, Plummer and Mackenzie (1974) assumed the extrapolated pHs represented the solution pH at metastable thermodynamic equilibrium with the dissolving magnesian calcite. Thorstenson and Plummer (1977) argued that thermodynamic equilibrium conditions cannot be calculated from congruent dissolution data, and that the extrapolated pH values of Plummer and Mackenzie and other investigators represent a state of stoichiometric saturation.

Thorstenson and Plummer (1977), in an elegant theoretical discussion (see section on The Fundamental Problems), discussed the equilibrium criteria applicable to a system composed of a two-component solid that is a member of a binary solid solution and an aqueous phase, depending on whether the solid reacts with fixed or variable composition. Because of kinetic restrictions, a solid may react with a fixed composition, even though it is a member of a continuous solid solution. Thorstenson and Plummer refer to equilibrium between such a solid and an aqueous phase as stoichiometric saturation. Because the solid reacts with fixed composition (reacts congruently), the chemical potentials of individual components cannot be equated between phases; the solid reacts thermodynamically as a one-component phase. The variance of the system is reduced from two to one and, according to Thorstenson and Plummer, the only equilibrium constraint is $IAP_{Mg\text{-}calcite} = K_{eq(x)}$, where $K_{eq(x)}$ is the equilibrium constant for the solid, a function of composition (x) (see the derivation in Thorstenson and Plummer, 1977). It should be noted that $IAP_{Mg\text{-}calcite} = K_{eq(x)}$ is also a necessary, but not sufficient, condition for thermodynamic equilibrium. The problem is that there is a continuous series of solid solutions between 0 and 30 mole % $MgCO_3$, and thus a series of well defined individual components. Reduction of the variance of the system is indirectly a violation of the phase rule. Experiments to test the validity of the hypothesis that magnesian calcite reactions are governed by stoichiometric saturation have been inconclusive (Marijns, 1978; Wollast and Marijns, 1980; Walter and Morse, 1984a).

If care is taken in sample preparation, then it appears that consistent values are obtained for solubility calculated as $IAP_{Mg\text{-}calcite}$. This observation leads us to conclude that $IAP_{Mg\text{-}calcite}$ is a valid way to represent the *relative solubilities* of the magnesian calcites. In Figure 3.7 we plotted the solubilities of the biogenic magnesian calcites in terms of -log $IAP_{Mg\text{-}calcite}$. All these values were obtained from dissolution experiments starting with undersaturated solutions and thus do not include results from experiments performed in surface seawater, which is usually oversaturated with respect to calcium carbonates. The curve through the biogenic experimental results indicates only a broad trend of increasing solubility with $MgCO_3$ content as evaluated from data presently available on the dissolution of biogenic magnesian calcites. For comparison the stability-composition curve for synthetic magnesian calcites and that of Plummer and Mackenzie (1974) are also shown.

The above discussion generates an obvious question: What are the "correct" values for the relative solubilities of the magnesian calcites? Bischoff et al. (1987) offer an explanation of the three curves in Figure 3.7 that has bearing on natural processes affecting calcites discussed later in this book. They suggest that each of the curves is applicable to the reactivity of magnesian calcites, but under different theoretical and environmental constraints:

    1. The results of Plummer and Mackenzie (1974) may best describe how biogenic materials react in nature. In nature, pieces of magnesian calcite skeletons will contain organic matter, have very fine-grained magnesian calcite particles adhered to larger grains, and have surface defects and dislocations. The apparent Gibbs free energies of formation for these materials are likely more dependent on kinetic considerations than on the true thermodynamic properties of the solid.

    2. The results of Walter and Morse (1984a) and Bischoff et al. (1987) on cleaned, annealed and size-fractionated magnesian calcite skeletons best describe the reactivity of biogenic materials with chemical and physical

heterogeneities. These materials do not properly belong in the simple two-component system $CaCO_3$-$MgCO_3$, and are compositionally and structurally more complex than the synthetic phases.

3. The results of Bischoff et al. (1987), and perhaps Mucci and Morse (1984), on synthetic phases may best describe the thermodynamic properties of the magnesian calcite solid solution in the system $CaCO_3$-$MgCO_3$. These results have been confirmed recently by the experimental work of Busenberg and Plummer (1989) It is these results that should be used in modelling the thermodynamic behavior of the solid solution, and provide a basis for evaluating the effect of heterogeneities on reactivity of calcite in natural environments. The implications of these three points to carbonate mineral diagenesis are discussed in subsequent chapters.

Hypothesis of a Hydrated Magnesian Calcite

Lippmann (1960) was the first to our knowledge to suggest that magnesian calcite could contain some water molecules owing to the difficulty of dehydrating the Mg^{2+} ion during precipitation of the carbonate. The possibility is also reflected in the fact that precipitation of magnesian carbonate at ordinary temperatures leads invariably to the formation of hydrated compounds. The hypothesis of poisoning of magnesian calcites by hydroxyl groups or water (Schmalz, 1965; Weber and Kaufman, 1965; Lahann, 1978; Wollast et al., 1980; Mucci and Morse, 1983) has been used by several authors mainly to explain the kinetics and mechanisms of precipitation of magnesian calcites.

Evidence supporting this hypothesis has been obtained only recently by infrared spectrometry of the magnesian calcites (Mackenzie et al., 1983) and spectrometry (Gaffey, 1988) of the magnesian calcites. Figure 3.15 shows the absorbance of the O-H bonds for various magnesian calcites dried at 110°C. Water in the magnesian calcite structure may be the cause of this absorbance. If so, the absorbance is proportional to the water content and is possibly correlated with the magnesium content. The natural samples containing 2 and 6 mole % $MgCO_3$ exhibit rather high water contents, and two samples synthesized hydrothermally by Bischoff et al. (1983) are characterized by low water contents. From this regression line and loss of weight measurements, it can be estimated that, on the average, there is about one molecule of water per Mg atom in the natural magnesian calcites.

The possible presence of water in the structure of the magnesian calcite solid solution must affect its free energy of formation and thus its solubility. Inspection of the free energies and enthalpies of formation (Robie et al., 1979) of nesquehonite [$MgCO_3 \cdot 3H_2O$], hydromagnesite [$Mg_5(OH)_2(CO_3)_4 \cdot 4H_2O$], artinite [$Mg_2(OH)_2CO_3 \cdot 3H_2O$], and brucite [$Mg(OH)_2$] indicates that the introduction of water molecules or hydroxyl groups into the lattice of magnesian calcites should increase the heat of dissolution of the hydrated structure and decrease the free energy of formation.

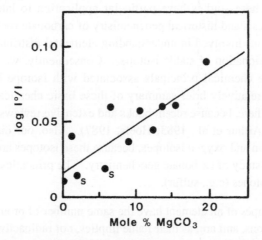

Figure 3.15. Infrared absorbance of the O-H bond as a function of the magnesium content in magnesian calcites after drying at 110°C. All samples are biogenic except for two hydrothermally synthesized magnesian calcites indicated by the letter S. (After Mackenzie et al., 1983.)

It is possible, therefore, that the magnesian calcites should be considered at least as a three component system: $CaCO_3\text{-}MgCO_3\text{-}H_2O$. It is not obvious, however, that the magnesian calcites can be described as a mixture of pure $CaCO_3$ and hydrated $MgCO_3$. The apparent high water content of biogenic magnesian calcites with low $MgCO_3$ contents (Mackenzie et al., 1983; Gaffey, 1988) seems to indicate that $CaCO_3$ may also be partially hydrated. Furthermore, it is likely that the water, at least in biogenic phases, occurs in three positions in the structure: hydrated H_2O, hydroxyl groups, and fluid inclusions. Thus, it is premature to try to estimate quantitatively the influence of water content on the free energy of formation and solubility of magnesian calcites. The possible presence of water, however, is certainly one important factor that may partially explain the high enthalpies of mixing (Mackenzie et al., 1983) of magnesian calcites, their high solubilities, and their mechanisms of precipitation.

Stable isotope chemistry

General Considerations

Over the last 30 years the study of the stable isotope composition of carbonates has been one of the more active areas of research in carbonate geochemistry. These studies have particular application to later discussion of carbonate diagenesis and historical geochemistry of carbonate rocks. Many of the same considerations involved in understanding elemental distribution coefficients apply to the fractionation of stable isotopes. Consequently, we have included a discussion of the chemical principals associated with isotope behavior in this chapter. Only a relatively brief summary of these basic chemical considerations will be presented here, because recent books and extensive reviews are available on this topic (e.g., Arthur et al., 1983; Hoefs, 1987). Also, our discussion will be restricted to carbon and oxygen isotopes, because these isotopes are by far the most important for the study of carbonate geochemistry. The principles, however, apply to other stable isotopes (e.g., sulfur).

Stable isotopes of an element have the same number of protons, but different numbers of neutrons, and are, as their name implies, not radioactive. The differing number of neutrons leads to atoms having similar chemical properties but different mass. Although their chemical properties are similar, they are not identical owing to quantum mechanical effects. The variations in properties of different isotopes are approximately proportional to the relative (percentage) difference in mass.

The fact that different isotopes of an element do not have the same physical-chemical properties means that kinetic and isotope exchange processes can lead to variations in isotopic composition. This phenomenon is usually referred to as isotope fractionation. Isotope fractionation, in most cases, leads to only small differences in isotopic composition. Consequently, isotope ratios are generally reported in terms of parts per thousand (per mil, o/oo) differences. While it is possible to report isotope compositions in absolute terms, it has been found most convenient to report them relative to a standard with a composition typical of common natural materials. These two considerations result in the commonly used "δ" notation:

$$\delta_x = \left(\frac{\text{Isotope ratio in X - Isotope ratio in std.}}{\text{Isotope ratio in std.}} \right) \times 10^3$$

(3.12)

where X is the unknown sample and std. is the standard.

The fractionation of isotopes between different phases is of primary concern to geochemists. For two different phases A and B, the fractionation factor (α_{A-B}) is defined as (isotope ratio in A) / (isotope ratio in B) and can be related to δ values by equation 3.13.

$$\alpha_{A-B} = (\delta_A + 10^3) / (\delta_B + 10^3) \tag{3.13}$$

Other useful and common relations are:

$$\varepsilon_{A-B} = (\alpha_{A-B} - 1) \times 10^3 \tag{3.14}$$

where ε_{A-B} is the enrichment factor, and:

$$\delta_A - \delta_B = \Delta_{A-B} \approx 10^3 \ln \alpha_{A-B} \tag{3.15}$$

Fractionation depends on temperature and reaction rate. A complication in carbonate geochemistry is that most calcite and aragonite are formed by organisms which exert a "vital" effect on fractionation. In spite of these complications, the study of oxygen and carbon isotope ratios in carbonates has proven an immensely useful geochemical tool in understanding diagenesis and paleoenvironments. Applications of stable isotopes to understanding geochemical processes are discussed in subsequent chapters of this book.

Oxygen Isotopes

Oxygen isotopes have been the most thoroughly studied isotopes in carbonate mineral systems. There are three stable isotopes of oxygen; $^{16}O = 99.763\%$, $^{17}O = 0.0375\%$ and $^{18}O = 0.1995\%$ (Garlick, 1969). The fractionation of ^{18}O relative to ^{16}O is commonly measured and reported as the value of $\delta^{18}O$ relative to a standard. Two standards have been widely used in reporting $\delta^{18}O$ values. The most common one in general usage now is "SMOW," which stands for standard mean ocean water.

It has an ^{18}O / ^{16}O ratio of 2005.20 (± 0.45) x 10^{-6} (Baertschi, 1976). In many studies involving carbonates, PDB is the standard relative to which values are reported. The PDB standard is based on the oxygen isotope ratio of a Cretaceous belemnite from the Peedee formation in South Carolina. Because it was composed of calcite, many limestones have isotopic ratios close to it. Values ("X") reported relative to the two standards are related by equation 3.16.

$$\delta^{18}O_{(X\ vs.SMOW)} = 1.03086\ \delta^{18}O_{(X\ vs.\ PDB)} + 30.86 \qquad (3.16)$$

One of the most important uses of oxygen isotopes in carbonates is as geothermometers. Isotope fractionation is pressure independent (Clayton et al., 1975), giving it an advantage over other geothermometers which are generally both temperature and pressure dependent. The equation for calculating the temperature at which calcite precipitates from water is:

$$T(^oC) = 16.0 - 4.14\ (\delta_c - \delta_w) + 0.13\ (\delta_c - \delta_w)^2 \qquad (3.17)$$

where δ_c is the oxygen isotope ratio in calcite relative to the PDB standard, and δ_w is the ratio in water relative to the SMOW standard (see Arthur et al., 1983).

Oxygen isotope fractionation in other carbonate minerals has also been studied. Magnesian calcites are about 0.06 per mil heavier than calcite for each mole % $MgCO_3$, and aragonite is also 0.6 per mil heavier than calcite at 25°C (Tarutani et al., 1969). Grossman and Ku (1981) found that aragonitic mollusks and foraminifera are about 0.7 per mil heavier than calcite over the temperature range of 3°C to 19°C. Oxygen isotope fractionation in dolomites is predicted to be about 6 per mil heavier than calcite at 25°C, based on extrapolations from high temperature measurements (e.g., Northrop and Clayton, 1966; O'Neil and Epstein, 1966). However, observations of natural dolomites have given highly variable isotopic results and been the source of considerable controversy. Oxygen isotope ratios in dolomites will be discussed in conjunction with the general problems of dolomite formation later in this book.

Studies of oxygen isotopic fractionation in inorganic precipitates provide conflicting evidence for a correlation between fractionation and precipitation rate. A correlation between the isotopic composition of the inorganic precipitates and the proportion of carbonate in the carbonate-bicarbonate solution (Figure 3.16) was observed by McCrea (1950). He hypothesized that, when first formed, the calcium carbonate has the same isotopic composition as the ion in solution, but if the solid is built up slowly, there is time for each layer to reach an equilibrium between the solid state and the solution. Tarutani et al. (1969) considered whether the variation they observed in measured ^{18}O fractionation factors could be attributed to precipitation rate. Their fractionation factors did not correlate with nitrogen flow rate, and thus were not considered to be related to precipitation rate. They did observe, however, a positive correlation between initial calcium concentration and fractionation factor. Because a greater initial calcium concentration corresponds to a greater initial ion activity product (IAP) and increased initial carbonate saturation, there may have been a positive correlation between initial precipitation rate and oxygen isotopic fractionation in Tarutani et al.'s results. This result is opposite to that of McCrea (1950). Oba (see Horibe et al., 1969) noted that the poor reproducibility in more than 50 experiments precipitating calcite and aragonite between 0° and 30^oC seemed to be dependent upon the speed of recrystallization (N_2 flow). Unfortunately, in none of these studies of oxygen isotope fractionation was precipitation rate directly measured, nor was it held constant throughout an experiment, so the evidence for or against a precipitation rate effect on isotopic fractionation is as yet circumstantial. However, as mentioned previously, studies of cation partition coefficients, where these kinetic factors were closely monitored, clearly show major kinetic influences for trace components (e.g., Lorens, 1981).

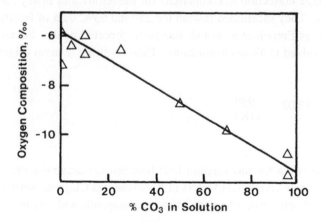

Fig. 3.16. Isotopic variation of calcium carbonate precipitate with solution composition at 25^oC. (After McCrea, 1950.)

Carbon Stable Isotopes

Carbon has two stable isotopes, ^{12}C=98.89% and ^{13}C=1.11% (Nier, 1950), whose fractionation has also been extensively studied. $\delta^{13}C$ values are usually reported relative to the previously described PDB calcium carbonate standard. The carbon isotopic composition of carbonates, dissolved carbonic acid species, methane and organic matter exhibit considerable variation, usually associated with biologic processes such as food webs and bacterial oxidation of organic matter. This observation means that they can be used to obtain information not generally available from oxygen isotopes. Consequently, both carbon and oxygen isotopes, particularly when used to complement each other, are important to the study of carbonate geochemistry.

The generally accepted values for carbon isotope equilibrium between calcium carbonate and dissolved bicarbonate are derived from the inorganic precipitate data of Rubinson and Clayton (1969), Emrich et al. (1970), and Turner (1982). Rubinson and Clayton found calcite and aragonite to be 0.9 and 2.7 $^o/_{oo}$ enriched in ^{13}C relative to bicarbonate at 25°C. Emrich et al. determined enrichment factors for undifferentiated calcium carbonate-bicarbonate fractionation (ε_{s-b}) over the temperature range 20° to 50°C. In the final analysis of their data, the authors questioned the accuracy of their 20°C datum, because it required a large correction to compensate for the removal of heavy carbon during precipitation. They substituted instead the 25° and 63°C data of Baertschi (1957), even though, as Emrich et al. noted, Baertschi reported neither the technique nor the errors involved in his determinations. Their resulting relation (Figure 3.17) is:

$$\varepsilon_{s-b} = 12.02 - \frac{2980}{T(K)}$$

(3.18)

Note that the curve for this relation falls between the aragonite-bicarbonate and calcite-bicarbonate enrichment factors of Rubinson and Clayton, consistent with the fact that the calcium carbonate was a mixture of aragonite and calcite.

In an attempt to compare isotopic fractionation in foraminifera to that in inorganic precipitates, Grossman (1984) presented tentative calcite-bicarbonate and aragonite-bicarbonate fractionation relations by combining the monomineralic results of Rubinson and Clayton with the temperature dependence obtained from the Emrich et al. relation. Grossman also redetermined the isotopic composition of the dissolved bicarbonate in Rubinson and Clayton's experiments. The $\delta^{13}C$ of bicarbonate must be calculated from the isotopic composition of the dissolved inorganic carbon, the relative proportion of carbonate ion, bicarbonate ion, and aqueous CO_2, and the fractionation factors between them. The results of Rubinson and Clayton are based on the carbonate ion-gaseous CO_2 enrichment factor of Thode et al. (1965), 13.3 to 16.6$^0/_{00}$. Recent redetermination of this enrichment factor has yielded a value of 6.54$^0/_{00}$ (Turner, 1982), a value in better agreement with the theoretically calculated value of 7.4$^0/_{00}$ (Thode et al., 1965). Grossman's revised Rubinson and Clayton values for aragonite-bicarbonate and

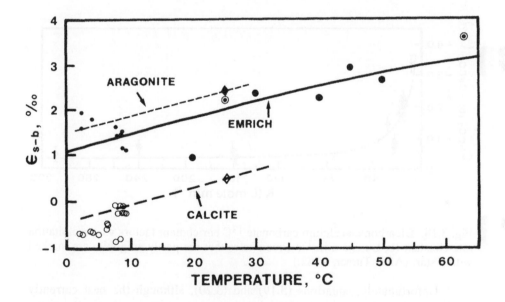

Fig. 3.17. Carbon isotopic enrichment factors for aragonitic (dots) and calcitic (circles) foraminifera and inorganic precipitates as a function of temperature. Shown are the inorganic precipitate data of Baertschi (1957)- circle with dot; Emrich et al. (1970)- large dots; and Rubinson and Clayton (1969) revised - see text; aragonite - solid diamond, calcite - hollow diamond. (After Grossman, 1984.)

calcite-bicarbonate fractionation at 25°C are 2.40 ± 0.23$^0/_{00}$ and 0.51 ± 0.22$^0/_{00}$, respectively (Figure 3.17). Applying Emrich et al.'s temperature dependence yields the tentative relations:

$$\varepsilon_{cl\text{-}b}\,(^o/oo) = 10.51 - \frac{2980}{T(^oK)} \tag{3.19}$$

$$\varepsilon_{ar\text{-}b}\,(^o/oo) = 12.40 - \frac{2980}{T(^oK)} \tag{3.20}$$

for calcite and aragonite (Figure 3.18).

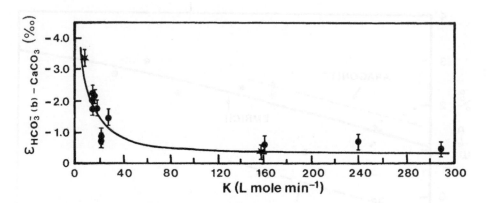

Fig. 3.18. Bicarbonate-calcium carbonate ^{13}C enrichment factors vs. precipitation rate constant (k, L mole min[-1]). Precipitates that are partly aragonite are marked with a star. (After Turner, 1982.)

Unfortunately, equations (3.19) and (3.20), although the best currently available for the data, can only be considered tentative. The weakness lies in the temperature dependence. As mentioned earlier, it is not known whether the results reflect a temperature dependence or a change in the proportion of aragonite to calcite. Another shortcoming in this relationship is that the temperature range of the experiments, 20° to 63°C (if Baertschi's data are included), is well above the temperature at which many organisms secrete carbonate. Because benthic foraminifera have proved very useful in carbon isotopic studies, it is important to have carbon isotopic equilibrium defined over their temperature range.

One of the first comprehensive investigations of kinetic effects in isotopic fractionation was conducted by Turner (1982). Noting that calcite and low magnesian calcite inorganically precipitated in freshwater lakes were enriched in

[13]C in excess of what the data of Rubinson and Clayton and Emrich et al. would predict, Turner examined the possibility that very slow precipitation could produce this enrichment. Calcium carbonate was precipitated at different rates by varying the ion activity product of calcite in the solution. Turner found that calcite-bicarbonate fractionation at 25°C was between 1.8 and 2.3°/$_{oo}$ at slow precipitation rates, and rapidly decreased to about 0.6°/$_{oo}$ as precipitation rate increased (Figure 3.18). Note that the values obtained at faster precipitation rates are almost identical to the revised Rubinson and Clayton (1969) value. Turner attributed the large fractionation at slow rates of precipitation to thermodynamic isotope fractionation, and the smaller fractionation at faster precipitation rates to the competing effects of thermodynamic fractionation and kinetic fractionation. Applying the surface diffusion crystal growth model of Burton et al. (1951), Turner suggested that kinetic fractionation occurs during diffusion of ions across the crystal surface. Because of its faster diffusion rate this process favors the light isotope-bearing ions, whereas thermodynamic equilibrium favors the precipitation of the heavier isotope-bearing ions. The results of Turner provide valuable insight into the complexity of carbon isotope fractionation. Unfortunately, the experiment was limited to carbon isotope fractionation; no data were presented on oxygen isotope fractionation. If the model used by Turner is correct, then kinetic effects on oxygen isotope fractionation should be twice that of carbon isotope fractionation.

Whether calcite-bicarbonate and aragonite-bicarbonate fractionation in biogenic material is dependent on precipitation rate is unclear. As mentioned earlier, an inverse relationship between precipitation rate and carbon isotopic composition has been observed in biogenic carbonate. If kinetic fractionation were the sole cause of disequilibrium $\delta^{13}C$ and $\delta^{18}O$ values, then $\delta^{13}C$ and $\delta^{18}O$ should always covary, with $\delta^{18}O$ showing twice the effect of $\delta^{13}C$. In nature this variance is usually not the case. In addition, the large carbon isotopic enrichment observed in the slow precipitation runs of Turner essentially are never seen in biogenic calcite. In fact, the $\delta^{13}C$ values of calcitic shells usually do not even show the 0.5°/$_{oo}$ enrichment that Turner observes in rapidly precipitated calcite.

Concluding remarks

At this stage in the development of the subject of the geochemistry of sedimentary carbonates, we have dealt primarily with the mineralogy and basic physical chemistry of the carbon system and some of its important phases. In the following chapters, this information and additional data and interpretations are utilized to explain the behavior of sedimentary carbonates in shallow water and deep marine environments, and during early and late diagenesis. The biogeochemical cycle of carbon and carbonates is also investigated, as is the use of carbonates in interpretation of the history of Earth's surface environment. It should be apparent to the reader at this stage that studies of the basic properties of sedimentary carbonates compose a voluminous literature; the application of these studies to geological and geochemical systems requires knowledge of a variety of other chemical, physical, biological, and geological considerations. These considerations are emphasized in the rest of this book.

Chapter 4

THE OCEANIC CARBONATE SYSTEM AND CALCIUM CARBONATE ACCUMULATION IN DEEP SEA SEDIMENTS

An overview of major processes

The processes responsible for the distribution of carbon dioxide in seawater and the accumulation of calcium carbonate in deep sea sediments are among the most intensely studied in earth sciences. They involve an intricate interplay among chemical, biological and physical components of the land, atmosphere and world ocean. Interest in this complex system has escalated as a result of growing concern about the impact of fossil fuel-derived carbon dioxide on global climate. A vast literature on this system has resulted, much of which is beyond the scope of this book. In so far as specialized books (e.g., Andersen and Malahoff, 1977; Williams, 1978; Broecker and Peng, 1982; Sundquist and Broecker, 1985) and major review articles are available that deal in detail with various aspects of this system, we will primarily emphasize those processes most relevant to the deposition and accumulation of carbonates in deep sea sediments. The early diagenesis of deep sea sediments is discussed in this chapter, and their long-term diagenesis is discussed in Chapter 8.

It is convenient to divide the deposition of carbonates in marine sediments into those being deposited in shallow (shoal) water (water depths of a few hundred meters or less) and those being deposited in deep sea sediments, where the water depth is on the order of kilometers. The primary reasons for this division are the differing sources, dominant mineralogies, and accumulation processes operative in these environments. Shoal water carbonates are the topic of Chapter 5. Naturally, there are "grey" areas of intermediate characteristics between these two extremes, such as continental slopes and the flanks of carbonate banks and atolls.

Before proceeding with an examination of the oceanic carbonate system and the accumulation of calcium carbonate in deep sea sediments, it is useful to consider briefly the general relationships among the different components and their most basic characteristics. The various components of first order importance to the system are shown in Figure 4.1. These can be divided into external components,

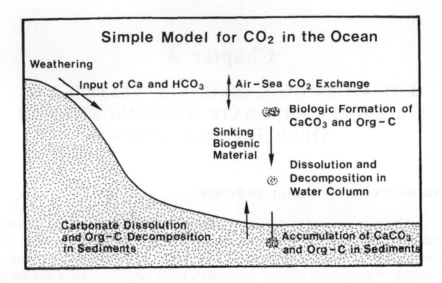

Figure 4.1. A simple model for the major controls on the carbonic acid system in the oceans.

such as continental waters and exchange of CO_2 across the air-sea interface, and internal components. The primary internal components of interest are those by which $CaCO_3$ is formed, dissolved and removed. A major portion of $CaCO_3$ formed in the oceans is precipitated by pelagic organisms in the upper ocean where the waters are supersaturated with respect to both calcite and aragonite. The carbonates formed in the upper ocean sink into the deep sea along with organic matter. The influences of decreasing temperature, and increasing pressure and P_{CO_2} all act to cause the deep waters of the oceans to become undersaturated, first with respect to aragonite, and then with respect to calcite. The increase in P_{CO_2} is primarily the result of oxidation of organic matter; the spatial heterogeneity in the extent of this reaction gives the deep oceans much of their variation in saturation state. These same factors also act on $CaCO_3$ which reaches the seafloor; as a result, only a small fraction of the total amount of $CaCO_3$ produced is eventually permanently buried in the sediment. It is important to keep in mind that the formation and preservation of $CaCO_3$ is intimately tied to oceanic productivity, circulation and the complex cycle of organic carbon in the oceans, as well as dissolution and burial.

Models of great complexity have been produced in recent years in an attempt to predict the probable influences of increased atmospheric carbon dioxide content on the oceans, and the oceans' ability to buffer these changes. These are discussed in Chapter 9.

The CO_2 system in oceanic waters

The Upper Ocean

The distribution of CO_2 and the associated carbonic acid system species in the upper ocean (here loosely defined as waters above the thermocline and generally only a few hundred meters in depth) is primarily controlled by the exchange of CO_2 across the air-sea interface, biological activity, and circulation of the ocean, mainly through vertical mixing processes. Other factors, such as the temperature and salinity of the water, can also contribute to variations by influencing the solubility of CO_2 in seawater and the equilibrium constants of the carbonic acid system.

The influences of temperature and salinity on the carbonate chemistry of seawater can be computed using the methods previously described in this book. Calculated values of ΣCO_2 and saturation state with respect to calcite are shown in Figure 4.2 for a reasonable range of conditions (T=0 to 30°C; S=33 to 37) for major ocean waters. The influence of salinity variation on ΣCO_2 and Ω_c is small ($\leq$ ~1%). Temperature variation can cause about a factor of 3 change in Ω_c, and about a 10 % variation in ΣCO_2. Pressure is of only negligible importance to carbonate chemistry variations in the upper ocean.

Revelle and Suess (1957) introduced a dimensionless factor (γ), commonly referred to as the Revelle factor, to describe the relation between changes in P_{CO2} and ΣCO_2 in seawater. The equation for this relation (based on the form of Takahashi et al., 1980a), where A = alkalinity, is:

$$\gamma = \left(\frac{\Sigma CO_2}{P_{CO_2}}\right)\left(\frac{\partial P_{CO_2}}{\partial \Sigma CO_2}\right)_{A,T,S} = \left(\frac{\partial \ln P_{CO_2}}{\partial \ln \Sigma CO_2}\right)_{A,T,S}$$

(4.1)

The Revelle factor is about 10 for typical surface seawater. The details of the chemistry of this general relationship and its derivation have also been discussed by Sundquist et al. (1979), who called it the "homogeneous buffer" factor. Of interest is the fact that using the Revelle factor one can calculate for an instantaneous change in the P_{CO2} of the atmosphere, the distribution of carbon between the atmosphere and seawater.

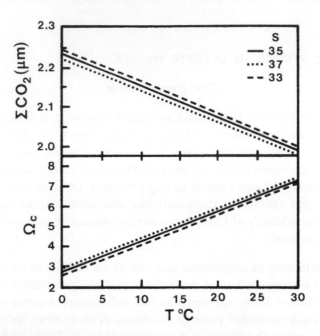

Figure 4.2. The variation of total carbon dioxide (ΣCO_2) and the saturation state of seawater with respect to calcite (Ω_c) with temperature for seawater with a total alkalinity of 2400 µeq kg^{-1} seawater and in equilibrium with atmospheric CO_2 (P_{CO_2} = 330 µatm).

Exchange of CO_2 across the air-sea interface can occur by diffusive processes and air injection into the water by breaking waves (e.g., Bolin, 1960; Broecker and Peng, 1974; Memery and Merlivat, 1985; Smith and Jones, 1985, 1986; Broecker et al., 1986). The former process has been extensively investigated, and a simple model generally in use is shown in Figure 4.3. In this model, CO_2 exchange is assumed to occur between the relatively rapidly and well-mixed reservoirs of atmosphere and seawater through a layer of water of thickness t, where molecular diffusion (D_{CO_2}) is the rate-limiting transport mechanism. The flux of CO_2 (J_{CO_2}) across the air-water interface is then given by equation 4.2:

$$J_{CO_2} = D_{CO_2} \; ([CO_{2(equilib. \, w \, atm)}] - [CO_{2(bulk \, sw)}]) / t \qquad (4.2)$$

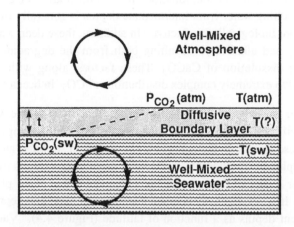

Figure 4.3. Model for the diffusive exchange of CO$_2$ across the air-sea interface. (Based on general model of Bolin, 1960; see also Broecker and Peng, 1974.)

The primary difficulty in employing equation 4.2 is determining appropriate values for the thickness of the boundary layer, t, which can be strongly influenced by wind (e.g., Kanwisher, 1963; Broecker and Peng, 1974; House et al., 1984). Typical values range from about 100 to 300 μm. D/t has units of velocity and is often referred to in the oceanographic literature as a "piston" velocity (e.g., Broecker and Peng, 1974). Furthermore, organic layers on the surface of the ocean can impede the transfer of CO$_2$ across the air-sea interface (e.g., Liss, 1973). It was suspected for some time that the enzyme carbonic anhydrase might catalyze the transfer of CO$_2$ into the ocean by accelerating the hydrolysis of CO$_2$, but more recent work (Goldman and Dennett, 1983) has largely discounted the importance of this process. An additional potential difficulty arises from the fact that the system may not be thermally homogeneous. If the air and seawater temperatures are greatly different, which is commonly the case, it is difficult to estimate the temperature in the diffusive layer and hence the appropriate values for the solubility of CO$_2$.

Two other factors contribute to major variations in the CO$_2$ content of upper ocean waters. The first is biologic activity. During photosynthesis CO$_2$ is consumed and O$_2$ is produced. Below the photic zone oxidative degradation of organic matter, primarily through bacterial attack, causes O$_2$ to be consumed and CO$_2$ to be released. Thus, major deviations of oceanic oxygen and carbon dioxide concentrations from those expected for equilibrium with the atmosphere, are generally inversely related. Biologic activity and ocean circulation are closely linked. As a result of the degradation of organic matter in deep water and sediments, upwelling water from intermediate depths is generally rich in nutrients and, hence, more biologically productive. In addition, these deep waters also have elevated P_{CO_2} and alkalinity, resulting both from the degradation of organic matter and the dissolution of $CaCO_3$. These factors, along with variable wind speeds can lead to extremely complex distributions of CO_2 in ocean surface waters.

Broecker and Peng (1982) pointed out that the surface waters of the Equatorial Pacific Ocean provide an excellent example of the importance of the interplay of these factors on the concentration of CO_2 in seawater and its exchange with the atmosphere. The high P_{CO_2} values of these waters results from upwelling. The deep water source of the excess CO_2 is clearly evident in the plot of potential P_{CO_2} (= P_{CO_2} if water were isochemically warmed to 25°C and depressurized to 1 atm) at different depths as a function of latitude (Figure 4.4). Variations in the relation between temperature and alkalinity of surface ocean waters also occur in different areas of the ocean (see for example Figure 4.5). The differences in alkalinity result from the formation of biogenic carbonates in the upper ocean, which removes alkalinity, and the subsequent dissolution of the carbonate in deep water, which adds alkalinity. Where older deep waters upwell, surface alkalinity is thus enhanced. In addition, alkalinity variations can be associated with processes which lead to variations in salinity. These variations can be accounted for by normalizing alkalinity to salinity. The resulting quantity is known as specific alkalinity.

It should be kept in mind that, in spite of these major variations in the CO_2-carbonic acid system, virtually all surface seawater is supersaturated with respect to calcite and aragonite. However, variations in the composition of surface waters can have a major influence on the depth at which deep seawater becomes undersaturated with respect to these minerals. The CO_2 content of the water is the primary factor controlling its initial saturation state. The productivity and temperature of surface seawater also play major roles, in determining the types and amounts of biogenic carbonates that are produced. Later it will be shown that there is a definite relation between the saturation state of deep seawater, the rain rate of biogenic material and the accumulation of calcium carbonate in deep sea sediments.

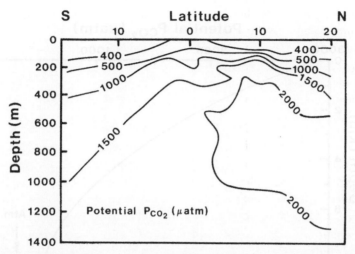

Figure 4.4. Potential P$_{CO_2}$ in Pacific Ocean waters as a function of depth at different latitudes (After Broecker and Peng, 1982, based on GEOSECS data.)

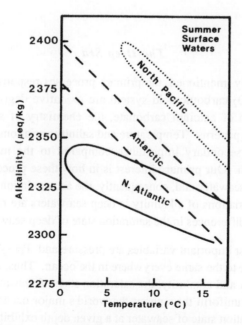

Figure 4.5. Variation in the alkalinity of surface waters versus temperature for different major ocean water bodies. (After Broecker et al., 1985.)

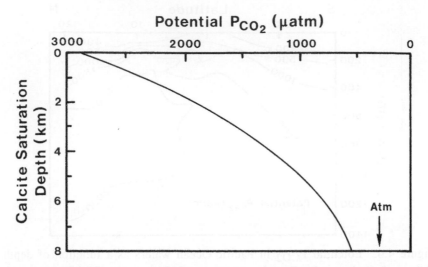

Figure 4.6. Variation in the saturation depth of seawater with respect to calcite as a function of potential P_{CO_2}, for seawater at 2°C, S=35, and A_t=2400 µeq kg^{-1} seawater.

The Deep Sea

As previously mentioned, the primary processes responsible for variations in the deep sea CO_2-carbonic acid system are oxidative degradation of organic matter, dissolution of calcium carbonate, the chemistry of source waters and oceanic circulation patterns. Temperature and salinity variations in deep seawaters are small and of secondary importance compared to the major variations in pressure with depth. Our primary interest is in how these processes influence the saturation state of seawater and, consequently, the accumulation of $CaCO_3$ in deep sea sediments. Variations of alkalinity in deep sea waters are relatively small and contribute little to differences in the saturation state of deep seawater.

The two most important variables are pressure and P_{CO_2}. The pressure at a given depth is close to the same everywhere in the ocean. Thus, if all deep seawater had the same P_{CO_2} and alkalinity, the saturation state of seawater at a given depth would be close to uniform throughout the world's major oceans. This is not the case, and the saturation state of seawater at a given depth exhibits major variability. The relation between the depth at which seawater is at equilibrium with calcite and the potential P_{CO_2} of the seawater is shown in Figure 4.6. It is important to note that if no CO_2 were added to deep seawater by respiration-oxidation that virtually none of the ocean floor would be overlain by undersaturated seawater.

The variation of alkalinity and ΣCO_2 in different major ocean basins is summarized in Figure 4.7. A simple general circulation model for oceanic waters, useful for understanding these distributions, is presented in Figure 4.8. Major source regions for deep waters are in the North and South Atlantic oceans. The general circulation can be roughly viewed as starting in the North Atlantic Ocean with flow to the south, where more water is added in the region of Antarctica. Flow then continues from west to east, with major portions of this flow breaking off and going northward into either the Indian or Pacific oceans. Thus, the youngest waters, which are the least influenced by organic matter oxidation or $CaCO_3$ dissolution, are in the North Atlantic, whereas the waters in the northern Indian and Pacific oceans are the oldest and show the largest alkalinity and ΣCO_2 values because of these processes. Broecker and Peng (1982) demonstrated that the ratio of the addition of $CaCO_3$-C to organic-C to deep waters from the upper ocean is not the same for all ocean basins. This is primarily the result of different input ratios and more extensive dissolution of $CaCO_3$ in the more undersaturated northern Indian and Pacific ocean waters.

The potential P_{CO2} values of northern, equatorial and southern Pacific and Atlantic ocean waters are presented in Figure 4.9. They were calculated by the methods presented in Chapter 2 using corrected (Bradshaw et al., 1981) GEOSECS data. They show similar, but much larger, variations than those observed for alkalinity and ΣCO_2. As previously discussed, this major variation in potential P_{CO2} is primarily the result of organic matter oxidation. This conclusion has been confirmed by the $\delta^{13}C$ studies of ΣCO_2 by Kroopnick (1985). The increasingly negative $\delta^{13}C$ values found in older waters of higher ΣCO_2 indicate a dominantly organic source for the CO_2 .

Similar attempts to evaluate the relation between changes in alkalinity and $CaCO_3$ dissolution have been made using dissolved Ca^{2+} concentrations in Pacific Ocean deep waters (e.g., Horibe et al., 1974; Brewer et al., 1975; Shiller and Gieskes, 1980; Tsunogai and Watanabe, 1981; Chen et al., 1982). This procedure has proven more complex than originally anticipated. In addition, the fact that the percent change in Ca^{2+} concentrations is about one tenth that in alkalinity means high (~0.1%) precision calcium data must be obtained.

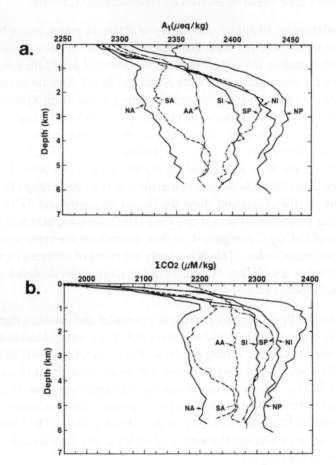

Figure 4.7. The mean vertical distribution of (a) alkalinity and (b) total CO_2 concentration normalized to the mean world ocean salinity value of 34.78. NA = North Atlantic, SA = South Atlantic, NP = North Pacific, SP = South Pacific, NI = North Indian, SI = South Indian, and AA = Antarctic region. (After Takahashi et al., 1980b.)

Figure 4.8. General circulation of deep ocean waters. (After Stommel, 1957.)

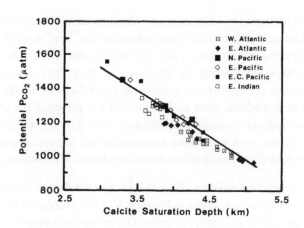

Figure 4.9. Potential P_{CO2} values for northern, equatorial and southern Atlantic and Pacific oceans (= NA, EA, SA, NP, EP, SP). (Based on GEOSECS data.)

Saturation State of Deep Seawater with Respect to CaCO₃

The calculation of the degree of saturation of deep ocean waters with respect to calcite and aragonite has been a continuously controversial topic. This calculation must be done with considerable precision, because saturation state typically changes only by about 20% per km change in water depth. In Chapter 1, the many constants which must be known as a function of temperature and pressure to accomplish this task were presented. Reevaluation, disputes over the "best" values, and incorrect usage of these constants (e.g., Li et al., 1969; see Pytkowicz 1970, for discussion) have led to calculations of substantially different values for saturation states of deep seawaters with respect to carbonate minerals. Attempts to perform *in situ* solubility measurements (e.g., Peterson, 1966; Ben-Yaakov and Kaplan, 1971; Sayles, 1980) unfortunately have not been successful at resolving the situation. As will be discussed later in this chapter, precise determination of the saturation state of deep seawater with respect to calcite and aragonite is pivotal to understanding the relationship between seawater chemistry and the accumulation of calcium carbonate in deep sea sediments. To understand $CaCO_3$ accumulation during geologic time, it is necessary to resolve today's situation.

Here, the methods for calculating the saturation state of seawater presented in Chapter 1 will be used. We feel that these values are the "best" at this time. Most of the saturation states were calculated using GEOSECS data (Takahashi et al., 1980b) and corrected as recommended by Bradshaw et al. (1981). Even using these corrections, serious questions remain about the reliability of the GEOSECS data, and the analytical methods used by the GEOSECS program for the carbonate system are now in doubt (Bradshaw and Brewer, 1988 a,b). However, this data set remains the only one with wide-scale coverage of the oceans. Because of these concerns, the values of saturation states should be viewed with caution.

Typical vertical saturation profiles for the North Atlantic, North Pacific, and Central Indian oceans are presented in Figure 4.10. The profiles in the Atlantic and Indian oceans are similar in shape, but Indian Ocean waters at these GEOSECS sites are definitely more undersaturated than the Atlantic Ocean. The saturation profile in the Pacific Ocean is complex. The water column between 1 and 4 km depth is close to equilibrium with calcite. This finding is primarily the result of a broad oxygen minimum-CO_2 maximum in mid-water and makes choosing the saturation depth (SD) where $\Omega_c = 1$ difficult (the saturation depth is also often referred to as the saturation level; SL).

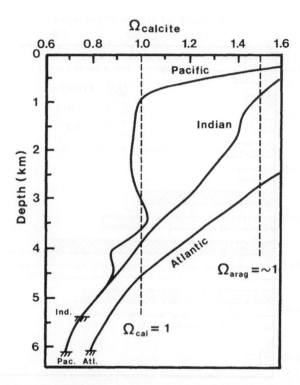

Figure 4.10. Saturation profiles for the northern Atlantic and Pacific oceans, and the central Indian Ocean (GEOSECS stations 31, 221, 450.)

It is impórtant to note that because of the greater solubility of aragonite, the water column becomes undersaturated with respect to aragonite at much shallower depths than the SD for calcite. In the Pacific and Indian oceans only the uppermost part of the water column is in equilibrium or supersaturated with respect to aragonite. Hypsographic data for the Atlantic, Pacific and Indian ocean basins are presented in Figure 4.11. The depth distribution of the seafloor in the different basins is strikingly similar. Only a small portion of the seafloor is within the depth range where the water is close to equilibrium or supersaturated with respect to aragonite. This is a primary reason for the preferential accumulation of calcite, relative to aragonite, in deep sea sediments.

We have calculated the saturation depth (SD) with respect to calcite for various regions of the Atlantic, Pacific and Indian oceans as shown in Figure 4.12. The saturation depth is deepest in the Eastern Atlantic ocean. In both the eastern and western Atlantic Ocean, the saturation depth becomes shallower to the south,

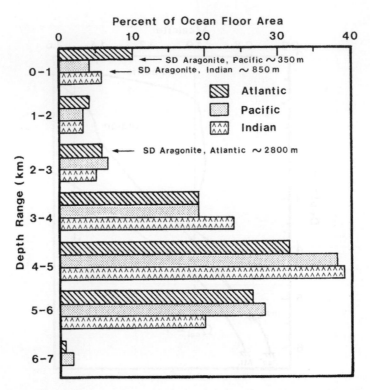

Figure 4.11. Distribution of ocean floor area by depth. (Based on data presented in Sverdrup et al., 1942.)

approaching the values for the Indian Ocean. In the Pacific Ocean the saturation depth (here picked as the depth below which the water column remains undersaturated) is highly variable for the reasons previously given.   These differences between and within oceans are primarily the result of the variations in P_{CO_2} of the waters.   Approximately two-thirds of the Pacific Ocean and one-half of the Indian Ocean seafloor are below the calcite SD, whereas only about one third of the Atlantic seafloor is below this SD (Berger, 1976). This positioning of the SD is a major reason for the relatively high abundance of calcium carbonate-rich sediments in the Atlantic Ocean.

Millero et al. (1979) summarized data of their own and of other investigators on the carbonate chemistry of the Mediterranean Sea. This sea is interesting, because its waters are supersaturated everywhere with respect to both calcite and aragonite. Both minerals are found in abundance in the sediments. The carbonate system in the Mediterranean Sea may be similar to that found in ancient epicontinental seas.

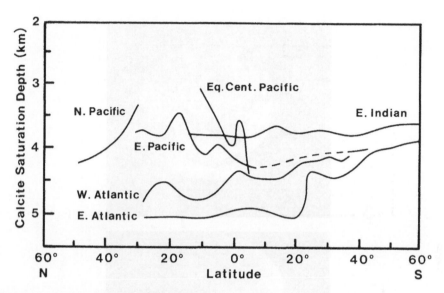

Figure 4.12. Saturation depths with respect to calcite in different areas of the ocean calculated from corrected GEOSECS data (see text).

Sources and sedimentation of deep sea carbonates

Sources

In the present ocean calcium carbonate formation is dominated by pelagic plants (coccolithophores) and animals (foraminifera, pteropods, and heteropods). Examples are presented in Figure 4.13. Although benthic organisms are important in shoal water sediments, and for dating and geochemical studies in the deep sea sediments, they constitute only a minor portion of the calcium carbonate removed from deep seawater. Shoal water carbonates are discussed in detail in Chapter 5.

The distribution of calcium carbonate-secreting pelagic organisms is primarily controlled by the fertility and temperature of the near surface ocean. The fertility of seawater is largely a result of ocean circulation patterns and, in particular, those processes leading to upwelling of nutrient-rich waters. In general, coccolithophores are common in temperate waters, but rare in high latitude cold waters where diatoms dominate. Coccolithophores are numerically much more

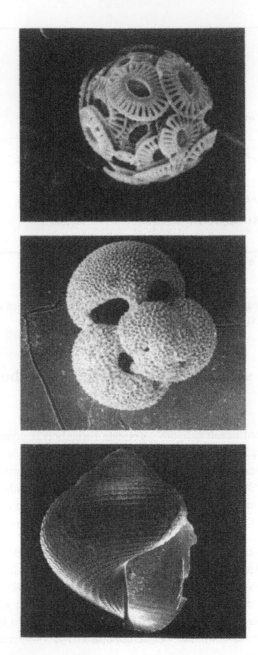

Figure 4.13. Photomicrographs of coccolithosphere (top, ~10 μm dia.), pelagic foraminifera (middle, ~100 μm dia.) and pelagic pteropod (bottom, ~1000 μm dia.).

abundant (usually ~ 10,000x) than foraminifera. They are commonly small (<10 μm), and form skeletal edifices termed coccospheres composed of over 30 disc-like structures (coccoliths). Planktonic forams range in size from about 30 μm to 1 mm. They and coccolithophores are composed of low magnesian calcite (< 1 mole % $MgCO_3$). Benthic forams can be formed of either aragonitic or high magnesian calcite. Because of their general scarcity in deep sea sediments, aragonitic pelagic organisms have received relatively little attention. Pteropods are the most abundant pelagic aragonitic organisms. They can reach over a centimeter in size. Their abundance, and rate of shell production have been sources of disagreement (e.g., Harbison and Gilmer, 1986). Knowledge of these variables is important, because they are necessary for assessment of the relative production rate of aragonite to calcite in the ocean. Recent estimates are primarily based on sediment trap results which will be discussed in the next section of this chapter.

Sedimentation

The mechanisms and rates of transfer of biogenic carbonate material from near-surface waters to deep sea sediments have been intensely investigated. Major questions have been concerned with the transition from living organism to carbonate test, rates of sinking, extent of dissolution in the water column and on the sea bottom, and the relation between life and death assemblages. Most major advances were made in this area after the extensive use of sediment traps was begun in the 1970's. Use of these traps has made direct collection of the sedimenting materials and their calculated vertical fluxes possible at different depths. Consequently, data and theories presented here will not generally include earlier and more hypothetical approaches to these processes.

Because of their small size, individual coccoliths should sink slowly (Lerman et al., 1974) and, consequently, spend long periods of time (on the order of 100 years) in the water column. This long residence time should lead to dissolution of the coccoliths in the undersaturated part of the water column. The origin of coccolith ooze on portions of the seafloor overlain by undersaturated waters, therefore, is difficult to explain by arguments based on settling rates of individual particles (e.g., Bramlette, 1958, 1961; Honjo, 1975, 1976; Roth et al., 1975). Schneidermann (1977) has summarized the major factors considered to be important in accumulation of coccoliths in sediments overlain by undersaturated waters. They are:

(1) Presence of organic material coating coccoliths which retards the rates of their dissolution.

(2) Rapid settling of coccoliths as large aggregate bodies, limiting the exposure for long periods of time of their high surface areas to the water column.

(3) Redeposition from shallower sites, both by mass transport (turbidity currents) and by resuspension.

(4) Rapid burial in the sediment.

The second factor, settling as aggregate bodies, has received a great deal of attention in recent studies. The primary mechanism for aggregation of coccoliths has been clearly shown to be grazing by zooplankton and encapsulation in fecal matter. The general model of Honjo (1976) for this complex process is presented in Figure 4.14. Notice that several processes are involved before discrete coccoliths are susceptible to solution in deep water. Honjo's paper and that of Schneidermann (1977) both present detailed discussions of coccolith distribution and sedimentation.

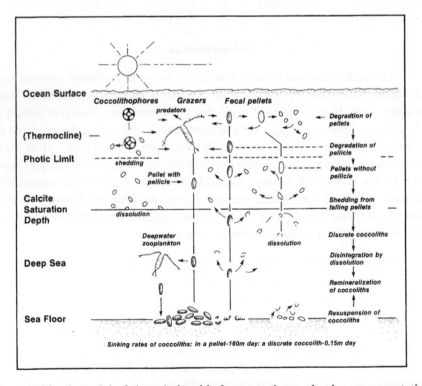

Figure 4.14. A model of the relationship between the production, transportation, dissolution and deposition of coccoliths in open, deep ocean waters. Scales are not in proportion. (After Honjo, 1976.)

Berger and Piper (1972) found the modal settling velocity for foraminifera was approximately 2.4 times less than that for a quartz sphere of equivalent maximum diameter. A good deal of scatter was observed from modal values, that the authors related to test shape and wall thickness. Berger and Piper (1972) found that the sequence of foraminifera observed in graded beds closely corresponded to their experimental results. Adelseck and Berger (1975) found evidence from net tows and the surfaces of box cores that larger foraminifera and pteropods experience little dissolution before reaching the sea floor. Similar experimental results and observations were obtained by Takahashi and Bé (1984). However, they emphasized the importance of shell weight and spines rather than simply size in controlling sinking speeds.

Thunell and Honjo (1981) conducted an extensive investigation of planktonic foraminifera in sediment trap samples from the tropical Pacific and central Atlantic oceans. At both study sites, the total foraminiferal flux and the carbonate flux tend to decrease with depth. In addition, the flux of individual species of planktonic foraminifera varies significantly with depth, with the number of small, solution-susceptible species decreasing with increasing depth. These results suggest that there is significant dissolution of small (<150 μm) foraminifera as they settle through the water column. Material collected from the sediment-water interface directly below the Pacific sediment trap array contained no planktonic foraminifera, suggesting that the residence time of an individual skeleton on the seafloor before it dissolves is extremely short (Thunell and Honjo, 1981).

Berner (1977) was among the first to emphasize the importance of pelagic aragonite sedimentation to the oceanic carbonate system. He estimated that about half of the $CaCO_3$ generated by pelagic organisms is aragonite, primarily in pteropod shells. Because of their large size, he also felt that it was likely that most reached the sediment-water interface prior to dissolution. In subsequent work (Berner and Honjo, 1981), the estimate for the percentage aragonite flux was revised downward to about 12%, based on PARFLUX (particle flux) sediment traps and examination of sediments from the Rio Grande Rise of the Atlantic Ocean.

More recent studies by Byrne et al. (1984) and Betzer et al. (1984, 1986) have contributed substantially to our understanding of aragonite sedimentation in the pelagic environment. In their work, short term deployments of large sediment traps were utilized to minimize the problem of dissolution of material within the trap. Byrne et al. (1984) carried out an elegant examination of pteropod dissolution in the water column in the North Pacific. Considerable variation of the expected extent of dissolution for different species was observed. Some of the pteropods, however, can undergo extensive dissolution in the water column of the North Pacific Ocean which is highly undersaturated with respect to aragonite.

Betzer et al. (1984, 1986) studied the sedimentation of pteropods and foraminifera in the North Pacific. Their sediment trap results confirmed that considerable dissolution of pteropods was taking place in the water column. They calculated that approximately 90% of the aragonite flux was remineralized in the upper 2.2 km of the water column. Dissolution was estimated to be almost enough to balance the alkalinity budget for the intermediate water maximum of the Pacific Ocean. It should be noted that the depth for total dissolution in the water column is considerably deeper than the aragonite compensation depth. This is probably due to the short residence time of pteropods in the water column because of their rapid rates of sinking.

The distribution of $CaCO_3$ in deep sea sediments and carbonate lithofacies

General Considerations

The distribution of $CaCO_3$ in deep sediments has been studied for close to 100 years. Murray and Renard (1891), as part of the Challenger Expedition, collected sediment samples showing that the abundance of calcium carbonate in deep sea sediments decreased with increasing water depth, commonly dropping to less than 30 percent at around 4000 meters, and with latitude, being rare in polar regions. They also made the general observation that calcium carbonate is more abundant in the Atlantic Ocean, and generally occurs in significant amounts to deeper depths than in the Pacific Ocean. The decrease in calcium carbonate content of the sediments was attributed to dissolution resulting from an increase in pressure and a decrease in temperature. With increasing depth, a survival sequence of calcareous tests was found where pteropods disappeared first, followed by the small and fragile types of foraminifera, until in the deeper parts of the oceans only badly damaged tests of large thick-shelled types of foraminifera remained in the surface sediments. A correlation was also found between surface productivity of calcium carbonate tests and the absolute amount found at equivalent depths in the underlying surface sediments. This observation subsequently led Murray to the idea of a calcium carbonate compensation depth (CCD) where the rate of calcium carbonate dissolution is balanced by the rate of infall, and the calcium carbonate content of surface sediments is 0 weight % (e.g., Bramlette, 1961). The CCD has been confused with the calcium carbonate *critical* depth (sometimes used interchangeably with the lysocline discussed later in this section), where the carbonate content of the surface sediment drops below 10 weight percent. Here the term "CCD" will be used to specify the depth at which the carbonate content is equal to 0 weight % because of its more specific relation to a balance between infall and dissolution of CaCO$_3$. A similar marker level in deep sea sediments is the aragonite compensation depth (ACD), below which aragonite is no longer observed to accumulate in sediments.

In the period between the Challenger Expedition and the early 1960's, a large number of investigators studied the distribution of calcium carbonate in deep sea sediments. Their observations have tended generally to confirm and add detail to the initial findings of Murray and his co-workers. During this period of time, there was also increased interest in the carbonic acid system, and the solubility of calcite and aragonite in sea water. Revelle and Fairbridge (1957) made a calculation indicating that sea water generally becomes undersaturated with respect to calcite at around 4000 meters. They suggested that dissolution should begin at this level. The data of Turekian (1964) showed a very definite decrease in the average carbonate content of sediments in the Mid-Atlantic oceanic region near this depth; however, Bramlette's (1961) data for the East Pacific showed the decrease at a slightly shallower depth. These data seemed to support the idea that in large regions of the ocean there was a "snow line" separating calcium carbonate-rich sediments from those deficient in CaCO$_3$. This "snow line" was believed to closely approximate the depth at which the ocean becomes undersaturated with respect to calcite.

In 1966 Peterson (1966) and Berger (1967) carried out experiments on the dissolution kinetics of calcium carbonate suspended at different depths along a taut mooring line in the central Pacific Ocean near Horizon Guyot. Peterson used polished calcite spheres from which he calculated the dissolution rate per unit surface area as a function of depth, after 4 months of suspension in seawater. His results showed dissolution beginning at about 300 to 500 meters, and a very slow increase in the rate of dissolution with depth until a depth of about 3800 meters was reached. Below this depth the rate began to increase very rapidly with increasing depth. Peterson attributed the sudden increase of the dissolution rate at depth to the water having reached a critical value of undersaturation sufficient to overcome surface inhibition caused either by adsorption of ions or organic matter from sea water. Berger (1967, 1970) used suspended tests of foraminifera from East Pacific Rise sediment to gather information on the relative resistance to dissolution of different species. His results indicated slight dissolution of the foraminifera between 300 and 4000 meters depth, and rapidly increasing dissolution below this depth.

These experiments demonstrated empirically that, although the saturation level was very shallow in this region of the Pacific Ocean, significant dissolution rates were not encountered for either natural or synthetic calcite until about 3500 meters below the saturation level. Berger (1968) applied the name "lysocline" to this region of rapid increase in dissolution rate. He suggested that it could be recognized in adjacent surface sediments as the level at which small thin-shelled foraminifera disappear or start to show strong signs of dissolution.

The foraminiferal lysocline (FL) was defined by Berger (1968) as the depth where the predominant type of foraminifera shifts in surface sediments, because of dissolution, from "soluble" to "resistant" species (~50% change in ratio). To determine the depth of the foraminiferal lysocline, three conditions must be met (Parker and Berger, 1971):

(1) There must be sufficient topographic relief so that sediment samples from both above and below the foraminiferal lysocline can be obtained.
(2) The correct starting assemblage of foraminifera must be present in order to apply the solution index.
(3) Well preserved samples from a nearby shallow area must be present for a standard.

In many regions of the ocean all three conditions are not met, yet the concept has had general utility, and represents a major step in our understanding of the behavior of calcium carbonate in the deep ocean.

Berger (1970) defined another useful marker level in pelagic sediments, which is also based on the change in the ratio of "soluble" to "resistant" species of foraminifera. This is the R_0 level at which the first significant (~10%) change in the ratio of these two types of foraminifera is observable in sediments. Adelseck (1978) carried out laboratory experiments on sediments to determine how much carbonate must be dissolved to produce assemblages of foraminifera that are characteristic of the FL and R_0 levels. He found that about 80 percent dissolution is necessary to produce the FL assemblage, and 50 percent dissolution is needed to produce the R_0 assemblage.

The distribution of calcium carbonate in sediments with ocean depth shows wide variations. In open ocean basins, where rates of detrital sedimentation are moderate to low, sediments above 3000 meters water depth are generally high in calcium carbonate, whereas sediments below 6000 meters generally have very low calcium carbonate content. Between these depths there is a poor correlation between the weight % calcium carbonate and depth (Smith et al., 1968). Turekian

(1964) has shown that even for a limited region such as the Mid-Atlantic, there is a large variation in weight % calcium carbonate at any depth, and that this variation increases with increasing depth. Bramlette (1961) found a similar trend in the sediments of the South Pacific.

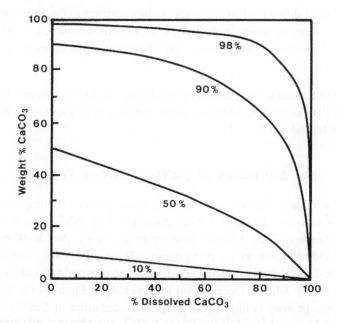

Figure 4.15. A plot of weight % calcium carbonate versus percent original calcium carbonate dissolved. For sediments of high initial calcium carbonate content, little change in weight % takes place until approximately 50 percent of the carbonate has dissolved. Small variations in initial calcium carbonate concentration can lead to large variations in calcium carbonate concentration at high dissolution values. (After Morse, 1973.)

Heath and Culberson (1970) and Berger (1970) have shown that percent original calcium carbonate dissolved is usually a more useful parameter than weight % calcium carbonate in problems concerned with calcium carbonate dissolution. The original weight % carbonate in the sediment can often be estimated from nearby sediments that are at depths substantially above those at which the water becomes undersaturated. Figure 4.15 shows an idealized relation between weight % calcium carbonate and percent calcium carbonate dissolved. It illustrates two important points for sediments of potentially high initial calcium carbonate content:

    (1) Dissolution of over half the calcium carbonate results in only a small change in the weight % of calcium carbonate, but dissolution of the last

10 percent of the original calcium carbonate can result in a decrease of over 80 weight % calcium carbonate. As the initial calcium carbonate concentration decreases, the effect of dissolution is more directly evidenced as a decrease in the concentration of calcium carbonate.

(2) Small differences in the initial calcium carbonate concentration can result in large variations in the weight % calcium carbonate, for a given degree of dissolution, when more than 75 percent of the calcium carbonate has dissolved.

It is important to keep in mind in the following discussions that it is difficult to observe signs of calcium carbonate dissolution in deep sea sediments unless major dissolution has occurred.

The Distribution of CaCO₃ in Surface Sediments

In 1976 three papers were published that presented comprehensive and internally consistent data sets for the distribution of calcium carbonate in the surface sediments of the Atlantic (Biscaye et al., 1976), Pacific (Berger et al., 1976) and Indian (Kolla et al., 1976) oceans. These papers will form the primary basis for the following discussion. The distribution of weight % $CaCO_3$ in surface sediments in the Atlantic Ocean basin is presented in Figure 4.16. Topographic control is the primary feature influencing the distribution of $CaCO_3$. The major topographic feature in the Atlantic Ocean basin is the Mid-Atlantic Ridge, whose general elevation above the CCD results in a central carbonate high running the length of the basin. Biscaye et al. (1976) pointed out that significant variations in carbonate content are found along the topographic highs. The high areas of $CaCO_3$ on the Mid-Atlantic Ridge from the equator to about 40°S and near 30°N are believed to be the result of less dilution by terrigenous material, whereas low carbonate content in the region of the Azores is the result of dilution from volcanic debris. These authors also noted that areas of higher productivity associated with the Gulf Stream and upwelling off west Africa have higher than usual carbonate contents, whereas in the south, near Antarctica, dilution by biogenic silica causes exceptionally low values of sediment carbonate content. The most common feature is a slow decrease in carbonate content to a depth of about 4 to 5 km, followed by a much more rapid decline in carbonate content with increasing depth. Biscaye et al. (1976) plotted the intercept of linear fits of the data for these two depth regions and the intercept of the region of rapid carbonate decrease with 0% carbonate content for the different regions of the Atlantic Ocean basin (Figure 4.17). The intercept of the two linear data fits generally closely corresponds to the depth of the

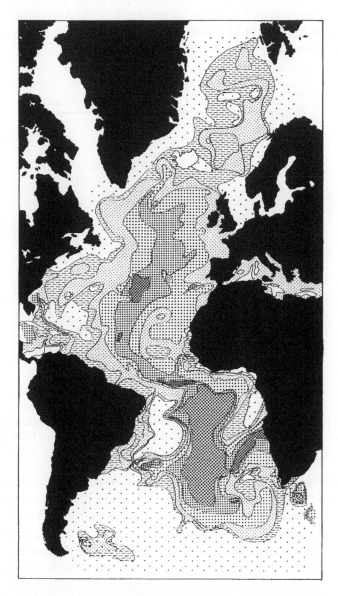

Figure 4.16. Variations in carbonate content of Atlantic Ocean sediments. (After Biscaye et al., 1976.) Percent CaCO$_3$:

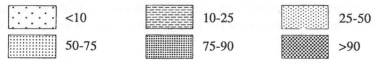

foraminiferal lysocline in th Atlantic Ocean (Berger, 1976), and the 0% intercept is the CCD. The slope (m / %CaCO3) of the region of rapid carbonate decrease is also presented in Figure 4.17. This slope may be indicative of the intensity of dissolution in these deep waters.

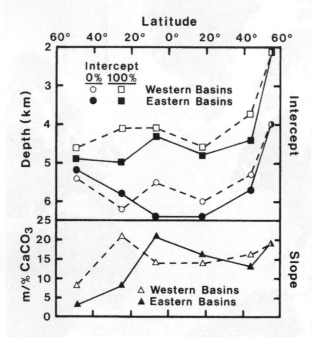

Figure 4.17. Variations in carbonate content versus depth characteristics as functions of latitude. See text for discussion. (After Biscaye et al., 1976.)

A similar approach was taken by Berger et al. (1976) in examining the distribution of carbonate in surface sediments in the Pacific Ocean basin. The general distribution of CaCO3 in surface sediments of the Pacific Ocean is presented in Figure 4.18. The East Pacific Rise is a central ocean topographic high similar to the Mid-Atlantic Ridge on which sediments with high carbonate contents are accumulating. It is interesting to note that the CCD is depressed in the equatorial area where upwelling results in enhanced biologic activity. The CCD never exceeds 5 km and is generally between 4 and 5 km. This depth range is approximately 1 km shallower than that observed in the Atlantic Ocean basin. Also, the rate of carbonate decrease with increasing water depth, in the region of rapid decrease, is more extreme in the Pacific than Atlantic ocean basins. As will be

Figure 4.18. A contour map of calcium carbonate percentages in Pacific surface sediments. Stippled areas are high (>80%) in CaCO₃.(After Berger et al., 1976.)

discussed later, this observation is consistent with the more undersaturated waters found in the Pacific Ocean. Berger et al. (1976) emphasized that interpretation of such plots should be done with caution, because many factors other than dissolution influence the shape of these curves.

The distribution of CaCO₃ in Indian Ocean basin sediments has been examined by Kolla et al. (1976). Their plot of surface sediment carbonate distribution is presented in Figure 4.19. Again the first order control of carbonate distribution is the topographic highs resulting from a central ocean ridge system (Mid-Indian Ridge). Dilution by terrigenous material is important in the Indian Ocean basin in areas of high sediment input such as the Bengal Fan, and dilution of carbonate content by siliceous biogenic material is important in the southern part of the basin.

Figure 4.19. Areal distribution of percent calcium carbonate in the Indian Ocean with 85%, 75%, 50%, 25% and <10% contours. (After Kolla et al., 1976.)

Kolla et al. (1976) plotted the lysocline and the carbonate critical depth (Figure 4.20) using methods similar to those of Biscaye et al. (1976). Unfortunately, they did not use the 0 weight % intercept of the CCD which appears to be generally 100 to 400 m deeper than the carbonate critical depth. Also, their use of the term lysocline is not the same as the FL, because it is entirely based on the relation between water depth and carbonate content of the sediment, not preservation of foraminifera tests. Nevertheless, given these caveats, it is interesting to note the major variations in water depth separating these two marker levels. It is possible that the large separation between the lysocline and the carbonate critical depth in the equatorial Indian Ocean is related to high productivity, but few supporting data are currently available to test this hypothesis.

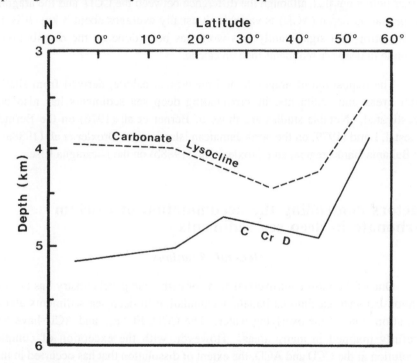

Figure 4.20. Carbonate lysocline and carbonate critical depth (CCrD) variations with latitude in the Indian Ocean. Data from Bengal and Arabian fans are excluded. (After Kolla et al., 1976.)

Schneidermann (1973), and Roth et al. (1975) emphasized that the distribution patterns of coccoliths and foraminifera differ significantly in deep sea sediments. In most sediments, the more easily dissolved coccoliths are missing. Even in areas where the overlying water is supersaturated with respect to calcite (e.g., 3000 m in the Atlantic Ocean, Schneidermann, 1973), significant dissolution of coccoliths is observed. Solution-resistant coccoliths, however, may survive in sediments in which all foraminifera tests have been dissolved. Schneidermann (1973) used these characteristics to create further subdivisions of the region near the CCD. Roth et al. (1975) emphasized that coccolith and foraminifera solution indices can be used in a complementary manner. Coccoliths are good indicators of dissolution above the lysocline, whereas foraminifera are better indicators of dissolution below the lysocline.

Berger (1978) summarized much of what is known about pteropod distribution, and the aragonite compensation depth in pelagic marine sediments. Berger points out that, although the difference between the CCD and the aragonite compensation depth (ACD) is variable, it usually averages about 3 km. It is also worth noting that significantly less work has been done on the distribution of aragonite in deep sea sediments than on calcite.

The deposition of aragonite and magnesian calcite, derived from shallow water areas and sediments, in surrounding deep sea sediments has also been investigated. Notable studies are those of Berner et al. (1976) on the Bermuda Pedestal, Land (1979) on the north Jamaican island slope, Droxler et al. (1988a) on the Bahama Banks region, and Droxler et al. (1988b) on the Nicaragua Rise.

Factors controlling the accumulation of calcium carbonate in deep sea sediments

General Relations

One of the most controversial areas of carbonate geochemistry has been the relation between calcium carbonate accumulation in deep sea sediments and the saturation state of the overlying water. The CCD, FL, R_0, and ACD have been carefully mapped in many areas. However, with the exception of complete dissolution at the CCD and ACD, the extent of dissolution that has occurred in most sediments is difficult to determine. Consequently, it is generally not possible to make reasonably precise plots of percent dissolution versus depth. In addition, the analytical chemistry of the carbonate system (e.g., GEOSECS data) and constants used to calculate the saturation states of seawater have been a source of almost constant contention (see earlier discussions). Even our own calculations have resulted in differences for the saturation depth in the Atlantic of close to 1 km (e.g., Morse and Berner, 1979; this book).

A reason that there has been so much controversy associated with the relation between the extent of carbonate dissolution occurring in deep sea sediments and the saturation state of the overlying water is that models for the processes controlling carbonate deposition depend strongly on this relation. Hypotheses have ranged from a nearly "thermodynamic" ocean where the CCD and ACD are close to coincident with calcite and aragonite saturation levels (e.g., Turekian, 1964; Li et al., 1969) to a strongly kinetically controlled system (e.g., Morse and Berner, 1972) where major differences in the CCD, FL, and saturation depth exist.

One of the more "aesthetically pleasing" relations was put forth for the eastern and western Atlantic Ocean basins by Berger (1977). In his plots (see Figure 4.21) the R_0, FL and CCD were generally widely separated and usually close to parallel. The saturation depth (SD) was close to coincident with the R_0 level. However, even this picture has problems. If the R_0 and SD are closely coincident, how can the 50% dissolution occur that is required to produce the R_0 level (Adelseck, 1978) ?

More recent calculations such as those in this book indicate substantially lower saturation depths. Those calculated here are plotted in Figure 4.21. The SD is generally about 1 km deeper than that presented by Berger (1977). Clearly the new SD is much deeper than the R_0 and appears only loosely related to the FL. Indeed, in the equatorial eastern Atlantic Ocean, the FL is about 600 m shallower than the SD. If these new calculations are even close to correct, the long cherished idea of a "tight" relation between seawater chemistry and carbonate depositional facies must be reconsidered. However, the major control of calcium carbonate accumulation in deep sea sediments, with the exceptions of high latitude and continental slope sediments, generally remains the chemistry of the water. This fact is clearly shown by the differences between the accumulation of calcium carbonate in Atlantic and Pacific ocean sediments, and the major differences in the saturation states of their deep waters.

It is also important to keep in mind that the relation between the saturation state of seawater and carbonate dissolution kinetics is not a simple first order dependency. Instead it is an exponential of about third to fourth order (e.g., Berner and Morse, 1974). Thus dissolution rates are very sensitive to saturation state. This type of behavior has not only been demonstrated in the laboratory (see Chapter 2), but also has been observed in numerous *in situ* experiments in which carbonate materials and tests have been suspended in the oceanic water column. The depth at which a rapid increase in dissolution rate with increasing water depth is observed usually has been referred to as the chemical or hydrographic lysocline. In some areas of the ocean it is close to coincident with the FL (e.g., Morse and Berner, 1972).

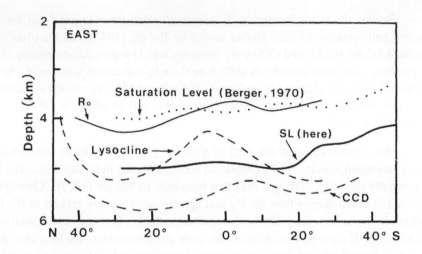

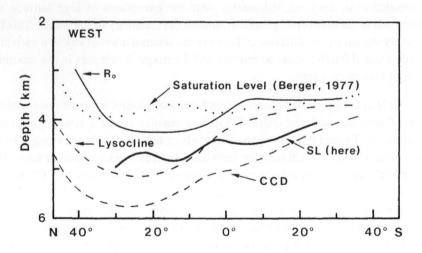

Figure 4.21. Latitudinal variation of saturation depths (SD) and carbonate sediment facies in the eastern and western Atlantic Ocean basins. (Modified after Berger, 1977.)

Factors Leading to Variability

Two major types of variability in the relationship between overlying water chemistry and carbonate accumulation in deep sea sediments occur. The first is the previously discussed relation of the saturation state of the water to the R_0, FL and CCD. The second is the relative separation of these different sedimentary features. In some areas of the ocean these relations can be influenced by transitions in water masses having different chemical and hydrographic characteristics (e.g., Thunell, 1982), but in many areas of the ocean the only major variable influencing the saturation state over wide areas is pressure, which leads to a nearly uniform gradient in saturation state with respect to depth.

Input rates have been recognized as primary factors causing variability in the relation between overlying water saturation state, the various sedimentary carbonate facies, and the thickness of the transition depths between these facies. The major solid inputs that are observed to vary in the ocean and influence calcium carbonate contents in sediments are detrital material, such as clays, biogenic silica, organic matter, total calcium carbonate and non-calcite carbonate minerals (aragonite and magnesian calcite). The clays and biogenic silica act primarily as diluents. Their presence influences not only the absolute weight percentage of carbonate in the sediment, but also the "sharpness" of transitions. The role of organic matter has recently been recognized to be of great importance, because its oxidation can release CO_2 and cause carbonate dissolution. This process will be discussed in the next section of this chapter.

The variability of calcium carbonate input to the sediment-water interface has long been recognized as an important factor influencing the carbonate content of deep sea sediments and its depth distribution. In simple terms, the basic idea is that if the input rate of calcium carbonate is high, the material is more quickly buried, and hence isolated from the chemical aggressiveness of the overlying undersaturated waters. Simple models have been set forth for this process (e.g., Broecker and Peng, 1982). The input rate of calcium carbonate is generally tied to the productivity of the overlying waters. This influence is most clearly seen in areas such as the equatorial Pacific Ocean (see Figure 4.22).

In addition to calcite, aragonite and occasionally magnesian calcites may also reach the ocean floor. They may play an important role even in sediments where they do not accumulate, by contributing carbonate ion to pore waters because of their dissolution, thus causing less calcite dissolution to occur than would be the case in their absence. This process is enhanced by the benthic process of bioturbation, whereby organisms cause a physical mixing of sediments down to depths of 5 to 30 cm in deep sea sediments (e.g., see Schink and Guinasso, 1977, for discussion of effects on carbonate dissolution). Recent calculations (Droxler et al., 1988a) indicate that the ACD is close to or even above the aragonite saturation level in the Atlantic Ocean. Droxler et al. (1988a) also observed that the dissolution behavior of high magnesian calcite (12 to 13 mole %) relative to aragonite was variable. They found that the magnesian calcites in the Bahamas dissolved before aragonite. Similar results were also found for the Nicaragua Rise by Droxler et al., (1988b). However, the reverse behavior has been observed in the Red and Sargasso seas. These relations may be the result of both transport processes and the relative solubility of the carbonate minerals. The many complex factors which control magnesian calcite solubility were discussed in Chapter 3.

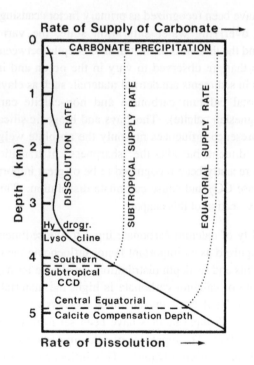

Figure 4.22. Simple conceptual model of the equatorial CCD depression. The dissolution rate curve is based on Peterson (1966). Decrease in supply rate at depth because of dissolution during settling is not quantitative. (After Berger et al., 1976.)

It is interesting to note that if the overlying water chemistry changes so that it becomes more undersaturated, areas of the ocean floor that at one time are above the CCD can move below the CCD. This change will lead to chemical erosion of the older carbonate from the sediments. This situation has been observed and documented in the equatorial Pacific by Keir (1984). Unless one knows that the carbonate in the sediments is not accumulating, but eroding, inclusion of such regions in maps based solely on carbonate content of sediment can severely mislead interpretations of the relationship between water chemistry and carbonate accumulation. These types of changes could also contribute to the variability of other carbonate facies relative to water chemistry.

Near Interfacial Processes

Numerous models have been proposed for the processes occurring near the sediment-water interface in deep sea sediments that lead to a balance between dissolution and retention of calcium carbonate in these sediments. Investigation of these processes is currently one of the most active areas of research in the study of calcium carbonate behavior in the oceans. A major difficulty in studying and modeling these processes is that many of the most important changes take place over distances of only a few millimeters in a highly dynamic environment.

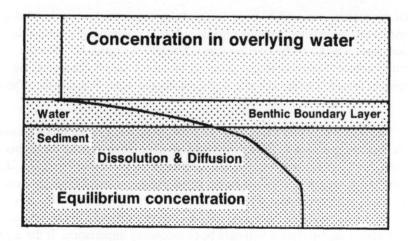

Figure 4.23. Schematic representation of concentration gradient across the sediment-water interface.

Early models for the process of calcium carbonate dissolution from deep sea sediments (e.g., Takahashi and Broecker, 1977) were based on simple diagenetic models in which calcium carbonate dissolved into the pore waters of the sediments.

The altered concentrations result in diffusion of the reactants and products, causing exchange with the overlying waters (see schematic Figure 4.23). The process is governed by the difference between equilibrium concentrations and those found in the overlying waters, the rate of dissolution as a function of disequilibrium, diffusion rates and the thickness of the stagnant benthic boundary layer above the interface. Because of the relatively rapid dissolution rates of calcium carbonate, the depth in the sediment calculated for equilibrium has generally been only a few millimeters.

A major benthic process, that had only casually been considered for its potential influence on carbonate accumulation in deep sea sediments, is the oxidation of organic matter. The general reaction for this process involving marine organic matter is (Emerson and Bender, 1981):

$$(CHOH)_{106}(NH_3)_{16}H_3PO_4 + 138\,O_2 + 124\,CaCO_3 \rightarrow$$

$$16\,H_2O + 16\,NO_3^- + HPO_4^{2-} + 124\,Ca^{2+} + 230\,HCO_3^- \qquad (4.3)$$

While this reaction was widely recognized as being important in continental slope sediments that are relatively rich in organic matter, it was generally ignored in deep sea sediments which usually have less than 0.2 weight % organic-C. With the advent of sediment traps, however, it rapidly became apparent that significant amounts of organic matter were reaching the sediment-water interface (see Table 4.1).

Based on these observations, Emerson and Bender (1981) introduced a model for the influence of organic matter on carbonate dissolution in deep sea sediments. In their model, they emphasized the importance of determining the depth distribution of organic matter oxidation. If the organic matter is rapidly oxidized after arrival at the sediment-water interface, the CO_2 generated will have little chance to interact with the calcium carbonate, and the influence of the oxidation of organic matter on dissolution will only be of secondary importance. If the organic matter is rapidly mixed by bioturbation into the sediment, the oxidation of the organic matter can be very important. In fact, it can potentially change the depth at which calcium carbonate dissolution starts to several hundred meters above the saturation depth in the water column.

Table 4.1. Carbonate and carbon fluxes from near bottom sediment traps. (After Emerson and Bender, 1981.)

Location, Bottom Depth	Trap Depth (m)	Fluxes ($\times 10^6$ mole $cm^{-2}yr$)		$\dfrac{F_{Org-C}}{F_{CaCO_3}}$	Reference
		Organic-C	$CaCO_3$		
N. Atlantic 31°N 55°W (5581 m)	4000	4.9	7.9	0.6	Brewer et al. (1980)
Sub. Trop. Atlantic 13°N 54°W (5288 m)	5086	7.6	11.1	0.7	
Sub. Trop. Pacific 15°N 150°W (5792 m)	5582	2.0	2.4	0.8	Honjo (1980)
N. Atlantic 38°N 69°W (3520 m)	~3470	18.7	12.2	1.5	Hinga et al. (1979)
N. Atlantic 38°N 69°-72°W					Rowe and Gardner (1979)
(3577 m)	3059	35	30.7	1.1	
(2816 m)	2316	19.2	15.7	1.2	
(2815 m)	2795	32.8	31.2	1.1	
(2192 m)	2162	52.5	67.7	0.8	
Eq. Pacific 0°N 86°W (2670 m)	2570	13.6	24.9	0.6	Cobler and Dymond (1980)

Because of the importance of this process for both carbonate accumulation in deep sea sediments and the cycling of CO_2 in the deep sea, several careful studies have been carried out recently (e.g., see Bender and Heggie, 1984; Emerson et al., 1985). The studies of Peterson and Prell (1985a) and Jahnke et al. (1986) are particularly relevant to the deposition of $CaCO_3$ in deep sea sediments. Peterson and Prell found that the saturation depth and the FL in the equatorial Indian Ocean were close to coincident, but that substantial dissolution occurred above the FL (see Figure 4.24). They concluded that dissolution occurring above the FL was primarily the result of oxidation of organic matter. Jahnke et al. (1986) made an extremely detailed millimeter-by-millimeter geochemical study of a deep sea core and demonstrated that benthic organisms were selectively feeding on non-carbonate particles. This process resulted in a carbonate-poor surface layer and is a mechanism for at least partially separating carbonate tests from the effects of organic matter oxidation.

One of the most powerful approaches to studying early diagenesis in sediments is that of extracting pore water from the sediment at different depths below the sediment-water interface (e.g., see Berner, 1980). This procedure is usually done by obtaining a core, sectioning it and extracting the pore water either by squeezing or centrifugation. Numerous studies (e.g., Murray et al., 1980;

Emerson et al., 1982) have demonstrated that this approach will not work for studying the carbonate chemistry of deep sea sediment pore waters. The reason is that the solubility of carbonates changes substantially with temperature and pressure, and they are reactive enough to change the pore water chemistry when the cores are recovered. Consequently, most recent studies have relied on extracting pore waters *in situ*. The major difficulty with this technique is that it usually is not possible to obtain closely spaced samples or samples very near the sediment-water interface. This interface is, unfortunately, the region where most of the chemical changes associated with the carbonate-CO_2 system take place.

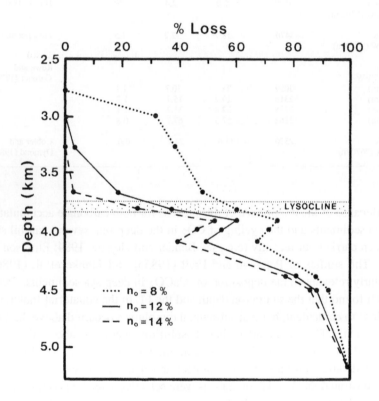

Figure 4.24. Estimates of Peterson and Prell (1985a) for the loss of carbonate to dissolution in the equatorial Indian Ocean. Depth profiles were calculated using the n_0(= initial weight % non-carbonate fraction) values given in the figure.

Most studies of carbonate chemistry in deep sea sediment pore waters have, therefore, focused on the the apparent solubility behavior of carbonates in these sediments. The basic hypothesis is that because of the relatively fast reaction kinetics of carbonate minerals and generally low sedimentation rates in deep sea sediments, the pore waters should come to equilibrium with the carbonate minerals within a few centimeters, at most, of the interface. The results of these studies have shown a surprising degree of variability and further demonstrated the complexities involved in calcium carbonate accumulation in deep sea sediments.

Several studies by different groups of investigators appeared using this approach at about the same time (Emerson et al., 1980; Murray et al., 1980; Sayles, 1980). The results of Emerson et al. (1980) and Sayles (1980) indicated that, in spite of considerable variability in the pore water chemistry of sediments at a number of different sites in the Atlantic and Pacific oceans, the pore waters are generally close to equilibrium with calcite. Emerson et al. (1980) noted higher ion concentration products in pore waters where Mn^{2+} was measurable, indicating possible control by a carbonate phase other than calcite. Similar changes were subsequently noted by Sayles (1985) below about 20 cm in many sediments. Sayles (1980) also found much higher ion concentration products in one sediment containing aragonite. Murray et al. (1980) conducted a similar study in the Guatemala Basin. The sediment had little (a few tenths of a weight %) $CaCO_3$. A sharp drop in pH below the sediment-water interface was observed as the result of organic-C oxidation. The pore waters were undersaturated in the top few centimeters, but came close to the predicted value for saturation at depth. These processes have been modeled by Emerson et al. (1982) at MANOP site C in the Pacific Ocean.

A detailed study of the chemistry of pore waters near the sediment-water interface of sediments from the equatorial Atlantic was conducted by Archer et al. (1989) using microelectrodes that were slowly lowered into the sediment. By modeling the resulting data they were able to confirm that calcite was dissolving above the saturation depth as a result of benthic oxidation of organic matter. The estimated *in situ* rate constant for calcite dissoluton was 1-100% day[-1]. This rate constant is 10 to 100 times slower than the one used in previous models, which was based on experimental data. If the slower rate constant proves to be correct, then dissolution of calcite by benthic metabolic processes will be of major importance.

Manganese appears to be capable of altering the apparent solubility behavior of calcite in deep sea sediments. This element has long been noted to be associated with calcium carbonate in deep sea sediments (e.g., Wangersky and Joensuu, 1964), and extensive experimental evidence exists for coprecipitation of Mn^{2+} with calcite (see Chapter 3). Both Pedersen and Price (1982) and Boyle (1983) have noted the close association of Mn with carbonate material in Panama Basin sediments. In fact, some of the pore waters approach equilibrium with $MnCO_3$, and mixed carbonate phases such as $(Mn_{0.48}Ca_{0.47}Mg_{0.05})CO_3$ have been observed (Pedersen and Price, 1982).

Submarine lithification and precipitation of cements in deep sea carbonate sediments are relatively rare processes in typical major ocean basin sediments. Milliman and his associates have summarized much of the information on these processes (Milliman, 1974; Milliman and Müller, 1973, 1977). The cements are of both aragonitic and magnesian calcite mineralogies, and are largely restricted to shallow seas such as the Mediterranean and Red seas, and sediments in the shallower parts of major ocean basins in which biogenic aragonite is also present. The formation of carbonate cements will be discussed in detail in subsequent chapters.

Mucci (1987; also see Garrels and Wollast, 1978) recently summarized much of the data on the composition of magnesian calcite cements from different environments, which indicate that many of the the shallow water and deep sea carbonate cements fall in the range of 10 to 15 mole % Mg with a strong maximum in abundance at about 13 mole % Mg (see Figure 4.25). Mucci notes that his experimental work at low temperatures, typical of the deep sea, failed to generate magnesian calcites with this composition. Mucci suggests that organic matter may play an important role in controlling magnesian calcite cement composition in the

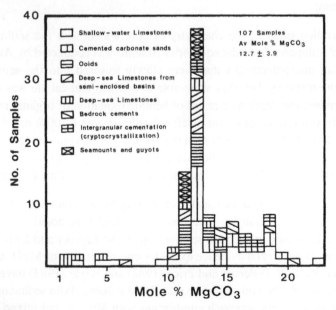

Figure 4.25. Distribution of Mg contents of magnesian calcite cements in marine sediments. (After Mucci, 1987.)

deep sea. This hypothesis was previously advanced for shoal water sediments by Morse and Mucci (1984).

Variability in calcium carbonate deposition in deep sea sediments with time

Influence of Glacial Times

A striking feature of many deep sea sediment cores is the major variability in calcium carbonate content within a given core. Much of this variability in calcium carbonate content has been shown to be associated with glacial-interglacial cycles of the last 2.4 million years (e.g., Zimmerman et al., 1981). Study of the factors leading to this association of carbonate content and the geochemistry of carbonates with climate has become a major area of research which has produced a voluminous literature (e.g., see Arrhenius, 1988, for summary and discussion). Here we will examine only the more general aspects of these relations, with particular emphasis on the geochemistry of deep sea sedimentary carbonates.

It appears that during interglacial times the carbonate content of deep sea sediments is higher than during glacial times in the North Atlantic Ocean. The opposite is true in the Eastern Equatorial Pacific Ocean. Explanations for these differences in both time and location have been the subject of extensive discussion and debate (e.g., Volat et al., 1980). The major factors leading to this variability are the same as those controlling carbonate variability in Recent deep sea sediments. They are the production rate of calcium carbonate, production ratio of calcium carbonate to organic matter, saturation state of deep seawater and input rate of noncarbonate material. In the Atlantic Ocean it is now reasonably well established that calcium carbonate production by coccoliths and foraminifera was higher during interglacial times by perhaps as much as a factor of 2, and that clay input was 2 to 5 times higher during glacial times (e.g., Bacon, 1984). In addition, there is evidence that during some glacial stages dissolution was more intense, owing perhaps to reduced circulation and increased deep water CO_2 concentrations (e.g., Curry and Lohmann, 1985, 1986). The relation between dissolution intensities in Atlantic and Pacific ocean sediments exhibits both in phase and out of phase correlations. Out of phase correlations are believed to be the result of changes in deep water circulation patterns, whereas in phase correlations are thought to result from shelf to basin carbon transfer (e.g., Crowley, 1985). Out of phase dissolution cycles have also been observed in Indian Ocean sediments, where variation in circulation patterns are also favored as the causal factor (Peterson and Prell,

1985b). One of the most interesting observations is the existence of a world wide aragonite preservation spike which is expressed as a pteropod-rich layer at the end of the last glacial period, about 14,000 y BP. This event appears to have been followed by a drastic rise in the compensation level in the North Pacific Ocean and other areas about 12,500 y BP. The precise reasons for these observations are still a matter of debate.

Boyle and his associates (Boyle, 1981, 1986; Hester and Boyle, 1982; Delaney et al., 1985) have carried out extensive work on the trace metal content of foraminifera tests that shows great promise for revealing considerable information about the paleo-phosphate concentrations in the ocean. Cadmium has been shown to be particularly useful in this regard. The relation of dissolved cadmium to dissolved phosphate in the modern ocean is close to linear. The concentration of Cd in foraminifera tests has been shown to reflect closely the Cd concentration in the waters in which the tests grew. An inverse relation has also been found between $\delta^{13}C$ of HCO_3^- and dissolved phosphate. It is thus possible to obtain information about paleoocean chemistry by examining Cd and $\delta^{13}C$ in foraminifera tests. By using both pelagic and benthic foraminifera, it is possible to ascertain differences between the near-surface ocean and deep sea waters in the past.

The Impact of Fossil Fuel CO_2 on the Ocean-Carbonate System

One of the major environmental issues of our time is the impact of anthropogenically generated CO_2 on the environment (see Chapter 9 for discussion). The major processes associated with fossil fuel CO_2 in the oceans are the uptake of fossil fuel CO_2 by the upper ocean, mixing and transport within the ocean, and reaction with calcium carbonate in sediments. In addition, biologic productivity may be influenced by increased P_{CO2} values. The two major reactions for uptake of CO_2 by the oceans, beyond those that would occur for a chemically unreactive gas, are in a simple form:

$$CO_2 + H_2O + CO_3^{2-} \rightarrow 2HCO_3^- \qquad\qquad (4.4)$$

$$CO_2 + H_2O + CaCO_3 \rightarrow Ca^{2+} + 2HCO_3^- \qquad\qquad (4.5)$$

Thus, the fossil fuel "neutralizing capacity" of seawater is largely dependent on the carbonate ion concentration and that of marine sediments depends on the calcium carbonate which is near enough to the sediment-water interface to react with the overlying water.

In general, the interaction of fossil fuel CO_2 with the oceans is divided into upper ocean processes that occur on time scales of a few years to decades, and deep sea processes that typically operate on time scales of hundreds to thousands of years. Because of the very rapid rise in the P_{CO_2} of the atmosphere owing to human activities, upper ocean processes will be the most important over the time span when it is anticipated most anthropogenic CO_2 input will occur. Evidence for this increase in the P_{CO_2} of upper ocean waters has been observed in the Sargasso Gyre of the Atlantic Ocean (Figure 4.26). Deep sea processes, however, will be by far the most important in eventually neutralizing the system and bringing it back to "natural" levels.

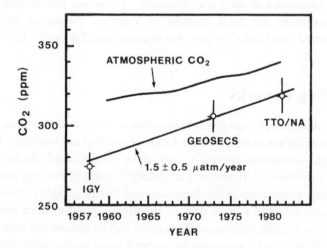

Figure 4.26. Comparison of mean P_{CO_2} values in surface waters of the Sargasso Sea Gyre with atmospheric values versus time. (After Takahashi et al., 1983.)

Nuclear weapons testing in the atmosphere introduced excess ^{14}C that has proven to be an excellent tracer for processes that will be important in the uptake of fossil fuel CO_2. For example, Peng (1986) mapped global uptake patterns of excess ^{14}C and used these to calculate fossil fuel CO_2 uptake by the oceans. His results indicate that about 10% of the fossil fuel CO_2 is "missing". In the next chapter we will discuss the possibility that this missing CO_2 may be the result of dissolution of highly soluble magnesian calcites from shallow water sediments. Also, Feely and

Chen (1982) pointed out that excess CO_2 has caused the aragonite saturation depth in the North Pacific Ocean to rise close to the sea surface.

A very simple model for the changes in deep water chemistry and carbonate accumulation because of the influence of changing atmospheric P_{CO_2} is to assume the increase in potential P_{CO_2} of deep water will be the same as that of the atmosphere. Because the oceanic reservoir is much larger, this almost certainly overestimates these changes and is, therefore, a "worst case" model. In Figure 4.9 the relation between potential CO_2 and the calcite saturation depth was shown. Based on this relation, a doubling of atmospheric P_{CO_2} from its preindustrial value would cause the calcite saturation depth to become about 2 km shallower. This would eliminate major areas of the seafloor (see Figure 4.11) as sites for carbonate accumulation, but much of the seafloor would then be available for neutralization of fossil fuel CO_2.

Major complications arise because of bioturbation and the complex reaction kinetics of carbonates in deep sea sediments. These two factors will, to a large degree, determine the total amount of calcium carbonate available for neutralization of fossil fuel CO_2 (e.g., see Broecker and Takahashi, 1977).

Concluding remarks

It is remarkable, and perhaps discouraging, that so very much work has been done on the oceanic carbon system and the factors governing accumulation in the deep sea. Pelagic carbonates comprise only about one-third of the total mass of sedimentary carbonates. However, particularly because of the long interest of the oceanographic community in the formation and fate of pelagic carbonates, these materials have received a great deal of attention. In the modern ocean we can evaluate, often directly, processes affecting the carbon system and carbonates, and one would think that in a century of research most major problems would be solved. As seen in this chapter, this is not the case; much remains to be done. The same statement may be made of the discussion of shoal water carbonates following in Chapter 5. The discouraging element is that when we study ancient continental carbonates of pelagic or shoal water origin, much information may have been lost because of diagenesis, and the oceanographic system in which these sediments were deposited may have been different from today's. Furthermore, the fluids found in the pores of these rocks may be substantially modified seawater or fluids and gases of other compositions. These factors should be kept in mind in ensuing discussions of early and late diagenesis and historical geochemistry of sedimentary carbonates. This chapter and Chapter 5 deal primarily with the modern situation in which processes can sometimes be directly observed, or at least closely inferred; as we progress further, the time factor will become more important, and resolution of the stratigraphic column more difficult, thus making a unique interpretation of a geological or geochemical question involving the sedimentary carbonates less likely.

Chapter 5

COMPOSITION AND SOURCE OF SHOAL-WATER CARBONATE SEDIMENTS

Introduction

In this chapter and the following, shoal-water ("shallow-water") carbonate sediments, and the seawaters they form from, are examined. Emphasis is on the biogeochemical processes affecting carbonate materials in this global marine environment, not on the general sedimentology of shoal-water carbonate deposits. The latter subject is discussed in innumerable publications such as "Carbonate Depositional Environments" (Scholle et al., 1983), to which the reader is referred.

Biotic and abiotic sources of shoal-water carbonates are discussed in this chapter to provide the reader with some idea of the starting carbonate compositions available for diagenesis of these sediments. Because modern and Phanerozoic carbonate sediments are predominantly skeletal in nature, some aspects of biomineralization are presented, with emphasis on the chemistry and mineralogy of skeletal calcite and aragonite. This latter discussion sets the stage for consideration of the chemical factors influencing precipitation of carbonates from seawater overlying modern shoal-water carbonate deposits. We emphasize the evidences and arguments for abiotic versus biotic processes of precipitation in the formation of aragonite needle muds and oöids. How these biotic and abiotic components of shoal-water carbonate sediments begin their transformation to rock is the subject of Chapter 6.

Shoal-water carbonates in space and time

From preceding chapters we have gained some idea of carbonate mineralogy and stability. We have seen that modern deep-sea carbonates are fairly homogeneous in composition, being principally low Mg- and Sr-bearing calcite and aragonite, respectively. In contrast, modern shoal-water carbonate sediments are more heterogeneous in composition, reflecting mainly the variety of carbonate mineralogies exhibited by shoal-water benthic organisms. In this section and the following, we look in more detail at shoal-water skeletal organisms and sediments, paying particular attention to organism mineralogy, chemistry and microarchitecture, and sediment composition. These factors play a dominant role in carbonate diagenetic processes.

It is informative to compare deep-sea production and accumulation rates with those of shoal-water sediments. Figure 5.1 is our present tentative interpretation of the fluxes involving carbonate carbon transport in the ocean; the following discussion is keyed to the figure. We estimate in Chapter 9 that about 18×10^{12} moles of Ca^{2+} and Mg^{2+} accumulate yearly as carbonate minerals, mainly as biological precipitates. Of this flux about 6×10^{12} moles are deposited as calcium and magnesium carbonates in shoal-water areas (cf. Milliman, 1974; Smith, 1978), and the remainder as calcareous oozes in the pelagic realm. The 12×10^{12} moles of carbonate accumulated annually in the deep sea are only about 17% of the annual carbonate production rate of 72×10^{12} moles of the open ocean photic zone.

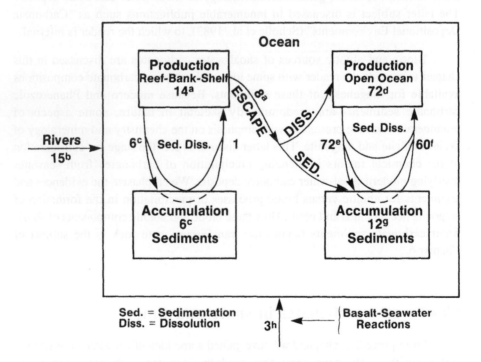

Figure 5.1. Tentative model of global carbonate cycle. Fluxes are in units of 10^{12} moles C y^{-1} as (Ca,Mg)CO$_3$. [a]Agegian et al. (1988); [b]Meybeck (1979); [c]Smith (1978) and Milliman (1974); [def]Broecker and Peng (1982); [g]Broecker and Peng (1982), Milliman (1974) and Hay and Southam (1977); [h]Wollast and Mackenzie (1983).

This efficient recycling of carbonate in the open ocean water column and at the sediment-water interface was discussed in Chapter 4. It is important to note that much shoal-water carbonate production ends up in sediments of reefs, banks, etc., so that, in contrast to the pelagic realm, production rate more closely approximates sedimentation rate. There is, however, escape of carbonate sediment from shoal-water areas to the deep sea where it is deposited or dissolved (e.g., Land, 1979a; Berner et al., 1976; Heath and Mullins, 1984). The magnitude of this flux is poorly known but may affect the chemistry of open-ocean regions owing to dissolution of the carbonate debris (Droxler et al., 1988; Agegian et al., 1988). Furthermore, its accumulation on the slopes of banks may act as a record of paleoenvironmental change (Droxler et al., 1983).

Whatever the case, it is this biogenic accumulation, which certainly has varied in amount and locus of deposition during geologic time -- shoal-waters versus deep sea -- that during the Phanerozoic provided the carbonate debris for diagenesis. In contrast to the Phanerozoic, for much of the Precambrian, the initial carbonate precipitates were likely inorganic in origin; perhaps their precipitation was influenced by biological processes, but certainly they were not truly skeletal. Thus, the starting materials for diagenesis were different in composition and particle morphology. As will be shown later, it is likely that during the Phanerozoic the composition of carbonate precipitates also varied, and thus even during this Eon, the starting materials for diagenesis varied, leading potentially to different pathways of diagenesis.

As a closing remark, if we accept the carbon isotopic record of carbonates and organic carbon through geologic time as an index of the relative accumulation rates of inorganic and organic carbon on the sea floor, the carbonate carbon accumulation rate has been remarkably constant (Chapter 10). This near constancy has been maintained despite changes in Earth's surface physical and chemical environment and biotic evolution.

Carbonate grains and skeletal parts

Overview and Examples

The carbonate grains of a carbonate sediment or rock may be subdivided into skeletal and non-skeletal components. The petrographic microscope has been the major tool used by the carbonate petrologist during this century to study the mineralogy and fabric of the agglomeration of grains that make up a carbonate rock. Geochemical (trace element and isotopic), X-ray, acetate peel, electron microscopic (both scanning and transmission), cathodoluminescence, and infra-red and Raman spectroscopic studies are now used to complement the standard light microscopy work. The principal aim of these studies is to determine the mineralogy and chemistry of the grains and cements of a carbonate rock, the texture of the grains as an aid to their recognition, the spatial relationships among grains and voids ("fabric"), and the paragenesis of cement types.

Non-skeletal carbonate grains have been divided into five major types (e.g., Scoffin, 1987): lithoclasts, mud, pellets and peloids, oöids, and relict. Lithoclasts are fragments of previously deposited, and usually somewhat lithified, carbonate sediment derived from outside (extraclasts) or within (intraclasts) the basin of deposition. Intraclasts formed by erosion of carbonate beachrock or semi-consolidated lime mud are commonly found in limestones. Carbonate muds of grain size less than 63 μm are common deposits of low-energy environments, such as the tidal flats and subtidal areas west of Andros Island, Great Bahama Bank. Pellets are formed by the ingestion of sediment by marine organisms, like worms and fishes, and excretion of faecal material. This material may rapidly harden in the marine environment and be preserved in a carbonate sediment. Often the term peloid is used for rounded, microcrystalline grains of equivocal faecal origin. Oöids are spherical to ovoid, 0.2 to 1 mm grains with an internal concentric or radial structure. Finally, relict grains are of older origin, having formed under previous environmental conditions. Recognition of these grains in limestone deposits can be a very important means of identifying periods of non-deposition in sequences that otherwise seem to represent continuous deposition. Further on in this chapter, we will discuss the origin of some non-skeletal grain types in more detail, particularly the lime muds and oöids.

The skeletal components of carbonate sediments represent the complete or partial skeleton, or the decomposed and disaggregated skeletal remains, of organisms extant at the time of deposition of the sediment. Older skeletal materials also may be incorporated in the sediment because of erosion of pre-existing, fossiliferous rocks, as well as, less commonly, younger materials derived subsequently to deposition, e.g., fragments of pulmonate snails transported from a soil horizon downward into a limestone by percolating waters. Thus, the skeletal grains of carbonate sediments are a record of the evolutionary history of life on Earth. Their original chemistry and mineralogy, and subsequent changes in composition during time, provide important information needed to determine the history of Earth's surface environment and the diagenetic history of a carbonate sediment.

Ever since the classic works of Lowenstam (1954a, b), Chave (1954 a, b) and Ginsburg (1956), geologists have been aware of the potential contribution of the skeletons of organisms to carbonate sediments and sedimentary rocks. Chave expanded his ideas in a series of widely read and oft-quoted papers (1960, 1962a, b, 1964, 1967). Even the fine-grained carbonate muds so characteristic of some modern carbonate environments like the Bahama Banks appear to be, at least in part, a product of the decomposition of skeletal organisms. The significance of the production of mud-size grains from disaggregation of organisms was beautifully demonstrated by Neumann and Land (1975). They showed that the production of lime mud by calcareous algae in the Bight of Abaca, Bahamas, exceeded its accumulation rate by 45-190 percent!

Sedimentary particles can be produced from calcareous organisms by disaggregation of their skeletons, by mechanical means related to wave and current energy, or by bioerosion of carbonate substrates like corals, molluscs and rocks. Bioerosional agents include such organisms as sponges (*Cliona*), molluscs (*Lithophage*), echinoids (*Tripneustes*), crabs, and parrot fish. These agents may act chemically (*Cliona*) and physically (crabs) in attacking the substrate.

The importance of organisms in supplying carbonate particles (aleochems) to a shoal-water sediment is seen in Table 5.1 and Figure 5.2. The sediment is a calcareous sand ("grainstone" if cemented) from a modern grass bed in Ferry Reach, Bermuda. The biogenic nature, as seen in Figure 5.2 for a grain size greater than 4 mm, is evident for all grain sizes in which constituents are identifiable using a binocular microscope. For many carbonate sediments, the use of SEM or TEM leads to identification of skeletal particles even at the micron-size level. Table 5.1 shows that the sediment is composed of a diverse assemblage of skeletal particles, the grains being mainly those of a few species of *Halimeda*, an aragonitic green algae. Notice that foraminifera, principally aragonitic and calcitic with various contents of magnesium, constitute nearly 40% of the species identified. Some typical organisms found in this sand and their break-down products are shown in Figure 5.3. The important point is that most modern shoal-water deposits are composed of the skeletons of marine organisms and their disaggregated tests. This mixture represents a metastable assemblage of minerals, with varying inter- and intragranular porosity.

In a further section, we emphasize the mineralogy and chemistry of carbonate skeletal grains, because of their abundance in Phanerozoic sediments, and their potential usefulness in paleoenvironmental interpretation. To provide the reader at this point with some idea of the complexity of skeletal carbonate grains, Tables 5.2 and 5.3 summarize the chemistry, mineralogy and structure of major

Table 5.1. Skeletal composition of a calcareous sand from a grass bed, Ferry Reach, Bermuda. (After Pestana, 1977.)

Taxa Identified	Number of Grains	Percent
Gastropods	162	6.4
Bivalves	81	3.2
Foraminifera	722	28.7
Grass fragments	205	8.2
Halimeda spp.	1117	44.5
Cymopolia spp.	26	1.0
Ostracods	36	1.4
Encrusting red algae	9	0.4
Serpulid (?) worm tubes	9	0.4
Agglutinate worm tubes	11	0.4
Caecum gastropods	10	0.4
Unidentified grains	90	3.6
Total grains counted	2515[a]	98.6

Type of Organism	Number of Species
Gastropods	9
Bivalves	5
Foraminifera	21
Grasses	3
Halimeda	3
Cymopolia	1
Other green algae	1
Encrusting red algae	1
Chitons	1
Ostracods	3
Echinoids	1
Agglutinated worm tubes	1
Crabs	1
Serpulid worms	1
Scaphopods	1
Caecum gastropods	1
Sponges	1
Total species counted	55

[a]The remainder included non-calcareous green algal fragments, chitons, echinoid spines, scaphopods, sponge spicules, and crab carapace fragments.

organism groups. It can be seen that modern skeletal organisms produce a variety of structures and that shoal-water organisms are composed principally of aragonite and intermediate- to high-magnesian calcitec ompositions. This composition contrasts sharply with that of modern, open-ocean pelagic, calcareous plankton, represented mainly by the calcitic Foraminiferida and Coccolithophyceae, and the aragonitic, Pteropoda. The original chemistry, mineralogy, and microstructure of skeletal particles are the principal factors governing the reactivity of these particles during the changing environmental conditions of diagenesis.

Figure 5.2. Grass bed calcareous sediment of Table 5.1, Ferry Reach, Bermuda; size grade > 4 mm. (After Pestana, 1977.)

Figure 5.3. Photographs of some modern calcareous organisms and breakdown products. A) *Penicillus*, an aragonitic green algae, these grains usually break up into microsize needles; B) *Halimeda*, an aragonitic green algae, these skeletal fragments may produce a bimodal distribution of grain sizes in the fine-sand and fine-silt size ranges; C) *Amphiroa*, a high-magnesian calcite red algae and its fragments; D) Peneropolid foraminifera of high-magnesian calcite; E) Various genera of mixed aragonite and calcitic gastropods; F) *Diploria*, an aragonitic hermatypic coral and coral fragments; and G) *Millipora*, an aragonitic hydrozoan and its fragments. (Photographs courtesy of Pestana, 1977, and Cavaliere et al., 1987.)

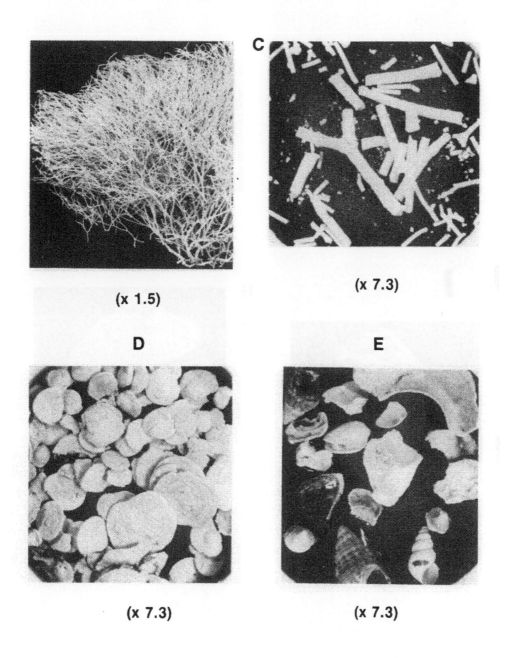

(x 1.5)

C

(x 7.3)

D

E

(x 7.3)

(x 7.3)

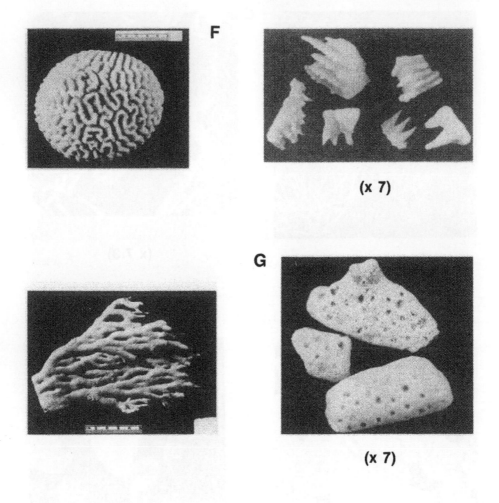

F

(x 7)

G

(x 7)

Table 5.2. Mineralogy and chemistry of major organism groups.*

| Type of Organism | Most Common Mineralogy | Range in Chemical Composition | | |
		$CaCO_3$ (mole %)	$MgCO_3$ (mole %)	$SrCO_3$ (mean mole %)
Foraminifera	Calcite	77->99	<1.2-19	0.249
Calcareous sponges	Calcite	71-85	0.08-0.7	0.079
Scleractinian corals	Aragonite	98-99	5.9-16.7	0.911
Octacorals	Calcite	73-99	0.4-19	0.264
Echinoids	Calcite	78-92	4.8-19	0.229
Crinoids	Calcite	83-92	8.3-19	0.167
Asteroids	Calcite	84-91	10.7-19	0.154
Ophiuroids	Calcite	83-91	0.2-13.1	0.262
Bryozoa	Calcite, aragonite	63-97	0.2-13.1	0.262
Calcareous brachiopods	Calcite	89-99	0.6-10.7	0.130
Phosphatic brachiopods	Chitinophosphatic	?-8	2.4-8.3	--
Annelid worms	Calcite, aragonite	83-94	7.1-20.2	0.484
Pelecypods	Calcite, aragonite	98.6-99.8	0-3.6	0.177
Gastropods	Calcite, aragonite	96.6-99.9	0-2.4	0.160
Cephalopods**	Aragonite	93.8-99.5	Trace-0.4	0.337
Crustaceans	Calcite, calcium phosphate	29-83	1.2-19	0.392
Calcareous algae	Calcite, aragonite	65-88	8.3-35	0.221

*Data from Clarke and Wheeler (1917), Chave (1954a), Thompson and Chow (1955), and Lowenstam (1961). **Egg case of *Argonauta argo* is calcite and contains 8.3 mole % $MgCO_3$. Table modified after Mackenzie et al. (1983).

Sediment Classification

In order to communicate further ideas on sediment diagenesis, it is necessary to consider briefly the sedimentary materials that make up carbonate sediments and rocks and classifications of these materials. For detailed discussion the reader is referred to Pettijohn (1957), Folk (1959), Dunham (1962), Ham (1962), and Blatt et al. (1980).

The grain size of the limestone formed the basis of early classifications of limestone lithic type; the same approach was used for dolostones but with different modifying terms. Since the 1950's combined genetic and descriptive limestone classifications have developed, based principally on the recognition that a carbonate sediment contains three major components: (1) a framework of discrete grains or

Table 5.3. Some common types of skeletal microstructure as seen in thin section under the microscope. (After Scoffin, 1987.)

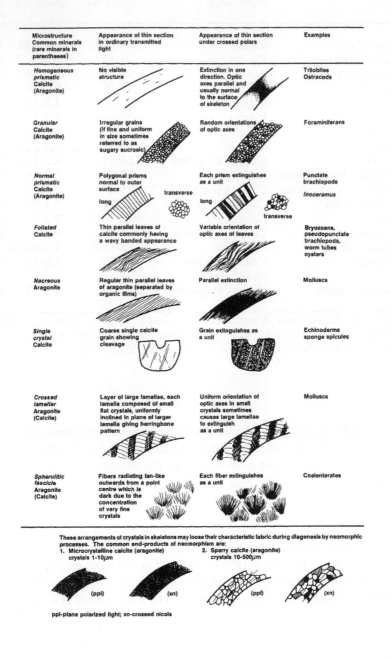

Microstructure Common minerals (rare minerals in parentheses)	Appearance of thin section in ordinary transmitted light	Appearance of thin section under crossed polars	Examples
Homogeneous prismatic Calcite (Aragonite)	No visible structure	Extinction in one direction. Optic axes parallel and usually normal to the surface of skeleton	Trilobites Ostracodes
Granular Calcite (Aragonite)	Irregular grains (if fine and uniform in size sometimes referred to as sugary sucrosic)	Random orientations of optic axes	Foraminiferans
Normal prismatic Calcite (Aragonite)	Polygonal prisms normal to outer surface long transverse	Each prism extinguishes as a unit long transverse	Punctate brachiopods *Inoceramus*
Foliated Calcite	Thin parallel leaves of calcite commonly having a wavy banded appearance	Variable orientation of optic axes of leaves	Bryozoans, pseudopunctate brachiopods, worm tubes oysters
Nacreous Aragonite	Regular thin parallel leaves of aragonite (separated by organic films)	Parallel extinction	Molluscs
Single crystal Calcite	Coarse single calcite grain showing cleavage	Grain extinguishes as a unit	Echinoderms sponge spicules
Crossed lamellar Aragonite (Calcite)	Layer of large lamellae, each lamella composed of small flat crystals, uniformly inclined in plane of larger lamella giving herringbone pattern	Uniform orientation of optic axes in small crystals sometimes causes large lamellae to extinguish as a unit	Molluscs
Spherulitic fascicle Aragonite (Calcite)	Fibers radiating fan-like outwards from a point centre which is dark due to the concentration of very fine crystals	Each fiber extinguishes as a unit	Coelenterates

These arrangements of crystals in skeletons may loose their characteristic fabric during diagenesis by neomorphic processes. The common end-products of neomorphism are:
1. Microcrystalline calcite (aragonite) crystals 1-10μm
2. Sparry calcite (aragonite) crystals 10-500μm

(ppl)   (xn)   (ppl)   (xn)

ppl-plane polarized light; xn-crossed nicols

in situ organic structures, (2) a fine-grained matrix (micrite), and (3) cement. Open spaces or pores constitute the rest of the sediment. Their arrangement and degree of connectiveness are major factors influencing fluid storage and movement in the sediment. The classification schemes of Folk (1962) and Dunham (1962) are given in Tables 5.4 and 5.5. A few words concerning these schemes are appropriate here.

Table 5.4. Classification of carbonate sediments based on texture and nature of allochems according to Folk (1962).

	OVER 2/3 LIME MUD MATRIX				SUBEQUAL SPAR AND LIME MUD	OVER 2/3 SPAR CEMENT		
Percent Allochems	0-1%	1-10%	1-50%	Over 50%		SORTING POOR	SORTING GOOD	ROUNDED &ABRADED
Representative Rock Terms	Micrite and Dismicrite	Fossili ferous Micrite	Sparse Biomicrite	Packed Biomicrite	Poorly Washed Biosparite	Unsorted Biosparite	Sorted Biosparite	Rounded Biosparite
1959 Terminology Terrigenous Analogues	Micrite and Dismicrite	Fossiliferous Micrite	Biomicrite			Biosparite		
	Claystone		Sandy Claystone	Clayey or Immature Sandstone		Submature Sandstone	Mature Sandstone	Supermature Sandstone

Lime mud matrix   Sparry calcite cement

Carbonate Textural Spectrum

Table 5.5. Classification of carbonate sediments according to Dunham (1962).

Depositional texture recognizable					Depositional texture not recognizable
Original components not bound together during deposition				Original components were bound together during deposition as shown by intergrown skeletal matter, lamination contrary to gravity, or sediment-floored cavities that are roofed over by organic or questionably organic matter and are too large to be interstices.	
Contains mud (Particles of clay and fine silt size)			Lacks mud		Subdivide according to classifications designed to bear on physical texture or diagenesis.
Mud supported		Grain supported			
Less than 10% grains	More than 10% grains				
Mudstone	*Wackestone*	*Packstone*	*Grainstone*	*Boundstone*	*Crystalline carbonate*

Both classifications are based on the depositional texture of the limestone; Folk's emphasis is on the similarity of carbonate sedimentation to that of sandstones, resulting in a scheme employing as variables the lime mud:spar cement ratio and allochem composition. Allochems may be intraclasts, oöids, fossils or pellets, and depending on their relative abundances, various descriptors are applied to the terms micrite and sparite. Dunham's fundamental criterion of subdivision is whether the framework of the sediment is mud-supported or grain-supported. The former implies little turbulence in the original depositional environment, the latter high turbulence.

It is now recognized that mixed siliciclastic and carbonate sediments are common in both modern and ancient sedimentary deposits. These mixed sediments have received little attention by geochemists or sedimentologists until recently (e.g., Arnseth,1982; Mount, 1984, 1985). Figure 5.4 shows one scheme proposed

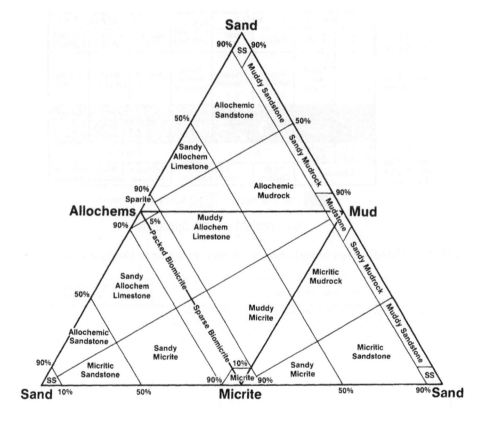

Figure 5.4. Classification of mixed siliciclastic-carbonate sediments proposed by Mount (1985). It is an unfolded tetrahedron in which the apices are siliciclastic sand, carbonate allochems, siliciclastic mud, and carbonate micrite. For comparison with Folk's (1962) classification, the classification of pure fossiliferous limestone is also plotted.

for the compositional and textural classification of these sediments. It is useful to remember that the initial carbonate materials available for diagenesis may be impure in many cases, and the ultimate pathway of diagenesis may be affected by these "dilutents", because of their influence on bulk composition and porosity and permeability. For example, the early diagenesis of a pure carbonate sediment containing little iron oxide will not lead to pyrite formation, whereas pyrite formation in an impure carbonate is almost invariably an important very early diagenetic reaction.

Depositional Environments

A large amount of the emphasis in carbonate research during the past 30 years has been on studies of modern depositional settings. These studies have been largely physical and biological in nature, although the work of Broecker and Takahashi (1966) on the chemistry of carbonate deposition on the Great Bahama Bank is an exception. These studies of modern depositional processes have led to development of analog models of carbonate deposition for ancient sediments. The investigation of diagenesis in carbonate sediments necessitates knowledge of the stratigraphy and sedimentology of the rock sequence of interest. The early diagenetic processes leading to cementation of a carbonate beach sand are certainly considerably different from those resulting in dolomite formation in a tidal flat sequence. Early marine cementation of a beach may inhibit reaction of the beach allochems with subsequent groundwater and other subsurface solutions, and thus, this cementation may affect the later stages of diagenesis. Early dolomitization of a tidal flat sequence may make it more susceptible to further dolomitization by fluids later in the history of the sediment. Whatever the case, diagenetic studies should be undertaken within a sound stratigraphic and sedimentologic framework. We refer the reader to the many excellent works on carbonate depositional environments and facies interpretation (e.g., Friedman, 1969; Bathurst, 1975; Reading, 1978; Scholle et al., 1983). In later sections we will employ the principle that stratigraphic and sedimentologic analyses should form the conceptual framework for diagenetic studies by discussion of the geology of carbonate sequences used as case histories of diagenesis.

Concluding Statement

The melange of carbonate particle compositions found in modern shoal-water carbonate sediments is to a significant extent the starting materials for diagenesis of these sediments throughout the Phanerozoic. Before this time, it is likely that carbonate sediments were principally abiotic in origin and probably because of the lack of carbonate skeletal materials of diverse compositions, more compositionally homogeneous. The exact percentages and compositions of calcitic, aragonitic and dolomitic phases in shoal-water sediments have certainly varied during the Phanerozoic, because of biological evolution and changes in the relative proportions of abiotic and biotic precipitates during this Eon. It is true, however, that during the Phanerozoic an initial metastable assemblage of carbonate minerals was produced in shoal-water environments and went through the diagenetic fence of Figure 5.5. This schematic diagram shows the overall processes of production,

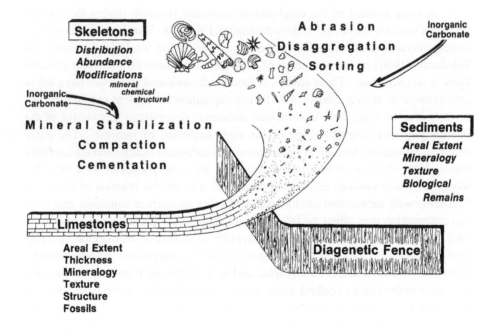

Figure 5.5. Steps in the development of a limestone. A myriad of processes are involved in the conversion of skeletons of organisms into sediments and the transformation of sediments into rocks. The diagenetic fence represents that complex of processes that lead to the conversion of a sediment into a rock. (After Chave, 1960.)

accumulation and diagenesis of carbonate materials, and summarizes well how a carbonate rock is formed. Organisms calcify and form carbonate skeletons composed of aragonite and a variety of calcites. These skeletal organisms form parts of a complex pelagic or benthic community. The skeletons of these organisms are mechanically or biologically disaggregated and altered by the processes of abrasion, disintegration, and sorting to form sediments. The sediments have variable properties such as areal extent, mineralogy, texture, and biological remains. These properties are modified by the diagenetic processes of mineral stabilization, compaction, and cementation to form limestones. Inorganic carbonate as cement may be added to the mixture early in its history or later. To become a rock, the skeletal sediment passes through a realm of complex processes termed the diagenetic fence. Our emphasis in the latter part of this book is on the diagenetic fence.

Biomineralization

General Aspects

We have demonstrated that biogenic carbonates are usually the dominant source of calcium carbonate in modern shoal-water marine environments. A wide variety of organisms utilize substantially different biologic processes to form carbonates that have distinctive mineralogies and compositions. As a broad generality, the complexity of an organism is reflected in the sophistication of the manner by which it precipitates carbonate minerals. The observation that environmental factors, such as temperature, salinity, water turbidity, and dissolved oxygen, can influence the form and chemistry of biogenic carbonates has resulted in a broad interest among carbonate geochemists in biomineralization because of its potential applications to interpretation of paleoenvironments.

We will not attempt to cover the specific mechanisms and complex biochemistry of calcification in different organisms, which have been reviewed in numerous articles (e.g., Degens, 1976; Watabe, 1981; Chave, 1984; Lowenstam and Weiner, 1989). Instead we will concentrate on observations of how biogenic carbonates reflect environmental conditions. These observations can be divided into the major areas of mineralogy, stable isotope ratios, and trace component concentration. The stable isotope and trace component chemistries have received substantially more attention during the last 25 years than has mineralogy. Inorganic phases are also discussed in this section as a means of comparison with skeletal materials.

Environmental Controls on Mineralogy

The major relations between the crystal form of calcium carbonate that is secreted by organisms and temperature were well described by Lowenstam in 1954. He grouped organisms into four categories based on increasing complexity and sensitivity to skeletal mineralogical responses to temperature changes (Figure 5.6). His first category was organisms whose skeletons consist entirely of aragonite, but in which the number of species increases with increasing water temperature (e.g., scleractinian corals). The second category consists of classes or subclasses of species that secrete either all calcite or all aragonite, but those organisms in orders secreting aragonite are confined to warmer waters (e.g., the *Florideae* red algae). A transition category is made up of organisms which secrete trace amounts of calcite in cold waters (e.g., the molluscan genera *Brachidontes* and *Littorina*).

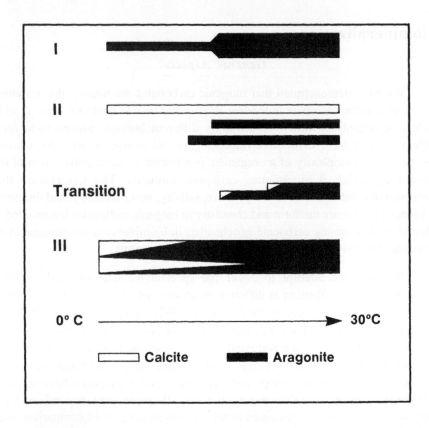

Figure 5.6. Influence of temperature on skeletal mineralogy for different categories (see text) of organisms. (After Lowenstam, 1954a.)

Lowenstam's final category is comprised of organisms which can secrete both calcite and aragonite, with the percentage of aragonite increasing with increasing temperature (e.g., *Mytilus*). This category of organisms has been most intensely studied, because of the potential for recognizing climatic variability from the mineralogy of shells. Mineralogic ratios, at a given site, usually have a reasonable relation to temperature variability. However, when mineralogic variability is compared between different sites, its relation to temperature is generally not as strong. Factors such as differences in the temperature response of different species, and the potential impact of other variables, such as salinity, diminish the general utility of using calcite to aragonite ratios to determine paleotemperatures.

Stable Isotopes

The stable isotope ratios of carbon and oxygen represent a potentially powerful geochemical means of determining environmental conditions, particularly temperature, under which biogenic carbonates were formed. The primary concern in interpreting carbon and oxygen isotope ratios in biogenic carbonates arises from the observation that organisms can significantly alter isotope ratios from those that would result from direct precipitation from solution (Figure 5.7). Differences between isotope ratios in biogenic carbonates and those that would result from abiotic formation are generally ascribed to "vital effects".

Vital effects are highly variable between different types of organisms, and even within an individual can be influenced by numerous environmental parameters, and factors such as age may also play a role. It is not always clear, for pelagic organisms, where in the water column carbonate secretion takes place. Benthic organisms may draw on pore water and overlying water carbon dioxide pools. In addition, different parts of a given organism (e.g., echinoderms) may exhibit significantly different isotope ratios.

McConnaughey (1989a,b) has divided vital effects into two categories. The first is due to "kinetic" effects associated with the rate at which skeletogenesis takes place, and the second is "metabolic" effects associated with photosynthesis and respiration. The importance of understanding the factors controlling stable isotope ratios in biogenic carbonates has resulted in major efforts to "calibrate" different organisms (e.g., Weil et al., 1981).

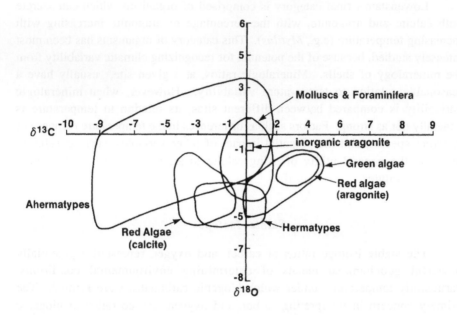

Figure 5.7. Isotopic composition of various groups of calcareous organisms. (After Swart, 1983.)

The stable isotope chemistry of scleractinian corals provides an example of some of the complexities encountered in studying isotope ratios in biogenic carbonates. Light, pH, and physiological factors, as well as temperature, all influence isotope ratios in these corals. Swart (1983) has reviewed the literature on this topic, and Figure 5.8 presents his model for carbon flow in the gastrodermis of a hermatypic coral. Given the complexity of this system, it is little wonder that stable isotope ratios can be significantly altered from those resulting from simple inorganic precipitation from solution.

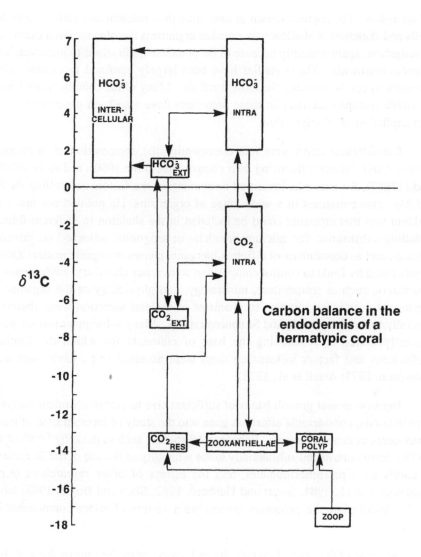

Figure 5.8. Schematic carbon flow in the gastrodermis of a zooxanthellate coral. (After Swart, 1983.)

Coprecipitation

Introduction The coprecipitation of ions other than calcium and carbonate in the shells and skeletons of shallow water marine organisms has also received extensive investigation, again primarily because of the potential applications to understanding paleoenvironments. These studies have been largely confined to cations, with strontium in corals receiving the most attention. Many of the complications found for stable isotope chemistry in these organisms have also been observed for the coprecipitation of "foreign" ions.

Considerable effort went into documenting the composition of calcareous skeletons and factors influencing their composition in the 1960's and early 1970's. Dodd (1967), for example, presented a summary of the factors controlling the Sr and Mg concentrations in a wide range of organisms. He pointed out that one problem was that elements could be included in the skeleton in different forms, including substitution for calcium in calcite or aragonite, adsorbed on mineral surfaces, and as constituents of different inorganic phases or organic matter. Other factors noted by Dodd to control composition were water chemistry, environmental parameters, such as temperature, mineralogy, and physiology of the organisms. Organisms with more complex mechanisms of skeletal secretion were observed generally to have lower Mg and Sr concentrations. Many subsequent studies were primarily aimed at broadening the base of elements for which distribution coefficients and factors influencing them were assessed (e.g., Livingston and Thompson, 1971; Ameil et al., 1973).

Because annual growth bands of sufficient size to permit chemical analysis form in corals, considerable effort has gone into the study of incorporation of trace components in these organisms. Experimental studies, such as those by Smith et al. (1979), have contributed substantially to the reliability of the use of the Sr content of corals as a paleothermometer, and the efforts of other researchers (e.g., Buddemeier et al., 1981; Swart and Hubbard, 1982; Shen and Boyle, 1988) have yielded insight into the processes controlling a variety of minor components in corals.

Because of the general complexity of factors controlling mineralogy, stable isotope composition and the chemistry of skeletal carbonates, it has been found that the best approach is to use these parameters in combination, rather than relying on any one of them (e.g., Lowenstam, 1961; Schifano and Censi, 1986). This approach is especially necessary in cases where early diagenesis may have taken place. Because, as will be discussed later in this chapter, this process can take place nearly concurrently with skeletal deposition, considerable care must be taken in making paleoenvironmental projections from skeletal carbonates. Nevertheless, it is among the best approaches we have for assessing how conditions may have varied during the past in shoal-water environments.

A variety of trace and minor elements may be found in skeletal carbonates. Veizer (1983) showed, however, that only magnesium, calcium, manganese, bromine, strontium and yttrium are concentrated preferentially in carbonate rocks and that phosphorus, sulfur, chlorine and iodine are found in these rocks within a factor of about one-half of their concentrations in shale. Furthermore he pointed out that, except for the above elements, the chemistry of sedimentary carbonates is principally controlled by the quantity and the chemistry of the non-carbonate constituents. Some trace elements that occur in significant concentrations in marine sedimentary carbonates are magnesium, strontium, sodium, zinc, iron, copper, uranium, barium, and to a lesser degree manganese, cadmium, and cobalt. Marine calcites tend to be enriched in magnesium, strontium, sodium, barium, and uranium whereas non-marine calcites contain greater concentrations of zinc, manganese, iron, cobalt, and copper (Veizer, 1983). This difference in trace element concentration reflects the difference in the chemistry of non-marine and marine waters. Because much effort has been invested in studies of magnesium in calcite and strontium in aragonite, and we know most about the processes and mechanisms controlling these elements in skeletal carbonates, these elements will be considered in detail in the following sections.

Chemistry and mineralogy of calcitic skeletal organisms The skeletons of modern organisms are composed predominantly of either aragonite or calcite containing a range of Mg concentrations or sometimes both minerals. The presence of Mg in biogenic hard parts was first noted by Silliman in 1846. Until 1952 wet chemical methods and stains were mainly used to determine the Mg concentration of calcareous skeletal materials. Clarke and Wheeler (1917), in their comprehensive study on the composition of marine invertebrates, demonstrated that carbonate skeletons can contain more than 20 mole % $MgCO_3$. At that time, however, and until the now-classic work of Chave (1952), knowledge of carbonate skeletal mineralogy was scant and mainly limited to staining tests as an index of mineral structure (e.g., Meigan, 1903; Clarke and Wheeler, 1917; Böggild, 1930). Interestingly, Böggild (1930) in his classic study of mollusc shells recognized three different compositional types of calcareous skeletons: low-Mg aragonite (<1-2 mole % $MgCO_3$), low-Mg calcite (2-3 mole % $MgCO_3$), and high-Mg calcite (12-17 mole % $MgCO_3$). He concluded that a chemical compositional gap existed in nature between the 10-15 mole % $MgCO_3$ found in the calcite of octacorals, echinoderms and algae, and the 1-2 mole % $MgCO_3$ found in calcite skeletons of brachiopods, barnacles and most molluscs. The further history of the quantitative relationship between the chemistry and mineralogy of carbonate materials, and environmental controls of these parameters, is tied strongly to the development and expanded use of X-ray powder diffractometry.

Chave (1952) reported a linear relationship between the position of the 1014 diffraction peak of skeletal calcites and their Mg concentrations, as determined by wet chemical techniques, over the range 0-18 mole % $MgCO_3$. At higher Mg concentrations, the relationship was observed to be nonlinear. He concluded "that Mg was replacing Ca in the calcite lattice, shrinking it, and forming a solid solution between calcite and dolomite" (Chave, 1981).

The most common organisms depositing skeletons of magnesian calcite today are the calcareous red algae, benthic foraminifera, bryozoans, echinoids, and barnacles, and in this order may be the most important contributors of skeletal magnesian calcite to shoal-water marine sediments (Chave, 1981). As mentioned previously, the pelagic Coccolithacae and Foraminiferida contain little magnesium (<1-2 mole % $MgCO_3$) in solid solution in their calcite skeleton. Thus, deep-sea calcite oozes are typically poor in magnesium, whereas shoal-water calcareous sediments contain an average of about 5 mole % $MgCO_3$, with a range of 0 to 10 mole % $MgCO_3$ (Chave, 1954b). The magnesium concentration of the magnesian calcite fraction of modern, shoal-water calcareous sediments averages about 14 mole %, with a range from 5-18 mole % (Garrels and Wollast, 1978; Agegian and Mackenzie, 1989) (Figure 5.9). This average reflects the range in magnesium content of the five most common magnesian calcite secreting organisms. Chave (1954a, 1981, 1984) estimates these ranges as calcareous red algae 10-25%, foraminifera 1-15%, bryozoans, highly variable because of mixed calcite-aragonite mineralogy, 0-11%, echinoderms 10-15%, and barnacles 1-15%.

Figure 5.10A illustrates the overall pattern of skeletal magnesian calcite compositions from the tropics to the poles. Data from major organism groups are plotted from Chave (1954a). This trend of compositional change with latitude may be interpreted as reflecting the influence of:

(1) *Temperature* -- Chave (1954a) and others (see Bathurst, 1975 and Milliman, 1974 for discussion) demonstrated a proportionality between the skeletal Mg content of organisms and their environmental temperature of growth. In general, the various organism taxa in modern seas show

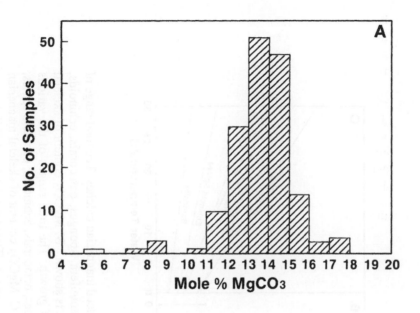

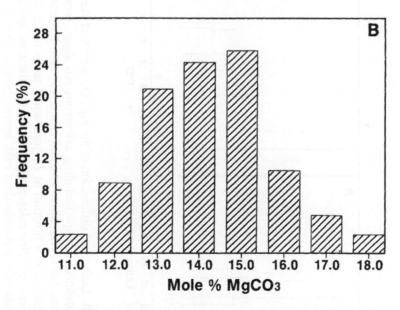

Figure 5.9. MgCO$_3$ content of magnesian calcite component of shoal-water calcareous sands. A. 161 samples compiled by Garrels and Wollast (1978); B. Samples of mid-depth bank sands (Agegian and Mackenzie, 1989.)

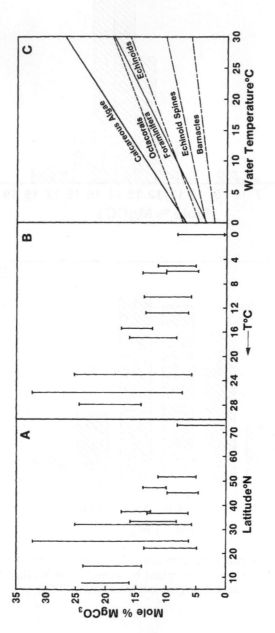

Figure 5.10. A. Latitudinal distribution of the MgCO₃ content of skeletal magnesian calcites. The total range of the various organism groups including calcareous red algae, foraminifera, sponges, octacorals, echinoids, echinoid spines, asteroids, ophiuroids, and crinoids at each latitude is shown. B. MgCO₃ content of skeletal magnesian calcites as a function of environmental temperature of growth. The total ranges of the various organism groups including calcareous red algae, foraminifera, sponges, octacorals, echinoids, echinoid spines, asteroids, ophiuroids, and crinoids at each temperature are shown. C. MgCO₃ content of skeletal magnesian calcites of various taxa as a function of environmental temperature of growth. (Data from Chave, 1954b.)

decreasing skeletal Mg concentrations with decreasing temperature (Figure 5.10b). The range of Mg concentration and rate of change of Mg with temperature are related to the taxonomy (physiology) of the organism (Figure 5.10c). For example, calcareous red algal skeletons vary from about 5 to 30 mole % $MgCO_3$ over a temperature range of 30°C, whereas barnacles contain less than 5 mole % $MgCO_3$ and exhibit small variations of only a few percent Mg over the same temperature range. Thus, temperature may be a reason for the latitudinal gradient in skeletal Mg content.

(2) *Growth rate* -- Moberly (1968) demonstrated that the magnesium content of coralline red algae and pelecypod skeletons depends on skeletal growth rates. These growth rates are not independent of temperature and physiology but depend on these variables as well as other factors. Calcification and growth rates and Mg content of calcareous organisms decrease with decreasing temperature. Thus, growth rate may be a reason for the latitudinal gradient in skeletal Mg content.

(3) *Seawater saturation state* -- The carbonate ion content of surface seawater decreases from the tropics to high latitudes, an exception being the equatorial upwelling area of lower CO_3^{2-} concentrations in both the Atlantic and Pacific Oceans. The latitudinal decrease in saturation state with respect to calcite would be expected to produce a magnesian calcite precipitate containing less magnesium. Lower CO_3^{2-} concentrations both kinetically and thermodynamically favor magnesian calcite precipitates of lower Mg content (see Chapter 3). Furthermore, Smith and Roth (1979) and Borowitzka (1981) demonstrated that the calcification rate in some coralline red algae is related to the dissolved inorganic carbon concentration in aqueous solution. In particular, Borowitzka's work appears to establish the fact that the CO_3^{2-} concentration and photosynthetic rate (see also Goreau, 1963, for example) strongly influence calcification rates in coralline algae. Calcification rates increase with increasing CO_3^{2-} concentration. It would be anticipated that the more rapid the calcification rate, the higher the Mg concentration in the skeletal magnesian calcite (Moberly, 1968). Thus, saturation state may be another reason for the latitudinal gradient in the magnesium content of skeletal magnesian calcite.

Recently, the calcareous red algae, *Porolithon gardineri*, was grown in seawater in controlled macrocosm experiments for periods of time up to three months (Agegian, 1985; Mackenzie and Agegian, 1988). The environmental conditions of light, temperature, and water chemistry were well regulated over this period of time. Figure 5.11 shows some results of these experiments.

It can be seen that with increasing temperature and carbonate saturation state, the growth rate of *Porolithon* increased, as did the magnesium content of the algae. Above about 30°C, the growth rate of the algae slowed, perhaps because the lethal temperature limit of the algae was being approached. The experimental results demonstrate that all three variables of temperature, seawater carbonate saturation state, and growth rate influence the magnesium content of *Porolithon*, and by analogy, perhaps that of other calcareous red algae and invertebrates. Agegian (1985) concluded, however, from a statistical analysis that most of the variability of the magnesium content of *Porolithon* stemmed from the saturation state variable. This conclusion, interestingly, is different from that drawn by Burton and Walter (1987) for inorganic calcite precipitates. They concluded temperature had more influence on magnesium content than did the carbonate saturation state of seawater.

Certainly for an aqueous solution with a nearly constant activity ratio of Mg^{2+} to Ca^{2+} like that of seawater, temperature and CO_3^{2-} ion activity ratio will play some role in governing the magnesium content of calcite precipitates. For skeletal precipitates, both of these variables influence growth rate similarly, leading to the aforementioned relationships between temperature, saturation state, growth rate, and skeletal magnesium content. For inorganic precipitates, like cements, it appears that seawater must be well oversaturated for precipitation to occur.

It can be demonstrated from the data of Videtich (1985) and others (see Mackenzie et al., 1983) that magnesian calcite marine cements and oöid compositions vary with latitude in much the same manner as skeletal calcite. Higher latitude phases contain less magnesium than lower latitude phases. Also, Videtich (1985) has nicely demonstrated that the magnesium content of calcite cements decreases with increasing depth in the ocean. She observed a strong positive correlation between temperature and the magnesium content of the cement. Furthermore, the average mole % $MgCO_3$ of naturally occurring magnesian calcite cements found at shallow depths in the ocean (Figure 5.12) is remarkably similar to the average $MgCO_3$ content of skeletal marine organisms secreting magnesian calcite and living in shallow marine water (Figure 5.9). This observation led Garrels and Wollast (1978) to conclude tentatively that "a 12 mole % magnesian calcite would be close to a composition in true equilibrium with surface seawater" (p. 1471).

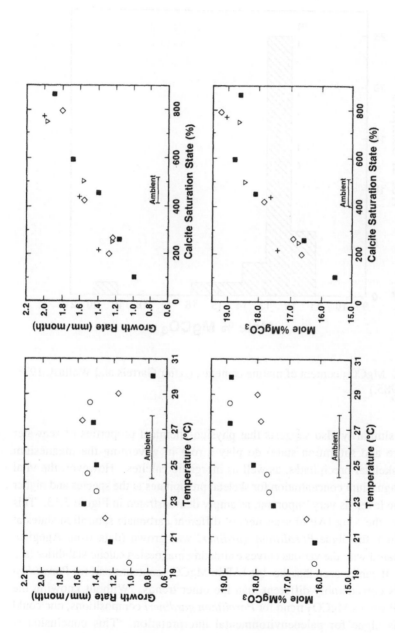

Figure 5.11. Results of macrocosm experiments on *Porolithon gardineri*, a coralline alga grown under controlled conditions of temperature, light and seawater chemistry. Different symbols simply represent different experiments. (After Agegian, 1985.)

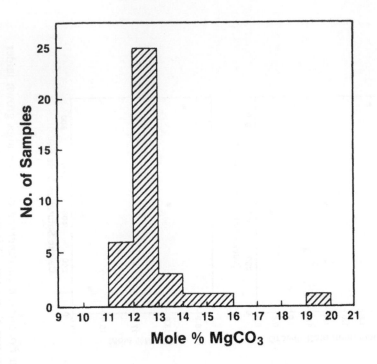

Figure 5.12. MgCO$_3$ content of marine cements. (After Garrels and Wollast, 1978; Videtich, 1985.)

This similarity also suggests that physico-chemical properties of seawater (temperature and saturation state) do play a role in governing the magnesium content of skeletal precipitates, as well as inorganic calcites. However, the vital effect on magnesium concentration for skeletal precipitates at the species and higher phylogenetic levels is very important, as amply demonstrated in Figure 5.13. This figure shows the -log IAP of seawaters of different carbonate saturation states at 25°C in which the algae *Porolithon gardineri* was grown (data from Agegian, 1985) compared with the various curves estimating magnesian calcite stabilities (see Chapter 3). It can be seen that the -log IAP in MgCO$_3$ content trend for *Porolithon gardineri* is considerably different than the other trends. Without knowing the specific IAP versus MgCO$_3$ trend for *Porolithon gardineri* compositions, one could not use this algae for paleoenvironmental interpretation. This conclusion is generally applicable to the use of most element concentrations in biogenic materials for environmental reconstructions.

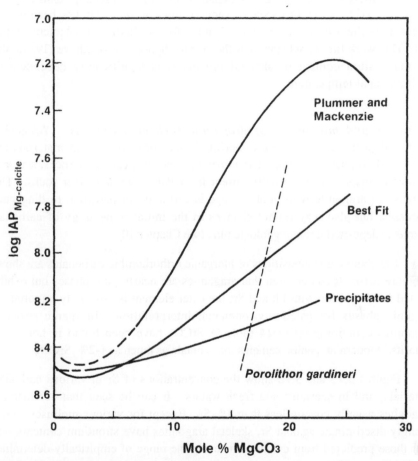

Figure 5.13. Experimentally determined trends of -log IAP magnesian calcite versus mole % MgCO3 compared with -log IAP of seawaters from which *Porolithon gardineri* skeletons of composition represented by the dashed line were deposited.

There is a strong need for more experimentation on the growth of calcareous marine organisms under controlled conditions to gain further insight into the environmental parameters controlling composition. As of now, we can conclude that temperature, seawater saturation state, and growth rate all affect the magnesium content of skeletal organisms; temperature and saturation state affect inorganic precipitates like cements, with temperature perhaps the most important overall variable. We are intrigued by the similarity between cement and skeletal magnesium concentrations with latitude and consequently temperature. It appears that temperature may be the most important variable governing the magnesium content of inorganic and skeletal precipitates in today's marine environment. The recent data of Izuka (1988) seem to support this assertion. He has been able to demonstrate the use of magnesium in benthic foraminifera as a paleotemperature tool. This work further substantiates the original hypothesis of Chave (1954a) that the magnesium content of skeletal organisms is significantly dependent on environmental temperature.

Chemistry and mineralogy of aragonitic skeletal organisms   Aragonites are found in modern calcareous deposits in the form of skeletal materials and cements. This phase is an important mineral of Cenozoic carbonate rocks, less so in Mesozoic rocks, and is virtually absent from Paleozoic and older rocks. This decrease in aragonite with geologic age is certainly, in part, due to diagenetic processes, but also may reflect changes in the initial mineralogy of carbonate sediments deposited through geologic time (see Chapter 10).

The chemical compositions of inorganic orthorhombic carbonates are shown in Figure 5.14. It can be seen that aragonites are nearly pure phases but exhibit limited solid solution with Pb and Sr; the latter element is especially important in biogenic phases for paleo-environmental interpretation. Inorganic strontian aragonites containing up to 14 mole % $SrCO_3$ have been found in hot spring deposits. Modern aragonitic cements may contain as much as 1-2% $SrCO_3$.

Figures 5.15 and 5.16 show the concentrations of Sr in various carbonate materials, and in seawater and fresh waters. It can be seen that sedimentary aragonites usually contain less than 1% Sr. Except for various molluscs, which strongly discriminate against Sr, skeletal aragonites have strontium contents very near those predicted from calculations using the range of empirically-determined distribution coefficients for strontium in inorganic aragonite at "equilibrium" with seawater. This observation suggests that for many skeletal aragonites, physio-chemical processes play a major role in the biomineralization process involving Sr incorporation. This statement is confirmed to some degree by the work of Onuma et al. (1979). They determined the enrichment factor, defined as concentration of element in shell divided by concentration of element in extrapallial fluid, for various cations incorporated in a marine bivalve, *Pinctada fucata*. Their data show a smooth parabolic relationship between enrichment factor and divalent cation radius (Figure 5.17). Onuma et al. concluded that the divalent cations, including Sr^{2+}, simply substitute for Ca^{2+}, and this process is controlled mainly by the difference between the ionic radius of calcium and that of the substituting divalent cation -- a physico-chemical control. In contrast to the divalent cations, no parabolic trend for the monovalent cations Li^+, Na^+, and K^+

was observed, suggesting their enrichment factors are not controlled by mineralogy, but that these elements are present as absorbed cations or in fluid or solid inclusions. Magnesium in skeletal aragonites is found in low concentrations (average weight % 0.1-0.2; Speer, 1983), as predicted from the distribution coefficient for Mg in inorganic aragonites (Figure 5.16); however, some Mg^{2+} in biogenic aragonites is present at sites where it is easily exchanged, or in organic matter (Amiel et al., 1973; Walls et al., 1977), and thus does not substitute directly for Ca^{2+}. Sodium is also found in low concentrations in aragonites (average weight % = 0.4; Speer, 1983), and appears to be in the aragonite structure with charge balance achieved by HCO_3^- substitution (Land and Hoops, 1973) or maintained by vacancies and interstitial substitutions (White, 1977).

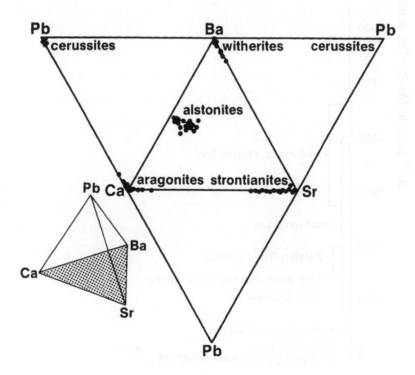

Figure 5.14. Chemical compositions of 106 samples of inorganic orthorhombic carbonate minerals. (After Speer, 1983.)

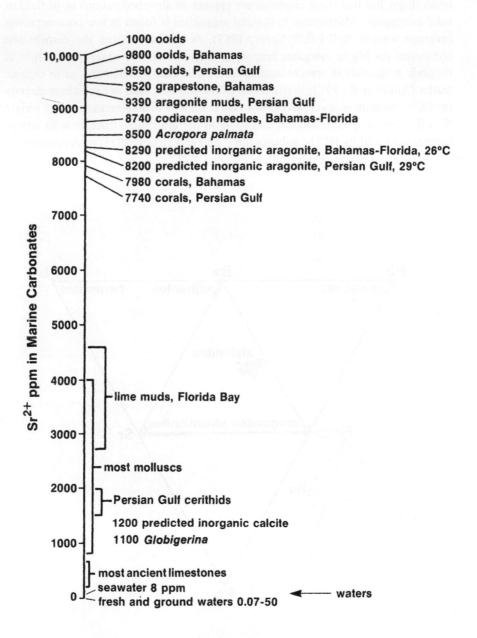

Figure 5.15. Concentrations of Sr^{2+} in natural phases. (After Kinsman, 1969 and Bathurst, 1975.) Compare with Figure 5.16.

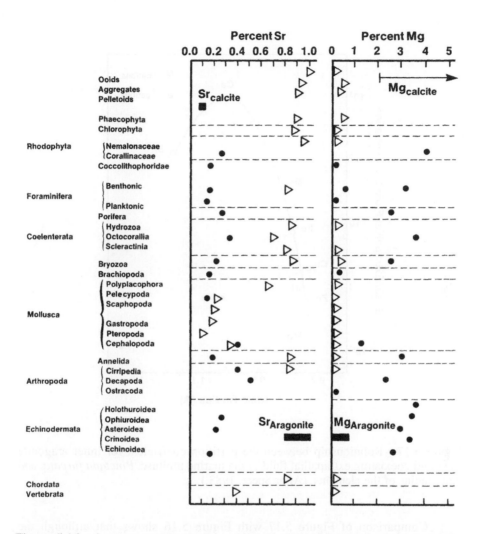

Figure 5.16. Average weight % Sr and Mg in various marine carbonate solids. (After Veizer, 1983). ● Calcite, ▷ Aragonite. Range of values of Sr and Mg in aragonite and calcite is also shown, based on calculations using distribution coefficients of Chapter 3 and an average seawater composition. Compare with Figure 5.15.

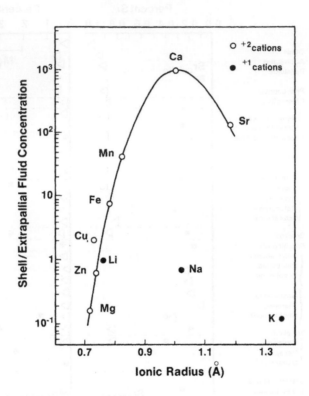

Figure 5.17. Relationship between the partition coefficient for inner aragonite shell and coexisting extrapallial fluid of the marine mollusc, *Pinctada furcata*, and ionic radius of the elements. (After Speer, 1983.)

Comparison of Figure 5.17 with Figure 5.16 shows that although the relationship between the enrichment factor for Sr^{2+} and ionic radius suggests a physico-chemical control for incorporation of Sr^{2+} in molluscan aragonite, the partitioning of this element between solid and fluid is not controlled by an equilibrium relationship with bulk seawater. The value of the distribution coefficient for inorganic aragonite predicts higher concentrations of Sr^{2+} in molluscan aragonite, as is observed for other marine organisms. This discrimination against Sr^{2+} by marine organisms is a result of the complexity of Sr^{2+} incorporation, in which the organism obtains Sr^{2+} for biomineralization from the extrapallial fluid via the blood which in turn receives its Sr^{2+} from seawater (e.g., Speer, 1983). In contrast, biomineralization in more primitive phyla is less complex and more extracellular; thus, the mechanism of incorporation of Sr^{2+} in aragonitic organisms of most phyla is closely related to that for inorganic aragonite precipitates.

Because the mechanism by which an organism precipitates its shell may vary from organism to organism, the "species" effect of Weber (1973), Sr^{2+} incorporation in aragonite shows a relationship to taxonomy. Apparently some taxa still exhibit a strong relationship between the Sr/Ca ratio of the shell and that of the ambient water (Figure 5.18). This relationship, however, need not be that predicted from an assumption of equilibrium. The distribution coefficient for the aragonitic molluscs of 0.24 shown in Figure 5.18, although constant, is not that predicted from the empirical inorganic aragonite values.

The above lines of evidence strongly suggest that for many marine aragonitic organisms, the Sr/Ca ratio of their shell will be related to the Sr^{2+}/Ca^{2+} ratio of the ambient water, which is essentially constant, and a distribution coefficient, which may or may not be the value for inorganic aragonite. The skeletal Sr/Ca ratio would be predicted to change with temperature. Smith et al. (1979), in a series of

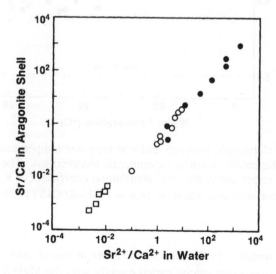

Figure 5.18. Relationship between the Sr/Ca ratio of aragonitic freshwater mollusc shells and the Sr^{2+}/Ca^{2+} ratio of the water in which they grew. The distribution coefficient fitting these data has a value of 0.24. (After Speer, 1983.)

careful experiments, demonstrated that these relationships exist for three genera of aragonitic corals (Figure 5.19). The relationship between coralline aragonite Sr/Ca ratio and temperature provides for a kind of thermometry known as sclerothermometry. Lorens' and Bender's work (1980) suggests that the difference in Sr/Ca versus temperature trends for inorganic aragonite and coralline aragonite shown in Figure 5.19 is a result of temperature controlling the physiological response of the corals, which in turn determines skeletal chemistry.

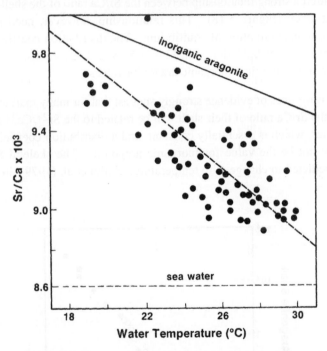

Figure 5.19. Relationship between ambient seawater temperature and the Sr/Ca ratio of aragonitic corals grown in experimental macrocosms. The dashed and full lines represent, respectively, the coral distribution coefficient of K = 1.30 - 0.0094 t (°C) and that for inorganic aragonite of K = 1.24 - 0.0045 t (°C). (After Smith et al., 1979.)

Concluding statement In Chapter 3 and earlier in this chapter, we saw that the Mg/Ca ratio of magnesian calcite varied directly with the Mg^{2+}/Ca^{2+} ratio of the ambient water, as seems to be the case for the Sr/Ca ratio. Also, it was demonstrated that the Mg/Ca ratio varied with temperature, as seems to be the case for the Sr/Ca ratio. For both magnesian calcites and Sr-bearing aragonites of biologic origin, the concentration of these elements in skeletal material is also related to the biochemical process of mineralization (Chave, 1954a; Lowenstam, 1963); as pointed out by Lowenstam (1963) "the tendency for both (Mg and Sr) to decrease as substituting ions for Ca from the less complex to the more highly differentiated tissue grades indicates an increase in power of discrimination by more advanced biochemical systems against ions from the external aqueous medium" (p. 167). This statement still holds true today, implying that both physico-chemical and "vital" processes play a role in governing the Mg and Sr contents of marine organisms. The morphologically and compositionally complex skeletons of these organisms form the major carbonate particles available for diagenesis in the Phanerozoic.

Precipitation of carbonates from seawater

Carbonate Chemistry of Shallow Seawater

Near-surface seawater is typically supersaturated by over six times with respect to calcite, and over four times with respect to aragonite (see Chapter 4). When it flows over shallow areas of the world's oceans, its chemistry can be modified by several different processes. The extent of modification is strongly dependent on the residence time of the water over the shallow areas. Morphologic controls on flow, such as embayments, barrier islands, and reefs, can often lead to significant restriction of the flow of seawater. In areas where these restrictions occur, major changes in the chemistry of the seawater are usually observed.

One commonly observed change is in the salinity of the water. This change does not involve any chemical reactions, unless such high salinities are reached that evaporite formation takes place. The two major factors influencing salinity are the balance between evaporation and rainfall, and inflow of nonmarine waters. In many environments the primary source of inflowing water is rivers, which act to dilute the seawater and lower its salinity. An excellent example of this situation is Florida Bay. Inflow of groundwaters may also alter salinity. Their chemistries vary extensively, and they usually have high P_{CO_2} values. Consequently, these waters may also alter relative concentrations of seawater components, as well as salinity. The balance between evaporation and precipitation controls salinity in areas where little or no inflow occurs. Such areas are typified by carbonate banks, atolls and coastal areas in arid regions. Evaporation often exceeds precipitation in these areas and elevated salinities are observed. In the absence of other processes, the saturation state of seawater will increase with increasing salinity and decrease with decreasing salinity.

Other major processes altering the chemistry of shallow seawater are precipitation of calcium carbonate, exchange of dissolved components between sediments and the overlying seawater, and biological activity. Both abiotic and biotic precipitation of carbonates leads to a lowering of the alkalinity and calcium concentrations, and, therefore, lowers the saturation state of the water. Exchange of dissolved components between sediment pore waters and the overlying seawater is a complex process, which will be discussed in Chapter 6. Generally, pore waters are less saturated with respect to carbonate minerals than overlying waters, and any exchange usually results in the immediately overlying water becoming less supersaturated. It should be noted that this situation is the opposite of that occurring in the deep sea, where carbonate mineral dissolution in sediments overlain by undersaturated waters acts to increase the saturation state of immediately overlying waters. Photosynthetic formation of organic matter leads to a decrease in the P_{CO_2} of the water, which causes the degree of saturation to rise. This effect is moderated by oxidation of organic matter and air-sea exchange of CO_2.

The waters overlying the Great Bahama Bank (Figure 5.20) have been extensively studied relative to their carbonate chemistry, and we will use them here to illustrate chemical modification of seawater in an important region of carbonate deposition. There is no major inflow of riverine or ground waters on the bank, and, consequently, salinity is primarily controlled by the balance between rainfall and evaporation, and the residence time of the water in different areas of the bank. The southern part of the bank is not protected by major obstructions to water flow. This situation results in residence times for the water that are short enough so that major chemical changes do not take place in the seawater as it passes over the southern part of the bank. The northern part of the bank is bordered by Andros Island, a topographic barrier that causes restricted flow and long residence times for the water. It is here where significant changes occur in the water chemistry.

Two major studies of the water chemistry of the northern Great Bahama Bank were done by Broecker and Takahashi (1966) and Morse et al. (1984), and our discussion will focus on their results. The variation in salinity observed by Morse et al. (1984), shown in Figure 5.21, is remarkably similar to that observed by Broecker and Takahashi (1966). Major increases in salinity, to values exceeding 44, were observed as Williams Island was approached. The ^{14}C distribution was also modeled in both studies, but by different methods, to obtain the residence time of the waters. Similar residence time values were obtained; Morse et al. (1984) estimated the highest salinity waters had been on the bank for

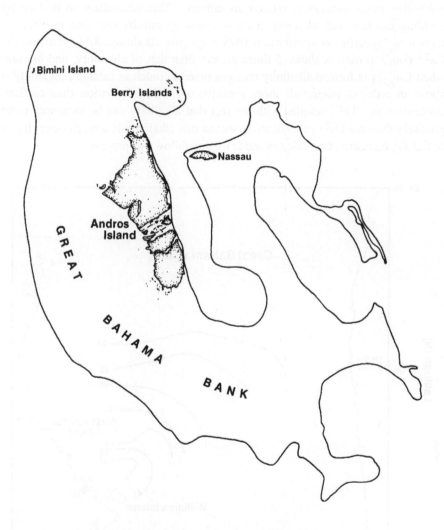

Figure 5.20. The Great Bahama Bank and Andros Island.

approximately 240 days. This residence time is in reasonable agreement with that which would be estimated from the net difference between evaporation and rainfall in the region.

Other than salinity changes, the major process altering the chemistry of the northern Great Bahama Bank waters is $CaCO_3$ production. Because changes in salinity will modify calcium concentrations and alkalinity, it is necessary to normalize these parameters relative to salinity. This normalization is done by dividing calcium and alkalinity concentrations by salinity resulting in what are known as "specific" concentrations (A/S = specific alkalinity, SA). Because the Ca^{2+} concentration is about 5 times greater than that of alkalinity, and because when $CaCO_3$ is formed alkalinity changes twice as much as calcium, alkalinity is about an order of magnitude more sensitive to $CaCO_3$ formation than calcium concentration. This, coupled with the fact that alkalinity can be measured more precisely than the Ca^{2+} concentration, means that alkalinity is a much better tracer of $CaCO_3$ formation than calcium and is used to follow this process.

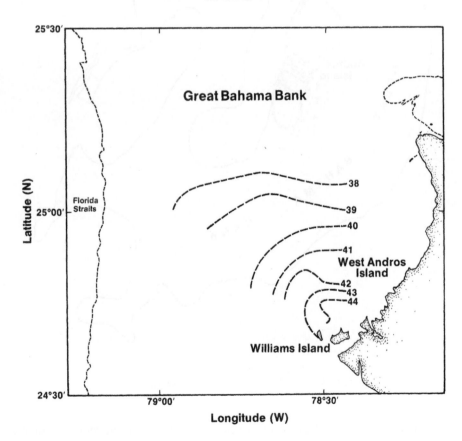

Figure 5.21. Salinity distribution on the northern Great Bahama Bank. (After Morse et al., 1984.)

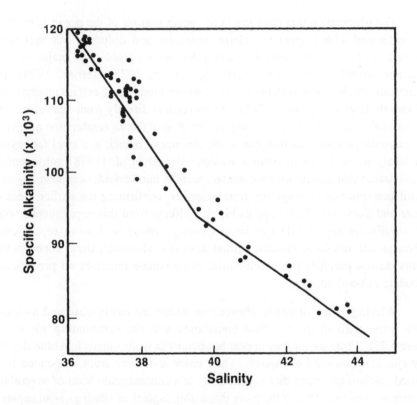

Figure 5.22. Variation of specific alkalinity with salinity on the northern Great Bahama Bank. (After Morse et al., 1984.)

The results of Morse et al. (1984) for specific alkalinity variation are shown in Figure 5.22. It can be seen that as salinity increases, the specific alkalinity drops significantly. In the highest salinity waters, about a third of the alkalinity has been lost because of $CaCO_3$ formation. The opposing influences on saturation state of increasing salinity and decreasing alkalinity are dominated by the alkalinity decrease so that in the higher salinity waters the saturation state is less than that of the incoming seawater. The relation of the water composition to processes of carbonate formation on the bank will be discussed later in this chapter.

Abiotic Precipitation of CaCO3 from Seawater

The observation that seawater in the upper regions of the ocean is strongly supersaturated with respect to calcite, aragonite and dolomite but that these minerals fail to precipitate directly from it, has been a problem of major interest. In experimental studies (e.g., Pytkowicz, 1965, 1973; Berner, 1975), the magnesium ion has been observed to be a strong inhibitor of calcite precipitation and largely responsible for its failure to precipitate directly from seawater. The presence of magnesium can also cause an elevation of the supersaturation necessary for aragonite precipitation to occur, in the absence of nuclei, to a level far beyond that likely to be found in natural waters. Berner et al. (1978) subsequently demonstrated that natural organic matter, such as humic acids, can also severely inhibit precipitation of aragonite from seawater, confirming the earlier ideas of Chave and Suess (1967). It appears highly unlikely from this experimental work that conditions appropriate for direct (homogeneous) nucleation of calcium carbonate will obtain in modern normal seawater. However, these experimental results do not preclude the precipitation of carbonate minerals on pre-existing carbonate mineral nuclei.

Whitings are a dramatic phenomena which are easily observed as milky white water with sharp to diffuse boundaries with the surrounding water (see Figure 5.23). From the air they appear as obvious as white clouds in a blue sky and they can be observed from space. Their color is derived from suspended fine-grained calcium carbonate that is commonly at a concentration level of several mg per liter in seawater. One of the more interesting aspects of whitings is an apparent ability to form suddenly, and to persist for days.

In spite of the strong experimental evidence to the contrary, controversy has continued to swirl around the possibility that whitings are caused by a sudden direct precipitation of aragonite from seawater. We will examine this controversy in some detail, because of its importance to the question of the origin of common aragonite needle muds, and also because it exemplifies the all too frequent problem in carbonate geochemistry of your eyes appearing to tell you something that is in major disagreement with the chemistry of the system.

There are three primary hypotheses for their mode of formation. The first is that they are simply sediment that has been resuspended by bottom feeding fish. The "mullet muds" in Florida Bay are probably an example of this process. The second is that they result from spontaneous precipitation of $CaCO_3$ which may be

Figure 5.23. Air photograph of whitings on the Great Bahama Bank.

triggered by algal blooms. Formation of whitings in the surface waters of the Great Lakes, U.S.A.,(where there is little Mg in the water) is a clear example of this mechanism, and whitings may possibly form in the Persian Gulf in this manner. An intermediate hypothesis is that they form on pre-existing carbonate nuclei and are concentrated by circulation. This mechanism for their formation may be operative in the Bahamas.

By far the most extensive investigations of whitings have been done in the Bahamas, primarily on the northern part of the Great Bahama Bank. These investigations have resulted in highly divergent explanations for their occurrence, and we will concentrate our discussion on whiting formation in Great Bahama Bank waters. Early work (e.g., Cloud, 1962 a,b) pointed out the difficulty in accepting the concept that whitings were produced by fish suspending bottom sediments, because many of the whitings occurred over coarse-grained sediments or hard-grounds. He argued that the whitings had to be inorganic precipitates. Subsequent extensive investigations of whitings on the Great Bahama Bank by Shinn et al. (1989) have also failed to demonstrate the association of schools of bottom feeding fish with whitings. Numerous attempts to produce whitings by resuspending bottom sediments by various means have failed to produce suspensions with the color or longevity of whitings. The concept that whitings are simply the result of fish stirring up bottom sediments appears unlikely as a general mechanism for producing whitings on the Great Bahama Bank.

The two most extensive investigations of the chemistry of whitings were conducted by Broecker and Takahashi (1966) and Morse et al. (1984) as part of their previously discussed examination of the chemistry of waters of the Great Bahama Bank. Both studies failed to find any changes between carbonate water chemistry within the whitings and outside of the whitings, as required to produce the observed concentrations of suspended calcium carbonate. Broecker and Takahashi (1966) conducted radiocarbon analyses of the whiting material and underlying sediments. They concluded that a minimum of 85% of the material had to be resuspended sediment. They also failed to find elevated concentrations of organic matter within the whitings which should occur if they are triggered by algal blooms.

Shinn et al. (1989) examined both the ^{14}C, and oxygen and carbon isotope composition (Figure 5.24) of carbonates collected from whitings on the Great Bahama Bank. They concluded from these data that 75 to 100% of the carbonates found in the whitings must be resuspended sediments. An examination was also conducted of the carbonate minerals found in the whitings and underlying sediments. More high magnesian calcite was found in the whitings (Figure 5.25) than in bottom sediments, but this may be the result of differential resuspension. The extensive study of Shinn et al. (1989) thus reinforces the paradox of the lack of a mechanism to suspend sediments to produce whitings in the Bahamas with the observation that they are dominantly composed of suspended sediment.

It has been observed that the whitings on the Great Bahama Bank are not randomly distributed, but form a halo on the northern and central part of the bank (Figure 5.26). Within this area the waters tend to be substantially more cloudy than on other areas of the bank, and the calculated rate of carbonate ion removal from these waters is substantially higher than in the surrounding region. It is possible that the higher carbonate ion removal rates are associated with precipitation of carbonate on the abundant suspended carbonate particles that act as nuclei in this area; a hypothesis favored by Morse et al.(1984) and Shinn et al. (1989). Through repetitive resuspension, this mechanism could produce a major portion of the fine-grained carbonate material forming on the Great Bahama Bank.

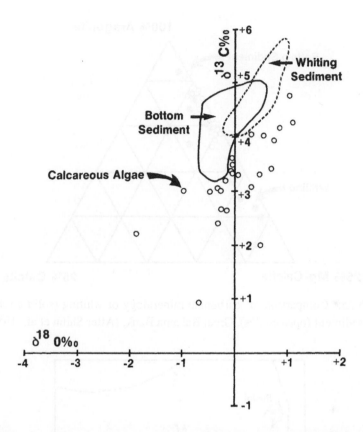

Figure 5.24. Isotopic composition relative to PDB standard of suspended sediment in whitings compared to the fine fraction of bottom sediment. (After Shinn et al., 1989.)

The best evidence to date for a marine whiting possibly being primarily the result of calcium carbonate precipitation from seawater comes from the observations of Wells and Illing (1964) in the Persian Gulf on the east side of the Qatar Peninsula. Here whitings up to 10 km across can form in only a few minutes, in waters up to 30 m in depth. The whitings always form near the water surface and clear water was observed between the whitings and the seafloor. The flocculation behavior of carbonates in the whitings is substantially greater than for suspended sediment, and the suspended sediment is much richer in aragonite than the bottom sediments. Wells and Illing (1964) provided inconclusive evidence that the calcium content of the seawater decreased in waters associated with whitings, but no pH changes were observed. They hypothesized that the whitings could be triggered by diatom blooms.

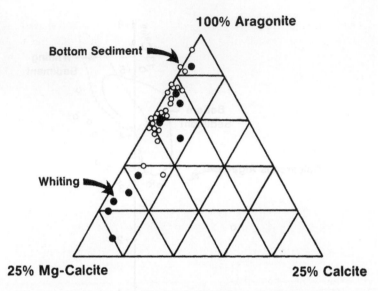

Figure 5.25. Comparison of carbonate mineralogy of whiting (solid circles) and bottom sediment (open circles), Great Bahama Bank. (After Shinn et al., 1989.)

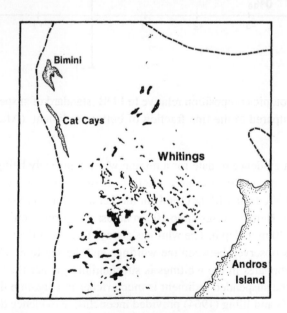

Figure 5.26. Distribution of whitings on the Great Bahama Bank. (After Morse et al., 1984.)

The effect of photosynthesis, which would be associated with diatom blooms, on inducing carbonate mineral precipitation is shown in the following reactions:

Photosynthesis $CO_2 + H_2O \rightarrow CH_2O + O_2$ (5.1)

Carbonate $Ca^{2+} + 2HCO_3^- \rightarrow CaCO_3 + CO_2 + H_2O$ (5.2)
Precipitation   _____

Net Reaction $Ca^{2+} + 2HCO_3^- \rightarrow CaCO_3 + CH_2O + O_2$ (5.3)

The uptake of CO_2 by the diatoms could raise the saturation state of seawater to a point where heterogeneous precipitation of aragonite takes place under conditions where "crystal breeding" occurs via dendrite fracture and microattrition (e.g., see Reddy and Gaillard, 1981, for discussion relative to calcium carbonate precipitation). This mechanism could account for the rapid spread and increase in carbonate nuclei in a whiting, without having to resort to a homogeneous precipitation mechanism that requires unreasonably high supersaturations.

Major problems concerning whiting formation still exist, not least of which are the distribution and longevity of whitings, their concentrated nature and lack of major water chemistry changes within whiting areas. The conclusions presented here have important implications for the production of fine-grained carbonate particles in modern and ancient depositional environments. Some further aspects of the production of these particles are discussed in the next section.

Sources of Aragonite Needle-Muds

The continued disagreement over the primary sources of common fine-grained aragonite needle-muds is closely connected to the controversy over whitings. The two major sources of aragonite needles generally considered are calcareous algae and non-biogenic precipitation. Numerous studies have been made on the source of aragonite for calcareous muds, and much of this literature has been reviewed by Milliman (1974), Bathurst (1975), and Scoffin (1987). As Milliman points out, the fine size of this material has resulted in most of the interpretation of its origin being largely based on chemical data, although needle morphology has also been used.

Attempts at making budgets of carbonate production and accumulation have also been used to answer the question of needle origin. In areas such as British Honduras and Florida Bay, the supply of biogenic material appears to be sufficient to provide the sediment for the carbonate muds, although some abiotic precipitation cannot be ruled out. However, in other areas such as the Great Bahama Bank and the Persian Gulf, the biogenic supply appears to be insufficient, and an abiotic source of aragonite needles is needed. The most commonly invoked source is whitings. If one accepts that whitings are not simply resuspended sediment, then they are potentially a major, if not dominant, source for the needed aragonite. Consequently, the question of what produces whitings is inseparable from the problem of the source of needle muds dominantly of aragonitic mineralogy.

A basic difficulty in determining the origin of these needles is that precipitation of aragonite from seawater produces aragonite needles with a very similar size and morphology to those resulting from breakdown of common codiacean green algae. It is necessary, therefore, to examine the chemistries of the needles to determine their origin. The most extensive research into the origin of the carbonate muds on the Great Bahama Bank using stable isotopes was made by Lowenstam and Epstein (1957). Their study included measurement of oxygen isotopes in samples of mud, algae and oöids. A summary of their results is presented in Figure 5.27. Interpretation of these results is clouded by problems associated with temperature and salinity variations. Bathurst (1975), in an excellent discussion of these difficulties, reaches the unfortunate conclusion that the results are inconclusive as to the origin of the muds. He notes, however, that oöid isotope ratios and mud isotope ratios do not overlap. Because oöids are believed to be of abiotic origin (see next section of this chapter), this lack of overlap favors an algal origin for the muds. In contrast, Milliman (1974) concluded that the Sr content of the aragonite in the muds of the Great Bahama Bank is close to that of oöids, and suggested a non-biologic origin for the aragonite.

An argument in favor of direct precipitation of the muds stems from the study of Broecker and Takahashi (1966). They show a decrease in the removal rate of carbonate ion with decreasing supersaturation of the water on the Great Bahama Bank. This correlation would not necessarily occur if the source of the needles were algae. The more recent work of Morse et al. (1984) also shows this trend. There are, however, two problems with this conclusion. The first is that Morse et al. (1984) found a continued strong removal of carbonate ion from the water in the inner part of the bank where whitings are not observed and clear water is present. The second is that nutrient availability is likely to decrease with increasing age of

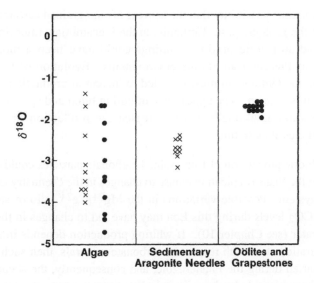

Figure 5.27. The range of $\delta^{18}O$ values for different carbonate materials from different areas after corrections for the $\delta^{18}O$ of water are made. (After Lowenstam and Epstein, 1957.)

the water, which correlates with the decreases in saturation state and removal rate. Thus, the biological removal rate also would be expected to decrease with decreasing saturation.

Loreau (1982), in a thorough work, examined grain morphology and crystallographic characteristics of modern aragonite needles. He found that many may not be derived from algae, and concluded that abiotic formation may be an important source of these allochems. His approach, along with new biochemical methods for distinguishing the sources of aragonite (e.g., Liebezeit et al., 1983), should contribute substantially to resolving the problem of the sources of aragonite in carbonate muds.

In summary, it is the case that aragonitic muds have two sources in the modern seawater environment: they are accumulations of fine-grained particles of abiotic and biotic origin. It certainly has been documented that organisms produce significant quantities of fine-grained aragonite, and that aragonite needles will precipitate directly from seawater of modified chemistry. Furthermore, it has been shown that bioerosional processes (e.g., Neumann, 1967) can lead to generation of fine-grained particles of diverse composition.

It is a distinct possibility that the sources of fine-grained carbonate particles varied during the geologic past. Certainly, in the Precambrian, micritic limestones must have had an abiotic origin. Whitings could have been a more common phenomenon of Precambrian marine environments. Evolution of the codiacean green algae in the Ordovician must have led to increased production of aragonite needles from this time on. The appearance of boring fungi and sponges also in the Ordovician should additionally have increased the production of fine-grained carbonate particles at these times.

The abiotic production of fine-grained carbonate particles could have varied as well during the Phanerozoic in response to changes in the chemistry of the ocean-atmosphere system. Possible variations in the Mg^{2+}/Ca^{2+} ratio of seawater and atmospheric CO_2 levels during this Eon may have led to changes in the saturation state of seawater (see Chapter 10). If whiting production depends in any way on seawater saturation state with respect to carbonate minerals, then such production could have varied during the Phanerozoic, and consequently, the accumulation of micritic carbonates could also have varied. Whatever the case, the production and accumulation of fine-grained carbonate particles from biotic and abiotic sources has not been constant through geologic time.

Formation of Oöids

General considerations For over 150 years subspherical carbonate grains generally referred to as oöids, because of their resemblance to fish eggs, have attracted the attention of investigators studying carbonate sediments and rocks. Sediments that are dominantly composed of such material are called oölites and are common members of modern shoal-water tropic and subtropic carbonate-rich sediments. Oölites are abundant in ancient carbonate sediments, where they are familiar host rocks for oil and gas deposits, and as recently as the last ice age, they were common allochems on many continental shelves.

Oöids are variable in both mineralogy and structure. Aragonite, high magnesian calcite and calcite have been observed to occur in oöids (e.g., Land et al., 1979). Radial, concentric-tangential (laminated) and micritic structures are commonly observed around a central nucleus of non-oöid material (Figure 5.28). The majority of modern, and probably most ancient, oöids are (or originated as) either aragonite with a laminated structure or high magnesian calcite with a radial structure. The laminated aragonitic oöids are by far the most abundant type of oöids

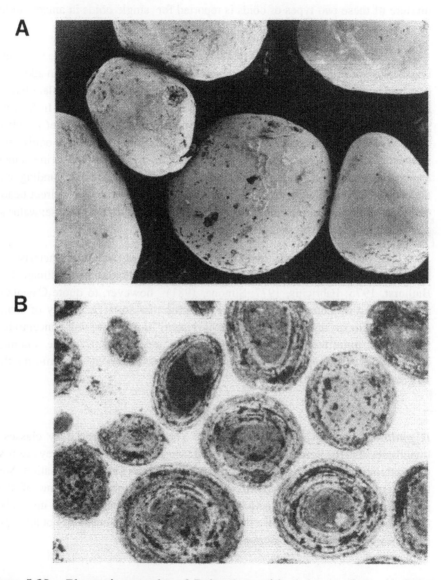

Figure 5.28. Photomicrographs of Bahamian oöids (~1 mm dia.). A) SEM photomicrograph of whole oöids. Note extensive boring and overgrowths. B) Thin section photomicrograph of oöids from same location. Note concentric structure, nuclei, and infilling of bored areas. (Courtesy of Tad Smith.)

in Recent sediments and have received the majority of attention. However, the mixture of these two types of oöids is reported for single oöids in ancient oölites (e.g., Tucker, 1984; Wilkinson et al., 1984).

Oöids are believed to be of non-biogenic origin and represent one of the most important modes of non-biogenic removal of $CaCO_3$ from the ocean. In addition, it has frequently been pointed out that their mineralogy and structure reflect those of marine carbonate cements (e.g., Fabricius, 1977; Land et al., 1979). In fact, they grade into each other in the case of Bahamian grapestone. Land et al. (1979) stressed the tie between carbonate cements and oöids, pointing out that oöids can be viewed as cements centrifugally deposited on grains, whereas submarine cements can be viewed as centripetally coated pores. Consequently, an understanding of the factors controlling oöid formation, mineralogy, and structure has a direct bearing on the overall question of non-biogenic carbonate precipitation from seawater and cementation of carbonate sediments.

We will not attempt to review the occurrence and general characteristics of oölites here, because this has been done by several authors (e.g., Milliman, 1974; Bathurst, 1975; Fabricius, 1977; Simone, 1981); however, to quote Cayeux (as translated by Bathurst, 1975), "it is remarkable that so large a body of writing should contain so small a measure of achievement." Instead, we will concentrate on the major hypotheses for oöid formation and summarize the chemical characteristics of oöids and experimental studies aimed at determining their processes of formation.

Hypotheses for the formation of oöids There are three primary classes of hypotheses for the processes responsible for the existence of oöids. They are based on bacterial-mechanical, algal, or chemical mechanisms for oöid formation. Some investigators have made hybrid models involving more than one of these mechanisms. Because the previously cited papers cover the literature up to about 1980, we will only present these hypotheses in a general manner and not attempt to reference the large number of papers relevant to the topic.

The "Snowball" or Bacterial-Mechanical Formation Hypothesis.

Sorby (1879) put forth one of the earliest non-biological hypotheses for the mechanism of oöid formation. His basic idea was that oöids with a concentric structure form by a process similar to that which forms a snowball. In his model, an organic coating is biologically deposited (the infamous "mucilagenous slime") on the surface of the oöid and then, as the oöid is rolled back and forth by wave action, small aragonite needles, which are common in most modern depositional environments where oöids are found, become imbedded in the organic matter. During subsequent quiescent times a new organic coating is added. This process is repeated many times (up to 200 lamellae have been observed in oöids) causing growth of the oöid.

This model has the attractive characteristics of not requiring biotic or chemical formation of $CaCO_3$ directly on oöid surfaces and offering an explanation for the observation that oöids appear to require "rest" between depositional cycles. It also provides a reasonable explanation for the concentric organic-rich layers generally observed in laminated aragonitic oöids. Researchers such as Oppenheimer (1961), Cloud (1962a), Wilson (1967), and Folk (1973) have accepted the general principles put forth in this model.

There are, however, some problems with the model. It clearly does not apply to radial oöids. When examined carefully, it fails to explain the general excellent sorting in grain size and mineralogy, tangential arrangement of the aragonite rods, or the formation of relatively thick lamellae on typical aragonitic oöids. It is also questionable how well such coatings could hold up on grain surfaces in the abrasive, high-energy hydrodynamic environments characteristic of laminated aragonitic oöid formation. Land et al. (1979) also pointed out the many failings of this model for laminated oöids in Baffin Bay, Texas. These oöids form in essentially terrigenous sediments, which should result in "dirty" snowballs. Such oöids are not observed in this environment.

It may be best to consider this bacterial-mechanical hypothesis for the formation of laminated aragonitic oöids as a precursor to more recent models, where adsorption of organic matter on oöid surfaces influences the precipitation of $CaCO_3$.

The Algal Formation Hypothesis.

As early as the last century (e.g., Wethered, 1890; Rothpletz, 1892), the ubiquitous association of marine algae with oöids had been observed, and much speculation developed concerning the possible formation of oöids by these algae. During much of the early part of the 20th century this hypothesis was in favor; however, the discovery that most of the algae in oöids are of the boring type, causing dissolution, not precipitation, resulted in this hypothesis being generally discarded.

The only investigator recently to argue forcibly in favor of the algal formation of oöids is Fabricius (1977). He observed that algae can make the aragonite "rod and granule" structure found in grapestones. He hypothesized that this was evidence that they can also make oöids. While his observations are interesting, he did not present a very strong connection between his observations in sediments and the formation of oöids. The lack of algal filaments in some laminated aragonitic oöids (e.g., Mitterer, 1968; Frishman and Behrens, 1969; Land et al., 1979) also has brought this hypothesis into question. In addition, Bathurst (1975) pointed out that there is no reason to conclude that the observation of organic residues on oöids mimicking their structure is other than a result of a time gap between algal colonization and oölitic growth.

The Direct Chemical Formation Hypothesis.

The hypothesis that oöids form by direct chemical processes has gained increasing popularity, and there is a growing body of experimental and observational evidence to support it. Link (1903) was the first person to base this hypothesis on direct experimental data. Subsequent observations of oöid formation in caves, hot springs, water purification plants and boilers have supported this mode of oöid formation (e.g., Berner, 1971). Also, stable isotope (e.g., Lowenstam and Epstein, 1957; Land et al., 1979) and Sr^{2+} concentration data are generally compatible with non-biogenic formation. In recent years, particular emphasis has been placed on the possible role of organic compounds (e.g., Mitterer,1968; Suess and Fütterer, 1972).

This hypothesis is also not without its problems. It fails to explain why it has been difficult to produce non-radial oöids by direct precipitation; it fails to give an explanation for organic lamellae or the general distribution of oöids; and it cannot explain why high magnesian calcites are formed in calcitic oöids.

Bathurst's Hybrid Chemical-Physical Hypothesis of Formation.

Bathurst has presented a model that solves some of the problems associated with laminated aragonitic oöid formation. His suggestion involves a means for growing inorganic precipitates of tangentially oriented rods. His arguments also offer plausible explanations for limitations on oöid size and other characteristics.

One of the difficulties in understanding typical aragonitic oöid formation is the fact that the aragonite needles, or battens, are arranged with their c crystallographic axes parallel to the solid surface. Bathurst (1974) observed that the radial-fibrous fabric found in calcitic oöids can be explained as the result of competition between crystals that grow most rapidly in a direction parallel to one crystallographic axis. The perpendicular orientation of the aragonite c axis is commonly observed for aragonite crystals on the walls of Foraminiferida, and in the utricles of *Halimeda*. It is also the preferred orientation of aragonite in beach rock. However, it is not the typical orientation of aragonite on free calcrete grains where the aragonite grains are arranged with their c axes parallel to the surface. Bathurst (1967) observed the growth of aragonite needles with their c axes parallel to the surface in 3 μm thick films on a variety of calcarenite grains outside of the oölitic shoal environment in the coralgal, stable sand, grapestone, and mud habitats at a number of locations. He suggested that the tangential orientation may result from the adsorption of colloidal sized aragonite nuclei on the solid surface, with their longest axis parallel the surface providing the most stable arrangement.

Contrary to the contention of many investigators of oöid formation, Bathurst (1975) suggests that there is not a strong correlation between the degree of water agitation and the orientation of aragonite needles. He cites his observation of the similarity between grains with oölitically structured films in the Bimini Lagoon and the oöids of Brown's Cay, Bahamas, which exhibit no petrographic differences, including the existence of lamellae, in spite of the radically different hydrographic environments. His ideas have been reinforced by the observations of Land et al. (1979) in Baffin Bay, Texas, where calcitic radial-fibrous and aragonitic tangential oöids coexist in the same sands and show no significant correlation with the probable wave energy differences at the different sampling sites.

Bathurst (1967) also examined the question of size and sorting of oöids and the thickness of individual oölitic coats. He feels that the thickness of the coats may reflect growth rates which are in turn a reflection of the degree of agitation. Higher energy waters will expose the oöids to more supersaturated water for a longer period of time. A problem with the sorting and size of oöids on banks is that the oöids are usually of similar size. Small ones should be winnowed from the oölite banks by the intense hydrodynamic conditions generally present. The source of the oöids, therefore, must be in deeper waters where hydrodynamic conditions are less intense. Bathurst suggested that the upper size limit on oöids may be governed by a balance between abrasion and growth rates.

Experimental studies of oöid formation   During the past 20 years relatively few experimental studies have been carried out on the formation of oöids. Most of these studies have emphasized the role of organic matter. The primary reason for this emphasis is the observation of abundant protein matrices in oöids (e.g.,

Mitterer, 1968, 1971, 1972; Trichet, 1968). Mitterer (1972) noted the abnormally high contents of aspartic and glutamic amino acids in these proteins, and stressed the similarity of this protein composition to that of fluids in carbonate secreting organisms. He suggested that there is an analogy between the two systems, and that this proteinaceous matter may play an important role in the precipitation of aragonite on oöid surfaces. Mitterer and Cunningham (1985) have discussed the possible mechanisms by which organic matter may interact with carbonate surfaces.

The study of Weyl (1967) was the first of the "recent" era. The primary contribution of his study, in which he observed carbonate precipitation in seawater flowing over a column of oöids, was that after an initial period of precipitation, the reaction between the oöids and supersaturated seawater appeared to stop. If the oöids were allowed to sit for a period of time with no water flowing, they would again become reactive. This experimental observation led to the idea that oöids need to "rest" between reactive periods. It was hypothesized that a high magnesian calcite coating initially formed, which later dissolved in pore waters of an oölite bank before more aragonite would precipitate. This hypothesis has been quite popular ever since Weyl's work, although it is far from proven.

Suess and Fütterer (1972) carried out the first major experimental investigation of the role of organic matter in the precipitation of oöids. By adding humates at concentrations up to 20 mg L^{-1} to seawater solutions, they were able to obtain a variety of aragonite morphologies, but when spherical precipitates were formed it was found that they had a radial-fibrous fabric. These precipitates also contained up to 7 weight % organic carbon (natural range 0.1 to 0.5), 20 weight % more Sr, and two to three times more Mg than that of natural oöids. Consequently, Suess and Fütterer questioned how meaningful these experiments were relative to natural conditions. Possible sources of the problems were the use of the spontaneous precipitation technique, in which unreasonably high supersaturations and reaction rates are produced, and the failure to control solution composition during precipitation (see Chapter 3).

By far the most extensive experimental study of the chemistry of oöid formation was done by Davies, Bubela, and Ferguson (see Davies et al., 1978 or Ferguson et al., 1978). Their primary emphasis was on the influence of different types of organic compounds and agitation conditions on the mineralogy and morphology of the precipitated carbonates.

In their non-agitated experiments, these authors used an experimental technique similar to that of Suess and Fütterer (1972) for generating spontaneous precipitation. The problems with this technique were previously discussed. After studying the influences of a variety of organic compounds on the mineralogy and morphology of the precipitates, it was found that humic acids were most effective in producing magnesian calcite (composition was not reported), with a fabric similar to that of radial oöids. It was also observed that, while low molecular weight humic acids were preferentially adsorbed relative to high molecular weight humic acids, the high molecular weight humic acids were most effective in producing oöid-like products. A lengthy discussion of the possible reasons for this observation was presented which noted that the high molecular weight humic acids had a higher Mg to Ca ratio, and that humic acids through hydrophilic-hydrophobic interactions could possibly make membrane-type structures.

Tangentially oriented batten-like aragonite crystals were synthesized on the surfaces of quartz grains and Bahamian oöids without the intervention of organic compounds. Again very high supersaturations were used (20 ml of 0.5 N Na_2CO_3 in 100 ml of seawater = an alkalinity ~40 times that of normal seawater). A high degree of agitation was found necessary, and, like Weyl (1967), Davies, Bubela and Ferguson observed the necessity of a "resting" period during which surfaces were reactivated, possibly by the release of Mg^{2+} or H^+.

Their studies are remarkable in indicating that organic matter may be important for the formation of magnesian calcite radial oöids, but not aragonitic tangential oöids. This observation is contrary to general concensus on the formation of these two types of oöids. It may also offer a major clue to the formation of aragonitic and calcitic carbonate cements.

Morse et al. (1985) and Bernstein and Morse (1985) have reported results of an investigation of the composition of Bahamian sediment pore waters and the factors controlling the calcium carbonate ion activity product in these sediments, discussed later in this chapter. As part of this study oölite banks were investigated. Porewater samples from the oölitic sediments were extracted *in situ* using a harpoon-type sampler from about 0.5 m below the sediment-water interface. The pH and total alkalinity of the oölitic sediment pore waters are not remarkably different from many other pore waters from other types of carbonate-rich sediments. They are significantly different from the composition of the overlying seawater, and supersaturated with respect to both aragonite and an 18 mole % magnesian calcite (based on the biogenic magnesian calcite solubility value of Walter and Morse, 1984a).

Experiments conducted on the solubility behavior of the oölitic sediments indicated that they were not in metastable equilibrium with their pore waters. On the basis of equilibration rates, Bernstein and Morse (1985) estimated that the residence time of the pore waters in these sediments was only a few days or less.

Given the fact that the pore waters also had a distinctly different composition from that of seawater, extensive chemical reactions must be taking place within these oölitic sediments. The increase in the alkalinity of the oölite pore waters indicates that carbonate dissolution may be occurring. Considering that the water is supersaturated with respect to aragonite and an 18 mole % magnesian calcite, the carbonate which is dissolving must be a magnesian calcite very high in Mg.

This idea was reinforced by Bernstein and Morse finding that the composition of oöid pore water remains close to constant in a closed system for periods ranging from a few days to months. The steady-state value for the calcium carbonate ion activity product is that of a high magnesian calcite. These results are again compatible with the hypothesis originally put forth by Weyl (1967) that the formation of a magnesian calcite "poisons" the surface of growing oöids. The major question is why should magnesian calcite form on a growing aragonite surface? The answer to this question has ramifications for the behavior of aragonite in seawater in general and awaits further research.

Concluding statement Seawater overlying shoal-water carbonate sediments is usually supersaturated with respect to calcite, aragonite and some high magnesian calcites. These minerals, therefore, have the potential to precipitate from such seawater. Precipitation does clearly occur in the case of oöids and probably occurs in at least some whitings. In both cases precipitation is heterogeneous, occurring on preexisting nuclei. The processes controlling the mineralogy and morphology of the precipitates are still not well understood, nor are the factors which control their distribution. These problems will be examined further in the next chapter as they relate to the formation of carbonate cements in shoal-water marine sediments.

Concluding remarks

Shoal-water carbonate sediments are made up of a variety of carbonate particles in a wide range of grain sizes. The particles have their origin in biotic and abiotic processes. However, on a global scale, these sediments are dominated in the modern environment by particles of skeletal origin. During the Phanerozoic shoal-water carbonate sediments, because of their organo-detrital source, are a record of the evolutionary history of biomineralization in skeletal organisms.

Many questions still exist concerning the physico-chemical and vital factors affecting biomineral composition in the modern environment. Because of these uncertainties, it is hazardous to use chemical, mineralogical, and isotopic data on fossil skeletal material without an understanding of these factors. These data, however, have proved invaluable in interpretation of environmental conditions, such as the Cenozoic temperature record (Chapter 10) derived from the $\delta^{18}O$ values of fossil foraminifera.

Oöids and aragonitic needle muds are important precipitates in today's marine carbonate environment; the former involve primarily abiotic processes, whereas the latter have both an abiotic and biotic source. Important problems concerning the formation of these allochems in the modern environment still exist, and thus their use in interpretation of past environmental conditions is somewhat limited. Oöid compositions during the Phanerozoic may have varied in response to changing global environmental conditions (see Chapter 10). Their abundance in the rock record certainly depended on the necessary environmental conditions of formation. With the advent of calcareous nannoplankton at the end of the Mesozoic and the general fall of Cenozoic sea level, oöids became less abundant allochems in shoal-water carbonate deposits. The aragonitic needle muds of biotic origin owe their existence to the evolution of the codeacean green algae in the Ordovician. Micrites prior to this time were formed principally by abiotic processes, with contributions from other sources such as the bioerosion of preexisting carbonate substrates.

In Chapter 10 the chemistry of Earth's surface environment is discussed in relation to the genesis of carbonate sediments. It is demonstrated that the composition of carbonate abiotic precipitates may have changed during geologic time because of global environmental changes in the ocean-atmosphere system. Assessment of the connection between global environmental change and carbonate composition is, in part, based on the relationships developed in this chapter.

Chapter 6

EARLY MARINE DIAGENESIS OF SHOAL-WATER CARBONATE SEDIMENTS

Introduction

Because this chapter begins our major discussion of carbonate diagenesis, we initially consider some thermodynamic and kinetic relationships. It is demonstrated that the ultimate end-point of the diagenetic processes affecting carbonate sediments is chemical stabilization of the compositionally heterogeneous mixture of carbonate particles initially deposited. That is, with increasing time (burial, temperature), carbonate particles dissolve, reprecipitate, and "recrystallize," gaining structural order and chemical purity. The product is a chemical system approaching a minimal free energy, and thus maximum entropy, state for the components $CaCO_3$-$MgCO_3$. It is not likely that such a thermodynamic total equilibrium state is ever reached for carbonate sediments because of many factors, not the least of which are time before exposure and destruction by weathering. The various diagenetic pathways followed by carbonates as they progress toward a stable assemblage of calcite and dolomite are significantly dependent on kinetic considerations, as will be discussed later in this book.

This chapter continues with a discussion of the very early diagenetic processes influencing carbonate sediments still exposed to marine waters. Emphasis is on the relative importance of the processes of carbonate mineral dissolution versus cement precipitation during the early stages of carbonate diagenesis. Chapters 7 and 8 continue with the story of carbonate diagenesis, emphasizing, respectively, early non-marine and deep subsurface diagenetic processes.

Some preliminary thermodynamic and kinetic considerations

At this junction it seems appropriate to consider some further thermodynamic and kinetic arguments that bear on carbonate diagenetic problems discussed in this chapter and in Chapters 7 and 8. Because most carbonate diagenetic reactions take place in an aqueous environment (seawater, freshwater, brines) over a range of temperature and pressure, it is rapidly becoming commonplace for fluid compositions to be analyzed or inferred. Seawater, shallowly-buried marine pore waters, and groundwaters in contact with carbonates are not difficult to analyze chemically. Subsurface brines and fluid inclusions can also be analyzed, although perhaps with more difficulty. Carbonate speciation and ion activity or concentration products of carbonate and other minerals are generally calculated from these analytical data (see Chapter 1). The calculations can provide some idea of the saturation state of the aqueous solution, and the potential of that solution to precipitate or leach mineral phases. As an illustration tool, equilibrium activity diagrams are commonly used to portray aqueous solution compositions found in a particular environment and their relationships to metastable and stable mineral phases (e.g., Garrels and Christ, 1965; Helgeson et al., 1969; Bowers et al., 1984; Nordstrom and Munoz, 1986). One such diagram that has particular interest at this point in our discussion was developed by A.B. Carpenter in 1962, and is reproduced and discussed in Garrels and Christ (1965). We will consider a modification of that diagram in some detail, because its development is important to further material in this book, and illustrates well a basic diagenetic principle -- time and water percolation through a carbonate sediment lead to a decrease in the number of carbonate phases originally present, and to a rock with a minimum number of minerals with broad ranges of stability (Garrels and Christ, 1965).

Figure 6.1 shows equilibrium relations between stable (A) and metastable (B) carbonate minerals. The boundaries between stability fields are most easily obtained by consideration of the mass action equations representing mineral compatibilities in terms of the variables shown in Figure 6.1. For example, consider the phase boundaries in Figure 6.1A; the reactions and the equilibrium constants for these boundaries in terms of the ratio of activity of Ca^{2+} to activity of Mg^{2+} and P_{CO_2} are as follows:

$$2CaCO_3 + Mg^{2+} \rightarrow CaMg(CO_3)_2 + Ca^{2+} \qquad\qquad (6.1)$$
$$\text{calcite} \qquad\qquad\qquad \text{dolomite}$$

$$K_{c\text{-}d} = \frac{a_{Ca^{2+}}}{a_{Mg^{2+}}}$$

$$\log K_{c\text{-}d} = 0.14 = \log \frac{a_{Ca^{2+}}}{a_{Mg^{2+}}}$$

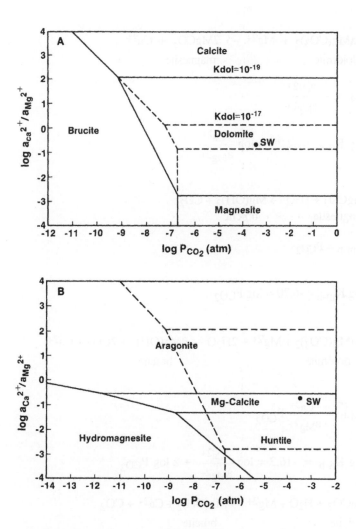

Figure 6.1. Equilibrium activity diagrams illustrating phase relations among calcium and magnesium carbonates at 25°C and one atmosphere total pressure as a function of the logarithmic variables $a_{Ca^{2+}}/a_{Mg^{2+}}$ and P_{CO_2}. (A) Equilibrium relations among stable phases showing two configurations; the expanded dolomite field is consistent with an equilibrium constant for dolomite dissolution of 10^{-19} ($\Delta G^o_{dolomite}$ = Gibbs free energy of formation of -2172.6 kJ mole^{-1}), whereas the contracted field is for a constant equal to 10^{-17} ($\Delta G^o_{dolomite}$ = -2161.7 kJ mole^{-1}). (B) Equilibrium relations among metastable phases, the magnesian calcite is of composition $Ca_{0.85}Mg_{0.15}CO_3$; dashed lines show calcite, dolomite, magnesite, and brucite relations from (A). Data used to construct diagrams from Table 6.1. SW is the point representing the composition of average seawater.

$$CaMg(CO_3)_2 + Mg^{2+} \rightarrow 2MgCO_3 + Ca^{2+} \qquad (6.2)$$
dolomite magnesite

$$K_{d-m} = \frac{a_{Ca^{2+}}}{a_{Mg^{2+}}}$$

$$\log K_{d-m} = -2.79 = \log \frac{a_{Ca^{2+}}}{a_{Mg^{2+}}}$$

$$MgCO_3 + H_2O \rightarrow Mg(OH)_2 + CO_2 \qquad (6.3)$$
magnesite brucite

$$K_{m-b} = P_{CO_2}$$

$$\log K_{m-b} = -6.70 = \log P_{CO_2}$$

$$CaMg(CO_3)_2 + Mg^{2+} + 2H_2O \rightarrow 2Mg(OH)_2 + 2CO_2 + Ca^{2+} \qquad (6.4)$$
dolomite brucite

$$K_{d-b} = \frac{a_{Ca^{2+}}}{a_{Mg^{2+}}} P_{CO_2}^2$$

$$\log K_{d-b} = -16.2 = \log \frac{a_{Ca^{2+}}}{a_{Mg^{2+}}} + 2 \log P_{CO_2}$$

$$CaCO_3 + H_2O + Mg^{2+} \rightarrow Mg(OH)_2 + Ca^{2+} + CO_2 \qquad (6.5)$$
calcite brucite

$$K_{c-b} = \frac{a_{Ca^{2+}}}{a_{Mg^{2+}}} P_{CO_2}$$

$$\log K_{c-b} = -7.07 = \log \frac{a_{Ca^{2+}}}{a_{Mg^{2+}}} + \log P_{CO_2}$$

Equations (6.1) - (6.5) give the slopes of the stability boundaries in Figure 6.1A. The equilibrium constants at 25°C and one atmosphere total pressure are calculated from the relationship:

$$\Sigma n_r \, \Delta G^0_{f,prod} - \Sigma n_r \, \Delta G^0_{f,react} = \Delta G^0_r = -5.709 \log K \qquad (6.6)$$

where $\Delta G^0_{f,react}$ and $\Delta G^0_{f,prod}$ are the standard Gibbs free energies of formation of the reactant and product minerals, respectively, n_r and n_p are, respectively, the reactant and product species stoichiometric coefficients, and ΔG^0_r is the standard Gibbs free energy of reaction. A self-consistent set of values of free energies of formation used to construct Figure 6.1 is given in Table 6.1. Figure 6.1B was constructed in a manner similar to that of 6.1A from the free energies of formation data for metastable phases given in this table.

Table 6.1. Standard Gibbs free energies of formation of minerals and aqueous species considered in development of mineral compatibility diagrams in Figure 6.1.

Mineral or Species	ΔG^0_f (kJ mole^{-1})	Source
Calcite ($CaCO_3$)	-1129.8	1
Aragonite ($CaCO_3$)	-1128.8	2
Dolomite [$CaMg(CO_3)_2$]	-2161.7 (-2172.6)	3(5)
Huntite [$CaMg_3(CO_3)_4$]	-4203.4	3
Magnesian Calcite ($Ca_{0.85}Mg_{0.15}CO_3$)*	-1114.5	4
Hydromagnesite [$Mg_5(CO_3)_4(OH)_2 \cdot 4H_2O$]	-5864.8	3
Magnesite ($MgCO_3$)	-1029	6
Brucite [$Mg(OH)_2$]	-833.5	3
Ca^{2+}	-553.5	3
Mg^{2+}	-454.8	3
CO_3^{2-}	-527.9	3
$H_2O_{(l)}$	-237.14	3
$CO_{2(g)}$	-394.4	3

(1) Plummer and Busenberg, 1982; (2) Morse et al., 1980; (3) Robie et al., 1979; (4) Bishoff et al., 1987; (5) Garrels et al., 1960; (6) Christ and Hostetler, 1970.
* Synthetic inorganic phase.

It should be emphasized that an equilibrium can be established between a metastable phase(s) and its environment. Aragonite, for example, can be precipitated from seawater at 25°C, but it is unstable at Earth-surface T and P, and can persist metastably because of kinetic reasons. This statement is illustrated by the following calculation. We can use the free energies of formation of Table 6.1, and calculate the Gibbs free energy of reaction for the mass action equation representing aragonite-calcite equilibrium:

$$CaCO_{3(aragonite)} \longleftrightarrow CaCO_{3(calcite)} \tag{6.7}$$

$$\Delta G_r = \Delta G_r^o + RT \ln \frac{a_{(calcite)}}{a_{(aragonite)}}$$

but $RT \ln \dfrac{a_{(calcite)}}{a_{(aragonite)}} = 0$ for the case of solid-solid reactions involving pure,

homogeneous phases ($a_{solid} = 1$), thus:

$$\Delta G_r = \Delta G_r^o = \Delta G_{(calcite)}^o - \Delta G_{(aragonite)}^o \tag{6.8}$$

$$= -1129.8 \text{ kJ mole}^{-1} \ (-1128.8 \text{ kJ mole}^{-1})$$

$$= -1 \text{ kJ mole}^{-1} \ (-239 \text{ calories mole}^{-1})$$

where ΔG_r is the Gibbs free energy of reaction, ΔG_r^o is the standard Gibbs free energy of reaction, $a_{(calcite)}$ and $a_{(aragonite)}$ are the activities of subscripted minerals, and $\Delta G^o{}_{(calcite)}$ and $\Delta G^o{}_{(aragonite)}$ are the Gibbs free energies of formation of subscripted minerals. The value of -1 kJ mole^{-1} indicates the reaction should proceed to the right at 25°C and one atmosphere total pressure, but the energy for conversion is not large. This small free energy difference, however, is sufficient for aragonite and calcite to coexist (Figure 6.1) in sedimentary environments for relatively long periods of time.

Another example of the "tricks" that metastability can play is that of the 15 mole % magnesian calcite, whose stability field is shown in Figure 6.1B. This phase is unstable relative to pure calcite and dolomite, but could precipitate from surface seawater of average composition. These conclusions are demonstrated by the following calculations. The reaction for dissolution of the 15 mole % magnesian calcite is written:

$$Ca_{0.85}Mg_{0.15}CO_3 \rightarrow 0.85\ Ca^{2+} + 0.15\ Mg^{2+} + CO_3^{2-} \tag{6.9}$$

and the stoichiometric saturation coefficient is:

$$K_{sat} = a_{Ca^{2+}}^{0.85}\ a_{Ca^{2+}}^{0.15}\ a_{CO_3^{2-}}$$

whose value at 25°C for the synthetic phase is $10^{-8.39}$ (Bischoff et al., 1987). In surface seawater of average composition, the ion activity product (IAP) for this mineral is:

$$IAP = a_{Ca^{2+}}^{0.85}\ a_{Mg^{2+}}^{0.15}\ a_{CO_3^{2-}}$$

$$= (10^{-2.62})^{0.85}\ (10^{-1.87})^{0.15}\ (10^{-5.25}) = 10^{-7.76} \tag{6.10}$$

Thus, surface seawater is $IAP/K_{sat} = 4.3$ times oversaturated with respect to this phase, but the magnesian calcite should spontaneously convert to nearly pure calcite plus dolomite according to:

$$Ca_{0.85}\ Mg_{0.15}\ CO_3 \rightarrow 0.7\ CaCO_3 + 0.15\ CaMg(CO_3)_2 \tag{6.11}$$

$$\Delta G_r = \Delta G°_r = 0.7\Delta G°_{calcite} + 0.15\Delta G°_{(dolomite)} - \Delta G°_{(Mg\text{-}calcite)}$$

$$= 0.7(-1129.8) + 0.15(-2172.6) - (-1114.5)$$

$$= -2.25\ kJ\ mole^{-1}\ (-538\ calories\ mole^{-1}).$$

It is the case that in surface seawater at 25°C, the ion activity products of all the minerals aragonite, magnesian calcite (15 mole %), calcite and dolomite are greater than their stoichiometric saturation constants, and therefore, these phases could precipitate from seawater. Figure 6.1 demonstrates, however, that in a solution of average seawater composition at 25°C, aragonite, magnesian calcite, and calcite are all probably metastable with respect to dolomite, and should convert to this phase. The energy associated with this metastability depends, in part, on the choice of the free energy of formation value for dolomite, whether one selects a value consistent with an equilibrium constant of dolomite dissolution of 10^{-17} (Robie et al., 1979) or 10^{-19} (Garrels et al., 1960). The value is still debated. Also, natural dolomites commonly contain $CaCO_3$ in excess of 50 mole % but rarely exceed 57 mole %. These non-stoichiometric phases have been termed protodolomites. Their stabilities certainly differ from stoichiometric, well-ordered phases. Whatever the case, it is likely that lack of conversion of carbonate phases to dolomite in the modern marine environment is a result of kinetic factors and is one of the principal observations leading to the long-standing "dolomite problem." This topic is discussed more extensively in Chapter 7.

As we have seen in the last chapter, a wide variety of calcium carbonate and calcium-magnesium carbonate minerals formed by both biologic and chemical processes is found in modern marine sediments. These phases, represented by the minerals shown in Figure 6.1, constitute a metastable assemblage; there are too many phases present for the number of components in the system. The calculations above demonstrate the metastability of aragonite with respect to calcite, and of magnesian calcite with respect to calcite plus dolomite. During geologic time when environmental conditions were appropriate, assemblages of carbonate minerals like those of Figure 6.1 were deposited, later to be transformed with time (temperature, burial) and water percolation to the stable assemblage of calcite plus dolomite. Thus, in a gross sense carbonate minerals undergo a stabilization process that is governed by thermodynamic considerations. The various diagenetic reaction pathways that carbonate minerals follow towards stabilization are to a significant degree controlled by kinetic processes, although thermodynamics still plays a role. These pathways in natural systems, and the relative role played by thermodynamics and kinetics in the processes and mechanisms leading to stabilization of carbonate minerals, are topics of this chapter and Chapters 7 and 8.

Very early diagenesis

Major Diagenetic Processes

Soon after a carbonate mineral is formed, a series of chemical, physical, and biological processes can start that may significantly alter it. These processes include dissolution and precipitation reactions, abrasion and transport, and ingestion and boring by organisms. The processes occurring after deposition fall into the category of diagenesis. Many shoal-water carbonates are repeatedly deposited, resuspended, and then deposited again during periods of elevated turbulence, which are common especially during storms. Consequently, carbonate minerals may experience several depositional environments before being permanently buried in a sediment. The study of carbonate diagenesis is further complicated by the fact that major amounts of shoal-water carbonates are commonly associated with reefs, for which traditional models of sediment diagenesis are not always appropriate. When addressing shallow-water carbonates it is therefore necessary to view diagenesis in a broad conceptual framework.

Diagenesis continues throughout the life of a sediment or sedimentary rock. It can involve early processes that began immediately after formation of a carbonate mineral and that may occur on a time scale of minutes, to processes occurring over hundreds of millions of years. We have chosen to subdivide the diagenesis of carbonates into three general areas. The first, which will be discussed in this chapter, involves processes usually occurring relatively close to the sediment-water interface and largely confined to seawater as the primary solution in the sediment pores. In Chapter 7 we will discuss diagenetic changes that may occur over relatively short geological time spans and do not involve deep burial. This stage of diagenesis is commonly associated with massive phase transformations and lithification. Diagenetic changes occurring over long time periods, and usually involving deep subsurface processes in carbonate rocks, will be the topic of Chapter 8. Although such divisions of a continuous process are largely arbitrary and many "gray" areas exist, this approach to diagenesis is a useful framework to employ as a focus for particular processes.

Before looking at aspects of very early diagenesis of particular interest for carbonate studies, it is useful to "set the stage" by briefly considering major properties of recent sediments and general processes involved in early marine diagenesis. We will not attempt to cover this large topic in a detailed treatment, as an excellent book entitled Early Diagenesis, A Theoretical Approach by Berner (1980), can be referred to by those wishing to pursue the topic further.

Soon after deposition, chemical and physical changes occur in carbonate sediments as a result of numerous diagenetic processes. Among the most important are dissolution-precipitation reactions and major alteration of pore water chemistry by bacterially-mediated oxidation of organic matter. These changes generally occur most rapidly in the upper few centimeters of sediment. Transport within the sediment of both solid and dissolved components is also of fundamental importance in diagenesis. It can be caused by physical stirring, which results from the movement of infaunal organisms (e.g., worms, bivalves, shrimp), and is called bioturbation. Closely associated with bioturbation is bioirrigation, which results from organisms pumping water through their burrows. Deeper within the sediment, advection of pore water and molecular diffusion in pore water usually become the dominant transport processes.

Since the early 1960's, much effort has gone into modeling the diagenesis of sediments. A central theme of most models is that diagenetic change results from the sum of transport and reactions of various types. Diagenetic models can be divided into those assuming constant conditions (steady-state models) and those taking into account factors that may vary with time (e.g., temperature, salinity, material input rates, water depth). The primary idea in the steady-state models is that a property in the sediment, at a given depth below the sediment-water interface, is time independent (Figure 6.2). Thus, if one were to measure the pore water alkalinity in a sediment at 20 cm depth and come back 10 years later and again measure alkalinity at the 20 cm depth it would be the same. This statement would be true even though the sediment was not physically the same at the 20 cm depth because of deposition during the intervening 10 years.

The general diagenetic equation from which many important models have been derived (as given by Berner, 1980) is:

$$\left(\frac{\partial \hat{C}_i}{\partial t}\right)_x = -\frac{\partial F_i}{\partial x} + \Sigma R_i$$

(6.12)

where $\hat{C}_i$ = concentration of solid or liquid component i in mass per unit volume of total sediments (solids plus pore water); F_i = flux of component i in terms of mass per unit area of total sediment per unit time; and R_i = rate of each diagenetic chemical, biochemical, and radiogenic reaction affecting i, in terms of mass per unit volume of total sediment per unit time. An example of application of this

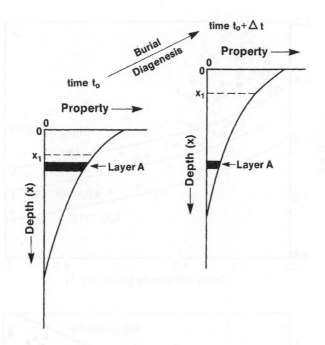

Figure 6.2. Illustration of the steady-state condition of a property in an accumulating sediment. (After Berner, 1971, 1980.)

equation would be the change in Ca^{2+} concentration, with time at a given depth in a sediment, as the result of diffusive transport and calcium carbonate dissolution.

For steady-state conditions this equation is set equal to 0, because at a given depth x, concentration does not change with time. Steady-state models are generally more amenable to mathematical solution than are non-steady-state models. Unfortunately, diagenesis in many shoal-water carbonate sediments is significantly influenced or even dominated by non-steady-state processes.

Pore Water Chemistry

The study of the chemistry of sediment pore water provides important information on chemical reactions occurring in sediments. In many cases, the reactions between pore waters and solids are not obvious from observations of the solids, but because of the large solid to solution ratio in sediments, major changes

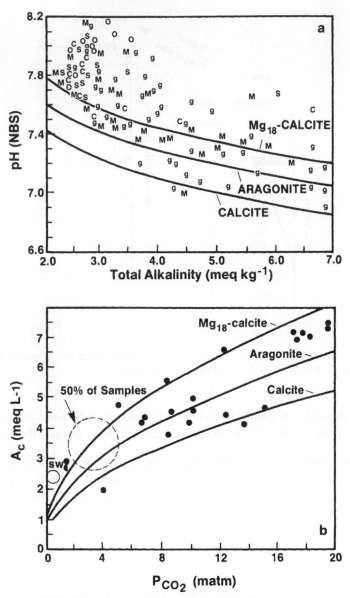

Figure 6.3. a) Plot of pore water pH versus total alkalinity for samples from different sediment types. O = oölite; M = mud; S = sand; C = coarse-grained grass bed; g = fine-grained grass bed. b) Plot of carbonate alkalinity versus P_{CO_2} for fine-grained sediments collected at depths > 6 cm below the sediment-water interface. (After Morse et al., 1985.)

can occur to the pore water during diagenesis. If an equilibration reaction alone were to take place in carbonate-rich sediments between the pore water and solids, precipitation would occur until equilibrium was reached between the pore water and a carbonate mineral. A major question is, what carbonate mineral would precipitate in a sediment containing calcite, aragonite, and a range of magnesian calcites, all of differing solubility? Furthermore, oxidation of organic matter is also an important process in these sediments. The carbon dioxide added by this process can be sufficient to cause undersaturation, resulting in dissolution not precipitation, as the dominant early reaction.

Several investigations of shallow water carbonate-rich sediments have been made to determine what factors control the calcium carbonate ion activity product. Two comprehensive studies we will focus on were made by Berner (1966) in south Florida and Bermuda, and by Morse et al. (1985) in the Bahamas. Berner's pioneering study was important in demonstrating through pore water chemistry that, in sediments exposed to normal seawater, there is little diagenetic alteration of sedimentary carbonates. Evidence of dissolution of magnesian calcites was found in pore waters from sediments in the northern part of Florida Bay where pore waters are fresher. Based on carbonate equilibrium constants available at the time, Berner concluded that most pore waters are close to equilibrium with low magnesian calcite. Calculations using new constants (see the following discussion) indicate that this state is usually not the general situation.

Morse et al. (1985) studied a number of pore waters in both coarse- and fine-grained sediments from the northern Great Bahama Bank and Little Bahama Bank. Figure 6.3a summarizes their findings for pH and total alkalinity in sediments from different environments. Lines are also given in this figure for equilibrium with calcite, aragonite, and an 18 mole % magnesian calcite, using the solubility constant of Walter and Morse (1984a). Clearly there is great variability in pore water composition, and many sediment pore waters are substantially supersaturated with respect to all these phases. Even in pore waters where such supersaturation is observed, elevated alkalinities are indicative of dissolution of carbonate from some more soluble phase because sulfate reduction is not important in these sediment pore waters. P_{CO_2} values are also high in these sediments, being typically at least 10 times the atmospheric value (Figure 6.3b). Morse et al. (1985) also found that even within a given core, calcium carbonate ion activity products were highly variable. Their results, along with saturation states for pore waters in south Florida and Bermuda recalculated from Berner's (1966) data, are summarized in Figure 6.4. Clearly, no simple equilibrium process dominates in any environment.

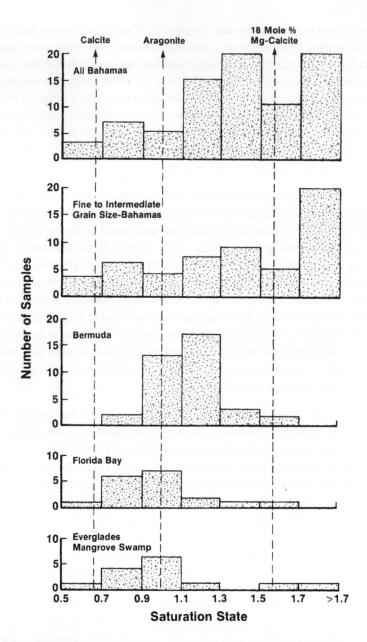

Figure 6.4. Histograms of the number of samples found within different ranges of saturation state with respect to aragonite for different shallow water calcium carbonate-rich sediments. Values for Bermuda, Florida Bay and the Everglades were calculated from the data of Berner (1966). (After Morse et al., 1985.)

However, there is a strong tendency to higher supersaturations with respect to aragonite in the Bahamian sediments, and lower saturation states in the south Florida sediments. Because aragonite is generally the dominate carbonate phase in these sediments, control of the IAP by the most abundant phase does not generally explain these observations.

Bernstein and Morse (1985) performed a number of solubility experiments on sediments from the Bahama sites at which pore waters were collected, in an attempt to understand the processes controlling the calcium carbonate ion activity product in these sediments. Equilibrium was approached from both supersaturation and undersaturation in these experiments, and for different time periods up to 50 days. Generally good agreement was found between the field and laboratory observations for fine-grained sediments, indicating at least a dynamic steady-state within the sediments between the pore water and some solid phase. Less agreement was found for coarse-grained sediments where flow of water through these sediments (e.g., oölite banks) may result in a residence time too short for a dynamic equilibrium to be reached. An interesting observation was that oöids, although composed almost entirely of aragonite, gave solubilities significantly higher than that of aragonite, in line with Weyl's original hypothesis of a magnesian calcite coating.

Both field and laboratory observations are consistent with the idea that dissolution can proceed faster than precipitation in carbonate sediments (also see Pytkowicz, 1971; Berner et al., 1978; Moulin et al., 1985; Burton and Walter, 1987), and that the pore waters reach steady-state ion activity products close to those of the most unstable phase (dissolution processes will be discussed later in this chapter). Carbonate ion may be "pumped" down to values at saturation with less soluble phases, as dissolution of the more soluble material eventually causes its removal. However, the persistence of high magnesian calcites in sediments for long periods of time indicates that this process does not involve a large amount of mass transfer under normal marine conditions.

Other factors which are important in understanding the pore water chemistry of carbonate-rich sediments include sulfate reduction, the influence of seagrass beds, and time variability of pore water chemistry. Rosenfeld (1979) found dissolved sulfate profiles in a Florida Bay sediment which were not typical of those found in fine-grained clastic sediments. In the Florida Bay sediments, the sulfate concentration decreased down to about 20 cm and then increased. Rosenfeld explained this profile in terms of mixing processes and varying rates of organic matter degradation. His results were in agreement with those of others, in finding little evidence for recrystallization of carbonates within the sediment.

Short et al. (1985) investigated the influence of seagrass beds on fine-grained sediments near San Salvador Island, and Morse et al. (1987) studied their influence on pore water chemistry in coarser-grained sediments near the Berry Islands. Their influence was less in the coarse-grained sediments, but even in these sediments, elevated P_{CO_2} and alkalinity values in sediments beneath the seagrass beds were apparent. This observation is probably the result of increased organic matter concentrations associated with roots and debris from the plants.

The major influence that seasonal temperature variation can have on pore water chemistry in carbonate-rich sediments was shown by Thorstenson and Mackenzie (1974) for sediments from Devil's Hole in Bermuda. They found that chemical gradients in the pore waters varied greatly between summer and winter down to a depth of ~1 m in the sediment. In the winter they were closer to the chemistry of the overlying water than in the summer. A major difficulty in understanding the change occurring in late summer or early fall was that there was no reasonable way to explain the change by chemical reactions, and calculations indicated that diffusion of ions was also too slow to produce the rapid and major changes which were observed. Bioturbation was also rejected by Thorstenson and Mackenzie as a possible mechanism, and they hypothesized that the change in chemistry was associated with the disappearance of the thermocline at this time of year. Penetration of thermal change into the sediment may cause mixing of water similar to that occurring in the overlying water. They also noted that sulfate reduction did not appear to have a major influence on carbonate mineral dissolution in these anoxic sediments.

Precipitation of Early Carbonate Cements

The formation of carbonate overgrowths and cements is certainly one of the most important and highly studied aspects of the behavior of carbonates in sediments and sedimentary rocks. It is largely through this process that not only carbonate-rich sediments, but also other sediments such as sands, begin to become lithified. Even during the early diagenetic stage, porosity, permeability, composition and fabric may be modified. Evidence of these changes may be found in carbonate sediments that are ancient and buried to great depths. The timing, mineralogy, and extent of cementation are all critical. In this section, we will primarily discuss the geochemical aspects of the formation of the earliest cements and overgrowths, which are generally produced near the sediment-water interface under close to normal marine conditions.

An extensive literature exists on the occurrence of early carbonate precipitates in marine sediments, where they are generally termed cements. Included in this literature are books devoted solely to carbonate cements (e.g., Bricker, 1971; Schneidermann and Harris, 1985) and numerous reviews (e.g., Milliman, 1974; Bathurst, 1974, 1975; Harris et al., 1985). Many investigations have been largely descriptive in nature, focusing primarily on the distribution, mineralogy, and morphology of the cements. Here we will briefly summarize the major aspects of these observations, and we will concentrate on the chemical aspects of the formation of these precipitates.

Carbonate cements in calcareous sediments fall into three major groups. The most common are those that occur in voids found in biogenic carbonates. This group is especially important in reefs, where it can represent a major portion of the total carbonate deposited (e.g., Land and Goreau, 1970; Ginsburg and Schroeder, 1973; Lighty, 1985). The common pattern of reef cementation is presented in Figure 6.5. Such cementation in reefs can lead to well-indurated structures in the subsurface of reefs currently growing in the modern marine environment (Ginsburg and Schroeder, 1973). Cements also occur on the exterior of carbonate particles where their intergrowth can cause formation of hardened pellets, grapestones, crusts, hardgrounds, and beachrock. The possible relation of this type of cementation to oöid formation has been discussed earlier (e.g., see also Fabricius, 1977). Micritic cements associated with boring algae comprise the third common type of cement. Photomicrographs of common forms of carbonate cements are presented in Figure 6.6, and examples of shallow marine cementation are presented in Table 6.2.

One of the most interesting questions about carbonate cements in the marine environment is why they are not more abundant. In carbonate-rich sediments there are abundant carbonate surfaces for overgrowths to form on. Also, both overlying seawater and, as previously discussed, many pore waters are supersaturated with respect to aragonite, calcite, and some magnesian calcites. The fact that precipitation of cements from pore waters of carbonate sediments is severely inhibited was directly demonstrated by Morse and Mucci (1984). They buried Iceland spar calcite crystals in a variety of carbonate sediments in the Bahamas for several months. After recovery and analysis using very sensitive depth-profiling Auger electron spectroscopy, they were able to detect overgrowths on only a few crystals. The calculated growth rates of the precipitates were much less than observed in laboratory experiments using seawater at the same supersaturation state. At least part of the answer to these observations must be that precipitation is

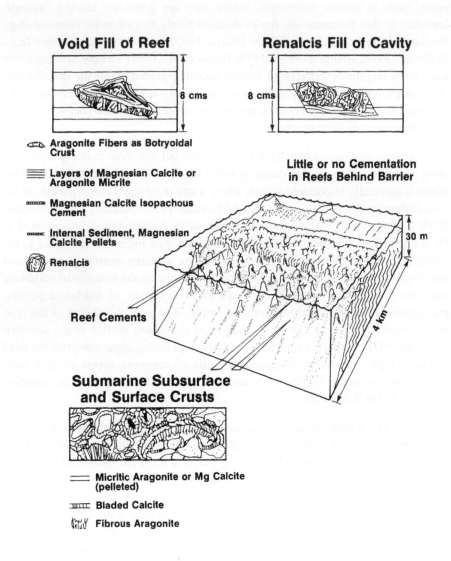

Figure 6.5. Examples of reef barrier cementation. Void filling cements are believed to be formed by direct chemical precipitation, while renalcis filling is assumed to be produced by organisms. (After Harris et al., 1985.)

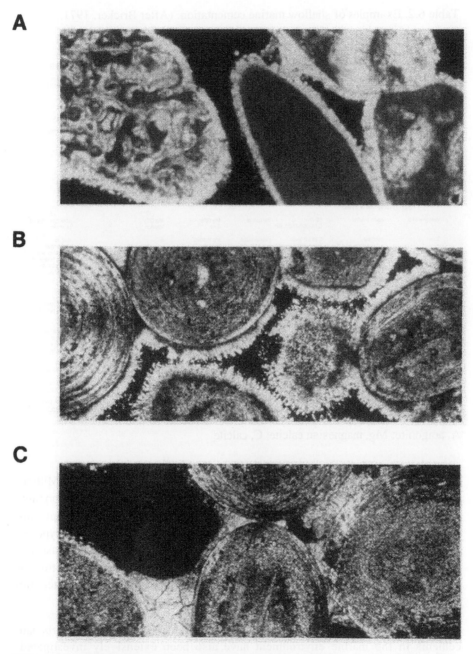

Figure 6.6. Photomicrographs of common marine cements. A) Whalebone Bay, Bermuda, 3m water depth. B) Beachrock from the Bahamas. C) Subaerial Holocene from the Bahamas. (Photomicrographs courtesy of R. Perkins.)

Table 6.2. Examples of shallow marine cementation. (After Bricker, 1971.)

Author	Environment	Evidence	Area	Micrite	Cements Fibrous	Granular drusy	Remarks
Friedman	shallow marine	79 m	N. Jersey	A	A		cement promoted
Ginsburg	subtidal	sed structures reefs, isotopes 3100 BP	Bermuda		Mg(16)		
Land	reef	75 feet	Jamaica	Mg	Mg(18)		reef talus
		C14 C&O isotopes					cemented
Hoskins	reef rampant	0 to 10 feet	Alacran		A?		
Pingitore	subaerial exposure Pleistocene reefs	?	Barbados		A?		
Glover, Pray	subtidal intraparticle	4 to 10 feet	Bahamas		Mg (15-19) A		shots shots
Roberts, Moore	subtidal (no other source of grains)	10 to 12 feet	Grand Cayman	A (partly algal boring)	A		
McIntyre et al.	open marine	60 feet C14-500-1500 BP	Barbados	Mg (20) rims	Mg (20) interior		stronger around burrows
Shinn	subtidal	5 to 90 feet 2000 BP to recent diagnostic structures C&O isotopes	Persian Gulf	Mg in mud A (Mg) mud 15% Mg	A		due to carbonate saturated sea water, related to grain size and rate of ppt.
Milliman	seamounts steep slopes supersaline	fossils O&C isotopes	everywhere except N. Pacific Red Sea, Mediterranean world	Mg (10-14) A Mg (12) in veins	A (some)	Mg(10-14)	
	volcanic seamounts			?		Ca	filling cavities, voids, pelletal matrix
McIntyre, Milliman	outer shelf	oolites, algal	SE U.S.		A		fine grained pelleted probably first a beachrock
	13,500-15,000 BP	calcarenite		Mg	Mg (12)		
							when sealevel lower
Friedman	1,300-1,700 m		Red Sea	A prismatic			indurated layers
Marlowe	340-380 M		Caribbean	minor in pellets	Mg (11-14) rhomb		minor A crystals of 2-5 microns

A, aragonite; Mg, magnesian calcite; C, calcite

severely inhibited by organics such as humic acids (Berner et al., 1978). Mitterer and his associates (e.g., Mitterer and Cunningham, 1985) have explored the possible role of organic matter in cement formation. Their general concepts are summarized in Figure 6.7. These authors suggested that, whereas some types of organic matter inhibit precipitation, other types, particularly those rich in aspartic acid, may favor precipitation by complexing calcium. Thus, inhibition of precipitation, coupled with slow transfer of fresh supersaturated seawater into sediment pores, seems to account for lack of extensive early cementation.

Factors controlling the mineralogy and chemical composition of carbonate cements in the marine environment have also been extensively investigated. Bathurst (e.g., 1975; 1987) summarized many of the observations (e.g., Glover and

Organic Matter Interaction with CaCO$_3$

TOC	O$_2$ Level	OM Type	Effect of OM	Typical Environments
High	Anoxic	Hydrophobic (lipid-rich)	Complexing of Ca by DOM; No ppt.	Pore waters in fine-grained sediments
Moderate			Lipid coating Isolation of grain No ppt.	Open marine basin
Low		Hydrophilic	Adsorption of OM and complexing of Ca Possible ppt. of CaCO$_3$ (epitaxy?)	Shallow marine grainstone and boundstone
Absent	Oxic		None: ppt. follows inorganic kinetics	Vadose

Increasing Oxygen (arrow pointing down, O$_2$ Level column)
Increasing Degree of Oxidation (arrow pointing down, OM Type column)

Figure 6.7. A model for the interaction of organic matter (OM) with calcium carbonate. Variables are total organic carbon (TOC), dissolved oxygen, and composition of organic matter. (After Mitterer et al., 1985.)

Pray, 1971) indicating that in some cases the host carbonate mineral is a factor in determining composition. Usually aragonite grains have aragonite overgrowths, whereas high magnesian calcite grains have high magnesian calcite overgrowths of similar Mg content. In some instances, syntaxial formation of cement is observed. Although this explanation of host control is a "comfortable" one, it does not explain the common occurrence of cements of mixed mineralogy or those that differ from their host grains.

Environmental factors also have been found to correlate loosely with cement formation, mineralogy and morphology. The energy (waves, currents, etc.) of the environment and rate of sedimentation are most often cited as being important in shallow water environments. The energy of the environment is important in supplying seawater, from which the carbonate cements can be derived. It clearly plays an important role in reefs, as shown by the work of Lighty (1985) in Figure 6.8. Because of the difficulties inherent in obtaining pore waters from inside reefs,

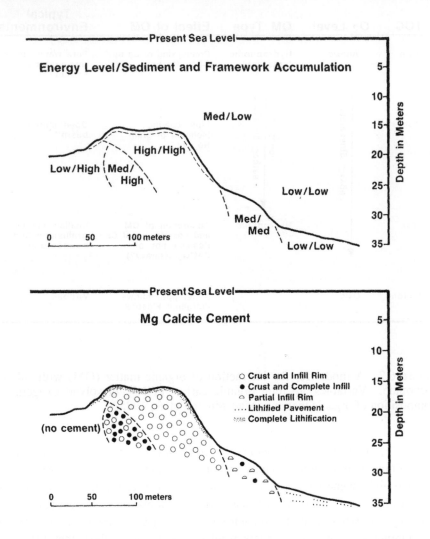

Figure 6.8. Schematic representation of energy environments and cements in a reef. (After Lighty, 1985.)

the chemical environment internal to reef structures has gone unstudied until recently. The studies by Sansone and associates (Sansone, 1985; Sansone et al., 1988a,b, 1990; Tribble, 1990), however, shed some light on the mode of origin of cement types shown in Figure 6.5, and the relationships between reef environments and cements illustrated in Figure 6.8. These studies of Checker Reef, Oahu, Hawaii (a lagoonal patch reef) and Davies Reef, Great Barrier Reef, Australia (a platform reef), show that microbial-mediated processes of oxic respiration, sulfate reduction, and to a lesser extent, nitrate reduction and methanogenesis are important in these reef structures.

The interstitial waters of these reefs are highly depleted in oxygen, exhibit lower pH values, and elevated concentrations of dissolved methane, sulfide, ammonium, phosphate, and silica relative to the external seawater surrounding the reef. In Checker Reef there is a gradient in the degree of alteration of interstitial waters from the margin to the center of the reef, but in Davies Reef, the most altered pore waters are at the margin of the reef. This difference between reefs most likely reflects framework structural differences, and differences in wave characteristics and tidal range in the two environments (Sansone et al., 1990). Anaerobic reactions in reef interstitial waters may not progress far if reef structures are open and well flushed. If, however, the systems are nearly closed, little fresh reactant will enter via seawater exchange, and mass transfer will be limited by the reactants trapped in the reef interstitial waters. This should be word enough to the wise that different reef systems will exhibit substantially different degrees of diagenesis in marine waters. Some may be extensively cemented whereas others may exhibit scant cement development.

As an example of the nature of interstitial water chemistry in reef frameworks, Figure 6.9 illustrates the mean concentration of dissolved constituents across Checker Reef. The interstitial waters of this reef are oversaturated with respect to calcite, near saturation with respect to aragonite (Figure 6.10), and elevated in dissolved calcium relative to those in seawater external to the reef. This suggests that dissolution of carbonates is occurring in reef interstitial waters because of oxic respiration and anaerobic decomposition of organic matter; the latter process, in particular, leads to the concentration gradients shown in Figure 6.9. The carbonate reactions involved may be the dissolution of high-magnesian calcite:

$$Ca_{1-x}Mg_xCO_3 + CO_2 + H_2O \rightarrow (1-x)\ Ca^{2+} + xMg^{2+} + 2HCO_3^- \qquad (6.13)$$

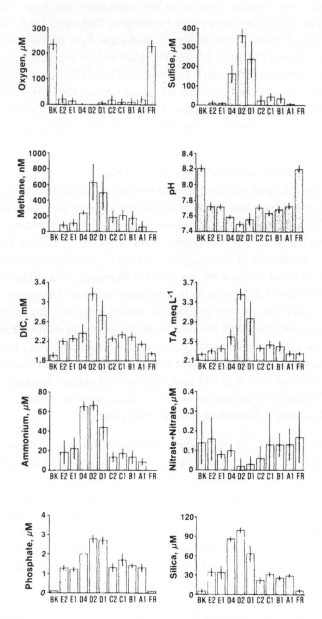

Figure 6.9. Mean concentrations of dissolved species in Checker Reef pore waters and overlying seawater. Vertical bars indicate standard deviations. "FR" and "BK" refer to seawater samples from the fore and back reef stations. Other symbols refer to 1-and-2 meter deep sampling stations across the reef (D is the midpoint of the reef). (After Sansone et al., 1990.)

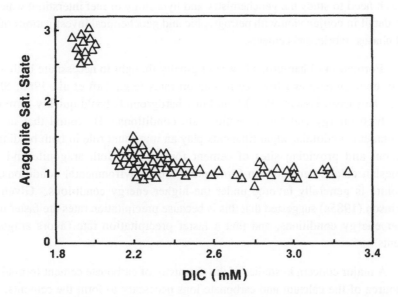

Figure 6.10. Aragonite saturation state versus DIC (dissolved inorganic carbon = ΣCO_2) in Checker Reef porewaters (lower points) as opposed to overlying seawater (upper points). (After Tribble, 1990.)

and precipitation of aragonite (or another magnesian calcite composition of solubility equivalent to aragonite)

$$Ca^{2+} + 2HCO_3^- \rightarrow CaCO_3 + CO_2 + H_2O \qquad (6.14)$$

The net of reaction (6.13) and (6.14) can result in addition of some dissolved calcium to Checker Reef interstitial water, and possible deposition of cement in the pores of the reef structure.

An important conclusion of Sansone's studies was that thermodynamic disequilibrium, among dissolved species like CH_4 and SO_4^{2-}, implies microzonation of chemical reactions. This microzonation resulting in slight differences in reef interstitial water compositions, as well as temporal changes in water composition, induced by a variety of factors including seasonal variables, may account for the co-existence of different cement mineralogies in reef structures, as well as the zonation of cements shown in Figures 6.5 and 6.8. Whatever the case, there is a strong research need to study the geochemistry and hydrology of reef interstitial waters in more detail in conjunction with petrographic and geochemical investigations of reef solid phases, fabric, and cements.

Formation of hardgrounds was originally thought to necessitate both a low energy environment and low sedimentation rates (e.g., Taft et al., 1968; Shinn, 1969). However, Dravis (1979) found that hardgrounds could quickly form even under high energy and sedimentation rate conditions. He found that in such environments endolithic algal filaments play an important role in both binding the sediment and providing sites of cement formation. Both aragonite and high magnesian calcite cements can form under most environmental conditions, but aragonite is generally favored under the higher energy conditions. Given and Wilkinson (1985a) suggested that this is because precipitation rates are faster under higher energy conditions, and that a faster precipitation rate favors aragonitic cements.

A major concern in studies of the chemistry of carbonate cement formation is the source of the calcium and carbonate ions necessary to form the cements. The obvious source is seawater, but large volumes of seawater are necessary if significant amounts of cement are to be produced. Cement formation is consequently favored near the sediment-water interface and in high-energy environments where water can be flushed through porous structures such as reefs. The observation that cements usually form only in thin crusts near the sediment-water interface also demonstrates the importance of normal seawater for cement precipitation. Further evidence for cement formation obtaining in normal seawater comes from stable isotopes. The $\delta^{13}C$ values of cements are usually close to those predicted for carbonates precipitating from seawater (e.g., see Given and Wilkinson, 1985b). Another possible source of the ions necessary for cement formation is from the dissolution of carbonate phases more soluble than the cements. High magnesian calcites could provide such a source, as has been demonstrated in periplatform oozes in the Bahamas (Mullins et al., 1985), and for coastal carbonate sediments of the Bay of Calvi, Corsica (Moulin et al., 1985). This process, leading to cement precipitation, is consistent with the previously discussed chemistry of many pore waters.

The precipitation of calcium carbonate (usually, but not exclusively, from petrographic evidence in the form of high-magnesian calcite, e.g., Alexandersson and Milliman, 1981) is generally inferred from a decrease in dissolved pore water calcium (e.g., Thorstenson and Mackenzie, 1974; Aller et al., 1986; Gaillard et al., 1986). Most studies of the impact of chemical diagenesis on the carbonate chemistry of anoxic sediments have focused primarily on the fact that sulfate reduction results in the production of alkalinity, which can cause precipitation of carbonate minerals (e.g., Berner, 1971). However, this process is complex and may also contribute to carbonate dissolution. It is discussed later in this chapter. Carbonate precipitation can also occur via methane oxidation instead of organic matter oxidation, resulting in distinctly "light" $\delta^{13}C$ values for authigenic carbonates (e.g., Ritger et al., 1987). These anaerobic processes also have been found to be important for the formation of cements in reefs (e.g., Sansone, 1985; Sansone et al., 1988a,b; Pigott and Land, 1986). The behavior of the pore water-solid carbonate system is, as previously discussed, complicated by the fact that in some sediments highly supersaturated pore waters can result, because of the inhibitory action of other components such as phosphate, organic matter (Berner et al., 1978), Mn and Fe (Dromgoole and Walter, 1987). Even in sediments from the same general area (e.g., Long Island Sound), examples of cores in which high supersaturation occurs, and cores where close to equilibrium conditions are reached with respect to calcite and aragonite can be found (Aller, 1982). The reasons for these differences are not well known.

Given the above considerations and the fact that waters overlying most shelf and upper slope sediments are substantially supersaturated with respect to calcite and aragonite, little or no dissolution of carbonate minerals generally is thought to occur in these environments (e.g., Broecker and Takahashi, 1977). However, as will be discussed later in this chapter, this assumption may be incorrect.

Attention has also been given to chemical controls on the cement phase that forms, and the Mg content of magnesian calcite cements (e.g., Given and Wilkinson, 1985a,b). Much of this attention has focused on the failure of laboratory coprecipitation studies to produce magnesian calcites, under conditions that are reasonable for sediment-interstitial water systems, with Mg contents as high as those observed in common shallow water magnesian calcite cements. Given and Wilkinson (1985a,b) argued that the Mg content of calcitic cements is kinetically controlled. However, Morse (1985) pointed out that over a 12 order of magnitude range in reaction rates, a close to constant ~8 mole percent magnesian calcite is produced from normal seawater. The influence of reaction kinetics appears to become important only at supersaturations greater than 100 with respect to calcite (see Chapter 3), which is unreasonably high for natural conditions. Although many hypotheses have been put forth to explain the compositional differences between synthetic and natural magnesian calcite precipitates, the problem remains unresolved. Dolomite cement is relatively rare in most shallow water marine environments where it is, as is the case for deep water sediments, generally associated with reducing environments (e.g., Baker and Burns, 1985;

Mitchell et al., 1987). A fundamental problem in resolving the chemical controls on cement formation is that these precipitates usually form in microenvironments from which it is difficult, if not impossible, to get good chemical data. Also, cement formation may be episodic, and even if such environments could be satisfactorily sampled, the observed chemistry may not be what was present at the time of cementation.

Finally, it should be noted that while most investigators have rejected the idea that carbonate cements are directly or even closely linked to biological mechanisms of formation, some investigators hold the position that organisms may be very important to cement formation processes. Marine peloids have received considerable attention in this regard because of their close association with bacterial clumps (e.g., Chafetz, 1986). In addition, Fabricius (1977) has presented a lengthy discussion in favor of the formation of grapestone cements and oöids by algae.

Dissolution of Carbonates

Continental shelves and slopes comprise approximately 10 percent of the Earth's surface, and contain over half the sediments in the ocean (Heezen and Tharp, 1965; Gregor, 1985). Recent estimates of marine carbonate burial rates (e.g., Hay and Southam, 1977; Sundquist, 1985) indicate that between about 35 to 70 percent of Holocene carbonate deposition has taken place on continental shelves. In spite of their importance for carbonate accumulation and the global CO_2 cycle, relatively few studies have been made on the chemical controls of calcium carbonate accumulation in these sediments, and most of these studies have been confined to near-shore environments.

Many studies of the impact of chemical diagenesis on the carbonate chemistry of anoxic sediments have focused primarily on the fact that sulfate reduction results in the production of alkalinity, which can cause precipitation of carbonate minerals (e.g., Berner, 1971). Although the many reactions involved are complex, this process can be schematically represented as:

$$1/53(CH_2O)_{106}(NH_3)_{16}H_3PO_4 + SO_4^{2-} \rightarrow$$

$$2HCO_3^- + HS^- + 16/53\, NH_3 + 1/53 H_3PO_4 + H^+ \tag{6.15}$$

which results in the commonly observed decrease in sulfate, and pH, and increase in alkalinity, hydrogen sulfide, phosphate, ammonia and total inorganic CO_2 in sediment pore waters. However, during the early stages of sulfate reduction (~2 to 35%), this reaction may not cause precipitation, but rather results in

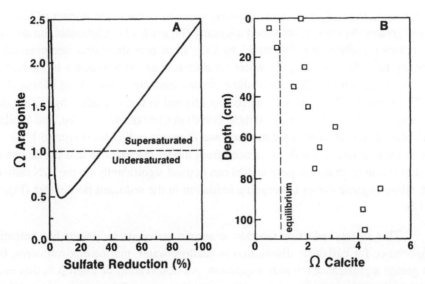

Figure 6.11. A. Saturation state of seawater with respect to aragonite as a function of sulfate reduction. Based on general model of Ben-Yaakov (1973), updated by using pK*$_a$ values and total ion activity coefficients of Millero (1982). B. Observed saturation state of Mangrove Lake, Bermuda pore waters with respect to calcite.

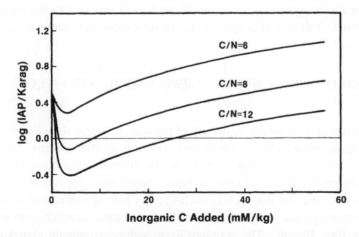

Figure 6.12. Calculated saturation states with respect to aragonite of marine pore waters undergoing anoxic diagenesis as a function of the original C/N ratio of labile organic matter. The extent of diagenesis is represented by the organic C oxidized in the sediment-pore water system, shown as inorganic carbon added to pore waters.

undersaturation with respect to carbonate minerals because the impact of lowered pH is greater than that of increased alkalinity (Figure 6.11). Carbonate ion activity decreases rapidly as it is "titrated " by CO_2 from organic matter decomposition leading to a decrease in pore water saturation state. This process is evident in observational data for Fe-poor shallow water carbonate sediments of Morse et al. (1985) from the Bahamas and has been confirmed in recent studies by Walter and Burton (1987) for Florida Bay, Tribble (1990) in Checker Reef, Oahu, and Wollast and Mackenzie (unpublished data) for some Bermuda sediments (Figure 6.11B). In the latter example it was demonstrated from model calculations that the degree of undersaturation in anoxic pore waters can depend significantly on the C/N ratio of the labile organic matter undergoing oxidation in the sediment pore water (Figure 6.12).

This sulfate reduction reaction in anoxic carbonate sediments has potential importance for carbonate dissolution in shallow-water, marine environments, but its global significance remains a question. An observation of interest is that even complete sulfate reduction returns the saturation state of the water to only about half its original value. Thus the sulfate reduction reaction by itself may not promote carbonate precipitation and partial sulfate reduction may result in carbonate dissolution.

The combination of the sulfate reduction reaction, and reactions primarily with iron and manganese oxide minerals can lead to significant calcium carbonate precipitation. This type of net process can be represented schematically as:

$$9CH_2O + 4SO_4^{2-} + 4FeOOH \rightarrow 4FeS_{(s)} + 9HCO_3^- + H^+ + 6H_2O \qquad (6.16)$$

Reaction 6.16 results in the pH of the waters, in the absence of carbonate precipitation, being buffered at higher pH values (e.g., Ben-Yaakov, 1973). It is, therefore, reasonable to expect that the effectiveness of sulfate reduction in producing carbonate dissolution or precipitation may depend in part on the availability of reactive iron. This conclusion has been demonstrated in a study of the pore water geochemistry of aluminosilicate and carbonate-rich sediments from Kaneohe Bay, Hawaii. The aluminosilicate sediments contain abundant pyrite whereas the pyrite content of the carbonate sediments in this bay is low. Pore waters collected from the aluminosilicate sediments have higher pH values than those collected from the carbonate-rich sediments. This observation is a result of the pH values in the pore waters of the aluminosilicate sediments being buffered at higher values than those for carbonate sediments because of reaction (6.16) (Ristvet, 1978; Mackenzie et al., 1981).

There are three primary processes that have been identified that can cause undersaturation, in addition to the previously discussed undersaturation that may result during the early stages of sulfate reduction. These are early post-death microenvironments within organisms, oxidation of organic matter by processes preceding sulfate reduction, and oxidation of sulfides. These processes commonly will be most important near the sediment-water interface.

Early studies by Hecht (1933) showed that dissolution of carbonate from mollusc shells can begin immediately after death. In laboratory studies he found shell weight losses of as much as 25% in two weeks. Other examples of this type of behavior come from the study of Berner (1969), who found calcium carbonate dissolution during bacterial decomposition of two types of clams, possibly as a result of the production of organic acids, in addition to CO_2, during early degradation of organic matter. Birnbaum and Wiremen (1984) have also pointed out that a proton gradient can be established across the cell wall of *Desulfovibrio desulfuricans* resulting in calcium carbonate dissolution in the nearby microenvironment.

Prior to the onset of sulfate reduction extensive organic matter degradation can occur by bacterially-mediated oxygen reduction (see Chapter 4). The influence of benthic bacterial activity under aerobic conditions on carbonate mineral dissolution has been nicely demonstrated by Moulin et al. (1985) for pore waters from sediments of the Gulf of Calvi, Corsica. Under aerobic conditions, the oxidation of organic matter may be written:

$$CH_2O + O_2 \rightarrow CO_2 + H_2O \rightarrow H_2CO_3 \tag{6.17}$$

and if carbonate minerals were to dissolve because of the CO_2 produced by organic matter decomposition, then the reaction:

$$CaCO_3 + H_2CO_3 \rightarrow Ca^{2+} + 2HCO_3^- \tag{6.18}$$

may also occur. Reaction (6.18) is possible only if the pore waters are undersaturated with a carbonate phase, and there are no other competing reactions, like significant ammonia release from decaying organic matter, to increase pore water saturation state. Figure 6.12 shows the increase in alkalinity versus increase in total CO_2 observed in shallowly buried pore waters in the Gulf of Calvi sediments. The slope of the regression line for this relationship is 1.04. This slope indicates that reaction 6.18 gives rise to the increased total alkalinity observed in these pore waters; that is, the H_2CO_3 produced by organic matter oxidation is stoichiometrically utilized in dissolution of a carbonate mineral phase in the

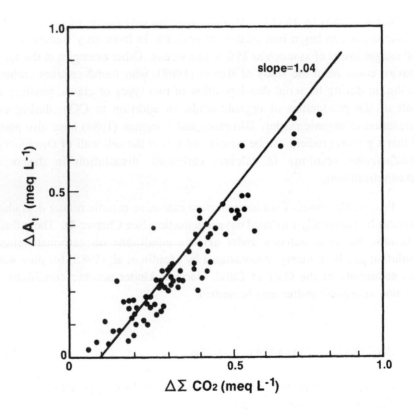

Figure 6.13. Increase in total alkalinity and total CO_2 over that observed in overlying waters, Gulf of Calvi, Corsica. (After Moulin et al., 1985).

sediment. If the dissolution were simply related to undersaturation with carbonate minerals, the reaction:

$$CaCO_3 \rightarrow Ca^{2+} + CO_3^{2-} \qquad (6.19)$$

would lead to an increase of alkalinity in Figure 6.13 of twice that of the total CO_2 increase.

Close inspection of the regression line in Figure 6.13 shows that it does not pass through the origin. Moulin et al. (1985) conclude that about 0.1 mmole L^{-1} of H_2CO_3 has been added to the pore waters from aerobic bacterial respiration prior to carbonate dissolution. This initial input of H_2CO_3 lowers the saturation state of the shallowly buried pore waters to that required for carbonate mineral dissolution. The authors have confirmed these pore water observations in experiments using a biological reactor, and from solid phase mineralogical studies. They have shown that high magnesian calcites are dissolved preferentially in these sediments, although the sediment contains a mixture of nearly pure calcite, aragonite, and magnesian calcites with up to 20 mole % $MgCO_3$. In deposits composed principally of the red alga *Lithothamnium* sp., this early diagenetic reaction has resulted in dissolution of 75% of the carbonate initially deposited. There is a strong need for more pore water and solid phase work of this nature to document the global significance of this aerobic carbonate dissolution process. It is possible that this is of more importance during early diagenesis of carbonate sediments than dissolution under anaerobic conditions, i.e., sulfate reduction.

Other reactions of probable less importance than those above leading to undersaturated conditions with respect to calcium carbonate near the sediment-water interface include nitrate reduction and fermentation (e.g., Aller, 1980). Such reactions may also be important near the sediment-water interface of continental shelf and slope sediments, where bioturbation and bioirrigation can result in enhanced transport of reactants. Generally, as water depth increases over continental slope sediments, the depth within the sediment at which significant sulfate reduction commences also increases. It is probable that the influence of reactions other than sulfate reduction on carbonate chemistry may increase with increasing water depth.

One major paper attacking the problem of the relationship between the preservation of calcium carbonate in shallow anoxic marine sediments and their chemistry was by Aller (1982). The study was conducted at sites in Long Island Sound. The calcium carbonate content of the sediments decreased with increasing water depth. At the shallow FOAM (Friends of Anaerobic Muds) site shell layers associated with storms resulted in irregular variations in the carbonate content of the sediment. Ca^{2+} loss from the pore waters, indicative of calcium carbonate precipitation, was found only at the FOAM site below ~20 cm depth. During the winter, elevated Ca^{2+} to Cl^- ratios were observed near the sediment-water interface at the FOAM site down to ~10 cm and at the other sites. The potential for dissolution was confirmed by saturation state calculations, which indicated that pore waters in the upper few centimeters of the sediments were undersaturated with respect to aragonite and possibly calcite. The extent of undersaturation increased with increasing water depth. It was hypothesized that dissolution may be occurring in the summer as well but that it is obscured by higher bioturbation rates. Aller modeled dissolution rates based on the pore water saturation profiles and concluded that the required dissolution could be accomplished through oxidation of only 2-14% of the yearly carbon supply.

The rates of sulfate reduction at the different sites were similar, but concentrations of reactant products varied owing to differences in exchange rates, which were controlled by the extent of benthic community development. More highly developed infaunal communities occurred at the deeper sites. The aggregation of shells into layers by storms also helped preserve shell material in the shallow FOAM site.

A major contribution of this paper was pointing out the importance of bioturbation and bioirrigation on chemical processes associated with carbonate dissolution. In the movement of sulfidic sediment from depth to near the interface by biological processes, oxidation of the sediment produces sulfuric acid which ends up titrating alkalinity, lowering pH, and thus lowers saturation state (e.g., Berner and Westrich, 1985). Actually this process is very complex, involving many reactive intermediate compounds such as sulfite, thiosulfate, polythionates, etc. Aller and Rude (1988) demonstrated an additional complication to this process. Mn oxides may oxidize iron sulfides by a bacterial pathway that causes the saturation state of the solution to rise with respect to carbonate minerals, rather than decrease as is the case when oxidation takes place with oxygen.

Burrows, and transport of solute in them, may contribute to dissolution by enhancing oxic degradation of organic matter near the burrow walls. However, the situation is complex, and depending on factors such as the type of burrow wall produced, cementation rather than dissolution of carbonates may be promoted. Aller's observation that the best carbonate preservation takes place in the most physically disturbed and biologically underdeveloped environments points to the need for studies of continental shelf and slope environments where carbonate dissolution could be even more intense than that observed at the sites studied in Long Island Sound.

Reaves (1986) described another study specifically aimed at relating calcium carbonate preservation to chemical diagenesis in sediments. He compared *Mercinaria* shells and pore water chemistry in intertidal mud flat and tidal creek samples from Connecticut and Georgia. The shells from Georgia showed little sign of dissolution, while even shells of living organisms from Connecticut exhibited extensive dissolution. He concluded that the more variable climatic conditions and reactive organic matter at the Connecticut site were responsible for the major differences in carbonate preservation. During the summer the rate of organic matter supply was high, resulting in formation of iron sulfides near the sediment-water interface. In the winter when the organic supply was substantially less, conditions became less reducing in the upper part of the sediment, and significantly undersaturated conditions resulted from oxidation of iron sulfides. He discounted bioturbation and bioirrigation as relatively unimportant processes at these sites.

Studies such as those of Berger and Soutar (1970) and Sholkovitz (1973) also point to the importance of chemical parameters in controlling calcium carbonate preservation. These authors noted that carbonate preservation is substantially greater in the sulfidic Santa Barbara basin sediments than in adjacent slope sediments that are overlain by oxic waters. This observation probably results from the fact that oxic degradation of organic matter and oxidation of sulfides are not likely to occur in this anoxic basin.

Concluding remarks

In contrast to modern pelagic carbonate sediments, which are bimineralogic, shoal-water sediments are compositionally more complex. They are dominated by Sr-bearing aragonite and Mg-rich calcite. The vast majority of modern shoal-water sediments are skeletal in nature, but commonly contain inorganic precipitates of oöids, cements, and needle muds. The alteration of these sediments begins early in their history. During transportation and in the early stages of accumulation, very soluble phases may be dissolved, and cements of aragonite and magnesian calcite precipitated in the pores of sediments. These processes modify carbonate sediment composition and result in the loss of information concerning original sediment chemistry, mineralogy, and biotic composition. Also, sediment characteristics such as porosity, permeability, and fabric are changed during these early diagenetic stages. These modifications will continue to occur during the later stages in diagenesis, but it should be remembered that the compositionally-complex and metastable assemblage of shoal-water sedimentary materials begins to undergo transformations very early in the history of the sediment. Future diagenetic pathways of a sediment will be determined to some extent by these early diagenetic transformations. Early cements or dissolution reactions may lead to infilling of pore space or development of secondary porosity, respectively. In contrast to aluminosilicate sediments, shoal-water carbonates, because of their greater solubility and chemical reaction rates, undergo major diagenetic change before these less reactive sediments, and these changes continue as the sediment is buried and subsequently uplifted. These diagenetic changes accompanying exposure of carbonate sediments and burial are the subjects of the following two chapters.

Chapter 7

EARLY NON-MARINE DIAGENESIS
OF SEDIMENTARY CARBONATES

Introduction

This chapter initiates a departure in the methodology employed to investigate problems in carbonate geochemistry. The departure stems from the fact that carbonate sediments age, and the geologic history of the carbonate rock becomes an important variable influencing the chemical, mineralogical, and fabric changes it undergoes. In the modern environment, access to little-modified marine waters and biotic and abiotic carbonate precipitates is relatively easy. Geologists can snorkel or scuba dive in modern shoal-water environments to view processes and collect samples by hand-coring, or other methods, of shallowly-buried sediments and interstitial waters. The same phases can be recovered in deeper waters by sophisticated coring methods and for pore waters by remote controlled *in situ* samplers. Observations and collections also can be made using remote observational vehicles (ROVS) or submersibles. The hydrogeochemical realm of this shallow skin of carbonate sediments is amenable, in many cases, to direct observation and experimentation, for the time scale of early chemical diagenesis (Chapter 6).

In the deeper stratigraphic column of the modern oceans and for rocks on land, interpretation of diagenetic pathways is more difficult than for shallowly-buried carbonate materials. The geologic history of the carbonate sediment becomes an important variable. Temporal changes in the burial parameters of pressure and temperature, and of geothermal gradient, interstitial fluid composition, and migration paths, and tectonic movement must be evaluated to decipher diagenetic pathways. Because experiments in carbonate geochemistry are short-lived, it is virtually impossible to simulate the many processes leading to transformation of a carbonate sediment to a rock; processes which generally occur over thousands to millions of years, or longer.

A few experiments have been successfully performed at low temperatures to simulate carbonate diagenetic processes; for example, cements have been precipitated on skeletal carbonate sands in experimental reaction chambers designed to mimic vadose and phreatic meteoric cementation (Thorstenson et al., 1972; Badiozamani et al., 1977). These cements are remarkably similar in composition and morphology to those found in rocks cemented in the meteoric environment. However, other experimentation to simulate, for example, low temperature dolomitization of skeletal carbonates (Land, 1966, 1967) or replacement of aragonite by calcite (e.g., Fyfe and Bischoff, 1965; Land, 1966; Taft, 1967) has been performed at temperatures elevated significantly above 25°C to obtain reasonable rates of transformation in time periods amenable to observation. In general, increased temperatures and/or unusual fluid compositions have been employed experimentally to evaluate carbonate diagenetic reactions. It is conceivable that this "manipulation" results in different mechanisms of transformation than those that operate in nature and their utility in understanding controls on carbonate diagenesis is, consequently, dubious at best.

We are in a quandary. In general, it has been difficult to design experiments to test mechanisms and rates of diagenetic processes affecting carbonate sediments. Thus, unlike the experimental chemical petrologists dealing with high temperature reactions pertaining to igneous and metamorphic rock genesis, sedimentary geologists do not have the necessary experimental data to constrain models of carbonate diagenetic processes. We must rely on quantitative petrographic, chemical, mineralogical, and isotopic data derived from a sediment at an instant in time, and inferences, based on stratigraphy, sedimentology, and structural setting, of the burial history of the carbonate sequence to obtain a model of carbonate diagenesis. The model requires application of the thermodynamic and kinetic approaches presented earlier in this book. The former require calculations of free energy changes and equilibrium constants to obtain information on the equilibrium state of the carbonate rock system and the direction in which diagenesis should proceed; the latter necessitate knowledge of chemical mechanisms and the rates of dissolution, precipitation and "recrystallization" reactions (see e.g., Berner, 1986). Using these approaches, geochemistry has the potential to unravel the diagenetic history of a carbonate rock unit. To do so, as Land (1986) so succinctly points out, requires "quantitative studies which integrate a variety of techniques ..., coupled with rigorous mathematical models" (p. 135).

The diagenesis of carbonate sediments before their destruction by high-grade metamorphism or weathering and erosion involves a myriad of chemical reactions and reaction pathways, the end-product of which is a system of lower free energy and increased entropy (e.g., Morse and Casey, 1988). Diagenesis includes the transformation of one phase to another, i.e., "recrystallization" by processes of microsolution-precipitation. Replacement involves gross chemical changes between the reactant and product phases, e.g., calcite $\rightarrow$ dolomite. Neomorphism refers to transformations involving polymorphs or members of solid solutions, e.g., aragonite $\rightarrow$ calcite. Recrystallization occurs when the reactant and product phase remain nearly the same compositionally and structurally, but grain growth occurs, e.g., Ostwald "ripening" of small carbonate grains to larger grains [see Morse and Casey (1988) for discussion of theory with regard to carbonates]. Land (1986) characterizes all stabilization processes as proceeding by replacement or wholesale dissolution. These transformations generally take place in the presence of a fluid; i.e., marine interstitial water, meteoric water, connate water, or subsurface brine. In some cases, pressures and temperatures are elevated above depositional P and T; in other cases, transformation takes place under near Earth surface conditions of P and T. Fluids move through the carbonate sediment package transporting dissolved reactant and product constituents centimeters to hundreds or thousands of kilometers. It is this fluid movement, or in some carbonate sequences lack of long-distance movement, that is responsible for significant mass transfer of carbonate components regionally, or locally, during diagenesis. Carbonate diagenesis also involves processes in which new minerals precipitate in original or secondary pore space, and carbonate minerals are leached from the sediment mass, as discussed initially in Chapter 6. Thus, the authigenesis or neoformation of mineral phases, as well as "recrystallization", can lead to changes in a sediment's ability to transmit fluids because of changes in porosity and permeability during diagenesis. Obviously these latter changes and their temporal and spatial distribution are important to studies involving oil and gas generation, migration, and entrapment (Moore, 1989).

Because of the complexity of carbonate diagenesis involving a multitude of chemical reactions and reaction pathways, much of the literature is environment specific, concentrating on mechanisms, processes, and geologic events affecting a hydrogeochemical environment, e.g., the normal marine, evaporitic marine, meteoric, mixed meteoric-marine, and burial diagenetic environments (see e.g., Moore, 1989). To some extent, this approach has proven profitable; it is necessary to study carbonate sequences that have been exposed to a single hydrogeochemical environment either in the modern marine depositional realm or later in the history of the sediment. Geologists have used this approach of study of local diagenetic processes to interpret the post-depositional history of carbonate rocks. It is essentially the approach we employ in this chapter and the following. It should be kept in mind, however, that there are larger-scale geologic controls on local environmental conditions that determine the direction and ultimate state of diagenetic change (Siever, 1979). These controls are to a significant degree linked to the plate tectonic history of the carbonate sediment sequence, and to the relationship between plate tectonics and the chemical evolution of Earth's surface environmental system (see Chapter 10). In the next section, the utilization of plate tectonic theory for relating diagenesis to geochemical and geophysical parameters will be discussed briefly with respect to carbonate rocks. For a detailed description, the reader is referred to the thorough review by Siever (1979).

Wherever possible in further analysis of carbonate diagenetic processes, we will consider the diagenetic history of the rocks in light of the framework of plate tectonics.

Plate-tectonic controls on diagenesis

Although the record is still in a continuous state of change, it appears that plate tectonics has played a major role in carbonate sedimentation and diagenesis. For example, marine carbonate shelf limestones are characterized by shoaling upward, cyclic sequences of basal marine, intermediate intertidal, and overlying supratidal units (e.g., Wilson, 1975; James, 1979). Different diagenetic processes may occur within the different units, including early marine cementation of subtidal units, dolomitization of supratidal units, and on exposure, meteoric diagenesis of the subjacent rock cycle. It is likely that the thickness of these shoaling upward cycles is controlled by the rate of shelf subsidence (Read et al., 1986). If the carbonate shelf were part of the plate tectonic environment of a trailing edge continental margin, the subsidence of the continental border, because of cooling and contraction of the plate, would be a major control on the rate of shelf subsidence, and conceivably on the type and extent of diagenesis.

On a broader scale, consider the effect of sea level rise and fall as induced by changes in plate tectonic regime on shoal-water marine carbonate sedimentation. Fluctuations in sea level may be driven globally by several forcing mechanisms, including changes in the volume of mid-ocean ridges and intraplate volcanic buildups. This plate tectonic mechanism has been persuasively argued as the cause of first-order cyclic changes in sea level (Vail et al., 1977) during the Phanerozoic. Even the second and third order changes in the Vail et al. sea level curve may be driven by plate tectonic processes. If so, then the evolution of some carbonate shelves, as depicted in Figure 7.1, may be the result of sea level change brought about by changes in the volume of submarine volcanism. In this idealized summary diagram, the rise and fall of sea level produce a stratigraphically-complex package of carbonate sediment deposited initially during rapid sea level rise on a carbonate ramp. With slowing of sea level rise and establishment of a sea level highstand, the ramp evolves into a rimmed shelf and then into a high energy shelf. Eventually, with sea level fall, the shelf becomes an emergent karst platform. The diagenetic history of such a mixture of carbonate lithofacies would be complex, with marine cementation dominating the early stages in diagenesis of this carbonate package and extensive vadose and phreatic meteoric diagenetic processes the later stages. Whatever the case, plate tectonic processes driving sea level rise and fall provide the large-scale geological control on diagenesis of the carbonate facies produced in this idealized scenario.

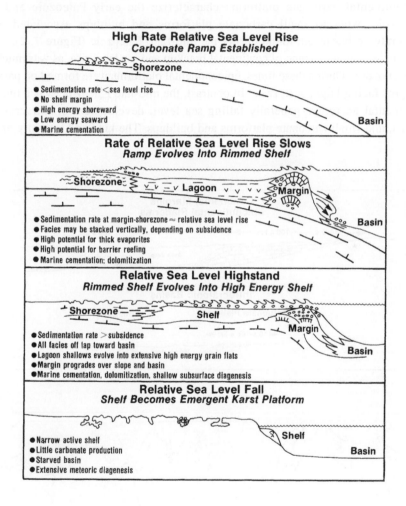

Figure 7.1. Diagram summarizing the general response of shallow marine carbonate sedimentation to various changes of relative sea level, assuming other environmental parameters remain unchanged. (Modified from Moore, 1989.)

It is likely that plate tectonic processes have also played a major role in the type and development of carbonate shelves during the Phanerozoic. Large epicontinental carbonate platforms characterize the early Paleozoic and the Mesozoic, whereas small carbonate platforms and buildups associated with intracratonic basins are found during the mid-Late Paleozoic (Figure 7. 2). The early Paleozoic was a time of continental breakup and sea level rise, as was much of the Cenozoic. During these times, epicontinental platforms were formed on passive margins facing large, open seas. In contrast, the mid-late Paleozoic was a time of continental accretion, generally falling sea level, development of intracratonic basins, and small carbonate platforms and buildups. The kind and intensity of the

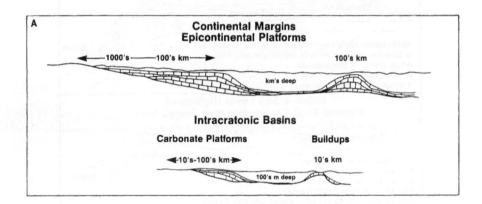

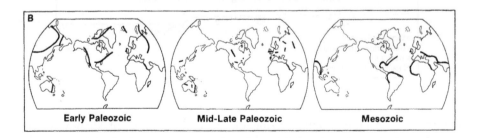

Figure 7.2.   A) Contrasting scales of continental margin and intracratonic margin associated carbonate platforms. B) The distribution of epicontinental (long lines) and cratonic-based (short lines) platform margins during different times in the Phanerozoic. (Based on James and Mountjoy, 1983; After Moore, 1989.)

diagenetic parameters (see below) associated with these two plate tectonic environments were certainly different and led to different styles of diagenesis of the two major types of carbonate shelf margins.

The diagenetic variables are shown in Figure 7.3, adapted from Siever (1979). In this figure, diagenesis is described as a function of mineralogy and chemistry of the initial mineral assemblage, the composition of early and late interstitial fluids, the burial parameters of P and T, the geothermal gradient, time, and the nature of tectonic movement. For epicontinental carbonate platform deposits on trailing plate continental margins, these deposits would be exposed, for example, to normal to low geothermal gradients (25-30°C km^{-1}), and moderately rapid burial to 10 kilometers or more over times of 50 to 200 million years. In contrast, the small, intracratonic carbonate platforms would be subjected to normal continental P and T gradients, and slow burial to depths rarely exceeding 3 to 4 kilometers over a 100-300 million year time span. Reversals of uplift and subsidence characteristic of the craton would subject this intracratonic carbonate package to oscillations of P and T, and probably, interstitial fluid composition. Certainly these two types of carbonate platforms would be affected by different post-depositional conditions induced by plate tectonic processes. These different conditions can lead to different pathways of diagenesis. Deep burial diagenetic processes could affect the epicontinental platform deposits, whereas those deposited in intracratonic basins, and not displaced from that tectonic realm, would be subject to lower P and T processes of diagenesis.

As a contrasting plate tectonic environment of diagenesis, consider calcareous oozes formed at mid-ocean ridges that move away from these features at rates of 1 to 6 centimeters per year. The basal sections of these deposits are commonly baked and well cemented because of the high heat flow and attendant temperatures encountered at ridges. As the deposits spread toward abyssal plains, they undergo a slow burial of 0.5 to 1 kilometer for periods of 100 to 200 million years. Temperature gradients relax from 50 to 60°C km^{-1}, or even higher near the mid-ocean ridge, to more normal sea-floor gradients of 30°C km^{-1}. Until reaching the subduction zone, burial is generally irreversible and depends on the cooling and contraction rates of underlying oceanic crust, as well as sediment accumulation and compaction rates. This plate tectonic environment of diagenesis contrasts sharply with that of the trailing plate continental margin or the craton. When the calcareous ooze reaches a subduction zone, it will enter another plate tectonic environment where diagenetic processes may dramatically change because of changes in the intensities of the diagenetic variables shown in Figure 7.3.

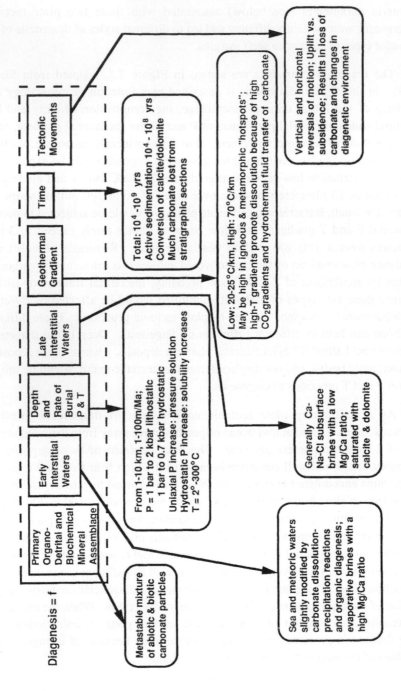

Figure 7.3. Schematic diagram of some diagenetic variables and their potential effects on the conversion of a carbonate sediment to a rock. (Adapted from Siever, 1979.)

Before concluding this section on plate tectonic controls of diagenesis, it is worth exploring briefly some other considerations. A number of authors have proposed relationships between plate tectonics and Phanerozoic ocean-atmosphere chemical change and the composition of marine carbonate precipitates (e.g., Mackenzie and Pigott, 1981; Sandberg, 1985; Wilkinson and Given, 1986; see additional references in Chapter 10). The important point here is that plate tectonic processes may lead to changes in the P_{CO_2} of the atmosphere and the Mg/Ca ratio and saturation state of seawater. It has been argued that these changes influence on a global scale the relative amounts of aragonite and calcites precipitated from the world's oceans during the Phanerozoic. If these contentions are true, then plate tectonics through its influence on the chemistry of Earth's surface environment may play a role in determining the composition of carbonate particles deposited and available for diagenesis. A more detailed discussion of this topic is provided in Chapter 10.

In summary, the major plate tectonic environments for diagenesis of carbonates and qualitative statements concerning the kind and intensity of the diagenetic variables for each environment are given in Table 7.1. Although the diagenesis of carbonates can be considered using plate tectonics as a framework, it should be kept in mind that in any one plate tectonic environment, stabilization of metastable carbonate phases may occur in a variety of waters, including marine, meteoric, evaporitic, and subsurface fluids, and over a range of P and T. For example, updip platform carbonates deposited on a trailing plate continental margin could be dolomitized early in their history, perhaps by reflux processes, and subjected to meteoric diagenesis during reversals in the general pattern of burial of these deposits. Burial-driven stabilization of the carbonate phases in the down-dip wackestones of these platform sequences could occur later in the burial history of the carbonate sequence. Thus, for any single plate tectonic environment, there may be a spectrum of diagenetic environments.

To understand the diagenetic processes affecting a carbonate sediment from its inception at deposition at the sediment-water or sediment-air interface to its present state, it is necessary to know the burial history of the carbonate sequence. The burial history of a carbonate rock largely depends on its plate tectonic setting. A sedimentation diagram (Figure 7.4) is commonly used to describe burial history. In this diagram, sediment burial depth with respect to sea level and the sediment-water or sediment-air interface is plotted against time (von Bubnoff, 1954; Siever, 1979). The mean elevation of the land surface, schematic subsurface isotherms, and values of lithostatic and hydrostatic pressure can be shown on such diagrams. This type of diagram is important in constraining models attempting to quantify

Table 7.1 Kind and intensity of diagenetic parameters associated with some major plate tectonic regimes. (Modified from Siever, 1979.)

Parameter	Mid-Ocean Ridge and Oceanic Intraplate	Trailing Plate-Continental Margin	Subduction (Trench)	Continent-Continent Collision	Rift Basins	Intra-Plate (Craton)
Detrital-Chemical Assemblage	Basalt+ pelagic carbonate & terrigenous detritus.	Terrigenous detritus (immature + mature) + platform carbonates.	Immature terrigenous detritus + pelagic; volcaniclastics, oceanic crust.	Terrigenous detritus: mature early, immature late (unroofing).	Immature alluvium, evaporites, volcanics.	Matrure, reworked sedimentary detritus. Platform carbonates.
Early Pore Waters	Carbonate: undersat if below CCD. Silica: undersat; affected by hydrothermal exchange with basalts.	Sea, brackish, or regressive fresh water. Carbonate: sat or supersat; silica: undersat.	Much like sea water. Carbonate: unsat. Amorphous silica: sat or undersat. High alkali/H+.	Fresh waters, modified by alteration of detritals. Marine turbidites: sea water.	Fresh to saline, alkaline. Amorphous silica: super-sat in some alkaline lakes.	Carbonate: sat or super-sat, amorphous silica: undersat. Fresh, brackish, or marine water.
Burial Parameters	Little burial: spreading to abyssal plains, where slow burial to 0.5-1km with normal oceanic gradients P,T.	Normal to low geothermal gradients: moderately rapid burial to 10 km.	Low heat flow: subnormal geothermal gradient; high-P, low-T of subducted material, rapid burial.	Low geothermal gradient; rapid burial in inter-montane basins or in offshore deltas and fans to 10 km.	Normal to high heat flow, rapid moderate burial to 4-7 km.	Normal continental P & T gradients; slow burial rarely exceeds 3-4 km.
Deep Pore Waters	May approach hydrothermal: silicate, alkali-rich; abyssal plains: like sea water.	Salt concentration inc. with depth and age to high ionic strength. Quartz: supersat; calcite: sat or undersat; dolomite: supersat. Fresh water infiltration from coast. High salinities near buried evaporites.	Silica, alkali-rich. May approach hydrothermal if near volcanic belt.	Continental facies: altered meteoric waters, some salinity inc. with depth. Marine facies: sea water with some salinity inc. with depth.	Some salt concentration; some hydro-thermal; may become alkaline.	Salt concentration inc. with depth and age to high ionic strength. Quartz: supersat; calcite: sat or undersat; dolomite: supersat.
Time (of Active Subsidence and Sedimentation)	Spreading rates: 1-6 cm/yr. Abyssal plains: slow burial: 100-200 m.y.	Moderate to great subsidence: 50-200 m.y.	Rapid burial: 10-50 m.y.	Rapid basin fill: 1-30 m.y. and longer	Active subsidence: 5-20 m.y. and longer	Slow subsidence: 100-300 m.y.
Oscillatory Movements	During spreading, some uplift along ridge flanks. Generally irreversible subsidence.	Mostly irreversible burial; some reversals locally and at craton edge of sediment wedge	Irreversible burial up to or more than 20 km followed by orogeny; subduction zone migration.	Many intermittent reversals as orogeny cannibalizes intermontane and marine basins.	Mostly irreversible burial; some reversals at borders.	Many reversals of uplift and subsidence: many unconformities + mild structural deformations.
Example	Mid-Atlantic Ridge.	North American Atlantic and Gulf Coast Mesozoic and Cenozoic.	Aleutian trench.	Siwalik series of India, Bengal submarine fan.	Triassic of Eastern North America.	Michigan and Eastern interior basins of North America.

Abbreviations: Sat = Saturated with respect to given mineral; Unsat = Undersaturated with respect to given mineral; Supersat = Supersaturated with respect to given mineral; P = Pressure; T = Temperature; Inc = Increases; Pptd = Precipitated; m.y. = Million years

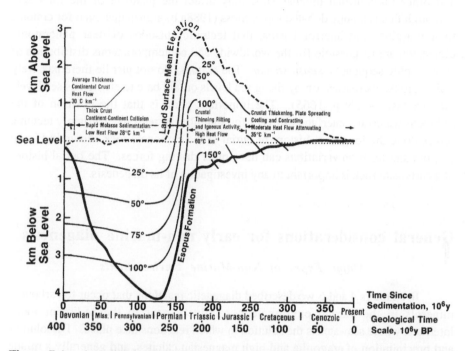

Figure 7.4. A sedimentation diagram for a hypothetical Devonian-age carbonate sequence. The curve for the sequence gives its inferred position relative to sea level and land surface elevation since the time of its sedimentation. Schematic subsurface isotherms are also presented based on estimates of crustal thickness and hypothetical heat flow values related to plate movements and igneous activity. [Adapted from Siever's (1979) representation of the Esopus Formation.]

carbonate diagenesis. To some extent, there is a feedback loop between the sedimentation diagram and diagenetic environments. Recognition of the latter can be used as additional evidence to geological and geophysical information to obtain the inferred position of a rock unit with respect to sea level and land surface elevation through geologic time. Knowledge of the former is important in interpretation of carbonate units undergoing burial diagenesis and potential uplift, for example, during continent-continent collisional tectonics.

These vertical movements can be linked to plate tectonic dynamics, but not all vertical movements may be a result of plate tectonics. It should be kept in mind that before the time of unidirectional burial of a sediment, as depicted for the carbonate rock sequence of Figure 7.4, for its first 125 million years of existence, the position of the sediment-water (air) interface may vary for reasons other than tectonics. Episodes of continental glaciation and deglaciation, as well as geoidal perturbations and other unidentified mechanisms, may affect the position of the interface. Although this is a hotly debated topic, Sloss (1988a,b) argues that even for cratons, their margins, and interior basins, that tectonic episodes of near pancratonic dimension are responsible for the worldwide and contemporaneous distribution of stratigraphic sequences, *sensu stricto*. The cratons "do not just lie there passively waiting to be encroached on by rising sea levels or laid bare to erosion as sea levels fall" (Sloss, 1988b; p. 1665). The important point is that the position of the sediment-water(air) interface can vary for a number of reasons. Certainly tectonic forces drive the long-term history of burial of the carbonate rock shown in Figure 7.4, but shorter term variations can have other driving forces. The burial history of a carbonate rock is important to any investigation of its diagenesis.

General considerations for early non-marine diagenesis

Major Types of Non-Marine Environments

In chapters 4 and 6 we described diagenetic reactions that occur to carbonate minerals in the marine environment, relatively near to the sediment-water interface, where seawater is the solution in which reactions take place. Dissolution and precipitation of aragonite and high magnesian calcites, and generally a minor amount of dolomite formation, are the primary reactions in the shoal-water marine environment. In the pelagic marine environment dissolution of aragonite and low magnesian calcite is the dominant reaction. However, a number of processes, such as sea level fluctuation, invasion of groundwaters, and formation of supratidal evaporitic ponds, can result in environments where carbonate sediments are exposed to solutions whose composition differs widely from that of seawater. Such changes in the chemical environment of carbonate sediments can occur on different time scales ranging from almost concurrent with deposition to lengthy burial.

The reactions resulting from these changes in chemical environment are among the most important for the diagenesis of carbonate minerals. They include such major processes as the destruction of high magnesian calcites, transformation of aragonite to calcite and large scale dolomitization, which may all combine to form primary lithification mechanisms whereby carbonate sediments become carbonate rocks. We will concentrate in this chapter on these processes because they occur relatively early in the life of carbonate bodies, with time spans on the order of a few million years at most. Also these processes are generally restricted to relatively shallow burial, where changes in temperature and pressure are not of primary importance. Chapter 8 will deal with carbonate diagenesis that occurs over longer time spans and with deeper burial. In this chapter and Chapter 8 we will first discuss major chemical processes and then examine the diagenesis of carbonates in some specific environments.

One convenient way to view diagenetic environments is in terms of the types of waters involved in reactions with carbonate sediments. There are four primary "end member" waters: seawater, relatively pure meteoric water (e.g., freshwater aquifers), hypersaline solutions derived either from seawater (e.g., sabkhas) or evaporite mineral deposits, and subsurface waters that may have developed unique chemistries through previous water-rock interactions (e.g., hydrothermal fluids). The mixing of these different classes of waters to form solutions of intermediate composition is of great importance to major diagenetic processes such as dolomitization and beachrock formation. Another division in diagenetic environments is between phreatic conditions, where the pores of the sediment or rock are saturated with water, and vadose conditions, where the pores are, at least episodically, unsaturated with water and gas exchange can occur. These terms are often combined with the type of water for a more specific description of the diagenetic environment (e.g., marine phreatic).

Water Chemistry

The chemistry of waters in which diagenetic reactions occur is of central importance in understanding massive carbonate diagenesis. Topics that have received the most attention are the influence of mixing waters of dissimilar chemistries and the ability of waters to transport the necessary masses of diagenetic reactants and products.

The chemical composition of waters resulting from mixing would appear to be a simple matter of averaging the compositions of the waters that mix in proportion to their volume contributions to the mixture. While this situation is true for concentrations of components such as calcium and total carbon dioxide, it is not true where chemical speciation is concerned. This is because important parameters controlling speciation, such as activity coefficients, do not vary in direct proportion to composition (see Chapter 2). This conclusion is of great importance to carbonate mineral diagenesis (Runnells, 1969). Badiozamani (1973) and Plummer (1975) provided clear examples of this effect for the mixing of dilute groundwaters with seawater. One of the most important observations made was that mixing dilute groundwater, in equilibrium with calcite, with normal seawater that is supersaturated with respect to calcite, can produce solutions of intermediate composition that are undersaturated with respect to calcite (Figure 7.5).

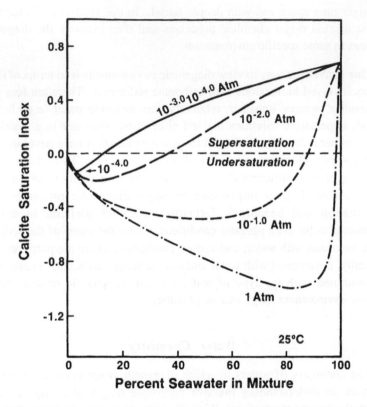

Figure. 7.5. Saturation state of mixtures of groundwater solutions in equilibrium with calcite, at different initial P_{CO_2} values, and seawater at 25°C. (After Plummer, 1975.)

Both authors' calculations also indicated that it is possible for solutions of reasonable compositions for natural waters to produce mixtures of freshwater and seawater that were undersaturated with respect to calcite but supersaturated with respect to dolomite. This observation is a cornerstone for some dolomitization models that are discussed later in this chapter. It is also important to note that the extent of undersaturation which results from mixing is strongly dependent on the initial P_{CO_2} of the dilute water when it is in equilibrium with calcite. Waters high in CO_2 can cause more extensive dissolution. If these waters enter a vadose zone where CO_2 can be degassed, they will become supersaturated and calcium carbonate can precipitate. This process provides an excellent mechanism for cementation near the water table. Because the water table can oscillate vertically, a considerable zone of cementation can result.

Questions concerning the ability of pore waters to transport reactants and products have largely focused on the removal of strontium, associated with primary aragonite, and supply of magnesium necessary for dolomitization. Such transport is also important in processes leading to the formation of secondary porosity and cementation. Because the moles per unit volume of water of most components of interest are small relative to the moles per unit volume in carbonate minerals, diagenetic models generally require the flow of massive amounts of water through carbonate bodies. Examples of these requirements will be presented later in this chapter.

Reactivity of Sedimentary Carbonates

A fundamental consideration in the diagenesis of carbonate minerals is the relative reactivity of different minerals under a given set of conditions. The primary factors that have been identified are the saturation state of the solution with respect to different carbonate minerals, and the presence of reaction inhibitors that may influence minerals differentially. The relative surface areas of different minerals and grains of similar mineralogy, derived from different sources of biogenic carbonates, that are available to react readily with a solution also have been identified as important influences on reaction rates (microstructural control). Degree of structural ordering, chemical heterogeneity, and lattice irregularities of carbonate grains are other factors involved in considerations of reactivity. All these factors combine to exert a major control on diagenetic pathways.

A basic concept is that a given carbonate mineral will not dissolve in a solution that is supersaturated with respect to that mineral or precipitate from a solution undersaturated with respect to that mineral. If a solution is undersaturated with respect to all carbonate minerals, they may all dissolve with their relative dissolution rates determined by grain size, microstructure, and solution composition, among other factors. The idea that under universally undersaturated conditions mineral solubility may not simply control dissolution rates, even for grains of the same size, was confirmed by Walter and Morse (1985). They observed that relative dissolution rates in seawater could not be normalized directly to total surface areas, but rather depended strongly on microarchitecture (Figure 7.6).

In Chapter 6, we discussed observations of marine pore water compositions in carbonate sediments that indicated that these waters usually come to a dynamic equilibrium with the most soluble carbonate phase present. Even though these waters are supersaturated with respect to other carbonate minerals, it appears that precipitation reactions are usually sufficiently inhibited that major diagenetic phase transformations do not generally occur. It should be kept in mind, however, that some carbonate phases, like high magnesian calcite, may undergo "recrystallization" even during the early stages of marine diagenesis (e.g., Moulin et al., 1985).

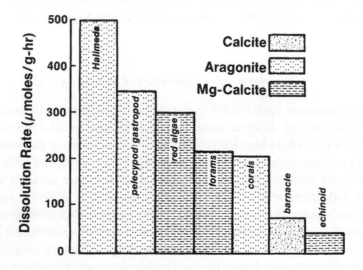

Figure 7.6. Relative dissolution rates of biogenic carbonates in seawater undersaturated with respect to calcite. (After Walter, 1985.)

The saturation state of waters can influence the diagenetic pathways by which sediments initially composed predominantly of aragonite and high magnesian calcites are transformed to calcite (e.g., Walter, 1985). In waters only undersaturated with respect to high magnesian calcite, it can dissolve and aragonitic or calcitic cements can precipitate. However, in waters undersaturated with respect to both high magnesian calcites and aragonite both mineralogical and microstructural factors control the relative dissolution rates of high magnesian calcites and aragonite. It is, therefore, possible to have either mineralogic or kinetic control of diagenetic pathways.

Major phase transformations

The Transformation of Aragonite to Calcite

Shoal-water carbonate sediments are commonly composed of 50 or more weight % aragonite. However, aragonite is rare in ancient limestones derived from such environments. The concept of thermodynamic stabilization of carbonate minerals with geologic time was discussed in Chapter 6, where the transformation of aragonite and high magnesian calcites to calcite and dolomite was shown to be a consequence of this overall process. The conditions necessary to accomplish such mineralogic transformations, and how these conditions influence reaction rates and diagenetic pathways, and resulting compositional and textural features of rocks are of interest in the study of carbonate diagenesis.

Uncertainty existed for a considerable period of time over whether, under shallow subsurface conditions, the aragonite to calcite transformation was accomplished via a solid state reaction (polymorphic inversion) or a dissolution-precipitation process that required water to proceed. The fact that sedimentary aragonite persists for hundreds of millions of years under conditions where exposure to aqueous solutions is limited argues against the polymorphic inversion mechanism. Also, a large body of experimental evidence indicates that the activation energy for this reaction is sufficiently large to make the kinetics of transformation too slow to be of importance below temperatures of 300° to 500°C (see Carlson, 1983, for review and discussion). Chemical analyses of the product calcite have shown that the reaction is not isochemical. Consequently, research on the factors controlling the diagenetic alteration of aragonite to calcite has focused primarily on the chemical attributes of aqueous solutions. These chemical parameters influence the rates at which the dissolution-precipitation reaction can obtain.

Numerous experimental investigations of the transformation of aragonite to calcite in solutions of various compositions were conducted in the 1960's and early 1970's (e.g., Brown et al., 1962; Taft, 1967; Metzger and Barnard, 1968; Bischoff and Fyfe, 1968; Jackson and Bischoff, 1971). Experiments generally consisted of stirring aragonite in aqueous solutions at temperatures greater than 50°C and observing the rate at which calcite formed. These moderately elevated temperatures were needed because of the slow rates of reaction. The major findings of these experimental investigations, relevant to common diagenetic environments, can be summarized as:

1) The rate of reaction is dependent on the nucleation and growth rates of calcite, not the dissolution rate of aragonite. Curiously, it has also been observed that absolute rates are strongly dependent on the aragonitic material used. This observation appears to contradict the generally held conclusion that rates are strictly dependent on calcite nucleation and precipitation rates, not the dissolution rate of aragonite.

2) The aragonite to calcite transition rate is very temperature dependent, with an activation energy in dilute solutions of 240 kJ mole^{-1} (Bischoff, 1969). However, this temperature dependence has only been established at temperatures above 50°C; below this temperature, results fell significantly off the general trend for higher temperature experiments. This observation raises the possibility that different mechanisms may be controlling the reaction rate at lower temperatures. It also raises the troubling question of how well the experimental results represent the mechanisms likely to be operating in shallow subsurface sediments.

3) The transformation of aragonite to calcite can be catalyzed in solutions containing electrolytes which are not inhibitors of calcite precipitation (e.g., NaCl). This observation has been interpreted to be due to increased concentrations of the calcium bicarbonate ion pair (e.g., Bischoff and Fyfe, 1968) in these solutions. Why this species should catalyze the reaction remains unexplained.

4) Mg^{2+}, SO_4^{2-} and some organic compounds (e.g., acidic amino acids, Jackson and Bischoff, 1971) can inhibit the transformation of aragonite to calcite. Mg^{2+} is a strong inhibitor and exerts a major control on the reaction at concentrations as low as 2 ppm (Bischoff and Fyfe, 1968). Mg^{2+} primarily inhibits the reaction by blocking the nucleation of calcite. Once calcite is nucleated and the transformation reaction starts, its rate of formation does not differ greatly from that in Mg^{2+}-free solutions. These results are consistent with a large body of data demonstrating the inhibitory capacity of Mg^{2+} on calcite nucleation. The need to lower the concentration of Mg^{2+} significantly below that of seawater is the most commonly cited control on the transformation of aragonite to calcite in sediments.

5) For natural sedimentary carbonates, the calcite produced diagenetically from biogenic aragonite precursors can contain over an order of magnitude less water (Gaffey, 1988). Furthermore, biogenic aragonite

precursors exhibit different structural ordering than synthetic aragonites commonly used in transformation experiments (Urmos et al., 1986).

6) Fyfe and Bischoff (1968) observed that the transformation of aragonite to calcite proceeds at a rate which is proportional to time squared, and that the final calcite crystals are close to equigranular in size. Carlson (1983) proposed a tentative model to explain these observations. He hypothesized formation of grains of close to the same size in a system where nucleation is constantly occurring, but growth stops when a particular grain size is reached. How this concept relates to the observation that once calcite is nucleated in solutions containing Mg^{2+}, the aragonite to calcite transition rate is close to that observed in Mg^{2+}-free solutions is unclear. There also is no explanation as to why growth of calcite crystals should stop at a given grain size in a supersaturated solution capable of producing new calcite nuclei.

A major factor influencing the morphology and trace component content of calcite that is produced as a result of the aragonite to calcite transformation is the relative volume of solution involved. This is largely controlled by both the porosity of the aragonitic solid (e.g., Martin et al., 1986) and movement of water during the transformation. The observation that both microstructures and modal compositions can be well preserved during the transformation has led to the idea that only a thin film (<100 Å) of water in a closed system may be necessary for the conversion of aragonite to calcite (e.g., Wardlaw et al., 1978).

Dolomite Formation

General considerations The "dolomite problem" has been one of the most intensely studied and debated topics in geology. The "problem" is that modern marine sediments contain only relatively rare and minor occurrences of this mineral, whereas it is a major component of sedimentary rocks. Questions about whether major dolomite formation from seawater could have occurred in the geologic past, the conditions necessary for dolomite formation, and many other aspects of the "dolomite problem" have resulted in a voluminous literature that includes entire books devoted to the topic (e.g., Zenger et al., 1980; Zenger and Mazzullo, 1982). The literature on dolomite formation is typified by the vigor with which contending hypotheses are supported and attacked. Many of the controversies have stemmed from attempts to find "the answer" to how sedimentary dolomite forms.

Land (e.g., 1985) has stressed the chemical and structural variations associated with natural dolomites, and has gone so far as to suggest that the name "dolomite" be used in the same way that the mineral name "feldspar" is used. The fact that dolomite is relatively unreactive compared to most other sedimentary carbonate minerals has severely limited experimental studies under temperature and pressure conditions that exist during shallow burial. Consequently, most information on the chemical behavior of dolomite must be obtained from observations of complex natural systems. Such observations are all too often open to multiple interpretations.

Intensive research on dolomite formation in sedimentary environments largely began during the 1960's. This research initially concentrated on the occurrence of dolomite in relatively young and unconsolidated sediments, which are generally bathed in highly saline waters. The fact that such environments apparently do not produce major volumes of dolomite today led to a refocusing of research on shallow burial environments where large scale circulation might lead to massive dolomitization of preexisting carbonate minerals. A fundamental consideration for all dolomitization models is a mechanism to supply the requisite amount of Mg. The observation that most meteoric waters and many subsurface brines contain relatively low concentrations of dissolved Mg^{2+} has led to the idea that seawater, or mixtures of seawater with groundwaters, must be involved in the dolomitization process.

Dolomite geochemistry   A primary consideration in the study of dolomite geochemistry is the importance of the Mg to Ca ratio and structural ordering. These two properties of sedimentary dolomites are commonly compared to the "ideal" for the mineral, and exert a large influence on the stability of dolomite and the incorporation of foreign ions. There is a general trend for the "ideality" of dolomite to increase with the age of dolomite over geologic time (Figure 7.7). Most dolomites that are currently forming in surficial sediments and that have been synthesized in the laboratory (e.g., Ohde and Kitano, 1978) are calcium-rich and far from perfectly ordered. Such dolomites are commonly referred to as "protodolomite", although the precise meaning of this term is rather ambiguous (e.g., see Gaines, 1977, for discussion). The influences of disordering and deviations in the Ca to Mg ratio from unity on dolomite stability have not been determined precisely but are probably large (e.g., Carpenter, 1980; Ohde, 1987; Oomori et al., 1988). Because non-ideal dolomite is metastable it can go through successive recrystallization reactions. Changes in trace component and stable isotope composition will occur as a by-product of the recrystallization process, and the phases will trend towards compositions of greater purity and isotopic values that are "re-set" during recrystallization.

The relationship between calcite (or aragonite) and dolomite solubilities has been extensively discussed in terms of reactions for the dolomitization of limestones. Although this relationship is relatively simple, it has been a source of controversy (e.g., Machel and Mountjoy, 1986; Stoessel, 1987). If both calcite and dolomite are in equilibrium with the same solution, they must be in equilibrium with each other. Under these conditions the following relations hold (e.g., Carpenter, 1980, see Chapter 6).

$$2CaCO_3 + Mg^{2+} \rightarrow CaMg(CO_3)_2 + Ca^{2+} \qquad (7.1)$$

$$K_{dol-cal} = \frac{K_{cal}^2}{K_{dol}} = \frac{a_{Ca^{2+}}}{a_{Mg^{2+}}} \qquad (7.2)$$

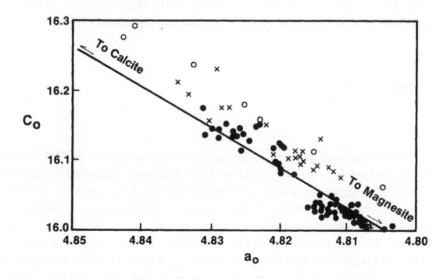

Figure 7.7. Unit cell measurements, in Angstroms, of synthetic (o), Holocene (x), pre-Holocene (●) and "ideal" (+) dolomites. (After Land, 1985.)

The observation that calcium and magnesium ion activity coefficients are generally similar in aqueous solution has led to the concept that the calcium to magnesium ion concentration ratio should be fixed at a given temperature and pressure in a solution in equilibrium with both calcite and dolomite. This ion concentration ratio can be calculated from the solubility products of calcite and dolomite, and is about 1.4 (see Chapter 6). Alternatively, the solubility product of dolomite can be calculated from observed ion concentration ratios. Extensive data on Ca to Mg ratios in subsurface waters have been collected. Unfortunately, these data scatter widely (see Carpenter, 1980, for summary and discussion), indicating that either equilibrium conditions are not reached or in some instances non-ideal dolomite may be controlling ion concentration ratios in these waters.

In order for dolomitization of a limestone to occur, dolomite must precipitate. This fact demands that a solution must be supersaturated with respect to dolomite. Consequently, the Ca to Mg ion concentration ratio of a dolomitizing solution is not fixed. It is probable that in many instances the dolomitizing solution is close to equilibrium with either calcite or aragonite owing to their rapid dissolution kinetics. If this is true, then the calcium carbonate ion activity product is fixed, but the relative concentrations of Ca^{2+} and Mg^{2+} are not (i.e. the larger the carbonate ion activity the smaller the calcium activity). The Ca^{2+} to Mg^{2+} ion activity ratio in a dolomitizing solution must be less than that which can be calculated using equation 7.2. Also, the activity of the Mg^{2+} ion must exceed the value that is obtained for a given carbonate ion activity via equation 7.3.

$$a_{Mg^{2+}} = \frac{(K_{dol} / K_{cal})}{a_{CO_3^{2-}}}$$

(7.3)

The formation of dolomite under conditions encountered in modern marine sediments and sediment burial to moderate depths is strongly controlled by reaction kinetics that are slow even at high supersaturations (e.g., Lippman, 1973; Morrow 1982a). A plausible explanation for the slow precipitation kinetics of dolomite is the requirement that cation ordering puts a major limit on the rate at which it can form (e.g., the "simplexity" principle, Goldsmith, 1953). It is likely that the dehydration kinetics of Mg^{2+} also play a role (e.g., Lippman, 1973). Because the precipitation kinetics of dolomite are too slow to be studied in the laboratory at near Earth surface temperatures, experiments on dolomite reaction kinetics have generally been conducted at elevated temperatures (typically between 100° and 300°C). While considerable interesting information has been gathered from these experiments, the applicability of the results to lower temperature and slower reaction kinetics remains highly questionable. Because of the uncertainties in how meaningful these results are for formation of dolomite under near Earth surface conditions, we will summarize only briefly the major observations. These are:

1) The reaction rate is very temperature sensitive (Land, 1967).

2) Dolomitization of aragonite proceeds faster than dolomitization of calcite. Dolomitization of high magnesian calcite may (Gaines, 1980) or may not (Katz and Mathews, 1977) be faster than that of pure calcite.

3) The reaction pathway can be complex with the formation of intermediate precursor phases (e.g., Katz and Mathews, 1977). An induction stage can be the slowest part of the reaction (Sibley et al., 1987).

4) Increasing the Mg^{2+} to Ca^{2+} ratio reduces the induction period (Sibley et al., 1987) and increases the reaction rate up to a Mg^{2+}/Ca^{2+} ratio of 5 (Gaines, 1980). Rates are also faster at high ionic strengths and elevated P_{CO2}.

5) A variety of dissolved components can greatly influence reaction rates. Strongly hydrated ions, such as Li^+, can catalyze the reaction (Gaines, 1974) but it is not a very abundant cation in natural waters. Organic compounds (Gaines, 1980) and sulfate (Baker and Kastner, 1981), that are abundant, however, can severely inhibit the reaction.

Observations of dolomite formation in natural systems also have been used as the basis for discerning additional factors that may influence the rate of dolomite formation. These include catalysis by certain clay minerals (e.g., Wanless, 1979) and production of organic by-products by bacteria (e.g., Gunatilaka et al., 1985). Mg^{2+} transport to sites of dolomite precipitation can inhibit the reaction in hemipelagic sediments (e.g., Baker and Burns, 1985). However, the true influence of reaction rates is largely speculative because the kinetic factors are generally deduced primarily from the presence or absence of dolomite in different environments. It has been often contended that the degree of nonideality of the dolomite reflects the rate at which it forms; however, the possibility that the degree of nonideality may reflect other influences, such as solution composition, cannot be ignored. Attempts also have been made (e.g., Folk and Land, 1975; Morrow 1982a), based largely on observational evidence, to construct "kinetic" phase boundaries between calcite and dolomite. One example is shown in Figure 7.8, where a dolomite-calcite "kinetic" field boundary is projected in a three-dimensional diagram where the axes represent the variables of Mg^{2+}/Ca^{2+} ratio, CO_3^{2-}/Ca^{2+} ratio, and salinity.

The stable isotope and trace component compositions of sedimentary dolomites have been intensely studied because of their potential utility in providing

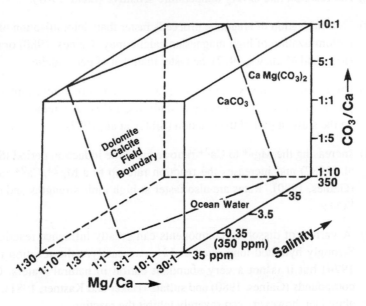

Figure 7.8.   Three dimensional "kinetic" phase diagram for the influence of solution composition on the formation of dolomite. (After Morrow, 1982a.)

information on the conditions under which dolomites form and their diagenesis. Land (1980) has summarized and evaluated much of the information on this topic. His analysis of existing data indicates that oxygen and carbon isotopes are of limited utility. The oxygen isotope composition of sedimentary dolomites covers a wide range (Figure 7.9) and in most cases may not reflect the original composition of dolomite because of recrystallization reactions. Most dolomites have carbon isotope compositions close to that of modern marine carbonates, although carbon derived from organic matter may be incorporated leading to variability (Figure 7.9). Strontium isotope ratios in dolomites have proven useful in distinguishing whether or not seawater was a major component of the solutions from which they formed.

The many problems that plague partition coefficients in calcite and aragonite (see Chapter 3) probably also are important for dolomite, but comparable experimental data are not available for dolomite at near Earth surface temperatures. Additional complications are likely to arise in dolomite in because of its degree

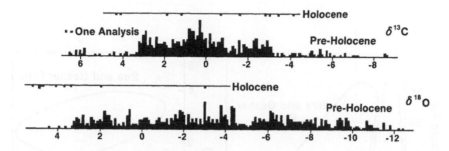

Figure 7.9. Summary of carbon and oxygen isotope ratios in sedimentary dolomites. (After Land, 1980.)

of ordering and Ca to Mg ratio, that may reflect solution Ca^{2+} to Mg^{2+} ratios. Because there are two cation sites in dolomite (the Mg and Ca layers), it may even be best to consider partitioning of other cations in dolomite in terms of two partition coefficients. Land (1980), in a pessimistic assessment of the utility of trace element geochemistry for interpreting sedimentary dolomites, pointed to many of the same problems associated with recrystallization that limit the usefulness of oxygen isotopes. He emphasized the importance of observed heterogeneities that range from within individual crystals to a regional scale and the increasing purity of dolomites with geological age.

In spite of these many difficulties in understanding the trace component and stable isotope geochemistry of dolomite, many investigators have attempted to use these variables to interpret the environments and mechanisms of dolomite formation and diagenesis (e.g., Sass and Katz, 1982; Veizer, 1985). While interpretations based on this type of data must clearly be taken with caution, it does appear that regional trends and large differences in composition can have substantial utility. This conclusion is well demonstrated by the recent work of Sass and Bein (1989). Their investigation emphasized the usefulness of combining Mg to Ca, Na to Ca, and $\delta^{18}O$ to Na ratios (see Figure 7.10) in interpreting the types of solutions (e.g., seawater, brines, etc.) involved in the formation of ancient dolomites. This paper is also interesting for its in-depth discussion of the factors influencing solution composition and the impact of varying solution compositions on partition coefficients.

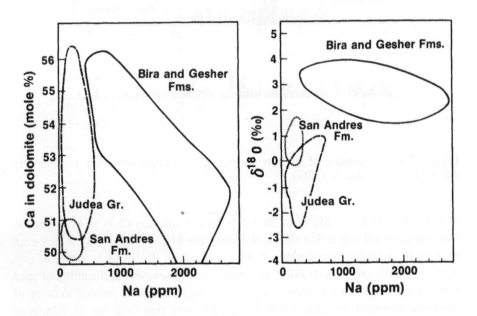

Figure 7.10. Distributions of mole percent Ca and $\delta^{18}O$ against Na for three dolomite assemblages. Note that although there is some overlap between data fields for different formations, they are generally distinctive. (After Sass and Bein, 1989.)

Dolomitization models The observation that major amounts of dolomite currently do not form directly from normal seawater has led investigators to look for diagenetic environments capable of producing dolomite. Such environments have usually been described in terms of system models which include hydrological components. A major consideration in the different models that have been proposed is their ability to produce regionally thick or massive dolomites. These models have been extensively described and strongly debated in numerous papers and review articles (e.g., Morrow, 1982b; Land, 1985; Machel and Mountjoy, 1986; Hardie, 1987). Here we shall summarize the more important aspects of the different hypotheses largely from a geochemical standpoint.

Before proceeding to the discussion of individual models it is useful to consider two general aspects of dolomitization. The first is the previously mentioned need for a supply of Mg. Two major types of dolomite formation are recognized, depending on if the dolomite is formed as a primary mineral in close association with overlying water or as a replacement product of preexisting carbonates in the subsurface. In the first case, Mg supply generally may not be a problem, whereas in the second case "plumbing" is necessary to transport Mg-bearing solutions to the site of dolomitization (Figure 7.11). This transport can involve the flow of large volumes of water. Land (1985), for example, calculated that for a typical carbonate sediment 648 pore volumes of seawater, or about 6500 pore volumes of mixed meteoric and seawater solution (10 to 1 ratio), would be required for complete dolomitization (see also Möller and Rajagopalan, 1976). The difficulty in obtaining the hydrological conditions necessary to produce such major flows of Mg-bearing solutions in a manner that will result in the formation of massive dolomite deposits lies at the heart of much of the controversy between competing models for dolomitization.

The second major consideration is the "volume problem". If most dolomites are replacements of previously existing carbonates the question arises as to whether or not a net gain or loss of material will occur during the dolomitization process. If dolomitization occurs with no net transport of carbonate (equation 7.1), a 13% reduction in the volume of the solid will occur if calcite is the precursor carbonate mineral. If the solution is initially undersaturated or supersaturated with respect to the precursor carbonate minerals, solid volume losses or gains can be roughly proportional to the degree of saturation. A complicating factor is that the Mg^{2+} to Ca^{2+} ratio in the dolomitizing solution can also play a role, because it will set the degree of supersaturation relative to dolomite in a solution in equilibrium with calcite or aragonite. However, Friedman and Sanders (1967) have observed that most dolomitized rocks appear to have undergone little volume change. This observation indicates that an approximate mole for mole replacement must have occurred (see also Morrow, 1982a, for discussion).

Most modern dolomite is forming from high ionic strength solutions that are typically derived from the evaporation of seawater or lakes in arid regions. These environments have been studied extensively because they provide an opportunity to observe directly systems in which substantial amounts of dolomite are currently forming. The associated hydrology and solution chemistry of the dolomitizing fluids can also be determined. Sites of particular note are the Persian Gulf sabkhas, Coorong district in South Australia, Bonaire Island in the Caribbean Sea, Sugarloaf

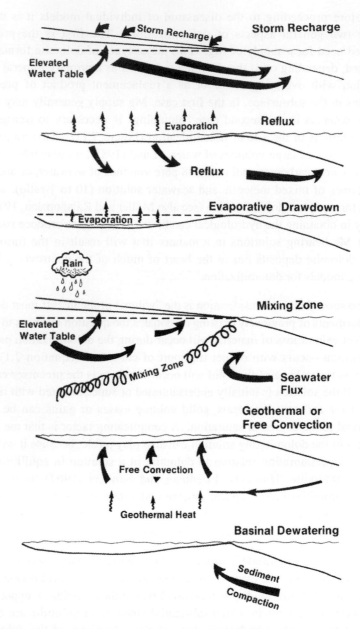

Figure 7.11. Various pumps that have been hypothesized for supplying the Mg necessary for dolomitization. (After Land, 1985.)

Key in Florida, supratidal sediments on Andros Island in the Bahamas and Deep Springs Lake in California. A general feature of many of these areas is that high Mg^{2+} to Ca^{2+} solution ratios appear to promote dolomitization. The importance of inhibition of dolomite formation by sulfate has been widely debated (e.g., see discussions of Hardie, 1987, and Gunatilaka, 1987). While dolomite formation is usually associated with sediments where sulfate reduction is active, the observation that dissolved sulfate concentrations can remain high, and dolomite still can form argues against dominant control by dissolved sulfate. This observation suggests that reaction products of sulfate reduction, such as increased alkalinity, may play a more important role in increasing the formation rate of dolomite than sulfate does in inhibiting it. This conclusion has been expanded upon by Compton (1988), who emphasized the general observation that highly elevated solution supersaturations with respect to dolomite are usually associated with sediments containing above average concentrations of organic matter. He termed dolomite that forms under these conditions "organogenic dolomite."

Studies of the hydrology of areas where dolomite is forming have resulted in models providing the flow necessary for dolomitization. These include storm recharge of supratidal waters and seepage reflux, in which brines with greater density than seawater sink through the system. Hardie (1987) has presented an excellent discussion of the chemistry of sabkha brines, illustrating the great complexity of these systems and their resistance to explanation by simple geochemical models. Friedman (1980) discussed the general importance of modern high salinity environments for dolomite formation and evidence in the rock record indicating a strong relation between evaporites and dolomite. He claimed that most sedimentary dolomite should be considered an evaporite mineral-- an opinion that is not universally accepted.

The idea, originally proposed by Hanshaw et al. (1971), that waters that are mixtures of meteoric waters and seawater could be effective solutions for producing dolomites, has enjoyed great popularity and considerable controversy. This concept was developed into the meteoric dolomitization model of Land (1973) and the "Dorag" model by Badiozamani (1973). At the heart of the Dorag model is the idea that mixing of meteoric waters and seawater can result in a water that is undersaturated with respect to calcite and aragonite, but supersaturated with respect to dolomite (see discussion of water chemistry and Figure 7.5). However, as previously discussed, it is not necessary for a solution to be undersaturated with respect to calcite or aragonite for dolomite formation to take place. Indeed, it is likely that waters that mix to produce an undersaturated solution with respect to these minerals more rapidly come to equilibrium with calcite or aragonite than dolomite precipitates. Hardie (1987) also pointed out that the "window" of water compositions that meet the requirements of the Dorag model is highly dependent on the solubility of the dolomite that forms. If non-ideal dolomite is precipitated, this window becomes quite small (Figure 7.12). Other thermodynamic and kinetic difficulties with the Dorag model include the fact that the mixed waters are less supersaturated with respect to dolomite than seawater (Carpenter, 1976), and that lowering ionic strength decreases the rate of dolomite formation (see previous discussion).

Another kind of mixing model, based on the original concept of Land (1973), is the "schizohaline model" of Folk and Land (1975). In this model, mixing of seawater or brines with freshwater results in solutions of lowered ionic strength, but these waters retain their high Mg^{2+} to Ca^{2+} ratios. A fundamental problem with these mixing models is the lack of unequivocal examples of dolomite formation in mixing zones (see Hardie, 1987, for extensive discussion). Various hypotheses have been put forth to explain this lack of supporting evidence (e.g., Pleistocene sea level variations, Land, 1985). However, given the lack of evidence for dolomite formation in waters of mixed composition and the extensive evidence of dolomite formation in high salinity waters, it is difficult to understand why there has been such a strong tendency to use the mixing zone models in favor of the high salinity models.

Difficulties with the high and low salinity models, along with their demands for solutions of "special" composition, have caused a resurgence of interest in seawater as a major dolomitizing fluid. The key elements that have been identified for dolomitization of carbonates by seawater (see Land, 1985; Hardie, 1987) are time, water movement, and perhaps temperature. Water movement can commonly result from burial compaction, tidal pumping, hydraulic heads, and thermally driven convection cells. It is interesting to note that dolomite formation in the carbonate sediments associated with the Floridan aquifer, which was the basis of the Hanshaw et al. (1971) mixing zone hypothesis for dolomitization, now is cited as one of the best examples of dolomitization by seawater. Dolomitization under near-normal seawater conditions also has been demonstrated for reef boundstone of north Jamaica (Mitchell et al., 1987) and supratidal sediments of Sugarloaf Key, Florida (Carballo et al., 1987).

Sperber et al. (1984) conducted an extensive examination of Paleozoic dolomites from North America that provides general insight into how different types of dolomitization processes may be reflected in carbonate rocks. Their data indicate that two separate processes may lead to two distinctive populations of sedimentary dolomite. The processes are divided into closed and open system

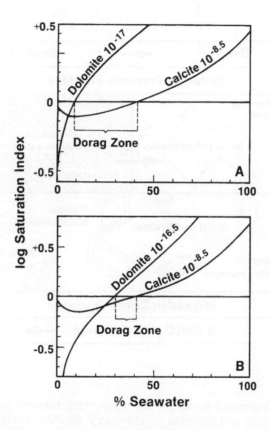

Figure 7.12. Saturation relations in mixtures of seawater and meteoric water for different solubilities of dolomite. A. Ideal dolomite; B. Calcium-rich dolomite. (After Hardie, 1987.)

dolomitization. Closed system dolomitization is associated with micrites that contained initially substantial amounts of magnesian calcites and results in dolomitic limestones. Dolomites in these limestones are generally calcium-rich, if they have undergone only shallow burial. Dolostones, where most of the carbonates have been converted to nearly stoichiometric dolomite, are believed to be the result of large scale circulation of fluids capable of providing the Mg necessary for dolomitization. The general concepts of Sperber et al. (1984) are summarized in Figure 7.13.

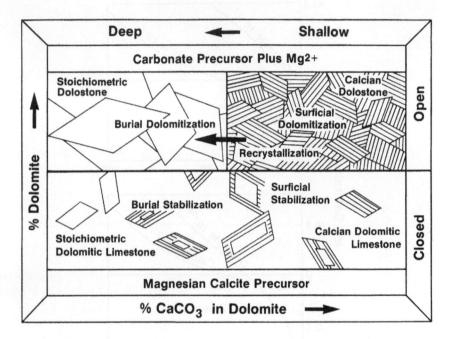

Figure 7.13. Conceptual model for the relationship between rock composition, openness of system, and dolomite stoichiometry. Shallow formation of dolomite produces less "ideal" dolomite. More open systems tend to produce more complete dolomitization. (After Sperber et al., 1984.)

Summary Remarks

Although the transformation of aragonite and high magnesian calcite to low magnesian calcite and the dolomitization of carbonate sediments and rocks are among the most important processes in the diagenesis of carbonates, major aspects of these processes are still poorly understood. One of the primary reasons for this is that although these reactions may occur over geologically short times, their kinetics are too slow to be studied in the laboratory under conditions closely approaching those in natural systems. Consequently, our understanding of the chemical processes and mechanisms that are reliably applicable to these major phase transformations is relatively limited.

Observations of natural systems have been subject to substantially different interpretations owing to the inherent complexity of these systems. There has been an unfortunate tendency to interpret observations using the popular hypothesis at a particular time in attempting to explain all such diagenesis. It should be kept in mind that just as there are many roads to heaven (or hell!), there are many pathways to thermodynamic stability.

Mass transfer during diagenesis

General Considerations

One of the most important aspects of carbonate diagenesis is the net movement of carbonate minerals. This mass transfer can be accomplished on a small scale by diffusive transport or on a large scale by the flow of subsurface waters. It is the basic process by which secondary porosity is created and cementation occurs. In most cases, it involves carbonate mineral dissolution at one site and precipitation at another. While this can be simply accomplished where mineralogic transformations from a metastable phase to a more stable phase are involved, more complex mechanisms may be required within mineralogically homogeneous carbonate bodies.

In many surficial systems the primary process responsible for dissolution is clearly associated with the increase in P_{CO2} that results from the oxidation of organic matter (e.g., see "Journey of a Raindrop" in Chapter 3). In near surface environments, this process can produce major dissolution features that lead to karst topography. Mixing of compositionally-dissimilar waters can produce undersaturated waters, with the degree of undersaturation being also related to the relative P_{CO2}'s in the mixing waters (see previous discussion, this chapter). Organic acids produced by the breakdown of organic matter may contribute substantially to carbonate mineral dissolution. Dissolution processes generally proceed relatively rapidly, and it is probable that in most carbonate mineral-rich environments, interstitial solutions become saturated with respect to the most soluble carbonate mineral phase soon after undersaturated solutions enter these systems. The sharpness of dissolution fronts is largely dependent on the relation between the initial degree of undersaturation of the solution and its flow velocity. Low flow velocities and high initial undersaturations usually result in a sharper dissolution front than high flow velocities and low initial degrees of undersaturation. The mode of fluid flow (pervasive versus channel) also exerts a major influence on dissolution patterns.

Buhmann and Dreybrodt (1985a,b; 1987), and their associates (Baumann et al., 1985), have conducted experimental studies and modelled calcium carbonate dissolution in porous systems primarily in regard to karst formation. Their research demonstrated the importance of turbulent versus laminar flow. The rates of dissolution under turbulent flow were about an order of magnitude faster than those under laminar flow. This "hydraulic jump" can be important in greatly increasing dissolution and precipitation rates when turbulent flow occurs, and may occur during the development of water conduits during karstification. Earlier work (e.g., Palciaukas and Domenico, 1976) also had shown that fluid flow rates were important in controlling both the distance and rate at which dissolution could occur. It is important, however, to note that in these models diffusion controlled kinetics were presumed to apply. This is true only in the case of major deviations from equilibrium (see Chapter 2).

Precipitation of cements is strongly dependent on the degree of supersaturation of the solution and presence of precipitation inhibitors. High degrees of supersaturation can be obtained near the water table where degassing of CO_2 can occur. This process is discussed subsequently with regard to beachrock formation. In the absence of a gas-water interface, the degree of supersaturation often will be primarily limited by the differences in solubility between the dissolving and precipitating minerals. During early diagenesis, where magnesian calcites and aragonite are dissolved, and calcite is precipitated, supersaturations in the range of 1.5 to ~3 may be typical. However, in calcitic limestones differences in solubility that result from differences in grain size and small compositional variations or pressure solution may be limited to a few percent at most.

The sharpness of cementation fronts will be influenced by factors similar to those for dissolution fronts, but also are probably more sensitive to solution composition and the surface chemistry of substrates. Although the factors influencing the formation of subsurface cementation fronts are not understood in detail, they are clearly of substantial practical importance in determining where features such as diagenetic petroleum traps may be emplaced. The formation of cements, of close to constant minor component and isotopic composition, over geographically wide areas implies that they formed from a solution of close to constant composition. If cements are to be derived from outside the region of cementation, the rate of cement precipitation must be extremely slow compared to the rate of flow of the cementing solution in order for the solution composition to be maintained at a nearly constant value. This concept is compatible with the idea that only very slight degrees of supersaturation may be obtained in the cementing solution.

Geochemical Constraints on Mass Transfer

Because of the importance of mass transport of carbonate components to problems of porosity development and cement formation, we will now discuss some of the more fundamental geochemical aspects of this process. A reasonable starting point is to consider the dissolving or precipitating capacity of a solution; that is, how much carbonate can be dissolved or precipitated by a solution owing to a change in saturation state. The dissolving capacity of a solution can cover a very wide range, because the initial solution composition may be acidic. This acidity can be viewed as a negative alkalinity and provide a reasonable estimate of the amount of calcium carbonate that the solution can dissolve. For undersaturated solutions close to equilibrium or supersaturated solutions, the methods presented in Chapter 2 (e.g., see "Journey of a Raindrop") can be used to determine the dissolution or precipitation capacity of a solution. It is important to keep in mind that the relationship between the amount of calcium carbonate that can be precipitated from or dissolved in a solution and its composition is not simple. The relationship depends on, among other factors: variation in activity coefficients, reactions within the carbonic acid system, the carbonate ion to calcium ion concentration ratio, and pressure and temperature.

Here we will use a simplified example to illustrate some basic aspects of the mass transport process for carbonates that avoids most of the more complex relationships. In this example, the calcium and carbonate ion concentrations are set equal, and values of the activity coefficients, temperature, and pressure are held constant. The carbonate ion concentration is considered to be independent of the carbonic acid system. The resulting simple (and approximate) relation between the change in saturation state of a solution and volume of calcite that can be dissolved or precipitated (V_c) is given by equation 7.4, where v is the molar volume of calcite.

$$V_c \text{ (cm}^3 \text{ calcite L}^{-1}) \cong v \left(\frac{K_{cal}}{\gamma_{Ca^{2+}} \, \gamma_{CO_3^{2-}}} \right)^{\frac{1}{2}} \left(\Omega^{\frac{1}{2}}_{cal(i)} - \Omega^{\frac{1}{2}}_{cal(f)} \right)$$

$$(7.4)$$

This relationship can be modified to calculate (equation 7.5) the commonly used concept of the number (N) of pore volumes of water necessary to cause a given change in porosity (ϕ). If an ion activity product of 0.4, a change in saturation state

$$N \cong \left(1 - \frac{\phi_f}{\phi_i}\right) \left[\nu \left(\frac{K_{cal}}{\gamma_{Ca^{2+}} \; \gamma_{CO_3^{2-}}}\right)^{\frac{1}{2}} \left(\Omega_{cal(i)}^{\frac{1}{2}} - \Omega_{cal(f)}^{\frac{1}{2}}\right) \times 10^{-3}\right]^{-1}$$

(7.5)

of 0.1 near equilibrium, and a change in porosity from 60 to 40 percent are used, close to 2 million pore volumes of water are needed. Even if the change in saturation state is increased to 0.5, about a half million pore volumes of water are required. The times necessary to accomplish this per km distance of rock to be cemented at a flow rate of 1 cm day^{-1} are, respectively, about 550 and 120 million years. It should also be noted that, by setting the calcium and carbonate ion concentrations equal, the maximum amount of precipitation occurs. Over reasonable ranges of differences in the concentration of these ions, at least an order of magnitude more time could be required.

The duration of time calculated to cause the change in porosity of 0.2 in this simple example is not geologically impossible, but it is orders of magnitude larger than some estimates of the time over which similar processes may occur in natural systems (e.g., Dorobek, 1987). Our longer calculated time, if even approximately correct, has some implications to the development of regionally extensive calcite cement zones in carbonate rocks (e.g., Meyers, 1978; Meyers and Lohmann, 1985). It would be difficult to have such geosynchronous cementation events over large areas by importing the cement components from considerable distances outside the carbonate body undergoing cementation. The time required for cementation by this process is too long. Thus it appears that a substantial portion of calcite cement must be generated within the rock body by dissolution of metastable precursor carbonate minerals and subsequent precipitation of calcite. Much of the acidity for this dissolution must also be supplied from sources near the region of cement formation.

Let us now consider the problem from the standpoint of calcite precipitation kinetics. At saturation states encountered in most natural waters, the calcite reaction rate is controlled by surface reaction kinetics, not diffusion. In a relatively chemically pure system the rate of precipitation can be approximated by a third order reaction with respect to disequilibrium $[(\Omega-1)^3$, see Chapter 2]. This high order means that the change in reaction rate is not simply proportional to the extent of disequilibrium. For example, if a water is initially in equilibrium with aragonite ($\Omega_c=1.5$) when it enters a rock body, and is close to equilibrium with respect to calcite ($\Omega_c \approx 1.01$), when it exits, the difference in precipitation rates between the two points will be over a factor of 100,000! The extent of cement or porosity formation across the length of the carbonate rock body will directly reflect these differences in reaction rates. The important implication of this rate behavior is that it is not possible to produce a uniform distribution of cements or porosity over a wide area, if the saturation state of the water changes much during its passage through the rocks. It should also be noted that the presence of common reaction inhibitors, such as phosphate and organic matter, increase the reaction order, making the process even more sensitive to saturation state change.

The results of these calculations for simple systems clearly point to the difficulties inherent in producing extensive carbonate cements or porosity for even a modest sized carbonate rock body, by transporting the calcium carbonate required into or out of the system. The need to have very large changes in saturation state during passage of a subsurface water through a rock body to produce the required transport of dissolved calcium and carbonate ions results in major changes in reaction rates causing most deposition or dissolution to occur soon after the water enters the rock body. Thus, as previously stated, there is a strong possibility that the components of most meteoric calcite cement have not been transported great distances and have been derived by dissolution of metastable carbonate precursor minerals close to the site of cement formation. Because of saturation state and kinetic constraints, porosity development and cement formation appear to be intimately coupled by dissolution-precipitation processes that do not involve long distance transport of chemical components. It may be possible to explain regionally coherent variations in trace component (e.g., iron and manganese) compositions in cements by obtaining the $CaCO_3$ from local sources, whereas the trace components are derived from the circulating waters or local sources as a result of regional changes in the redox chemistry of the meteoric waters. This topic will be discussed further in Chapter 9 in relation to "cement stratigraphy."

Beachrock Formation

The formation of beachrock will be examined as an example of carbonate cement formation, because it has been extensively investigated and because it represents a chance to study carbonate cement emplacement under conditions where the rate of cement precipitation is relatively rapid and the associated solutions can be analyzed directly. It also differs from the cementation process in our model in that carbon dioxide can be degassed to the atmosphere, resulting in major changes in the saturation state of the cementing solution.

As the name implies, beachrock generally forms in the intertidal to supratidal zone. Magnesian calcite and aragonite are the common cement-forming minerals in modern beachrock (e.g., see Milliman, 1974, for summary). Early explanations of the formation of beachrock usually invoked direct biologic processes or the breakdown of organic matter. However, during the past quarter of a century, direct precipitation from seawater or groundwater has been the primary mode of formation most favored, with controversy existing over whether mixing (e.g., Schmalz, 1971; Moore, 1973) or degassing processes (e.g., Hanor, 1978; Meyers, 1987) are responsible for cement precipitation.

The study by Hanor (1978) of beachrock formation in Boiler Bay, St. Croix, U.S. Virgin Islands, stands out as one of the most detailed investigations. It was designed specifically to test the competing hypotheses of whether mixing of waters or degassing of CO_2 was responsible for precipitation of cements. A series of wells were emplaced across the beach from which waters were sampled. Well-water samples was mixed with both seawater and distilled water and allowed to slowly degas to the atmosphere. Crust-like precipitates of low magnesian calcite (<5 mole percent $MgCO_3$) were observed to form from unmixed beach groundwater at the air-water interface. Both the time of first appearance and the quantity of precipitate decreased with increasing fraction of seawater or distilled water, suggesting that mixing inhibited rather than promoted precipitation.

Hanor calculated the calcium carbonate ion activity product for mixtures of groundwater with seawater for a closed system and after exchange with the atmosphere (Figure 7.14). Results indicated that for the closed system, the saturation state of the initial groundwater solution was close to equilibrium with respect to aragonite, and decreased with increasing fraction of seawater, up to ~75% seawater, then increased to the supersaturation of normal seawater. Over a range of about 30 to 90% seawater, the solution was undersaturated with respect to both calcite and aragonite. Upon degassing to atmospheric P_{CO2} values, supersaturations of all mixtures with respect to aragonite were calculated, with pure groundwater reaching a supersaturation with respect to aragonite in excess of an order of magnitude. Thus, both experiments and theoretical calculations support the idea that degassing of CO_2, rather than mixing of waters, is responsible for beachrock formation at this site.

Hanor (1978) also discussed the mechanisms that may be responsible for degassing of groundwaters in the supratidal zone. His ranking of processes in order of probable importance was: 1) tidally-induced oscillations of the groundwater table, 2) tidal pumping at the sediment-atmosphere interface, 3) a decrease in the average depth of the water table, and 4) a seaward increase in porosity .

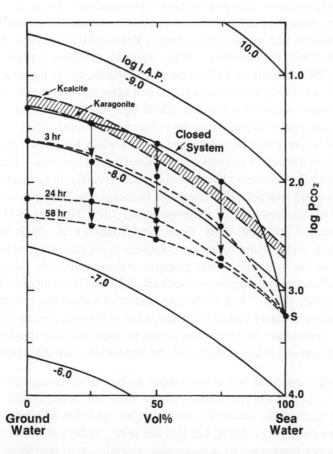

Figure 7.14. Calculated P_{CO_2} and calcium carbonate ion activity products for mixtures of groundwater and seawater. Arrows indicate degassing with time. (After Hanor, 1978.)

Lithification in the meteoric environment

Introduction

There is an extensive literature that exists concerning carbonate diagenesis in the vadose and phreatic meteoric environmental settings (see e.g., Bathurst, 1975, 1980; James and Choquette, 1984; and Moore, 1989 for reviews). Much of this work has dealt with the stabilization of aragonite and magnesian calcite to calcite in the present meteoric diagenetic realm associated with Pleistocene limestones. As

Land (1986) points out, many of the early studies dealt with processes and products of meteoric diagenesis of Pleistocene rocks that are subaerially exposed on tropical and subtropical land masses; for example, Bermuda (Gross, 1964; Land, 1967; Land et al., 1967; Plummer et al., 1976), Barbados (Matthews, 1968; Steinen and Matthews, 1973; Matthews, 1974; Allan and Matthews, 1977), the Red Sea (Gavish and Friedman, 1969), and Jamaica (Land and Epstein, 1970; Land 1973). These sediments were deposited during interglacial high stands, particularly at +5 meters 120,000 years ago and have been in the vadose zone for most of their history (Land, 1986). It is likely that phreatic meteoric diagenesis is a, if not the, dominant collection of processes responsible for limestone diagenesis. However, this diagenetic realm has received less attention, until recently, in carbonate diagenetic studies involved with both Pleistocene and Holocene sediments (for exceptions see Steinen and Matthews, 1973; Plummer et al., 1976; Halley and Harris, 1979; Buchbinder and Friedman, 1980; Budd, 1984; Chafetz et al., 1988; Vacher et al., 1989). Most authors today agree that diagenesis in the phreatic zone is more rapid than that in the vadose zone and produces textures more like those of ancient limestones than vadose diagenesis does(Land, 1970, 1973; Matthews, 1973; James and Choquette, 1984). Most of the case studies of vadose and phreatic meteoric diagenesis are certainly biased by investigation of these processes in Pleistocene carbonate sediments. Much caution should be taken in using these examples of meteoric diagenesis as actualistic models for ancient limestone diagenesis.

Further potential bias is introduced by two other considerations: that of secular changes in the composition of the skeletons of organisms and abiotic carbonate precipitates during Phanerozoic time, and that of changes in ocean-atmosphere composition during this Eon and prior. Today's shoal-water carbonate sediments are dominated by a metastable assemblage of aragonite, magnesian calcite, and calcite (see Chapter 5), in order of decreasing abundance. Holocene and Pleistocene shoal-water carbonates have mineralogies and chemical compositions similar to the carbonate deposits in modern seas. However, as pointed out by Lowenstam (1963), Milliken and Pigott (1977), Wilkinson (1979), and Mackenzie and Agegian (1989), there is a long-term increase in the production of biogenic carbonates dominated by a varied mineralogical assemblage of aragonite, magnesian calcite, and calcite with decreasing rock age in the Phanerozoic. In general, early and middle Paleozoic sediments were dominantly calcite or magnesian calcite, whereas those of the modern and Cenozoic are, or were, mostly composed of aragonite with lesser magnesian calcite, and generally, little pure calcite. Mesozoic skeletal carbonates of shoal-water origin are characterized by about equal abundances of aragonite and calcite (either high-or low-magnesian calcite). Furthermore, it appears that the abundance of benthic, carbonate-secreting algae may have been more important during the early Paleozoic and the middle to late Mesozoic than during the late Paleozoic and Cenozoic (Bischoff and Burke, 1989).

The history of the mineralogy of non-skeletal particles during the Phanerozoic is more complex; both oöid and cement mineralogy may have varied in an oscillating manner in the Phanerozoic (e.g., Mackenzie and Pigott, 1981; Sandberg, 1983; Wilkinson and Given, 1986; Mackenzie and Agegian, 1989). Oöids and cements of original calcite compositions may have been more important than aragonite abiotic precipitates in the Ordovician through Devonian and late Jurassic through early Tertiary seas.

The second consideration is that of changes in the chemistry of the ocean-atmosphere through time. The driving force for these changes on a long-time scale is plate tectonics. Some variables that may have changed during the Phanerozoic are the saturation state of seawater with respect to calcite and its Mg^{2+}/Ca^{2+} ratio, and the P_{CO2} of the atmosphere (see e.g., Mackenzie and Pigott, 1981; Berner et al., 1983; Wilkinson and Given, 1986). These chemical variations may be linked to the oscillatory behavior of abiotic carbonate precipitates during the Phanerozoic. Also, changes in P_{CO2} can lead to global temperature variations that can affect carbonate dissolution rates in the meteoric environment (see Chapter 10 for further discussion of the above).

The important point is that the pathways of meteoric diagenesis are partly dependent on the original mineralogy of the sediments undergoing alteration. Those sediments with a great deal of aragonite will be altered and cemented most in the meteoric environment, whereas those with abundant calcite will be less affected (James and Choquette, 1984). Also, the rate of diagenetic alteration in the meteoric environment can be affected by temperature; in the past higher atmospheric CO_2 levels may have led to higher global temperatures and consequently higher rates of carbonate mineral dissolution in the meteoric realm. Furthermore, if the saturation state of seawater has changed during geologic time, it is possible that the chemistry of the mixing zone between the freshwater phreatic and the shallow marine phreatic realms may have been chemically different from that of today. Thus studies of meteoric diagenesis in Holocene and Pleistocene carbonate sediments are biased not only because of Pleistocene sea-level changes but also because the composition of these deposits is different from that of most of geologic time, and ocean-atmosphere composition is not a static property of Earth's surface environment.

The Meteoric Environment and Cement Precipitates

The meteoric realm   The meteoric diagenetic environment and the hydrologic conditions present in the meteoric realm are sketched in Figure 7.15. The major settings, or hydrologic regimes, are the vadose, freshwater phreatic, and the mixing zone of brackish water. The water table marks the boundary between the vadose and freshwater phreatic zones. Near present-day continental coastlines and on oceanic islands, the mixing zone generally represents a mixture of Ca^{2+}-HCO_3^- high P_{CO_2} waters with normal seawater. Within continental interiors or at the borders of freshwater lenses extending many kilometers into sedimentary basins or seaward from the strandline beneath continental margins, the mixing zone can represent a mix of meteoric water with altered seawater or subsurface solutions of various compositions.

The most common phreatic aquifers are unconfined and open to the atmosphere, with a well-defined water table between the phreatic and vadose zones. In the vadose zone both a gas ("soil air") and a water phase are present. Water is concentrated by capillarity at grain contacts, and vadose carbonates are alternately wetted and dried as rain falls, infiltrates the vadose zone, evaporates, or is transpired by plants, and moves by vadose seepage and flow into the phreatic zone. The zone of infiltration in the upper vadose zone is an intense region of evapotranspiration, soil formation, and caliche development. There are many sites for nucleation of cements and a short time for carbonate crystals to grow from solution in the vadose zone. Water in this zone tends to concentrate where grains are in near contact. Precipitates tend to thicken along grain bottoms, and where two grains are in contact, the water rims coalesce to form a rounded contact with the air phase of the pore. In contrast, in a phreatic zone of continuous saturation where water moves by diffuse flow, the water phase is continuous, surrounding the grains and filling all pores. This type of phreatic zone generally characterizes young, freshly exposed carbonates undergoing chemical stabilization. In mature, stabilized karstic carbonate terrains free, or advective, flow via conduits and integrated systems of water discharge are important and groundwater tables may be difficult to recognize.

Cement morphology, composition, and origin in meteoric freshwater As might be anticipated, initial cement morphologies in the vadose and phreatic zones are distinctly different because of the different nature of the water-wetted surfaces in these regimes (Figure 7.15). The vadose percolation of water is not uniform, and thus cements generally are distributed irregularly throughout the

Meteoric Diagenesis

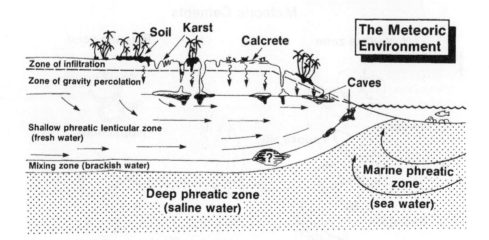

Controls

Intrinsic
Mineralogy
Grain size
Porosity and permeability

Extrinsic
Climate
Vegetation
Time

Settings

Vadose zone

Water table

Phreatic zone

Mixing zone

The Meteoric Environment

Soil   Karst   Calcrete

Zone of infiltration

Zone of gravity percolation

Caves

Shallow phreatic lenticular zone
(fresh water)

Mixing zone (brackish water)

Marine phreatic zone
(sea water)

Deep phreatic zone
(saline water)

Figure 7.15.   Major settings and principal controls of meteoric diagenesis, and a schematic drawing of the meteoric environment. (After James and Choquette, 1984.)

rock. These calcite cements are concentrated at grain boundaries and exhibit a curved surface, reflecting growth outward to capillary water-air curved interfaces of partially water-filled pores (the *meniscus cement* of Dunham, 1971). In deeper regions of the vadose zone, where more fluid is available, droplets of water can accumulate on the bottoms of grains. This leads to formation of pendant cement shapes (microstalactitic cements; Longman, 1980), reflecting the shape of the gravity-induced, convex-downward and elongated water-air interface. Epitaxial overgrowths on echinoderm grains and some benthic foraminifera may also develop in this realm (James and Choquette, 1984), as well as in the freshwater phreatic realm. Vadose silt derived from cements detached from pore ceilings and from fine-grained detritus infiltrating with water from above may line the bottom of pores to form geopetal layers (Dunham, 1969). Such cement features have been duplicated in experiments designed to simulate the meteoric environment (Thorstenson et al., 1972; Badiozamani et al., 1977). Observations of cement morphologies (e.g., Bricker, 1971) and knowledge of the nature of fluid flow in modern vadose environments and experimental evidence leave little doubt that the composition and morphology of vadose cement shown in Figure 7.16, if preserved in ancient limestones, are excellent criteria for identification of this

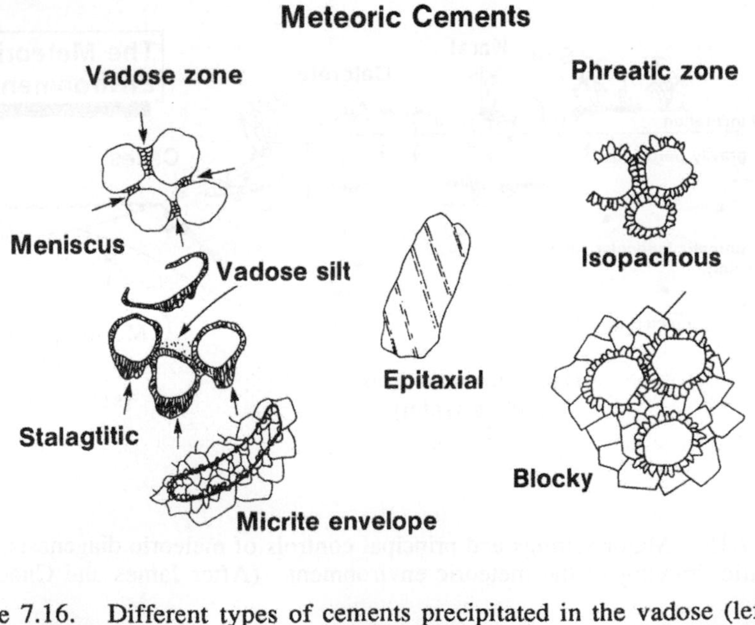

Meteoric Cements

Figure 7.16.   Different types of cements precipitated in the vadose (left) and phreatic (right) parts of the meteoric diagenetic environment. Epitaxial cements may be precipitated in either environment. (After James and Choquette, 1984.)

hydrogeochemical realm. However, if vadose cementation obtains for a long period of time and pores become filled, the above characteristic cement shapes may be lost (Budd, 1984). Also some vadose cementation patterns characterize cements in the marine environment (Moore and Brock, 1982), but these cements are not compositionally low magnesian calcites (Chapter 6).

In the phreatic environment, because pores are always filled with water, calcite cement crystals can grow unimpeded by water-air interfaces, and until they contact another carbonate grain or cement crystal. Phreatic meteoric cements may grow faster than vadose cements, because new dissolved constituents may be brought rapidly and in larger quantities by flowing and continuous phreatic freshwater to sites of cement nucleation and growth. Freshwater phreatic cements form isopachous layers around grains or occur as blocky calcite. The calcite cement generally is of larger crystal size than the fine-grained calcite spar found in the vadose zone, and commonly coarse crystals abut directly against grain boundaries. Once again such cements have been reproduced in laboratory experiments mimicking meteoric diagenesis.

It is not surprising that freshwater vadose and phreatic meteoric cements in carbonate rocks are low-magnesian calcite in composition. Observations of the composition of waters bathing modern near surface sediments and experimental evidence show that low magnesian calcite is the phase anticipated to form from such waters. Meteoric waters (Table 7.2) in carbonate sediments have low salinities, low Mg^{2+}/Ca^{2+} ratios, low SO_4^{2-} concentrations, variable, but measurable, concentrations of PO_4^{3-} and generally low dissolved organic carbon concentrations. These waters are generally undersaturated to slightly oversaturated in local regions of the vadose zone, and at saturation or several times supersaturated with respect to calcite in the phreatic zone, but undersaturated with respect to dolomite, and near saturation with aragonite. Obviously the saturation state of meteoric waters favors precipitation of calcite, but also kinetic considerations favor its formation.

Studies of the kinetics of nucleation and growth of carbonate phases from aqueous solution provide evidence for the nature of carbonate cements in the freshwater meteoric environment (e.g., Wollast, 1971; Nancollas and Reddy, 1971; Kazmlesczak et al., 1982; Walter, 1986; Busenberg and Plummer, 1986b; Burton and Walter, 1987; Zhong and Mucci, 1989). Wollast (1971) in a simple, but elegant, theoretical treatment demonstrated that regardless of supersaturation or temperature changes in a solution, homogeneous nucleation of carbonate cement is very unlikely in natural environments. This statement is particularly true of mineral-interstitial water systems which are notoriously heterogeneous. It can be

Table 7.2. Some representative meteoric water compositions from limestone terrains.

SiO$_2$	Ca^{2+}	Mg^{2+}	Na$^+$	K$^+$	HCO$_3^-$	SO$_4^{2-}$	Cl$^-$	Sr^{2+}	δ^{18}O ‰	δD ‰	P$_{CO_2}$ (10^{-3} atm)	pH	Alk (meq kg^{-1})	Ref.
<-------------------------- μM -------------------------->														
148	1200	239	174	18	2750	67	136	-	-	-	-	-	-	(1)
-	1070	-	913	-	-	-	-	5.3	-5.65	-38.6	-	-	-	(2)
-	349	-	-	-	-	-	-	-	-	-	0.8-1.3	-	-	(3)
-	2000	292	609	84	4150	19	960	1.7	-	-	19.4	7.1	4.15	(4)
-	2410	350	-	169	3860	-	215	68.5	-	-	12	7.26	3.36	(5)
267	1450	670	265	18	2670	740	254	-	-	-	2.8	7.75	2.67	(6)
34	1423	260	1465	295	6600	130	1305	7.5	-	-	30.9	7.15	6.29	(7)

(1) Miocene limestone (White et al., 1963); (2) Lincolnshire Limestone (Smalley et al., 1988); (3) Vadose flow in Lower Carboniferous rocks, Kentucky, U.S.A. (Thrailkill and Robl, 1981); (4) Pleistocene reef limestone, Yucatan Pennisula, Mexico (Back et al., 1986); (5) Pleistocene limestone, Bermuda, 1.05% seawater (Plummer et al., 1976); (6) Floridian aquifer, U.S.A., δ^{13}C = -10.8 (Plummer et al., 1983) (7) Fresh groundwater, Pleistocene limestone, Bermuda, also contains 3.07, 47.8, 1.3, 1.1 and 0.7 μM L^{-1} of NO$_3^-$ + NO$_2^-$, NH$_4^+$, PO$_4^{3-}$, Fe and Mn, respectively (Simmons, 1987).

demonstrated that supersaturations with respect to calcite in calcium carbonate solutions of several thousand times the product of a_{Ca2+} a_{CO32-} at equilibrium, and temperature changes of many tens of degrees centigrade, are insufficient to cause homogeneous nucleation of calcite from aqueous solutions. These supersaturations are never attained in natural waters. For example, the critical radius needed to stabilize an initial calcite domain nucleating homogeneously from aqueous solution is 2.5 times greater than that needed for heterogeneous nucleation. The activation energy necessary to achieve this critical size, however, is 180 times greater! Thus, cement formation requires heterogeneous nucleation of precipitates on grain surfaces and in cavities within and between grains.

Wollast (1971) showed that heterogeneous nucleation of calcite on a surface with a surficial energy close to calcite could begin at a state of supersaturation of four times the product of a_{Ca2+} a_{CO32-} at equilibrium with calcite. For a solution initially at saturation at 0°C with calcite, because of the retrograde solubility of calcite, a temperature variation of only +10°C is necessary to induce heterogeneous calcite precipitation. Furthermore, it is obvious, if calcite crystals are already present, nucleation of crystal cements will not be the rate determining step, but the rate of growth will control the reaction.

It can be demonstrated that grain boundaries or microcracks in sediments are preferred environments of cement nucleation. Consider the environment between two grains of carbonate separated by a void area of radius r (Figure 7.17); the energy (ΔF_{cav}) required for the formation of calcite cement domains in such a cavity of height h and radius r is:

$$\Delta F_{cav} = \pi r^2 h \Delta F_v + 2\pi h (\sigma_{12} - \sigma_{a2}) \tag{7.6}$$

where ΔF_v is the free energy change per unit volume, and σ_{12} and σ_{a2} are, respectively, the interfacial energy between the calcite precipitate and the original carbonate grain, and the interfacial energy between the original carbonate grain and the medium of the original void space. Because σ_{a2} is generally greater than σ_{12}, inspection of equation 7.6 shows that heterogeneous nucleation of a cement in a cavity is accompanied by a surface energy gain. This relationship leads to the interesting conclusion that precipitation of calcite on original allochems or in microcracks in grains can take place at calcite saturations less than those at equilibrium with the bulk solution.

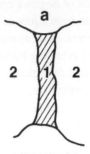

Figure 7.17. Schematic representation of nucleation of calcite in a cavity. (After Wollast, 1971.)

Once a cement has begun to nucleate on a grain its rate of growth is controlled by the difference between the rate at which constituent molecules enter the growing phase from the aqueous phase and the rate at which constituent molecules depart the phase. Two factors that influence calcite cement growth rates are degree of supersaturation and temperature (Figure 7.18). The differences between calcite and aragonite dissolution and precipitation rates also play a role in calcite cement precipitation in meteoric waters. Figure 7.19 demonstrates that in waters undersaturated with respect to calcite, aragonite dissolves faster than calcite, whereas in supersaturated waters calcite grows significantly faster than aragonite. Thus, we would anticipate that in the waters of the meteoric realm, which are generally undersaturated with respect to aragonite, that aragonite would dissolve and calcite would precipitate. Because most potential inhibitors of carbonate dissolution or precipitation also favor this reaction path (see following discussion), calcite precipitation is favored in the meteoric realm. Initial precipitation should occur in cavities and microcracks, as has been observed (Land et al., 1967), and the rate of growth will depend significantly on the degree and maintenance of calcite supersaturation states. The curved nature of the cement-pore boundary, the meniscus cement of Dunham (1971), is a result of equilibrium between the interfacial forces σ_{a1}, σ_{a2} and σ_{12} (Figure 7.17) acting on the surface of a growing cement precipitate:

$$\sigma_{a2} = \sigma_{12} + \sigma_{a1} \cos\theta \qquad (7.7)$$

where θ is the contact angle between phases 1 and 2. This angle is close to $30°$ for meniscus cements, and is that predicted for a cement precipitating from a dilute carbonate solution.

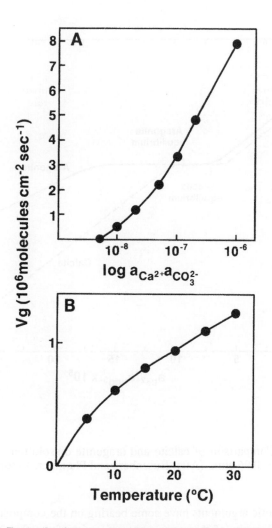

Figure 7.18. A) Rate of calcite growth (V_g) as a function of supersaturation. B) Rate of calcite growth as a function of temperature. (After Wollast, 1971.)

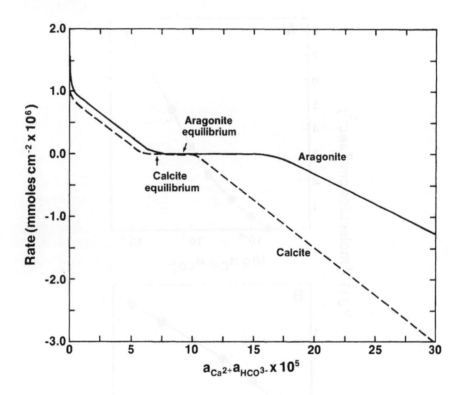

Figure 7.19.   Comparison of calcite and aragonite dissolution and precipitation rates at a $P_{CO_2} = 0.96$ atm. (After Busenberg and Plummer, 1986b.)

Other kinetic arguments have some bearing on the composition of meteoric cements. Some results of the experiments of various investigators are shown in Figure 7.20. The figure illustrates data for calcite precipitation rates on calcite seeds at 25°C and a P_{CO_2} of $10^{-2.5}$ atm (a reasonable meteoric realm value), as a function of calcite saturation state expressed as log (Ω-1). The supersaturations

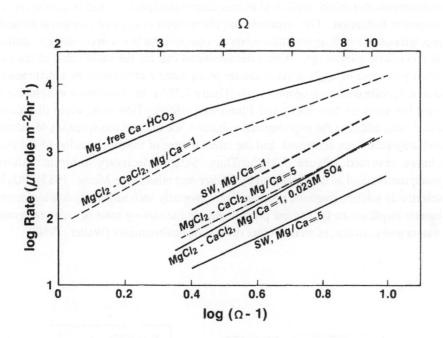

Figure 7.20. Trends in calcite precipitation rates on calcite seeds in different aqueous solutions at 25°C and a P_{CO_2} of $10^{-2.5}$ atm. as a function of calcite saturation state. Solution compositions are indicated on the diagram. (Data from Reddy et al., 1981; Mucci and Morse, 1983; Walter 1986.)

shown range from about 2 to 15 times calcite saturation values. The lower values are characteristic of meteoric environments; for example, calcite saturation states in Bermudian meteoric environments range from 0.5 to 2 (Plummer et al., 1976). The most straightforward interpretation of these experimental kinetic data in terms of meteoric calcite cement formation is that these low ionic strength, low dissolved Mg^{2+}, and low dissolved SO_4^{2-} waters are favorable, with respect to more compositionally-complex aqueous solutions, for the relatively rapid precipitation of calcite. This statement, of course, necessitates supersaturation of the waters with respect to calcite, and implies that in mixing zones of meteoric water with compositionally-different waters, calcite precipitation rates will change.

This latter statement is supported by further laboratory experiments of Walter (1986), in which calcite was precipitated on calcite seeds and aragonite deposited on aragonite seeds. It has been demonstrated in a variety of other experiments that calcite seeds will induce calcite precipitation, and aragonite seeds, aragonite formation. This experimental observation is not true of natural cement precipitates where it appears that substrate composition is not necessarily a control of precipitate mineralogy. Walter demonstrated that for the same value of the ion activity product of $a_{Ca^{2+}}\ a_{CO_3^{2-}}$, calcite precipitated more rapidly on calcite seeds than aragonite does on aragonite seeds (Figure 7.21A), an observation in agreement with the work of Busenberg and Plummer (1986b). However, when dissolved SO_4^{2-} was added to the experimental solutions, the rates of precipitation of calcite and aragonite were decreased, and the relative rates of the precipitation of the two phases reversed (Figure 7.21B). Thus, SO_4^{2-} selectively inhibits calcite precipitation, and as demonstrated by Walter and others (see Morse, 1983), PO_4^{3-} selectively inhibits aragonite precipitation. Obviously such selectivity has important kinetic implications for cement precipitation in the mixing zone between meteoric waters and seawater, as well as other diagenetic environments (Walter, 1986).

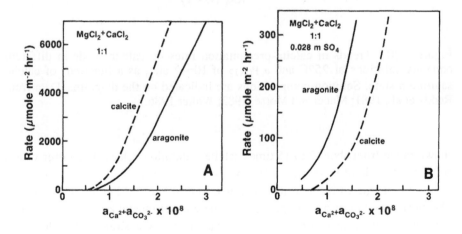

Figure 7.21. A) Relative rates of calcite and aragonite precipitation in solutions with Mg:Ca = 1. B) Relative rates of calcite and aragonite precipitation in solutions with Ca:Mg = 1 and containing dissolved sulfate. Note that in solutions without sulfate calcite precipitates faster, whereas in solutions containing sulfate aragonite precipitates faster. (After Walter, 1986.)

The reasons for the equant sparry habit of cements in the meteoric realm are probably more debatable than the fact that these cements should be calcite. There are two principal models in the literature used to explain carbonate cement morphology; these models are shown schematically in Figure 7.22. In the Folk (1974) model of Mg-poisoning, the Mg^{2+}/Ca^{2+} ratio of the water plays an

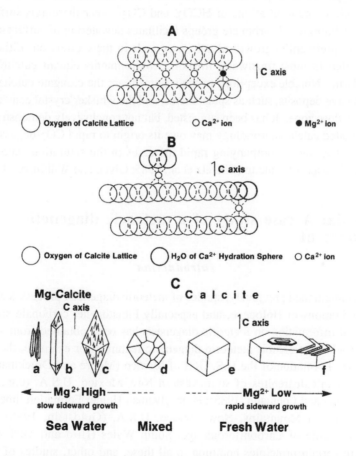

Figure 7.22. Mechanisms of ion incorporation into a calcite crystal and mechanisms of morphology control. In the Folk (1974) model (A), lattice distortion is induced by incorporation of the Mg^{2+} ion on the calcite edge face, whereas in the Lahann (1977) model (B), the mechanism of Ca^{2+} ion adsorption onto faces parallel and perpendicular to the calcite c-axis differs, resulting in differential charge development on crystal faces and morphologic variation. In (C) the postulated change in the habit of calcite with variation in the Mg^{2+} content of water is shown. (After Folk, 1974; Lahann, 1978.)

important role in governing cement morphology. In the Lahann (1978) model, differences in surface potential between growing crystal faces of calcite govern morphology. In the Folk model, where Mg^{2+} is abundant in solutions like seawater, it selectively poisons sideward growth of calcite so that fibrous crystals or elongate rhombs develop from these solutions (Figure 7.22). Lahann argues for a chemical mechanism promoting growth of calcite crystals elongated parallel to the c axis. Higher positive potential on a calcite c-axis surface than on its edge faces leads to higher concentrations of HCO_3^- and CO_3^{2-} near the c-axis surface. The greater availability of carbonate groups facilitates dewatering of surface-adsorbed Ca^{2+}, and preferential growth of calcite crystals in the c direction. Either model predicts that in most freshwater meteoric environments equant calcite crystals should form. Notable exceptions to the statement are the elongate calcite crystals found in cave deposits, such as speleothems, and the whisker crystal cements of the shallow vadose zone. It has been suggested, but not conclusively demonstrated, that this elongated calcite morphology may owe its origin to rapid CO_2-degassing from vadose waters and accompanying rapid increases in the saturation state of these waters with respect to calcite (Chafetz et al., 1985; Given and Wilkinson, 1985a).

Bermuda: A case study of a meteoric diagenetic environment

Introduction

As mentioned previously, studies of meteoric diagenetic processes are biased by investigations of Holocene, and especially Pleistocene, carbonate sequences. Additional information on meteoric diagenesis has been obtained from studies of ancient rocks that exhibit meteoric diagenetic features: for example, the Jurassic Smackover Formation of the U.S. Gulf of Mexico (Moore and Druckman, 1981); Mississippian calcarenites of southwestern New Mexico, U.S.A. (e.g., Meyers, 1978; Meyers and Lohmann, 1985); Pennsylvanian Holder Formation limestones of the Sacramento Mountains, New Mexico, U.S.A. (Goldstein, 1988); and the Brofiscin Oölite of Carboniferous age, South Wales (Hird and Tucker, 1988). There are certain principles common to all these, and other, studies of meteoric diagenesis. To illustrate these principles, the meteoric diagenesis of Bermudian calcarenites is selected as a case study. The reasons for this selection are three-fold: (1) the stratigraphic, sedimentologic, and structural setting of Bermudian limestones is well known, (2) Bermudian meteoric water hydrology and geochemistry, as well as the petrography and geochemistry of the calcarenites, have been intensively studied, and (3) the authors have spent considerable time on

Bermuda investigating the geology and geochemistry of its waters and sediments. In the following discussion, we will expand on the Bermuda case history by integration of observational data from other areas, and theoretical and experimental arguments.

Bermuda is a well-known group of originally 150 limestone islands and islets located 1000 kilometers east of Cape Hatteras, North Carolina, U.S.A., near the western boundary of the North Atlantic Sub-Tropical gyre, the Sargasso Sea. Because of its easy access from the East Coast of the U.S.A., and the location of a marine station on the islands, the Bermuda Biological Station for Research, many geologists and geochemists have studied the geology and biogeochemistry of the modern calcareous sediments and Pleistocene limestones of this locale. There are more than 200 articles written on these subjects. Thus, there is a wealth of information; the case study of Bermuda is based on the works of Bretz (1960), Gross (1964), Land (1966, 1967, 1970), Land et al. (1967), Mackenzie (1964a,b), Ristvet (1971), Lafon and Vacher (1975), Plummer et al. (1976), Harmon et al. (1978, 1981, 1983), and references therein. This case study of Bermudian limestones will illustrate the variety of geological and geochemical data needed to construct models of meteoric diagenetic processes.

Geological Framework

General aspects   Bermuda is a limestone cap overlying weathered volcanics that form the summit of a Cretaceous-to-Paleogene, basaltic volcanic pinnacle rising 4000 meters above the Bermuda abyssal plain. The exposed portion of the 20 to >300 meter thick limestone cap is a complex mosaic of Pleistocene strandline- and shoreline-dune (eolianite) grainstones and interlacing paleosols, some of island-wide extent. The limestones are principally biosparites deposited at times of high sea level (Interglacial ages), and are composed of the fragments of skeletal organisms that lived offshore at the time the presently-exposed beach and dune sands were deposited. The original mineralogy and chemistry of the allochems composing the calcarenites were very similar to the skeletal particles accumulating in the present-day platform environments of Bermuda. The "fossilized" dunes are beach dunes that commonly intercalate seaward with beach deposits. Recognition of this intercalation is one means of locating Pleistocene strandlines and indicates the source of sand for the Pleistocene dunes was the beaches, and indirectly, the subtidal marine deposits extant at that time. The island-wide red soils (terra-rossa paleosols) formed during times of low sea level (Glacial ages) and mark periods of extensive soil formation and development of solutional unconformities.

The alternation of limestone and island-wide terra-rossa paleosols has formed the principal basis for subdivision of the stratigraphic column. However, Pleistocene sea level oscillations of finer time and elevation scale have resulted in intercalation of marine and eolian grainstones, and in some cases, formation of poorly developed, brownish soil horizons of local extent ("accretionary soils"). These latter stratigraphic features and their interpretation are to some extent the reason for the differences in stratigraphic subdivisions shown in Table 7.3.

In the most recent interpretation of the stratigraphy of Bermuda (Vacher and Harmon, 1987; Vacher and Hearty, 1989; Vacher et al., 1989), five rock-stratigraphic units are recognized: the Southampton (~85,000 y BP), Rocky Bay (~125,000 y BP), Belmont (200,000 y BP), Town Hill, and Walsingham (>350,000 y BP), in order of increasing age. The units are distinct time-stratigraphically, and the four older units are bracketed by paleosols ("geosols") of island-wide extent. Each formation is predominantly of eolian facies ("eolianite"), but several units contain one or more marine tongues, and accretionary soils. A geologic map and simplified cross-section are shown in Figure 7.23.

Because of its stratigraphy and relative tectonic stability, Bermuda has been regarded as a "Pleistocene tide gauge" (Land et al., 1967; Harmon et al., 1983), requiring no tectonic correction for deducing Pleistocene sea level history. One evidence for this assertion is that Substage-5e of the last interglacial is recorded at 5 to 6 meters above present sea level in Bermudian calcarenites. This elevation is approximately coincident with other sea level estimations during Substage-5e, as indicated by widespread, coastal marine deposits found along stable shorelines around the world. The late Pleistocene paleosea-level curve for Bermuda is shown in Figure 7.24. The rise and fall of global sea level has alternately wet and "dried" Bermudian limestones. However, much of the rock volume currently above sea level has spent most of its history in the vadose meteoric zone. The present groundwater lenses of Bermuda bathe the Pleistocene calcarenites in freshwaters of the phreatic meteoric realm.

During major sea level falls of the Pleistocene sea, much of the Bermuda Platform was emergent. A Pleistocene eolianite outcrop at the edge of the Bermuda Platform some 12 kilometers north of the islands across the current lagoon floor attests to coastal dune formation at the edge of the Platform during sea level stands lower than present. However, during these low stands, little carbonate skeletal material was produced on the bank, and the constructional processes of dune building gave way to the destructional processes of weathering, chemical erosion, and development of solutional unconformities in underlying calcareous sands and grainstones.

Table 7.3. History of stratigraphic nomenclature in Bermuda. (After Vacher and Harmon, 1987.)

Verrill (1907)	Sayles (1931)	Land et al. (1967)	Vacher (1971, 1973)	Vacher and Harmon (1987)
Paget Formation	Southampton eolianite	Southampton Formation	Paget Formation / Upper Paget Member	Southampton Formation
	McGall's soil			
	Somerset eolianite			
	Signal Hill soil			
	Warwick eolianite			
	St. George's soil	St. George's Soil		
		Spencer's Point Fm	Lower Paget Member	Rocky Bay Formation
	Pembroke eolianite	Pembroke Formation		
	Harrington soil	Harrington Formation		
		Devonshire Formation		Devonshire Member
	Devonshire limestone			Shore Hills Geosol
	Shore Hills soil	Shore Hills Soil	Shore Hills Soil	Belmont Formation
Devonshire Formation				Ord Road Geosol
	Belmont limestone	Belmont Formation	Belmont Formation	Town Hill Formation
				Upper T.H.Member
				Harbour Road Geosol
				Lower T.H. Member
		Soil?	Sub-Belmont soil	Castle Harbour Geosol
Walsingham Formation	Walsingham eolianite	Walsingham Formation	Walsingham Formation	Walsingham Formation

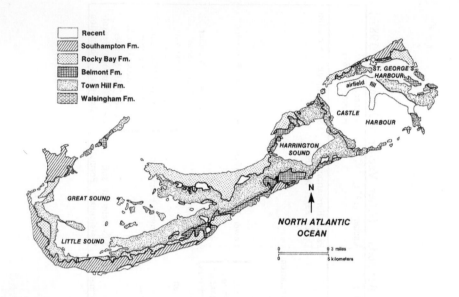

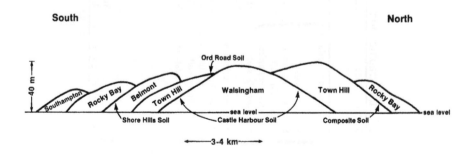

Figure 7.23.   Generalized geologic map and cross-section of Bermuda. (After Vacher, Rowe and Garrett, 1989.)

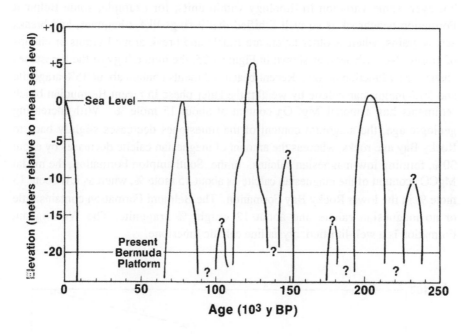

Figure 7.24. Late Pleistocene paleosea-level curve for Bermuda. (After Land et al., 1967, as modified from Harmon et al., 1983, and Vacher and Hearty, 1989.)

Limestone lithology The Bermudian sparites are composed principally of the skeletal remains of organisms living at the time of deposition. These include calcareous corals, molluscs, algae, foraminifera, echinoids, and other less important groups. These are the same skeletal organisms, with few exceptions, that are living in present-day Bermudian marine environments, and whose skeletal remains are found in the beaches and coastal dunes of modern Bermuda. Figure 5.2 and Table 5.2 of Chapter 5 showed an example of the skeletal composition of a calcareous sand found in one modern marine environment on the Bermuda Platform.

The source materials for both the modern calcareous sands and young calcarenites of Bermuda are characterized by a complex metastable carbonate assemblage of skeletal aragonite, magnesian calcite, and calcite. The limestone units shown in Figure 7.25 differ lithologically primarily because of a progressive change in mineralogy, cementation, and porosity with increasing geologic age, a result of the varying extent of diagenetic alteration. There is, however, some variation in lithology within units; for example, some Belmont Formation beachrock is so well lithified that it rings like a hammer, and breaks across grains, whereas other layers are friable and break around grains or clumps of grains. Nevertheless, as shown in Figure 7.25, the mineralogy of the limestones changes as a function of age. Recent beach sediments contain about 35% aragonite and 50% magnesian calcite by weight; the latter phase in recent Bermudian beach sediments has a modal $MgCO_3$ content of about 15 mole %. With increasing geologic age, the aragonite content of the limestones decreases slightly back to Rocky Bay age rocks, whereas the amount of magnesian calcite decreases by about 30%, forming low-magnesian calcites. In the Southampton Formation, the modal $MgCO_3$ content of the magnesian calcite is about 15 mole %, whereas it is about 13 mole % in the lower Rocky Bay Formation. The Belmont Formation contains little or no magnesian calcite and about 12 weight % aragonite. The Walsingham Formation is a well-lithified, crystalline calcitic limestone.

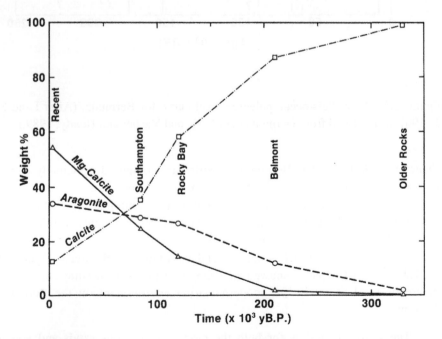

Figure 7.25. Trends in the mineralogical composition of Bermudian grainstones with geologic age. (Modified from Ristvet, 1971.)

The samples used to determine the mineralogy of the limestone units shown in Figure 7.25 were all collected from the present-day vadose zone of Bermuda. The sea level curve of Figure 7.24 shows that limestones younger than 125,000 years exposed on the Bermuda islands have had little contact with phreatic zone waters, whereas the Belmont and Walsingham sediments have spent more time in the saturated zone. Thus the trends in mineralogy with geologic age of Bermudian limestones are dependent to a considerable extent on sea level change. In 125,000 years little has happened to the bulk of Bermudian limestones above present-day sea level because they have had little contact with phreatic zone waters. Those limestones exposed to phreatic meteoric water, now or in the past, however, are extremely altered.

The older limestones, principally the Walsingham limestone, exhibit karst features and are cavernous. Groundwater flow in the Walsingham is essentially unrestricted. The Rocky Bay Formation, however, is a well sorted grainstone, and its permeability is related principally to development of intergranular porosity and increase in effective pore size. The Belmont has considerable moldic porosity, and in the present-day saturated zone contains many pencil-size and larger solution channels. This variation in porosity type and permeability exerts a strong influence on the distribution of groundwater on present-day Bermuda (Vacher, 1978).

Paleosols   The modern-day soil and the paleosols ("geosols") of Bermuda are important in the diagenesis of the limestones. These soils are derived from alteration of subjacent limestones, as well as from eolian dust (Bricker and Mackenzie, 1970). It has been demonstrated that the composition of meteoric water reaching the phreatic zone depends, in part, on the overlying soil type, and the mineralogy of vadose zone sediments through which it passes (Plummer 1970; Plummer et al., 1976). Water passing through calcareous sandy soils may dissolve considerable amounts of calcium carbonate on route to the vadose zone. The amount dissolved depends significantly on the rate of production of CO_2 by oxidation of organic matter in the soil zone and the residence time of water in this zone. An increase in the P_{CO_2} of the soil air in a calcareous sand soil of 100 times increases the solubility of calcite by a factor of 5 (Table 7.4). Calcite and aragonite dissolution rates in essentially Ca^{2+}-HCO_3^- solutions in the soil zone are relatively rapid (cf. Figures 7.19 and 7.20), so depending on the rate of water flow, meteoric water passing through the sandy soil zone can be slightly subsaturated or supersaturated with respect to calcite as it enters the subjacent grainstones. Those waters close to calcite saturation will be most aggressive toward high magnesian calcites and less corrosive toward aragonite and calcite. In contrast to sandy soils, meteoric water passing through the terra-rossa soils of low carbonate content and peat marshes is subsaturated with respect to calcite and highly aggressive to all carbonate phases.

The modern and Pleistocene soils of Bermuda exhibit an array of diagenetic features characteristic of soils developed on young carbonate terrains in subtropical climates throughout the world (James and Choquette, 1984). In general, the upper portion of Bermudian red soils is a zone of dissolution and accumulation of inert materials derived from external sources and weathering of underlying limestone. The lower portion exhibits features characteristic of both dissolution and precipitation of calcium carbonate, e.g., soil pipes, chalky limestone with moldic porosity, soil bases of fine-grained calcite precipitate, and calcite-crystal veining.

Table 7.4. The solubility of calcite in soil solutions as a function of P_{CO_2}.

P_{CO_2} (atm)	Ca^{2+} Concentration	
	moles kg^{-1}	ppm
$10^{-3.5}$	$10^{-3.30}$	20
$10^{-3.0}$	$10^{-3.13}$	29
$10^{-2.5}$	$10^{-3.03}$	43
$10^{-2.0}$	$10^{-2.80}$	63
$10^{-1.5}$	$10^{-2.63}$	93
$10^{-1.0}$	$10^{-2.47}$	136
10^{0}	$10^{-2.13}$	294

Hydrology   There are two major factors influencing the distribution of fresh groundwaters on Bermuda (Vacher, 1978). The first is that the limestones of Bermuda represent a lateral succession of stratigraphic units (Figure 7.23), rather than a layer-cake arrangement. The second is the diagenetic history of the

biosparites. Because of the depositional pattern of lateral accretion, the upper
saturated zone, including the phreatic freshwater lens, is laterally continuous across
island slices (Figure 7.26) of different formations with different diagenetic
characteristics. Because more than 80% of the saturated zone lies within the Rocky
Bay or Town Hill Formations, it is these limestones that are most important in the
occurrence of groundwater on Bermuda. Their relative permeabilities play a
major role in the configuration of the groundwater lens. The Rocky Bay
Formation is of lower permeability than the Town Hill; the Town Hill/Rocky Bay
ratio of hydraulic conductivities is 14, as derived from a fit of a Ghyben-Herzberg-
Dupuit model to the observed distribution of relative salinities in the Devonshire
water lens (Figure 7.26). This lens covers an area more than twice that of the
combined area of all other Bermudian freshwater lenses. The non-axisymmetric
distribution of freshwater in the Devonshire lens results from the effects of the
across-the-island variations in the permeability of Rocky Bay and Town Hill
formations on the Ghyben-Herzberg lens and the mixing zone.

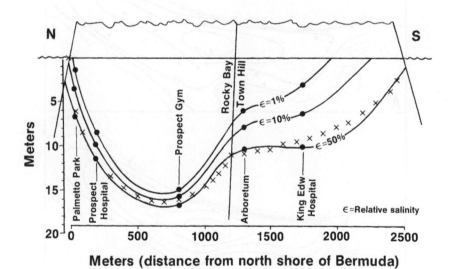

Figure 7.26. Hydrogeologic cross-section of the Devonshire groundwater lens.
The X's denote a calculated fit to the 50% isoline of relative salinity (ε) using the
Ghyben-Herzberg-Dupuit model. (After Vacher, 1974, 1978.)

Figure 7.27 illustrates the three-dimensional geometry of the Devonshire freshwater lens. Figure 7.27A shows the thickness of the Devonshire Ghyben-Herzberg lens, bounded by the water table and the freshwater-seawater interface, and Figure 7.27B illustrates the thickness of the freshwater nucleus, defined as less than 1% relative salinity, contained within the Ghyben-Herzberg lens. Comparison of Figures 7.26 and 7.27 shows that the bulk of the fresh groundwater of the Devonshire lens is limited to the Rocky Bay Formation, because the lens tends to swell in that unit, and the zone of mixing with underlying seawater is thinnest there. The water chemistry of this lens, and the relationship of this chemistry and the geohydrology to the diagenesis of Bermudian limestones, are discussed in a further section.

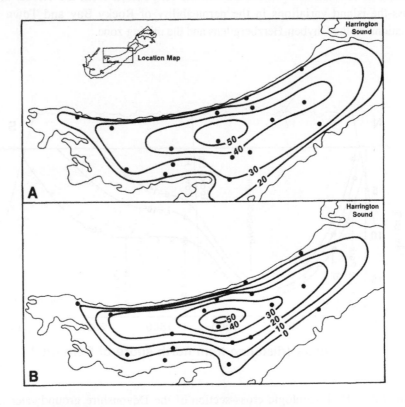

Figure 7.27. Maps showing the thickness of (A) the interface-bounded lens in feet below the water table to the iso-surface of 50% relative salinity, and (B) the fresh-water nucleus in feet below the water table to the iso-surface of 1% relative salinity for groundwaters in the central region of Bermuda. (After Vacher, 1974.)

Limestone Chemistry and Isotopic Composition

As might be anticipated from the mineralogical trends observed in Bermudian limestones, these rocks also exhibit bulk chemical and isotopic changes with increasing geologic age (Figure 7.28). The Mg, Sr, K, total Fe, Na and Cl

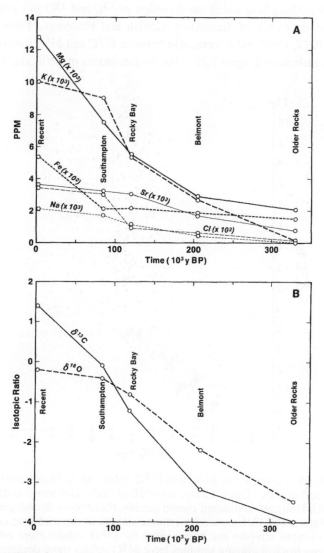

Figure 7.28. Trends in (A) chemical composition and (B) carbon and oxygen isotopic composition of Bermudian grainstones with geologic age. (Modified from Ristvet, 1971.)

contents all decrease with age, a result of the alteration of original aragonite and magnesian calcite to calcite. ^{13}C and ^{18}O are depleted in the older limestones relative to recent Bermudian skeletal sands. This depletion is the result of the precipitation of soil derived calcite as cement and veins in the biosparites, as well as the replacement of magnesian calcite and aragonite by calcite.

Many authors have noted the depletion of ^{13}C and ^{18}O in carbonate rocks with increasing extent of diagenesis (Gavish and Friedman, 1969; Allan and Matthews, 1977, 1982) and a correlation between $\delta^{13}C$ and $\delta^{18}O$ for Holocene and Pleistocene carbonates (Figure 7.29). For the limestones of Bermuda, the depletion

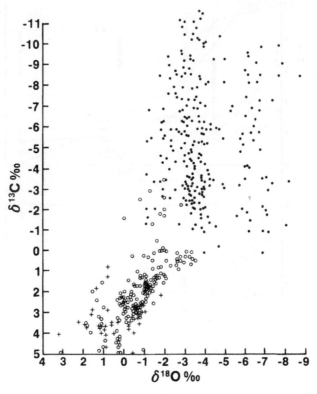

Figure 7.29. A scatter plot of $\delta^{13}C$ and $\delta^{18}O$ values of Holocene and Pleistocene carbonate sediments. Open circle, unlithified bulk Holocene sediments; +'s, lithified bulk Holocene sediment; closed circles, Pleistocene limestones altered by meteoric water. For Pleistocene limestones altered in the meteoric environment, temperate climate samples are represented by $\delta^{18}O$ values more enriched than -5‰, whereas tropical climate samples have $\delta^{18}O$ values more depleted than -5‰. The large depletion in ^{13}C is because the sediment samples are primarily from vadose environments. (After Land, 1986.)

of ^{13}C implies alteration in a relatively open system in which soil-gas CO_2 derived from oxidation of organic matter and depleted in ^{13}C is produced in the vadose zone. This "light" CO_2 dissolves in vadose water mixing with "heavier" carbon derived from dissolution of the limestones. The resultant vadose water contains HCO_3^- of intermediate $\delta^{13}C$. When this HCO_3^- precipitates during cementation or replacement reactions, the ultimate product is a cement, vein, or a fossil cast of ^{13}C depleted calcite. The process is shown schematically in Figure 7.30. It is not surprising that the Bermudian sparites are depleted in ^{13}C with increasing geologic age; most of the volume of Bermudian limestone has been subaerially exposed for much of its existence, and in this sub-tropical climate, which has not changed drastically during the late Pleistocene, soil plant and humus materials are a relatively abundant source of vadose zone CO_2 because of processes of organic decay. It should be kept in mind, however, that in a phreatic meteoric environment where the rock:water ratio is high, the carbon reservoir in the carbonate minerals can be a more important source of carbon for replacement reactions.

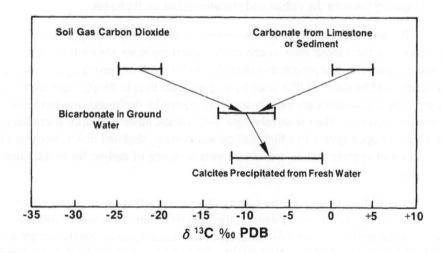

Figure 7.30. Diagram illustrating the effect of mixing carbon from metastable marine carbonate sediments with carbon derived from soil-gas CO_2 on the $\delta^{13}C$ composition of limestones undergoing freshwater diagenesis. (After Allan and Matthews, 1982.)

This latter statement is supported by carbon isotope data on limestone samples collected from the Pleistocene carbonate sequences of Barbados Island in the western Atlantic Ocean. Figure 7.31 shows two profiles of $\delta^{13}C$ for borehole limestone samples. All the data are whole-rock values obtained from analyses of the calcitic matrices surrounding corals. The profile of $\delta^{13}C$ versus depth for limestone samples collected wholly in the vadose environment shows a gradual enrichment in ^{13}C with depth (Figure 7.31A). The depletion in ^{13}C at the top of the profile presumably represents "light" carbon in the limestone calcite derived from decay of organic matter in the soil and shallow vadose environment. With increasing depth, replacement reactions take place in a vadose regime where more of the HCO_3^- is derived by dissolution of limestone enriched in ^{13}C. In Figure 7.31B the $\delta^{13}C$ profile shows a sharp inflection at the boundary between the vadose and phreatic regime. Water in the Barbados freshwater lens has only a short period of contact with soil gas CO_2 as it initially percolates through the soil zone. The water flows laterally down the hydrologic gradient and is relatively isolated from additional soil CO_2. As it moves, it dissolves metastable carbonate phases and precipitates calcite. The HCO_3^- in the water becomes progressively enriched in carbon derived from the limestones, and the $\delta^{13}C$ of the calcites in the limestones in the present-day phreatic zone increases down the hydrologic gradient. The inflection of $\delta^{13}C$ of limestones in the borehole of Figure 7.31B is coincident with the boundary between the vadose and phreatic realms on Barbados.

Because the bulk of the limestone volume of Bermuda has spent most of its existence in the vadose zone, much of the limestone mass shows the imprint of vadose diagenesis, and progressive depletion in ^{13}C with increasing age. However, limestones older than 125,000 years have spent more time in the phreatic meteoric zone. These limestones are more extensively altered by freshwater and freshwater-seawater mixtures. Their relatively light $\delta^{13}C$ values, however, imply alteration in a relatively open system in which soil carbon dioxide, depleted in ^{13}C because of oxidation of organic matter, was an important source of carbon for replacement reactions.

The $\delta^{18}O$ signature of the calcarenites is controlled by the temperature of deposition of replacement carbonates, the $^{18}O/^{16}O$ ratio of the water from which they precipitated, the original $\delta^{18}O$ values of the skeletal grains, and the water/rock ratio, among other factors. Gross (1964) showed that calcite precipitates from freshwater, stalactites, soil bases, and secondary calcites, have $\delta^{18}O$ values ranging between -2.7‰ to -4.7‰. This range is different from that of -0.2‰ to 1.2‰ for recent marine sediments. Thus it appears that the age trend of Figure 7.28B is best interpreted as reflecting dissolution of original metastable grains of the Bermudian limestones and precipitation of calcite in the meteoric freshwater environment, the waters of which are depleted in ^{18}O. Rocks younger than 125,000 years

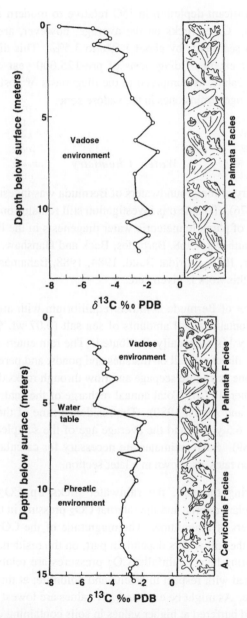

Figure 7.31. Depth trends in the bulk $\delta^{13}C$ values of two cores of Barbados limestones. (A) Profile through the vadose environment, and (B) profile through the vadose-phreatic environments and the water table contact. (After Allan and Matthews, 1977.)

do not show a significant depletion in ^{18}O relative to modern marine calcareous sands of Bermuda. Older rocks on the average, however, are depleted in ^{18}O relative to modern sediments by about 1.5‰ to 3.5‰. This difference probably reflects the greater extent of diagenesis of pre-125,000 year-old grainstones in phreatic meteoric realms, as compared to the diagenesis of post-125,000-year-old Rocky Hill and younger grainstones in the vadose zone.

Water Chemistry

The chemistry of the groundwaters of Bermuda was investigated in detail by Plummer et al. (1976). To date this investigation still remains one of the few major integrative studies of phreatic meteoric water diagenesis in the literature (see also, e.g., Harris and Matthews, 1968, Barbados; Back and Hanshaw, 1970, Florida and Yucatan; Plummer, 1977, Florida; Budd, 1984, 1988, Bahamas). A summary of the conclusions of this work is given here.

The rainwater of Bermuda is in near equilibrium with atmospheric $P_{CO_2} = 10^{-3.5}$ atm., and contains small amounts of sea salt (0.07 wt. % seawater). The rainfall of 147 cm y^{-1} is seasonally distributed. The rain enters the saturated zone by two main paths: direct rainfall on marshes and ponds, and percolation downward from the vadose zone as vadose seepage and flow through rocks during times of soil water excess (Vacher, 1978). Total annual recharge of the saturated zone is about 40 cm y^{-1}(Vacher and Ayers, 1980). The residence time of the groundwater has been calculated as 6.5 years, and the average age of the sampled water as 4 years (Vacher et al., 1989). Such estimates are necessary for calculations of carbonate mineral stabilization rates, as shown in a later section.

Upon entering the soil, the rainwater picks up CO_2 from oxidative decomposition of plant debris, and the internal CO_2 pressure in the water increases by nearly two orders of magnitude. The magnitude of the CO_2 increase and the saturation state of the soil waters depend, in part, on the residence time of water in the soil zone. During heavy rainfalls, CO_2 pressures are relatively low, and the water is subsaturated with respect to calcite and dolomite; at times of low rainfall, the converse is true. As might be expected, pH values are lowest in soils containing little carbonate, and buffered at higher values in soils containing calcite.

Vadose water moving by vadose flow through fractures has been sampled from the roofs of caves (Frantz, 1971) in the Town Hill and Walsingham formations. This water is lower in dissolved CO_2 and contains less Ca^{2+} and HCO_3^-

than soil waters, implying cementation of the open fractures as CO_2 evades and calcite precipitates. The sub-soil cavernous areas of Bermudian limestones are laced with microcracks and fractures partially filled with calcite cement.

The groundwaters of Bermuda are a mixture of calcium bicarbonate water with a seawater component. Plummer et al. (1976) calculated the seawater addition to the groundwaters sampled, and the contribution of dissolved calcium, magnesium and strontium from dissolution of carbonate minerals to the groundwaters. The seawater component amounts to 0.6 to 79% of the range of groundwater compositions sampled. Details of the geochemistry of the upper portion of the saturated zone within the Devonshire freshwater lens of Bermuda are shown in Figures 7.32 and 7.33.

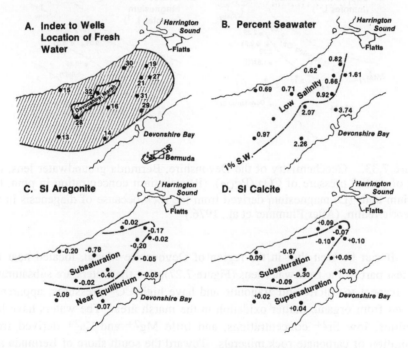

Figure 7.32. Geochemistry of the Devonshire, Bermuda groundwater lens. (A) Well samples and location of freshwater, (B) percentage of seawater in the groundwater, (C) saturation with respect to aragonite, (D) saturation with respect to calcite. SI (= Ω) is the saturation index. (After Plummer et al., 1976.)

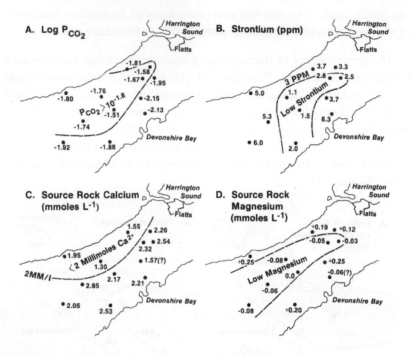

Figure 7.33.   Geochemistry of the Devonshire, Bermuda groundwater lens. (A) Log of partial pressure of CO_2 (P_{CO2}), (B) strontium concentration in ppm, (C) calcium and (D) magnesium derived from the rock because of diagenesis in the meteoric realm. (After Plummer et al., 1976.)

It can be seen that in the region of Devonshire Marsh, located near the thickest part of the freshwater lens (Figure 7.27), that the waters are subsaturated with respect to calcite and aragonite and have high CO_2 pressures, apparently derived from organic matter oxidation in the marsh area. The waters have low salinities, low Sr^{2+} concentrations, and little Mg^{2+} and Ca^{2+} derived from dissolution of carbonate rock minerals. Toward the south shore of Bermuda and eastward from Devonshire Marsh, the salinity of the waters increases, and the saturation state approaches near-equilibrium with calcite, and supersaturation with respect to aragonite. Lower P_{CO2} values characterize the waters farther away from Devonshire Marsh.

Based on these observations of the chemistry of the Devonshire groundwater lens, and on calculations whose results are portrayed in Figure 7.34, Plummer et al. (1976) concluded the following:

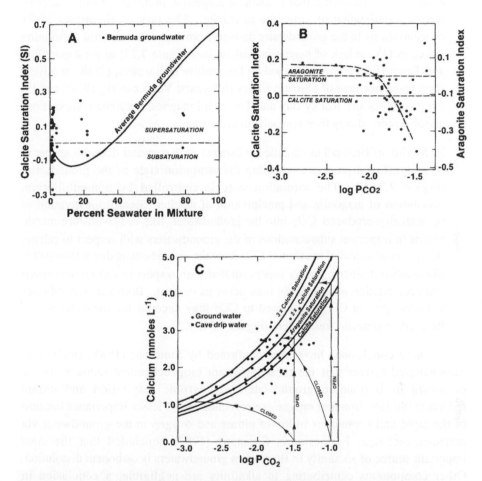

Figure 7.34. The results of a variety of calculations aimed at determining the processes controlling the geochemistry of Bermudian groundwaters. (A) Calculation of mixing the average composition Bermuda groundwater with seawater showing that such mixing is not a dominant process in controlling saturation state. (B) Comparison of the calcite saturation state of groundwaters with their computed internal CO_2 pressures (P_{CO2}) illustrating a weak, but likely, trend towards subsaturation with increasing P_{CO2}. (C) Total groundwater dissolved calcium as a function of P_{CO2} at 25°C. Different theoretical trends are shown for different saturation conditions as indicated. Theoretical reaction paths (arrows) are also shown for systems both open and closed to CO_2 as they approach equilibrium with aragonite. (After Plummer et al., 1976.)

1. Sr^{2+} in the groundwater is derived from the dissolution of aragonite at higher than atmospheric CO_2 levels. The high Sr^{2+}/Ca^{2+} ratios of the

groundwater, three-fold that of skeletal aragonite, indicate the importance of the recrystallization of aragonite to calcite. The lack of significant Mg^{2+} concentrations in the groundwater derived from carbonate mineral dissolution indicates (1) the lack of magnesian calcite (see Figure 7.25) in the Rocky Bay and Town Hill limestones bathed by Devonshire lens waters, (2) the relatively short residence time of phreatic waters (6.5 years; Vacher et al., 1989), and (3) the possibility that Rocky Bay and Town Hill magnesian calcite compositions dissolve more slowly than aragonite in these waters.

2. Mixing of Bermudian phreatic freshwaters with seawater does not appear to be a significant process controlling the saturation state of the groundwater (Figure 7.34A). The saturation state is controlled by non-equilibrium dissolution of aragonite and precipitation of calcite. Significant invasion of biologically-produced CO_2 into the groundwater, like at Devonshire marsh, results in important subsaturations of the groundwaters with respect to calcite. In regions of potential evasion of CO_2 from the groundwaters down flow paths, like the South Shore, waters supersaturated with respect to calcite may result and precipitation of this phase may occur as cement. Both soil-groundwater systems open to CO_2 and closed to CO_2 may account for the evolution of Bermudian phreatic freshwaters (Figure 7.34B,C).

These conclusions have been confirmed by Simmons (1983, 1987), who demonstrated further that the most important biogeochemical redox reactions occurring in Bermudian groundwaters are aerobic respiration and nitrate reduction. Sulfate, iron and manganese reduction are of lesser importance because of the rapid and continuous influx of nitrate and oxygen to the groundwater via meteoric recharge. Furthermore, Simmons (1987) concluded that the most important source of alkalinity in Bermudian groundwaters is carbonate dissolution. Other components contributing to alkalinity are negligible, a conclusion in agreement with Plummer et al.'s (1976) observations.

Carbonate Mass Transfer

The rate at which metastable phases dissolve or are replaced is an important problem in carbonate diagenesis. Carbonate mineral assemblages persist metastably in environments where they should have altered to stable assemblages. The question is "what are the time scales of these alterations"? They are certainly variable ranging from a few thousand to a few hundreds of millions of years. Even calcites in very old limestones show chemical and structural heterogeneities, indicating that the stabilization of these phases is not complete. Unfortunately, it is difficult, but not impossible, to apply directly the lessons learned about carbonate mineral dissolution and precipitation in the laboratory to natural environments.

Because of all the data available concerning Bermudian limestone diagenesis, several authors (Lafon and Vacher, 1975; Plummer et al., 1976; Vacher, 1978; Vacher et al., 1989) have attempted to determine rates of carbonate mineral stabilization in Bermudian calcarenites and the mass transfer involved in the process.

One approach used to date (Plummer et al., 1976; Halley and Harris, 1979; Budd, 1984, 1988) is to assume that the rate of conversion of aragonite to calcite (dA/dt) in the phreatic freshwater zone is directly proportional to the mass of aragonite present at a given time, and that the surface area of aragonite per unit mass of aragonite present in the phreatic zone does not change with time:

$$dA/dt = kA \qquad\qquad (7.8)$$

where A is the amount of aragonite present in the phreatic zone (cm^3 m^{-3}) at time t in years, and k is the first-order rate constant for the replacement process in reciprocal years (y^{-1}). In integrated form, the equation is:

$$A = A_0\, e^{-kt} \qquad\qquad (7.9)$$

where A_0 is the amount of aragonite present at the beginning of the conversion process in the initial rock mass. This first-order equation is a simplification of a very complex series of processes, but it appears to work well.

The rate of aragonite stabilization in the Bermudian groundwaters has been evaluated by Vacher et al. (1989) using a revised estimate of the average age of the sampled groundwater of 4 years (Plummer et al., 1976, estimated 100 years). Using this age, and chemical data from Plummer et al. (1976)-- 2.1 mmoles L^{-1} as the average Ca^{2+} derived from dissolution of carbonate minerals in the groundwaters, the observation that the Sr/Ca ratio in the groundwater is three-times that in aragonite, and a porosity of 20% -- they calculated that 7.2 cm^3 y^{-1} of aragonite is converted to calcite for each m^3 of the Bermudian phreatic freshwater zone. For South Joulters Cay, Bahamas, 1000-year old oölite, Halley and Harris (1979) calculated a conversion rate of 27.5 to 55 cm^3 m^{-3} y^{-1}. For comparison, Budd (1988) estimates the yearly rates of calcite precipitation resulting from the aragonite-to-calcite transformation in freshwater lenses on two small, Holocene oöid sand islands in the Schooner Cays, Bahamas, as ranging from 11 to 276 cm^3 m^{-3}.

Solution of equation 7.9 for Bermuda, where $dA/dt = 7.2$ cm^3 m^{-3} y^{-1}, and A $= 80,000$ cm^3 m^{-3} gives a value of $k = 9 \times 10^{-5}$ y^{-1}. This k applied to the aragonite stabilization rate of calcarenite in Bermuda and Joulters Cay gives a half-life of:

$$T_{\frac{1}{2}} = \frac{-\ln\left(\frac{1}{2}\right)}{k} = 7700 \text{ years} \tag{7.10}$$

In 51,000 years of continuous phreatic diagenesis of calcarenite in Bermuda and Joulters, 99% of the original aragonite should be stabilized. Budd (1988) obtained values for complete conversion of an aragonite precursor to calcite of 4,700 - 15,600 years in the phreatic freshwater zone, and 8,700 - 60,000 years in the upper portion of the mixing zone at Schooner Cays. These estimates from various localities are remarkably close, and suggest that phreatic meteoric diagenesis of metastable aragonite can result in a calcite limestone in less than about 50,000 years.

In contrast, Lafon and Vacher (1975) estimated the half-life of aragonite in the vadose zone of Bermuda as about 230,000 years, and the half-life of high-magnesian calcite as only 60,000 years. Minor corrections need to be made to these estimates, because of new estimates of limestone ages for Bermuda, but the fact still remains that stabilization in the vadose freshwater zone is much slower than in the phreatic zone (Land, 1970; Steinen and Matthews, 1973; Steinen, 1974; Vacher et al., 1989).

In conclusion, it can be calculated that 850 m^3 of the Bermuda islands are lost each year by dissolution of calcarenites and transported to the sea by groundwater. This rate is compared with estimates from other areas in Table 7.5.

Table 7.5. Calculated solution rates for small limestone islands. (After Anthony et al., 1989.)

Island⇒	Majuro	Enewetak	Bermuda
latitude (^{o}N)	7	12	32
area* (km^2)	1.4	1.06	14
rainfall (mm y^{-1})	3,560	1,470	1,460
recharge (mm y^{-1})	1,780	500	370
flushing (L y^{-1})	2.5×10^9	3.0×10^8	5.2×10^9
source rock (SR) (kg y^{-1})	3.7×10^{-4}	4.0×10^{-4}	3.3×10^{-4}
SR flushed (kg y^{-1})	9.3×10^5	1.2×10^5	1.7×10^5
volume dissolved (m^3 y^{-1})	465	60	850

*Area underlain by freshwater lens.

A brief synthesis of meteoric diagenesis

Diagenetic Stages

The case study of Bermuda should provide the reader with some appreciation of meteoric diagenetic processes operating in one geologic environment. In this environment, because of the original presence of the metastable mineral assemblage of low and high-magnesian calcites and Sr-rich aragonite, the pathway of diagenesis is strongly controlled by mineral alteration reactions in the phreatic zone. However, as shown in Figure 7.15, the intrinsic controls on meteoric diagenesis involve not only original mineralogy but also the grain size, porosity and permeability of the carbonate sediments undergoing diagenesis. Extrinsic controls are climatic conditions, vegetation type, standing crop and time. In this section we discuss the diagenesis of Bermudian calcarenites in relationship to other studies of meteoric diagenetic processes and arrive at some generalizations.

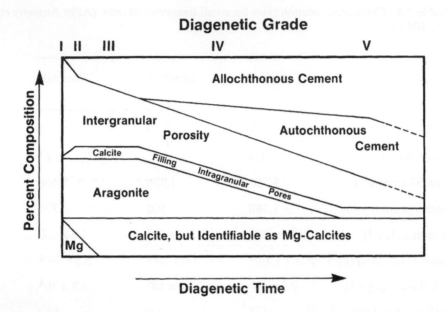

Figure 7.35. Schematic diagram of mineral balance versus diagenetic time for vadose and phreatic meteoric alteration of Bermudian limestones. These trends are mainly based on stratigraphic analysis and hand specimen petrography and geochemistry. (After Land et al., 1967.)

The idealized scheme of meteoric diagenesis of Bermudian grainstones is illustrated in Figure 7.35. Five diagenetic stages are recognized: (I) the initial sediment of unconsolidated biogenic sand which was composed mainly of a mixture of molluscan (aragonite and calcite), coral and *Halimeda* (aragonite) debris, forams, coralline algae, and echinoid (high-magnesian calcite) fragments, and low magnesian calcite lithoclasts of Pleistocene limestone; (II) a sediment in which the first cement of low-magnesian calcite is found at grain contacts and in the micro-pores of skeletal grains; (III) a sediment with only low-magnesian calcite and aragonite, the Mg^{2+} in the original magnesian calcite grains has been lost to solution; (IV) dissolution of aragonite and secondary porosity development are exhibited at this stage; and (V) this stage represents a limestone composed almost wholly of calcite and remaining porosity. In stages II and III, most of the calcite cement is derived from outside the cemented grainstones (allochthonous cement). The components of this cement have their origin in the leaching of carbonate minerals during chemical weathering and soil formation in overlying carbonate sediments. In stages IV and V, much of the calcite cement is derived internally from the grainstones (autochthonous cement), mainly by dissolution of aragonite. The components of this cement are transported over small distances. The geochemical changes associated with this procession of diagenetic stages are shown in Figure 7.28.

These same diagenetic changes have been observed in studies of meteoric diagenesis of young, mineralogically immature carbonate sediments from other areas, e.g., Mediterranean coast of Israel (Gavish and Friedman, 1969), Florida and the Bahamas (Stehli and Hower, 1961), Barbados (Pingitore, 1976), and Majuro atoll, Marshall Islands (Anthony et al., 1989). All of these studies demonstrate that, in general, the first diagenetic event to take place in the meteoric regime is the loss of magnesium from the highly chemically reactive high-magnesian calcites. This event is then followed by the gradual disappearance of aragonite and replacement by calcite during the $CaCO_3$ dissolution-precipitation process. During this latter event, Sr^{2+} is released to the meteoric water. The remarkable similarity between the trends in diagenetic texture, mineralogy, and geochemistry with progressive geologic age exhibited by Bermudian biosparites and those of carbonate sediments of the Mediterranean coast of Israel (Figure 7.36) supports the argument that the Bermudian diagenetic sequence is not an exception. As shown on pp. 291-292, this general diagenetic pathway may be modified because of relative differences in the chemical reactivity of carbonate skeletal mineral compositions resulting from surface area and microstructural effects.

The observed progression in diagenetic stage found in young polymineralic carbonate sediments may be related to geologic age (the original interpretation for the carbonates of Israel), but also may be a "snapshot" of the progress of diagenesis throughout the meteoric environment (Figure 7.37). For Bermudian biosparites, a large volume of post-125,000-year-old limestone has never been exposed to phreatic waters and has undergone only minor alteration in the vadose zone. Older limestones, however, have been bathed in phreatic meteoric waters and have been altered to nearly pure calcitic limestone. Thus, the diagenetic stages exhibited by Bermudian biosparites are not simply the result of progressive chemical and mineralogical changes with increasing geologic age, but also reflect the duration of time that a particular rock unit has spent in the various zones of meteoric diagenesis. This diagenetic time for Bermuda is mainly a consequence of the rise and fall of Pleistocene sea level.

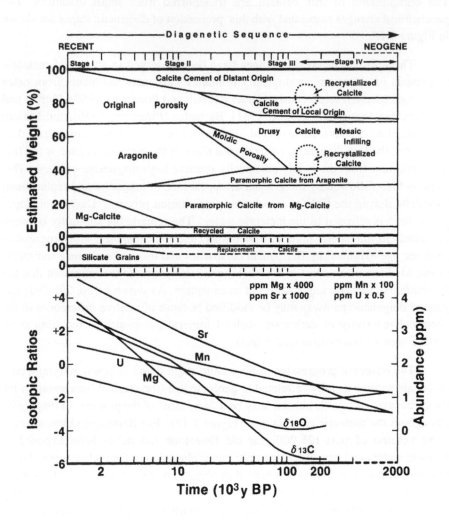

Figure 7.36.   Diagenetic sequence of textural, mineralogical and geochemical changes for the carbonate sediments of the Mediterranean coast of Israel. In this case the diagenetic sequence is interpreted to be related to the progressive aging of the carbonate sediments. (After Gavish and Friedman, 1969.)

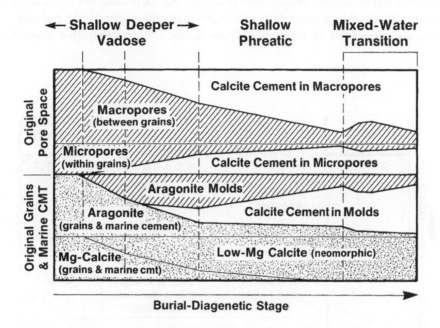

Figure 7.37. Idealized burial diagenetic evolution of a polymineralic assemblage of carbonate minerals as they pass through meteoric diagenetic zones. (After James and Choquette, 1984.)

In the case of the burial diagenesis of a mineralogically immature carbonate sediment (Figure 7.37), the rate of burial, and consequently the time spent in the various meteoric diagenetic zones, will determine to some degree how strong is the imprint of the various diagenetic stages on the lithic unit. A relatively long residence time in the phreatic meteoric zone will result in extensive diagenesis and a strong imprint of the textural, mineralogical, and geochemical features characterizing this environment. Because many shallow-water carbonate sequences "shoal up" to sea level, it is inevitable that these sequences will probably undergo diagenesis in the meteoric realm at some time in their history.

Effect of Original Mineralogy

The Bermudian biosparites are composed of a mixture of calcite, magnesian calcite and aragonite. The diagenetic pathway of these sediments is controlled to a significant degree by mineral-controlled alteration, especially the dissolution of aragonite and subsequent precipitation of calcite. For carbonates like chalk, lacking the very reactive high magnesian calcite and aragonite minerals, the pathway of meteoric diagenesis is different (Figure 7.38). Water-controlled alteration

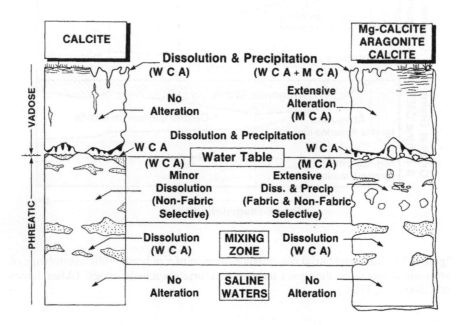

Figure 7.38. A schematic representation illustrating the differences between diagenetic processes affecting a carbonate composed of calcite and one composed of a metastable polymineralic assemblage of magnesian calcite, aragonite and calcite. Water-controlled alteration (WCA) processes are driven principally by invasion and evasion of CO_2, whereas mineral-controlled alteration (MCA) is mainly governed by the differences in the chemical reactivity of the carbonate minerals. (After James and Choquette, 1984.)

of these sediments appears to be due to invasion and evasion of CO_2. The major distinguishable process that operates during alteration of these calcitic carbonates is calcite dissolution. Calcite saturation seems to be very quickly obtained down the path of flowing groundwater, as has been demonstrated, for example, for the Lincolnshire Limestone of southwest England (Smalley et al., 1988). It is likely, but has not been demonstrated, that mineral controlled processes may also play a role even in this type of meteoric diagenesis. Precursor calcites, because of chemical heterogeneities, and structural disorder, may be more soluble and kinetically reactive than calcite cements precipitated in this environment. This difference would provide a driving mechanism, similar to the aragonite to calcite transformation, for diagenesis.

An example of the meteoric diagenesis of a calcitic limestone is the Eocene to Miocene age, bryozoan-rich, carbonate sands of the Gambier Limestone of southern Australia (Reeckman, 1988; James and Bone, 1989). These sediments were originally composed almost entirely of calcite containing variable amounts of magnesium. During deposition on the sea floor, the Gambier sediments were little altered. Subsequently, they appear to have been buried to depths no greater than 200 meters, and have been in the vadose and phreatic meteoric environments for more than 10 million years. Despite this considerable exposure to freshwater, these sediments are poorly cemented; the only cement of any consequence is epitaxial on echinoid debris. The most obvious diagenetic feature of the sediments is meteoric dissolution. Karst features, such as caves and speleothems, are numerous. There appears to be an order of chemical reactivity of the specific bioclasts from most to least reactive: bryozoans > miliolid foraminifers > echinoids > planktonic foraminifers, other benthic foraminifers, and brachiopods. The latter three bioclasts show no evidence of dissolution. Compaction has resulted in mechanical readjustment and fracturing of grains and interpenetrating grains indicative of pressure solution. Some of the $CaCO_3$ for cement precipitation in the Gambier Limestone probably came from pressure solution. The original high-magnesian calcite echinoid, benthic foraminifera, and bryozoan bioclasts have lost magnesium, and a part of the $CaCO_3$ liberated by pressure solution may have precipitated in the subsequent low magnesian calcite skeletons.

It is important to recognize that originally calcite-rich carbonate sediments show a different style of meteoric diagenesis than aragonite-rich sediments. The latter today and in the recent past are found in subtropical and tropical environments, whereas the former are more characteristic of higher latitude environments. James and Bone (1989) argue that temperate calcite sediments are a better model of the diagenesis of sediments of all ages than the varied mineralogical assemblage of calcite, magnesian calcite, and aragonite composing tropical carbonates on which most models of meteoric diagenesis are based. As mentioned previously, the depositional composition of carbonate sediments varied throughout the Phanerozoic, because of evolutionary changes in skeletal mineralogy and because of changes in the chemistry of Earth's surface environment. Low original aragonite content in Phanerozoic shallow-water carbonates was probably not unusual, and thus, it is no wonder that many of these sediments that have been exposed subaerially exhibit little or no petrographic record of meteoric diagenesis.

There is a need to explore the chemical and isotopic inhomogeneities and structure of calcite as it continues to stabilize in the diagenetic environment. These features may provide valuable clues to the processes of stabilization of calcite sediments in the meteoric, and other, diagenetic environments.

Climatic Effects

The Bermudian biosparites are undergoing meteoric diagenesis in a sub-tropical climate. Their previous diagenetic history also reflects this climatic regime. The climatic variables of rainfall and temperature influence both the intensity and rate of carbonate alteration, particularly in the vadose zone. Rainfall and rate of evaporation determine the availability of water in the meteoric zone, and temperature influences the rates of mineral alteration reactions and the rate of organic matter decomposition leading to CO_2 production. Figure 7.39 shows one example of the effects that different climates can produce during meteoric diagenesis.

Harmon et al. (1975) demonstrated a highly correlated (0.966) positive linear relation between the concentration of HCO_3^- in North American ground-waters resulting from weathering of carbonate rocks and the temperature of the water. Because groundwater temperature is a good index of mean annual air temperature, the HCO_3^- content of the groundwaters is directly related to the climatic variable of temperature. High temperatures promote rapid decomposition of organic matter in the soil zone and rapid production of soil gas CO_2. This process drives the schematic carbonate dissolution reaction

$$CaCO_3 + CO_2 + H_2O \rightarrow Ca^{2+} + 2HCO_3^- \tag{7.11}$$

to the right, and along with the fact that the intrinsic rate constants for carbonate mineral dissolution increase with increasing temperature, leads to relatively rapid production of HCO_3^- in soil and groundwaters. Provided the residence times of different groundwaters are not vastly different, their HCO_3^- content will be directly correlated with temperature. Thus alteration of a metastable carbonate mineral assemblage in the meteoric environment of a warm climate with sufficient availability of water will be faster than in a cold, wet climate. Obviously in a dry climate the availability of water will become a limiting factor in the rate of alteration of metastable carbonate minerals.

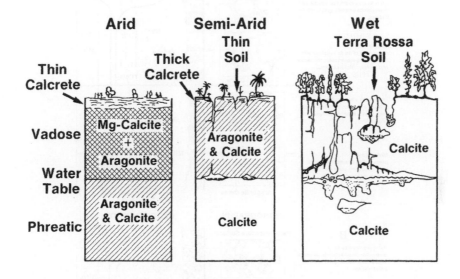

Figure 7.39. Schematic diagram illustrating the effects of climate on meteoric diagenetic pathways for coeval carbonates originally composed of a polymineralic assemblage of calcite with various amounts of magnesium and strontium-rich aragonite. (After James and Choquette, 1984.)

Some hypothetical examples of the vertical distribution of petrography, mineralogy and geochemistry developed on a complex of metastable carbonate sediments under different climatic regimes are shown in Figure 7.40. Little is known about meteoric diagenetic pathways under hot arid conditions. Karst features and thin calcretes probably will develop slowly. Little mineral alteration will occur in the vadose zone. In the phreatic zone, depending on the size of the catchment area and magnitude of interbasinal flows, water flow may be slow because of low recharge and cementation will be limited. An original calcite-rich carbonate exposed to such a meteoric environment may exhibit little evidence of diagenesis, except perhaps for pressure solution and scant formation of calcite cement.

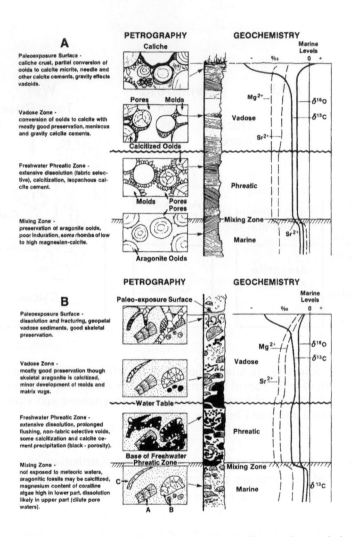

Figure 7.40. Schematic diagram illustrating meteoric diagenetic trends in terms of petrographic and bulk rock geochemistry for two hypothetical cores through (A) an oöid grainstone composed of Sr-rich aragonite and altered in a semi-arid climate, and (B) a reef limestone originally composed of coralline algae (A-magnesian calcite), bivalves (B), and coral (C-aragonite) altered in a warm, moderately humid climate. Both climate and original mineralogy are important factors affecting the meteoric diagenesis of these rocks. (After James and Choquette, 1984.)

In a warm and semi-arid climate thick calcretes can develop. Alteration in the vadose zone is relatively rapid but not as rapid as in the phreatic zone. Surface karst is present locally, and caves are small and rare. The mineralogical changes follow the pattern of Figure 7.25, but at a rate slower than that for a warm, sub-tropical climate like Bermuda. Under the extreme of a warm, wet tropical climate, extensive terra-rossa soils can develop, and dissolution features, such as caves, solution pipes and fractures, should be prevalent. Mineralogical stabilization should occur rapidly.

In a recent study, Hird and Tucker (1988) demonstrated a difference in diagenetic style of two Carboniferous oölites of inferred calcite composition from South Wales. They attributed these differences to the climatic regime present at the time of alteration of these limestones. The younger unit, the Gully Oölite, has a paucity of early meteoric cements, whereas the older unit, the Brofiscin Oölite, has abundant cement. The lack of cement in the Gully is interpreted to be the result of meteoric diagenesis in an arid climate. The abundant calcite cements of various types in the Brofiscin reflect a more humid climatic regime. Interestingly, during later burial diagenesis, the early cements of the Brofiscin inhibited grain-grain solution and pressure solution, whereas these processes were more effective in the Gully Oölite.

Although they were originally compositionally somewhat different, the diagenesis of the Gully bears some resemblance to that of the Cenozoic Gambier Limestone discussed previously. Both units were originally mainly calcitic in composition and pressure solution was important in development of cements in both carbonates. One can conclude that the rate of freshwater meteoric diagenesis is slow for a calcite limestone undergoing alteration in a dry climate. In a wetter climate, such a limestone will probably undergo dissolution and calcite cementation, with the net result that these two processes are a function of mean annual temperature and water residence time. The diagenetic history of calcite limestones of this nature may be such that they enter the realm of burial diagenesis with little cement to support the grain framework. Thus, pressure solution may play an important role during shallow burial compaction, before late diagenetic poikilotopic calcite spar develops. Obviously the rigid, low porosity grade V limestones of Bermuda, altered predominantly in the phreatic meteoric zone, if buried, probably would be less susceptible to significant grain-grain pressure solution during shallow burial compaction.

From the discussion in the previous two sections, it can be seen that the rate of alteration of carbonate sediments during meteoric diagenesis is strongly influenced by the original mineralogy of the sediments and the temperature and residence time of meteoric water. Carbonate sediments originally composed of the highly chemically reactive minerals aragonite and high-magnesian calcite in a warm, wet climate will alter rapidly in the meteoric environment. In contrast, calcitic carbonates undergoing meteoric diagenesis in a cold, dry climate will alter slowly and may show little petrographic evidence of alteration.

Rock-Water Relationships

The Bermudian calcarenites are composed of a polymineralic assemblage of aragonite and calcite with a range of magnesium contents. The effect of the contrasting reactivity among these co-existing phases leads to dissolution of the more soluble and rapidly reacting phases like aragonite and supersaturation of Bermudian groundwaters with respect to calcite. Thus, there is an intrinsic drive for rock-water reaction until the biosparites are completely converted to calcite limestone. Even though the groundwaters are saturated with respect to calcite, unstable phases may continue to dissolve supplying constituents to the waters. These dissolved constituents may accumulate in the waters or be removed in precipitating phases. Dissolution can continue until all metastable phases are exhausted and the rock is dominantly composed of calcite.

The concentrations of dissolved constituents like Mg^{2+} and Sr^{2+} in meteoric waters are controlled principally by the rates of mineral reactions versus the rate of water flow. Thus meteoric water chemistry, aside from initial constituent concentrations found in the rain water source, reflects the composition and availability of reactive mineral phases, the duration of rock-water interaction, and the rates of precipitation of secondary mineral phases (Pingitore, 1978). The extent of rock-water interaction, that is the number of dissolution-precipitation reactions seen by an individual volume of meteoric water (Lohmann, 1988), governs the change in composition of the meteoric fluid. For example, as aragonite and high-magnesian calcite dissolve, Mg^{2+} and Sr^{2+} are added to the groundwater. If calcite precipitates from this water, the Mg/Ca and Sr/Ca concentration ratios of this diagenetic phase are less than their ratios in the water, because the distribution coefficients for these elements in calcite are less than one (see Chapter 3). The concentration of these elements in an individual volume of water will increase as the extent of rock-water interaction increases, provided calcite is the only secondary phase formed. In relatively open systems of low rock-water interaction, the Sr/Ca ratio of diagenetic phases will decrease from original values as Sr^{2+} is lost via water flow, whereas in effectively closed systems with high rock-water interaction, the Sr/Ca ratios of these phases approach the values of the original precursor phases (Figure 7.41).

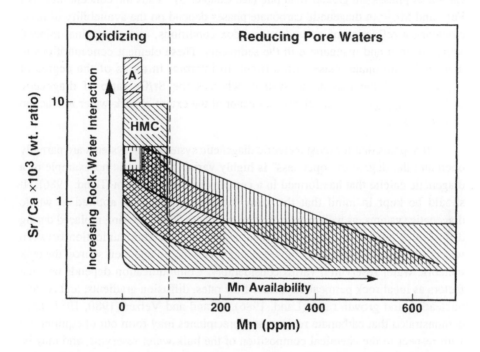

Figure 7.41. Diagenetic trends for aragonite (A), high-magnesian calcite (HMC), and low magnesian calcite (L) stabilized in meteoric waters. Sr^{2+} is lost to the water in systems with low water-rock interaction, and the Sr/Ca ratio of diagenetic phases will be less than their original ratio. In contrast, in nearly closed systems with high water-rock interaction, the Sr/Ca ratios of the diagenetic phases can approach those of the original precursor phases. The Mn contents of the diagenetic carbonates depend on the availability of manganese. (After Brand and Veizer, 1980, as modified by Lohmann, 1988.)

For elements like Fe^{2+} and Mn^{2+}, the situation is somewhat more complex because the aqueous chemistry of these species is controlled by redox equilibria. For example, in an oxidizing groundwater passing through a limestone, the small quantities of Fe^{2+} and Mn^{2+} released to solution from dissolution of mineral phases will rapidly precipitate as highly insoluble ferric and manganoan oxyhydroxides, and the water will contain vanishingly small concentrations of these elements. However, in groundwaters where dissolved oxygen and other oxidants like NO_3^- are depleted because of the oxidative decomposition of organic matter, Fe^{2+} and Mn^{2+} concentrations in these anaerobic waters may be on the ppm level. Diagenetic calcite precipitating from these waters will be enriched in these elements relative to water concentrations, because the distribution coefficients for these elements in water concentrations, because the distribution coefficients for these elements in carbonate phases are greater than one (see Chapter 3). Thus the concentrations of Fe^{2+} and Mn^{2+} in diagenetic carbonate phases depend on the availability of these elements, a reflection of groundwater redox conditions, and an original mineral source of iron and manganese in the sediments. These element concentrations in diagenetic carbonate phases are difficult to interpret in terms of the degree of "openness" of the diagenetic system, whereas the Sr/Ca ratio of diagenetic carbonates appears to be a valuable indicator of the extent of rock-water interaction (Figure 7.41).

It appears that for most meteoric diagenetic systems, the systems are partially open and the degree of "openness" is highly variable. There is no example of a diagenetic calcite that has formed in a completely closed system (Land, 1986). It should be kept in mind that the term "openness" has been applied to whole diagenetic systems, as well as to individual allochems that are being replaced during diffusive or advective diagenesis. The degree of fluid communication between waters causing dissolution and those causing precipitation and their source, the bulk meteoric water reservoirs, varies considerably. Communication depends on such factors as local rock permeability, fluid flow rates, diffusion gradients, and cement nucleation and growth rates (Land, 1986). Brand and Veizer (1980, 1981) have demonstrated that carbonate replacement precipitates may form out of equilibrium with respect to the chemical composition of the bulk water reservoir, and may be affected strongly by the chemical composition of the local phases undergoing dissolution.

Rock-water interactions also control to a significant extent the carbon and oxygen isotope compositions of calcite phases. It should be kept in mind that in most meteoric diagenetic systems, the reservoir mass of carbon and cations in carbonate minerals overwhelms their respective reservoir masses in the pore waters. This statement is not true for oxygen in which case the reservoir masses are more nearly the same. Thus, as Land (1986) so succinctly points out, "it is impossible to convert strontium-rich aragonite or Mg-calcite to strontium- and magnesium-depleted calcite and rigorously conserve the oxygen isotope composition of the precursor phase" (p. 131). However, the oxygen isotope composition of marine aragonite or magnesian calcite which have undergone alteration to calcite in marine water of the same composition as that in which they formed need not change. Diagenetic stabilization of aragonite to calcite in the marine environment results in loss of Sr^{2+}, but stable isotopic values of the diagenetic phase are certainly not those found for meteoric diagenetic calcite (see Chapter 8).

As demonstrated for Bermudian biosparites, soil-gas derived carbon imprints a distinctive light carbon isotopic signature on precipitates in the vadose environment. This signature is not characteristic, however, of most ancient carbonates, except adjacent to subaerial exposure surfaces where extensive vadose meteoric cementation has occurred. In Bermudian grainstones, even phreatic meteoric diagenetic calcites exhibit depletion in ^{13}C. This depletion is probably due to the fact that the Bermuda diagenetic system is relatively open, water residence time is short, and soil-gas CO_2 can mix far down the groundwater flow paths. In general, the larger the meteoric diagenetic system and the farther down the groundwater flow regime, the less chance that the DIC of the water will be depleted in ^{13}C derived from organic decomposition reactions in the recharge zone. In regions proximal to vadose water recharge, replacement carbonates should be depleted in ^{13}C, and in ^{18}O; their meteoric diagenetic signatures should be negative. As the meteoric water flow becomes more distal, because of the different sizes of the carbon and oxygen reservoir masses in the rock-water system, the $\delta^{13}C$ of carbonate cements will mimic more rapidly the $\delta^{13}C$ of the host carbonate phases, whereas the $\delta^{18}O$ may be little affected. Thus, phreatic cements proximal to vadose water recharge in large meteoric diagenetic systems may have highly variable carbon and nearly invariant oxygen isotope compositions (Lohmann, 1982). As the extent of rock-water interaction increases in the distal regions of large phreatic meteoric realms, and as the system becomes more closed, the $\delta^{18}O$ composition of the groundwater will become more like that of the rock. Thus distal phreatic carbonate cements may have somewhat variable $\delta^{18}O$ composition, and nearly invariable $\delta^{13}C$ compositions like that of the host carbonate rock. Lohmann (1982, 1988) has shown that for an idealized meteoric diagenetic system, a scatter plot of the $\delta^{13}C$ versus $\delta^{18}O$ compositions of calcite cement precipitates has the form of an "inverted J". An idealized plot of the variation of $\delta^{18}O$ and $\delta^{13}C$ characteristic of vadose and phreatic meteoric carbonates is shown in Figure 7.42. This figure summarizes the general trends in the stable isotope composition of meteoric carbonate precipitates discussed above.

As a final word of caution in applying stable isotope geochemistry to interpretation of meteoric diagenetic processes, it is worth recalling that the $\delta^{18}O$ content of seawater may have changed during the geologic past (see Chapter 10). If so, the meteoric calcite line of Figure 7.42 may be displaced in situations involving meteoric diagenesis of carbonates in the past. For example, Permian meteoric water may have been depleted in ^{18}O by several ‰ (Holland, 1984). Such depletion would lead to a displacement of the meteoric calcite line shown in Figure 7.42 to more negative values of $\delta^{18}O$. An additional complication arises from the fact that the isotopic signature of ^{13}C and ^{18}O-depleted vadose carbonates may be nearly identical to that found for some carbonate sequences undergoing burial diagenesis (Land, 1986). To truly understand meteoric diagenetic processes, it is necessary to obtain a variety of geochemical data on a carbonate rock unit, including carbon, oxygen and strontium isotopic values, and the concentration and distribution of the trace elements Mg, Sr, Mn and Fe. These data must then be used in models of the meteoric diagenetic system that quantitatively duplicate processes within the constraints imposed by the geology of the unit. This is a difficult task because of the data base needed, and because modeling of large meteoric diagenetic systems is in its infancy.

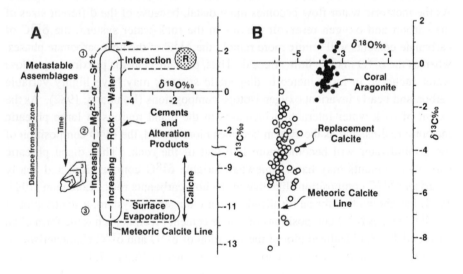

Figure 7.42. Comparison between (A) an idealized plot of variation in $\delta^{18}O$ and $\delta^{13}C$ for carbonates subjected to vadose and phreatic meteoric diagenesis (after Lohmann, 1988) with (B) the meteoric alteration trend observed for the Key Largo Limestone, Florida, U.S.A. (after Martin et al., 1986). The critical trend in isotopic composition is termed the *meteoric calcite line*. This trend may be modified at the water recharge surface where evaporation is an important process, caliche is formed and the diagenetic phases are depleted in ^{13}C derived from soil-gas CO_2. Another modification can occur distally to the recharge area where precipitating carbonate cements may have isotopic ratios nearly equivalent to dissolving phases.

The geochemical reaction model developed by Plummer et al. (1983) for the Floridan carbonate groundwater system is one step in this direction. Down the groundwater flow path of this aquifer, the net reactions are dolomite dissolution with calcite precipitation (incongruent reaction driven irreversibly by gypsum dissolution), with accompanying sulfate reduction and conservation of iron and sulfide in iron hydroxide and iron sulfide precipitation. Modeling calculations indicate that the system is open initially to atmospheric CO_2 and reacts with fossil organic carbon in the rock reservoir farther along the flow path of the water. The predicted reaction path that corresponds to observation is shown in Figure 7.43.

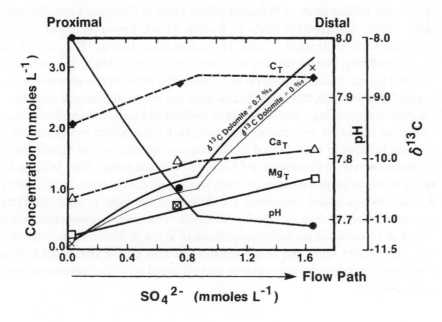

Figure 7.43. Predicted reaction path for the meteoric diagenetic system of the Floridan aquifer. These reaction path calculations agree reasonably well with observations, as shown by variousl symbols. Reaction path calculations of this nature can be applied to other modern meteoric diagenetic systems, and perhaps, with modifications to ancient systems now removed from original meteoric water. (After Plummer et al., 1983.)

Unfortunately in the rock record meteoric waters that caused diagenesis are not always present. However, if sufficient geohydrologic, chemical, mineralogical, and isotopic data can be gleaned from the rock record, it is likely that modified geochemical reaction models akin to that of the model of Plummer et al. (1983) can be applied to interpretation of ancient meteoric diagenetic regimes. More will be said about these types of models in the next chapter.

Mixed Meteoric-Marine Water Regime

Before concluding this section on synthesis of meteoric diagenesis, a few further statements concerning the mixing zone are appropriate. Dolomite has not been observed forming in the present-day mixed meteoric-marine water regime of Bermuda, but dissolution cavities and vugs have been observed. These dissolution features are generally found in most modern day mixing zones, including, for example, the mixing zones of Holocene oölitic sands of Schooner Cays, Bahamas (Budd, 1988), and coralgal sands of Engebi Island, Enewetak (Goff, 1979), Pleistocene rocks of Jamaica (O'Neil, 1974), Barbados (Harris, 1971), and of the Yucatan Peninsula, Mexico (Back et al., 1986; Stoessell et al., 1989), and carbonates of Andros Island, Bahamas (Smart et al., 1988). The cause of this dissolution was discussed previously in this chapter. Dolomite has not been observed forming in most present-day mixing zones, as would be anticipated based on theoretical mixing calculations discussed previously. However, both replacement and cementation dolomites inferred to have formed in Pleistocene meteoric-marine mixing zones have been observed in Pleistocene and Neogene carbonate rocks. Thus, because the kinetics of dolomite precipitation are notoriously slow, and modern mixing zones are relatively young and short-lived, the absence of dolomite in modern mixing zones may reflect inadequate time for formation. However, it is our opinion that changes in Phanerozoic seawater composition in terms of its carbonate saturation state, Mg^{2+}/Ca^{2+} ratio, and SO_4^{2-} content can also play a significant role in enhancing or retarding dolomitization in ancient mixed meteoric-seawater regimes (see Chapter 10).

Concluding remarks

Plate tectonics is a major driving force affecting Earth's surface environment. The kind and intensity of the physico-chemical parameters that control diagenetic pathways are determined to some degree directly or indirectly by plate tectonic processes. The plate tectonic environment exerts a direct control on diagenesis, whereas changes in Earth's surface chemistry induced by changes in the intensity of plate tectonic processes may modify the initial carbonate mineral compositions subjected to diagenesis and the waters that bathe them.

Under the modest temperature and pressure conditions characteristic of the environments in which meteoric diagenesis typically takes place, many of the most important reactions are slow. This has severely constrained the study of the chemical mechanisms and kinetics involved in such fundamental processes as the aragonite to calcite transformation and dolomite formation. Information on these processes obtained under conditions not typical of the meteoric realm (e.g., elevated temperatures) are of questionable applicability to "real world" carbonate diagenesis.

Meteoric diagenetic processes and their products are well known, and models are currently available to describe these systems quantitatively. However, problems still remain. Little is known about the continuous process of diagenesis of carbonates originally composed of calcite. Difficulties exist in unravelling the diagenetic history of a carbonate sequence that has seen multi-cycles of diagenesis in the meteoric environment, or in which meteoric diagenesis has been overprinted by later burial diagenetic processes. Furthermore, continuous and deep burial diagenesis may produce a carbonate rock product like that developed during severe meteoric diagenesis.

Kinetic and thermodynamic considerations show that the mass transfer leading to extensive alteration of carbonates in the meteoric realm is difficult to account for by transport of chemical components over long distances. This conclusion has important implications for the formation of regionally-distributed cements in carbonate rocks. These implications and further problems of mass transfer are subjects of the next chapter.

Chapter 8 continues on with the diagenesis theme, emphasizing processes and products in systems for which much less is known than for meteoric diagenesis. In all carbonate diagenetic studies, however, it should be emphasized that there is a strong need for synthesis of geologic, petrographic, chemical, and isotopic data coupled with development of geochemical models that are rigorous, mathematical, and hydrologically reasonable. In meteoric diagenetic studies in particular, the science has advanced far beyond the descriptive stage and into the realm of mathematical modeling of process and product, a reflection of the evolution of the science.

Chapter 8

CARBONATES AS SEDIMENTARY ROCKS IN SUBSURFACE PROCESSES

Introduction

In this chapter, we will consider the diagenesis of carbonate rocks that are "out of sight," and perhaps "out of mind" (Scholle and Halley, 1985). These are the carbonates that have undergone burial diagenesis. Some of these rocks show evidence of multi-diagenetic events, perhaps starting with early cementation in the marine environment and progressing through shallow-burial and deep-burial stages of diagenesis of varying intensities. Some rocks may be uplifted after deep burial, and all previous diagenetic features may be overprinted by a new cycle of diagenesis. A variety of geochemical and petrographic techniques (light microscopy, cathodoluminescence petrography, trace element, stable isotope, and fluid inclusion studies) employed within the context of a solid stratigraphic framework and basin history of the rock units under investigation are necessary to unravel the diagenetic history of rocks that have entered the deep subsurface. To quantify the subsurface processes of diagenesis, rigorous mathematical models need to be constructed of fluid movement in the basin and mineral-organic matter-aqueous solution reactions.

Studies of the burial diagenesis of carbonates are biased owing to the fact that most investigations are of Cenozoic carbonate rocks that have undergone unidirectional burial. This bias is analogous to that for meteoric diagenesis, in which most of our information comes from Holocene and Pleistocene carbonates containing a metastable assemblage of minerals (see Chapter 7). The geochemist-petrographer is faced with a difficult task today in interpreting the diagenetic history of deeply buried, and in some cases, subsequently uplifted, carbonate rock sequences. In this chapter, some general geochemical considerations of the subsurface environment are presented first, followed by discussion of several case histories of carbonates that have entered the increased pressure and temperature and changing fluid composition zone of burial diagenesis. To understand diagenetic processes in these P-T realms, it is first necessary to consider further the stability of carbonate phases as a function of P, T, and X (composition). Initial considerations were developed in Chapters 1-3.

P, T, and X and carbonate mineral stability

Carbonate minerals subjected to high P and T conditions will decarbonate; that is, the metal carbonate will decompose into a metal oxide and CO_2:

$$MeCO_3 \rightarrow MeO + CO_2 \tag{8.1}$$

where Me represents Ca, Mg, Sr, etc. Figure 8.1 shows the stabilities of magnesite, dolomite, and calcite as a function of P and T. The decarbonation of these metal carbonates requires conditions of P and T that are well outside the diagenetic realm. At 1 bar total pressure, magnesite decomposes at a temperature of about 500°C; dolomite and calcite require higher temperatures, with calcite decomposing at a temperature as high at 800°C. In the shallow crust, because of the availability of SiO_2, calcite would react with this component to form a calc-silicate at a temperature lower than its decarbonation temperature. Similar reactions with SiO_2 occur for magnesite and dolomite.

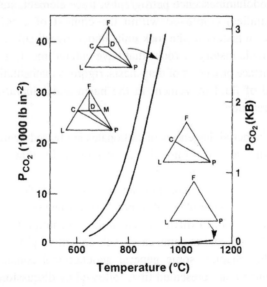

Figure 8.1. Univariant curves for the dissociation of calcite, magnesite, and dolomite in system CaO-MgO-CO_2. Compatibility diagrams are shown for each divariant area.   F=CO_2, C=calcite, D=dolomite, M=magnesite, L=lime, P=periclase. (After Harker and Tuttle, 1955.)

For the magnesian calcites, the situation is somewhat more complex. Thermogravimetric analytical studies of the decarbonation of these phases indicate two steps in the reaction (Koch, 1989). The component $MgCO_3$ in magnesian calcite decarbonates at a lower temperature than the $CaCO_3$ component. This initial decomposition reaction can occur at a temperature as low as 250°C under ambient experimental CO_2 fugacity. Further decomposition involves the loss of CO_2 from the calcite component. Because of the nature of the decarbonation reactions involving magnesian calcites, care must be taken when determining the organic carbon content of carbonate sediments by heating to 550°C. Under such conditions, some of the CO_2 may be lost from the magnesian calcite phase and not exclusively from the oxidative decomposition of organic matter. Also, at extreme diagenetic conditions, if magnesian calcites have survived earlier burial diagenetic reactions, they could begin to stabilize first by decarbonation.

Most diagenetic reactions involving carbonates occur in the presence of an aqueous solution. The number of chemical components is increased and the stability relationships of the metal carbonates are influenced by the composition of the diagenetic system. Figure 8.2 shows phase relationships in the system CaO-MgO-CO_2-H_2O-HCl as a function of P and T. In Figure 8.2, the equations for the phase boundaries are written conserving CO_2 so that phase relationships are portrayed in terms of the logarithmic activity ratios $a_{Mg^{2+}}/a^2_{H^+}$ and $a_{Ca^{2+}}/a^2_{H^+}$. The general approach used to construct Figure 8.2 was discussed in Chapter 6. An additional consideration is the effect of P and T on the equilibrium constants for the reactions. Pressure and temperature effects are calculated using molal volume and heat capacity data for the reactant and product phases, respectively (see Bowers et al., 1984, and references therein for detailed explanation). Because the phase relationships between magnesite, calcite, and dolomite are independent of CO_2 (see Figure 6.1, p. 243), the diagrams in Figure 8.2 are a valid representation of the stabilities of these carbonate phases in contact with aqueous solutions over a range of P and T.

Several water compositions are plotted on the stability diagrams in Figure 8.2. It can be seen that at shallow Earth surface pressures and temperatures, seawater plots in the stability field of dolomite whereas solutions of average river water composition and most shallow groundwaters plot in the field of calcite. With burial of carbonate sediments and elevated P and T, the dolomite field shrinks, but subsurface fluid compositions evolve toward a composition in equilibrium with dolomite. This conclusion is probably one of the most important arguments for the formation of dolomite during deep burial diagenesis (see also Hardie, 1987). Thermodynamic considerations favor this reaction path, as well as the fact that dolomitization reaction kinetics are enhanced at elevated temperatures (e.g., Land, 1966, 1967).

The subsurface situation, however, with respect to dolomitization is more complex, as shown by the work of Land and Prezbindowski (1981). It appears that in the Cenozoic Gulf Coast basin of the United States, subsurface, Ca-rich saline formation waters may be undersaturated with respect to dolomite, and as the brine moves updip, and is progressively diluted by shallow subsurface waters, the brine may cause dedolomitization of carbonates encountered in the subsurface.

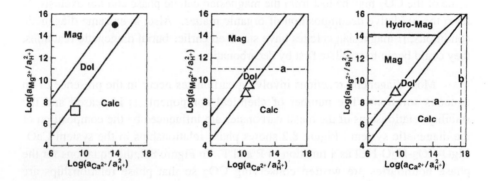

Figure 8.2. Phase relations in the system $CaO-MgO-HCl-H_2O-(CO_2)$. At A. 25°C, 0.001 kb; B. 150°C, 1 kb; C. 300°C, 3 kb. Mag, magnesite; Dol = dolomite; Calc = calcite; Hydro-Mag = hydromagnesite. Saturation lines for brucite (a) and lime (b) are also shown. • Average seawater composition, ˙ river water composition, Δ some subsurface water compositions. (After Bowers et al., 1984.)

The stability diagrams of Figure 8.2 are drawn for dolomite in its stable state of substitutional order/disorder. If the dolomite phase is disordered, its reaction to the stable dolomite phase entails a Gibbs free energy of reaction of -8.78 kJ mole^{-1}, or -2.10 kcal mole^{-1}, at 25°C. Figure 8.3 shows the change in the value of the equilibrium constant (log K) for the reaction

$$CaMg(CO_3)_2 \longleftrightarrow CaMg(CO_3)_2 \qquad (8.2)$$
$$\text{(disordered dolomite)} \qquad \text{(dolomite)}$$

as a function of temperature at various pressures. With increasing temperature the log K of the reaction becomes smaller, the two phases approach equilibrium, and the formation of the stable state of dolomite from a disordered dolomite precursor becomes energetically more favored. The Gibbs free energy of reaction at 300°C is -8.15 kJ mole^{-1}, or -1.95 kcal mole^{-1}. Thus the trend in the ideality of dolomite to increase with geologic age of the enclosing rocks (Chapter 7, p. 296) probably represents both the necessary time for metastable, Ca-rich, disordered dolomite to convert to stable dolomite and the probability that the rocks had been subjected to elevated temperatures.

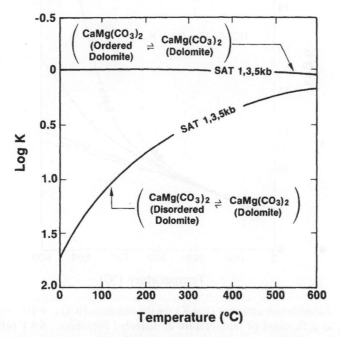

Figure 8.3. Equilibrium constants (log K) for the metastable coexistence of completely ordered and disordered $CaMg(CO_3)_2$ with dolomite in its stable order/disorder state as a function of temperature at constant pressure. SAT refers to the vapor-liquid curve for pure H_2O. (After Bowers et al., 1984.)

Aside from mineral stabilities, the behavior of the CO_2-H_2O system with increasing P and T is also important to an understanding of the deep burial diagenesis of carbonate rocks. One reaction of interest, which represents the summation of K_0 and K_1 (see Chapter 1) for the carbonic acid system, is

$$CO_2 + H_2O \rightarrow H^+ + HCO_3^- \tag{8.3}$$

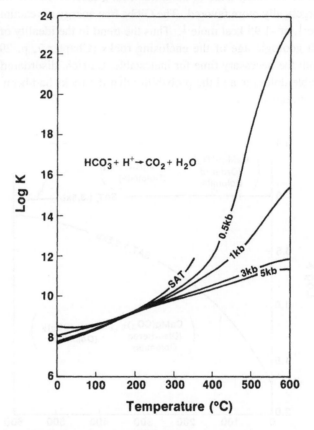

Figure 8.4. Equilibrium constant (log K) for the reaction $HCO_3^- + H^+ \rightarrow CO_{2(gas)} + H_2O_{(aq)}$ as a function of temperature at different pressures. SAT refers to the vapor-liquid equilibrium curve for pure H_2O. (After Bowers et al., 1984.)

Figure 8.4 illustrates the equilibrium constant for this reaction as a function of temperature at several constant pressures. Except at lower temperatures, where the converse is true up to several kilobars pressure, increasing pressure and temperature on the reaction have opposite effects. Thus, if the dissolution and precipitation of carbonate phases in the subsurface depend on reactions like

$$CaCO_3 + CO_2 + H_2O \rightarrow Ca^{2+} + 2HCO_3^- \tag{8.4}$$

it is likely that these reactions are more a function of subsurface water composition than changes in P and T. Nagy and Morse (1990) demonstrated that for typical diagenetic geothermal and lithostatic pressure gradients, calcite solubility is little influenced by changes in pressure and temperature (Figure 8.5). However, increasing the ionic strength of a solution from that of a meteoric water to a 0.5 m NaCl solution increases calcite solubility by a factor of 30. Further increases in ionic strength up to 6 m in NaCl solutions do not increase the solubility of calcite

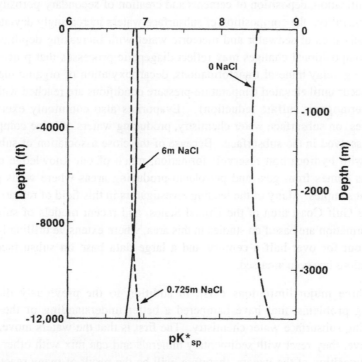

Figure 8.5. The influence of dissolved NaCl on pK^*_{sp} (-logarithm of stoichiometric solubility constant for calcite) under typical burial conditions. (After Nagy and Morse, 1990.)

greatly (~3x) and are approximately equal to the combined effects of pressure and temperature for diagenetic conditions typical of a sedimentary basin. Nagy and Morse (1990) conclude that even small variations in subsurface fluid composition can cause much greater changes in calcite solubility (and presumably other carbonate phases) than solubility changes owing to increasing pressure and temperature associated with sediment burial.

Subsurface water chemistry in sedimentary basins

The composition of subsurface waters is highly varied and exerts a primary influence on carbonates in sedimentary rocks. Subsurface water composition plays a dominant role in major late-stage diagenetic processes such as dolomitization-dedolomitization, deposition of cements and creation of secondary porosity. As a broad generality, the composition of subsurface waters increasingly deviates from simple mixtures of seawater and meteoric water with increasing depth and age. These compositional changes may reflect diagenetic processes that proceed very slowly (e.g., clay mineral transformations, decarboxylation of organic matter) or do not occur until elevated temperature-pressure conditions are reached with burial (e.g., thermogenic sulfate reduction). Evaporites also commonly exert major influences on subsurface water chemistry, producing waters of brine composition when dissolved in the subsurface. Because of the close association of subsurface waters with hydrocarbon reservoir formation, much of our knowledge of these solutions comes from gas- and petroleum-producing areas where wells provide abundant samples. Many of the leading investigators in this field of research come from the Gulf Coast area of the United States, and recent models of subsurface water formation are based on studies in this area, where extensive drilling has been carried out for over half a century and a large data base on subsurface water compositions has been amassed.

Three major limitations exist, in addition to the previously discussed sampling problems, that have hampered a better understanding of the factors controlling subsurface water chemistry. The first is that the waters move, and as they move, they react with sedimentary minerals and can mix with other waters. The composition of the waters, therefore, will be the result of many reaction and mixing events, under a variety of temperature-pressure conditions. The waters are not necessarily derived from the host rock in which they are found. This can lead to highly divergent interpretations of the factors responsible for water compositions (e.g., Stoessell and Moore, 1983; Land and Prezbindowski, 1985). The second factor is the current lack of chemical models, for brines in particular, that can be used to make good predictions of mineral solubility under subsurface conditions. In fact, even the Pitzer equation approach (Chapter 1) is inadequate at present to predict accurately calcite solubility in simple $NaCl-CaCl_2$ brines at 1 atm pressure and 25°C (Nagy, 1988). Finally, an equilibrium thermodynamic approach to understanding the controls on subsurface water composition is inappropriate for modeling processes where nonequilibrium processes usually dominate (e.g., Morse and Casey, 1988). Our woefully limited knowledge of the reaction kinetics of most of the important interactions involving subsurface waters and minerals, especially in brines at elevated temperatures and pressures, means that good predictive models are probably far in the future.

In spite of these major limitations, considerable progress has been made in understanding many of the important factors that influence subsurface water chemistry. Na^+, Ca^{2+} and Cl^- account for the major portion of dissolved components in most brines (Figure 8.6). Ca^{2+}, which can comprise up to 40% of the cations, usually increases relative to Na^+ with depth (Figure 8.7). Br^- and organic acids are commonly found at concentrations of 1 to 2 g L^{-1} (Land, 1987). The bicarbonate concentration is largely limited by carbonate mineral solubility, and sulfate is generally found in low concentrations as a result of bacterial and thermal reduction processes.

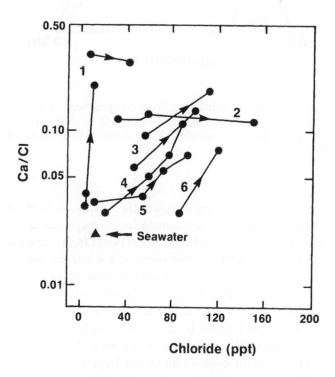

Figure 8.6. Ca:Cl ratios in brines from sedimentary basins in North America. 1-California; 2-West Virginia; 3- Alberta; 4- Illinois; 5-Texas; 6-Louisiana. (After White, 1965.)

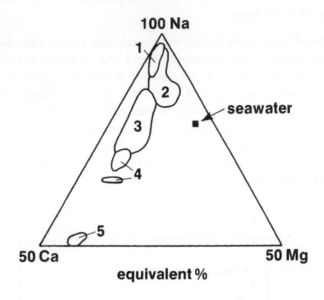

Figure 8.7. Representation of subsurface brine compositions on a ternary diagram for brines from U.S. sedimentary basins. 1- Texas; 2- California; 3- Kansas and Oklahoma; 4- Appalachia; 5- Arkansas (After Hanor, 1983, based on data from Desitter, 1947.)

Land (1987) has reviewed and discussed theories for the formation of saline brines in sedimentary basins. We will summarize his major relevant conclusions here. He points out that theories for deriving most brines from connate seawater, by processes such as shale membrane filtration, or connate evaporitic brines are usually inadequate to explain their composition, volume and distribution, and that most brines must be related, at least in part, to the interaction of subsurface waters with evaporite beds (primarily halite). The commonly observed increase in dissolved solids with depth is probably largely the result of simple "thermo-haline" circulation and density stratification. Also many basins have basal sequences of evaporites in them. Cation concentrations are largely controlled by mineral solubilities, with carbonate and feldspar minerals dominating so that Ca^{2+} must exceed Mg^{2+}, and Na^+ must exceed K^+ (Figures 8.8 and 8.9). Land (1987) hypothesizes that in deep basins devolatilization reactions associated with basement metamorphism may also provide an important source of dissolved components.

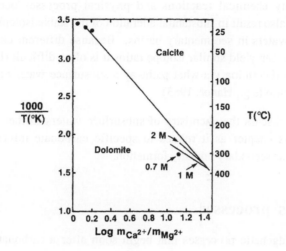

Figure 8.8. Log of the molar $Ca^{2+}:Mg^{2+}$ ratio of a solution in equilibrium with calcite plus dolomite. (After Land, 1987.)

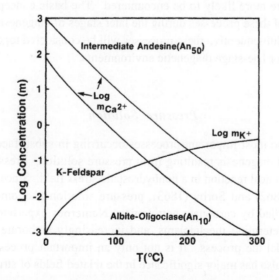

Figure 8.9. Log of molar concentration of K^+ in equilibrium with K-feldspar in 3m NaCl solution, and the log of the molar concentration of Ca^{2+} in equilibrium with two kinds of typical detrital plagioclase in 3m NaCl solution. (After Land, 1987.)

The many chemical reactions and physical processes occurring in the subsurface can also result in substantial alteration of the stable isotopic composition of subsurface waters in sedimentary basins. Because different combinations of these processes can yield similar isotope ratios it is often difficult (Figure 8.10) to unambiguously determine via what pathway a subsurface water has obtained its stable isotope ratio (e.g., Hanor, 1983).

We will discuss the chemistry of subsurface waters further in subsequent sections of this chapter as it relates to specific carbonate mineral diagenetic processes such as secondary porosity formation.

Continuous processes

Many diagenetic processes that begin soon after a carbonate sediment is deposited or during shallow burial can continue throughout the life of a carbonate unit. The relative importance of some processes, such as dolomitization, during early and late diagenesis is debatable, whereas others, such as pressure solution, are clearly dominant during the later stages of diagenesis where elevated temperatures and pressures are more likely to be encountered. The basic concepts and general characteristics of these processes during the later stages of diagenesis are discussed in this section. Subsequently, these processes will be considered together in models of different major late-stage diagenetic environments.

Pressure Solution

One of the most important processes occurring in subsurface carbonates is compaction and diagenesis resulting from pressure solution. Pressure solution is the result of chemical reaction in a nonhydrostatic stress field. Since the early work of Thomson (1862) and Sorby (1863), pressure solution phenomena have been extensively studied by carbonate petrologists. Numerous experiments have been conducted to determine mechanisms, and increasingly elaborate theories have evolved to model the process. It is not only an important process in carbonate diagenesis, but also has major significance in the related fields of structural geology and geophysics.

Major models for the mechanism by which pressure solution occurs can be divided into those that favor dissolution in a thin film of solution as a result of

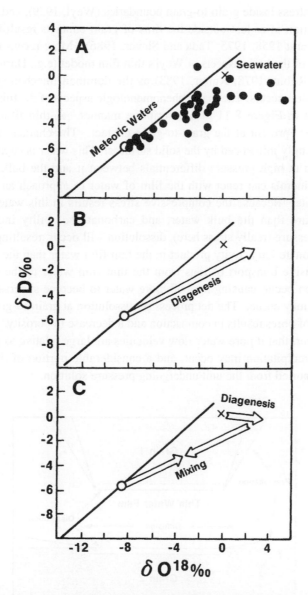

Figure 8.10. The isotopic composition of subsurface waters in the Illinois basin.
A. Data of Clayton et al. (1966). B. Interpretation of Clayton et al. (1966) that the
isotopic values are the result of progressive diagenetic alteration of meteoric water.
C. The alternative interpretation of Hanor (1983) that the isotopic values are the
result of mixing of meteoric water and diagenetically altered seawater. (After
Hanor, 1983.)

compressive stress inside grain-to-grain boundaries (Weyl, 1959), and those which favor dissolution at or just outside the rims of grain contacts resulting in under-cutting (Bathurst 1958, 1975; Tada and Siever, 1986). More recent models have favored most of the basic aspects of Weyl's thin film model (e.g., Durney, 1976; de Boer, 1977; Robin, 1978; Rutter, 1983) as the dominant mechanism by which pressure solution occurs. The major phenomenologic aspects of the thin film model are presented in Figure 8.11 in a descriptive manner. A thin film of water is assumed to be present at the grain-to-grain contact. The characteristics of this water are strongly influenced by the solid surfaces. This water is capable of being retained even at high pressure differentials between it and the bulk pore water. Carbonate minerals can react with the film of water to approach an equilibrium saturation state. Because the compressive stress results in this water being at a higher pressure than the bulk water, and carbonate solubility increases with increasing pressure (really stress here), dissolution will occur, resulting in a higher calcium carbonate ion activity product in the thin film water than the surrounding water. Diffusive transport of ions from the thin film water to the surrounding water can then occur, causing the bulk pore water to become supersaturated, and precipitation may ensue. The net process of dissolution at grain-to-grain contacts, and infilling of pores results in compaction and a decrease in porosity. It should be noted, however, that if pore water flow velocities are large relative to precipitation rates, little precipitation may occur, and a considerable portion of the carbonate may be transported from the unit undergoing pressure solution.

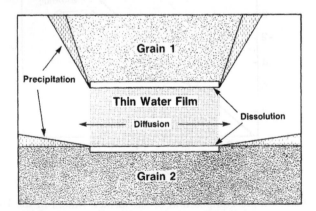

Figure 8.11. A general schematic model for pressure solution.

Many factors can influence the pressure-solution process, including temperature, the physical nature (e.g., roughness, grain size, etc.) of the solids, fluid composition, which may particularly influence precipitation kinetics, and fluid flow. However, the presence or absence of minor amounts of clay minerals generally is thought to be an important control on whether or not pressure solution occurs in carbonate rocks. Although the exact mechanism is unclear, it appears that clay minerals, which can contain appreciable amounts of loosely bound water, may act to create conduits for enhanced transfer of ions by effectively increasing the thickness of film between grain-to-grain contacts (see de Boer, 1977). Baker et al. (1980), however, have reported results of experimental pressure solution studies on natural carbonate sediments that indicate that clays and other siliceous sediment components may significantly inhibit the recrystallization processes. The reasons for the differences between the observations and experimental results are unclear, but they may involve the vastly different time scales over which the experimental studies were made and natural processes occur.

Pressure solution can cause major alterations in carbonate rock structures on megascopic to microscopic scales. Numerous papers and reviews deal with this topic (e.g., Bathurst, 1975; Choquette and James, 1987). We feel that one of the best attempts to bring an orderly picture out of the many complex features that are observed was that by Wanless (1979), who also emphasized the importance of pressure solution for subsurface dolomitization (see next section). Figure 8.12 presents his general model for the characteristics and controls on pressure solution types in limestones. The primary variables that Wanless considered were the clay content of the limestone, the concentration of structurally resistant elements, and variations between different units or beds. Temperature, pressure and fluid composition are also likely to play an important role in determining the timing and extent of pressure solution.

Dolomitization

The timing of the dolomitization of carbonate rock bodies and emplacement of dolomite cements has been one of the more controversial aspects of the "dolomite problem." Most of the basic factors controlling dolomite formation, where were discussed in Chapters 6 and 7, also apply to dolomite formation during the later stages of diagenesis. However, the extended periods of time, the solution compositions likely to be encountered, and the elevated temperature and pressure that occur during deep burial provide highly favorable conditions for dolomite formation.

Solution cannibalization (e.g., Goodell and Garman, 1969) and pressure solution (e.g., Wanless, 1979) have been suggested as mechanisms for late-stage dolomite formation. The major problem with these models is that the magnesium required for dolomitization must be largely derived from precursor calcite. Because the conversion of high magnesian calcite to low magensian calcite generally occurs earlier than this type of dolomitization, massive amounts of calcite must be dissolved to produce the dolomite (probably hundreds of volume units of calcite per volume unit of dolomite). It is unlikely, therefore, that these mechanisms are capable of producing massive dolomite bodies, although they can be locally important.

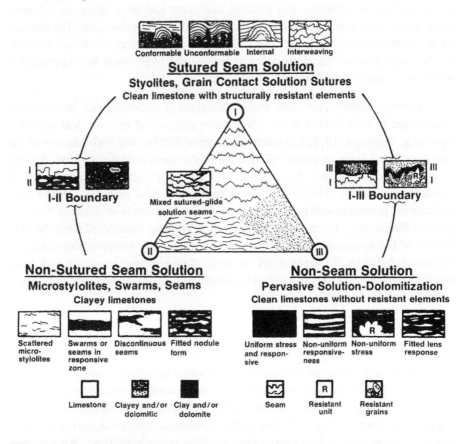

Figure 8.12. A summary diagram of characteristics and controls on pressure solution types with sketch examples. *Clean* means without significant platy silt or clay. *Responsiveness* refers qualitatively to how resistive the limestone is to change. (After Wanless, 1979.)

Clay mineral diagenesis also may play a role in dolomite formation during burial. The commonly observed conversion of smectites to illite can result in the release of the magnesium necessary for dolomite formation (e.g., McHargue and Price, 1982). Dolomite formation is observed near and within shale beds; however, this process again appears to be a localized mechanism and probably is incapable of producing large quantities of dolomite.

As with early dolomite formation models, most models of late-stage dolomite formation emphasize the need for major circulation of waters of the proper composition to provide the magnesium needed to transform calcite to dolomite (e.g., Wood, 1987). Interaction of these moving waters with subsurface evaporite minerals can contribute to their ability to dolomitize limestones. Several mechanisms have been proposed for major movement of dolomitizing subsurface waters. Water flow can result from compaction of sediments, and cellular and thermal convection. Topographically driven flow also has been suggested as a means of producing regional dolomitization (e.g., Gregg, 1985). While hydrothermal waters also may cause dolomitization, controversy exists over their importance, because many white sparry dolostones found in the vicinity of faults, and originally attributed to formation under hydrothermal conditions, are not found associated with other hydrothermal minerals.

Hardie (1987) has emphasized that with increased time and temperature the need for "special waters" with a narrow range of chemical composition for dolomitization decreases. An important point that he stresses is that as temperature increases the dissolved Mg^{2+} to Ca^{2+} ratio required to produce dolomitization decreases (Figure 8.13). Consequently, with deep burial and associated elevated temperatures, many subsurface waters can become potential dolomitizing solutions.

During extended periods of burial, two other important reactions involving dolomites may occur. The first is that dolomitic sediments may encounter waters whose compositions are undersaturated with respect to dolomite, and dedolomitization can occur. Land and Prezbindowski (1981), for example, have documented dedolomitization in Lower Cretaceous carbonates from south-central Texas by movement of saline formation waters that are generated deep in the Gulf of Mexico basin. The second is that epigenetic dolomites can form from non-ideal, metastable precursor dolomites. Land (e.g., 1985), in particular, has emphasized the importance of this process. He notes the increasing ideality of dolomites with geologic age. The generality of this relationship means that in older rocks of Paleozoic and Precambrian age, the dolomite is a secondary recrystallization

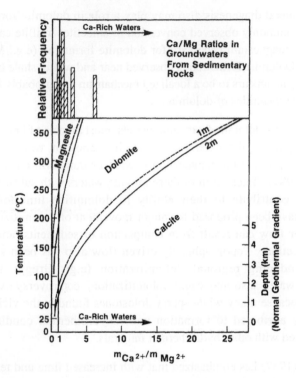

Figure 8.13. Plot of expected stability field for ordered dolomite as a function of temperature, and Ca^{2+}:Mg^{2+} ion ratio in 1 m and 2 m chloride brines. At the top is a histogram of Ca^{2+}:Mg^{2+} ion ratios of groundwaters in contact with a variety of sedimentary rocks, showing that at temperatures above 60-70°C many Ca-rich groundwaters could be dolomitizing fluids. (After Hardie, 1987.)

product of earlier dolomites, and is likely to reflect the conditions under which recrystallization occurred rather than the primary conditions of dolomitization.

Interpretation of the timing of dolomitization is important in a variety of studies, including hydrocarbon maturation and migration and porosity development. In Chapter 10 we discuss further the dolomitization process as a clue to ocean-atmosphere evolution.

Mud to Spar Neomorphism

It has been observed in many limestones that small grains (lime muds) appear to be replaced by larger grains (sparry calcites). This general process, when it occurs on a local scale, without fluid transport playing a major role, has been termed aggrading neomorphism (Folk, 1962). Bathurst (1975) presents an extensive discussion of observations of the different types of fabrics that can result from this process. There are many difficulties in clearly establishing that these fabrics have resulted from this recrystallization process. Bathurst states that submicron grains are typically replaced by grains that usually average 3 to 4 μm in diameter.

This process is, at least in a conceptual manner, similar to Ostwald ripening (see Baronnet, 1982; Morse and Casey, 1988, for discussion with regard to carbonates). The basic idea is that small grains have a greater solubility than large grains, because of the increasing contribution of surface free energy with decreasing grain size (see Chapter 2). Smaller grains should consequently dissolve and large grains grow, when in contact with the same solution. Because the influence of grain size on solubility decreases exponentially with increasing grain size, this process will be most important for submicron grains, but will result in only very small differences in solubility for grains larger than a micron.

Submicron carbonate grains persist in limestones for hundreds of millions of years, so reaction kinetics must play a dominant role in controlling the neomorphic process in natural systems. We have carried out a simple calculation for this process in seawater and Mg-free seawater at 25°C for different grain-size micrites. It is assumed in the calculation that the dissolution rate of the micrite is faster than the precipitation rate of the spar so that the solution composition is set by the solubility of the micrite. The reaction kinetics used are discussed in Chapter 2. Results are presented in Figure 8.14. The major influence of solution composition is reflected in the fact that reaction rates in the Mg-free seawater are about two orders of magnitude greater than in seawater. According to these calculations, a 1 μm grain would take on the order of 1 billion years for complete dissolution, but a 0.1 μm grain could dissolve in about 100,000 years in seawater.

Three major factors can alter these predicted rates by orders of magnitude. The first is that reaction inhibitors, such as organic matter, phosphate, and trace metals, are ubiquitous in pore waters of carbonate sediments. Even in highly supersaturated pore waters from modern sediments (see Chapter 6), these inhibitors are capable of effectively blocking precipitation. Because their inhibitory influence increases with decreasing supersaturation, and only small supersaturations are generated for grain sizes greater than about 0.1 μm, the presence of inhibitors in limestone pore waters is probably a primary reason that aggrading neomorphism is not more common. A second factor is that as this process proceeds, the growth of sparry calcite can cause a loss of porosity and an increase in sparry calcite grain contacts which may severely slow the rates of reaction. Temperature acts in opposition to these inhibiting influences and increases reaction rates. Although there is considerable petrographic evidence for the micrite to sparry calcite transformation, and the basic geochemical mechanisms are understood in a qualitative manner, we are still a very long way from understanding the details of the process. Such an understanding is essential to our ability to predict where neomorphism will occur in carbonates, and how extensive the process will be.

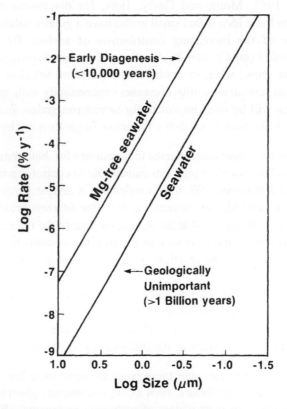

Figure 8.14. The dissolution rate of calcite as a function of initial grain size in seawater and Mg-free seawater, when dissolution is controlled by precipitation of large (>10 μm) grains. Times given are for complete dissolution.

Secondary Porosity

Secondary porosity results from the dissolution of carbonates in the subsurface environment. It can occur both in limestones and in sandstones where carbonate cements of original labile detrital minerals are dissolved. Because the formation of secondary porosity can substantially enhance the reservoir properties of sediments, it has received considerable attention from the petroleum industry.

Some of the basic processes in the formation of secondary porosity are similar to those for formation of carbonate cements. A solution of proper composition must be generated by subsurface processes, and this solution must also flow through the formation in which the dissolution reaction takes place in sufficient quantities to transport the dissolved carbonate. The primary differences between cement and secondary porosity formation are that an undersaturated solution must be generated rather than a supersaturated solution, and that while cement formation reduces porosity and can inhibit flow, formation of secondary porosity increases porosity and can result in enhanced flow of subsurface fluids.

The primary focus of research on secondary porosity formation has been on mechanisms for generating undersaturated formation waters. Because reactions that may result in undersaturation of waters with respect to carbonate minerals by consumption of calcium are unlikely to be quantitatively important, emphasis has been placed on reactions that may lower the carbonate ion concentration. Although not clearly documented in deep subsurface environments, mixing of waters of dissimilar composition can result in undersaturation with respect to calcite (see Chapter 7), and lead to secondary porosity formation. Acidic waters associated with igneous intrusions and thermal metamorphism can also cause carbonate dissolution that results in secondary porosity (e.g., deep Jurassic carbonates in Mississippi, U.S.A.; Parker, 1974).

The mechanisms for formation of secondary porosity that have received the most attention, based largely on studies in the Gulf Coast petroleum-producing areas of Texas and Louisiana, are thermal maturation of organic matter and clay mineral diagenesis. The decarboxylation of organic matter is thermally driven and yields carbon dioxide that can cause undersaturation of subsurface waters with respect to carbonate minerals (e.g., Schmidt and McDonald, 1979; Al-Shaieb and Shelton, 1981). In addition, major quantities of organic acid anions (e.g., acetate) can be produced during the thermal maturation of organic matter (Carothers and Kharaka, 1978; Surdam et al., 1984). Surdam et al. (1984) have discussed the importance of different kinds of naturally occurring organic acids in the formation of secondary porosity. These anions may buffer the pH of formation waters, and when CO_2 is added to the waters, they may precipitate or dissolve calcite. For most Cenozoic basin waters with high organic acid alkalinities (several hundred to 1000 mg L^{-1} acetate), addition of CO_2 will lead to undersaturation of these waters with respect to calcite and promotion of secondary porosity (Lundegard and Land, 1989).

Thermal dewatering of clays and the smectite to illite transition can produce dilute solutions, which lower saturation state and cause fluid flow (e.g., Schmidt, 1973; Klass et al., 1981). The dewatering of clays may produce overpressured zones (although depositional loading on low-permeability sequences is probably a generally more important mechanism for producing such zones), which are frequently correlated with the occurrence of secondary porosity (e.g., Lindquist, 1977). The thermal maturation of organic matter and clay mineral dewatering generally proceed concurrently and provide an attractive general mechanism for formation of secondary porosity (e.g., Hiltabrand et al., 1973).

Franks and Forester (1984) have discussed this mechanism in detail for Gulf Coast sediments, with particular emphasis on pre- and post-secondary porosity mineral assemblages. For many localities they found strikingly similar mineral assemblages (Table 8.1). Early carbonate cements had precipitation temperatures in the range of 40° to 75°C. Quartz overgrowths were observed to precipitate

Table 8.1. Mineral assemblages from different localities in the Gulf Coast. (After Franks and Forester, 1984.)

Lower Tertiary, Texas Gulf Coast	Oligocene Frio, South Texas	Eocene Wilcox, Central Texas	Eocene Wilcox, South & East Texas	Pennsylvanian Strawn, North Texas	Triassic Ivishak, North Alaska
Clay Calcite (Fe-poor) Calcite and/or dolomite (Fe-rich)	Calcite		Clay	Clay	
Quartz overgrowths Fe-poor calcite (Oligocene) Fe-rich calcite or dolomite (Eocene)	Quartz overgrowths Calcite	Quartz overgrowths Calcite	Quartz overgrowths Calcite	Quartz overgrowths Calcite	Quartz overgrowths Siderite
Dissolution Kaolinite Fe-rich dolomite or ankerite	Dissolution Kaolinite Fe-carbonate	Dissolution Kaolinite Fe-carbonate	Dissolution Kaolinite Fe-rich dolomite or ankerite	Dissolution Kaolinite Fe-carbonate	Dissolution Kaolinite Fe-calcite

before and during secondary porosity formation. The silica for these overgrowths was believed to be derived from the smectite to illite transition. Oxygen isotope data indicated that quartz overgrowths formed at temperatures above 80°C, which is consistent with the temperature range at which the smectite-illite transition occurs. Kaolinite was the phase most intimately associated with major dissolution events. Isotopic data and phase equilibria considerations indicate that the kaolinite was probably precipitated between 100° and 140°C from solutions in the pH range of 5 to 6. This pH range is consistent with subsurface pH measurements for this temperature range in the region under consideration. Albitization of detrital plagioclase overlapped and postdated secondary porosity formation. This reaction, based on oxygen isotope data, occurred at 135° ± 15°C. The reaction requires a low pH because it consumes H^+. Carbonate cements rich in iron were deposited subsequent to secondary porosity formation. The mineral assemblages and stable isotope data are thus all consistent with secondary porosity being produced in the temperature range of 100° to 140°C from waters in the pH range of 5 to 6, which were largely derived from the smectite to illite transition.

Franks and Forester (1984) have observed that there is a good correlation in the Gulf Coast region between high secondary porosities and high carbon dioxide concentrations in the Wilcox Formation of Texas (Figure 8.15). Major increases in carbon dioxide are found at 100° to 113°C in South Texas and 95° to 105°C in North Texas, in good agreement with the peak temperature for carbon dioxide generation from kerogen at about 100°C. This temperature is also consistent with the estimated temperature at which secondary porosity was produced through carbonate dissolution.

While the thermally driven diagenesis of shales remains the most attractive general model for secondary porosity formation, attempts to carry out mass balance calculations have yielded disturbing results. Franks and Forester (1984) found, for the previously discussed area, that the escaping shale fluids were capable of producing only 5 to 10 percent of the observed porosity. Lundegard et al. (1984) also calculated that decarboxylation of organic matter could only account for about 10 to 20 percent of the secondary porosity in the Frio Formation, unless very long range transport was involved. Carbon stable isotope data from carbon dioxide in natural gas, formation waters, and cements indicated that CO_2 derived from the decarboxylation of organic matter was of minor importance. The reaction kerogen + water $\rightarrow$ methane + CO_2 was consistent with the isotope data but was also insufficient to produce the observed secondary porosity.

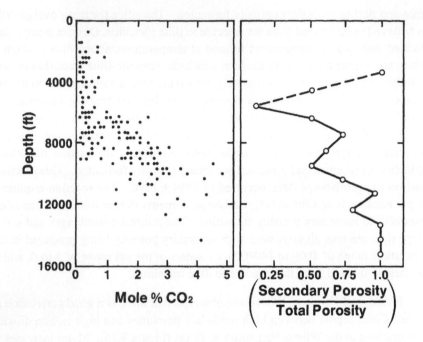

Figure 8.15. Comparison of mole % CO_2 in natural gas and ratio of secondary porosity to total porosity with depth for the Wilcox Formation in Texas. The shallow depth points probably have abundant secondary porosity as a result of meteoric water diagenesis, while deep secondary porosity is associated with the increase in CO_2. (After Franks and Forester, 1984.)

Cementation in the Subsurface

Although cementation is a process that can occur throughout the life of a sedimentary carbonate body, the dominant processes and types of cements produced generally differ substantially between those formed in the shallow-meteoric and deep-burial environments. Mineralogic stabilization (i.e., dissolution of magnesian calcites and aragonite, see Chapter 7) commonly drives cement formation during the early shallow-burial period, whereas the previously discussed processes of pressure solution and neomorphism are more important in the deep-burial environment. The pore waters in which cementation takes place also tend to differ substantially between the two environments. In shallow subsurface environments, cementation usually takes place in dilute meteoric waters that are oxic to only slightly reducing. Waters in the deep-burial environment generally have elevated salinities that usually increase with depth. These waters are reducing owing to their long isolation from the atmosphere and because of burial diagenesis of organic matter. An important consequence of the reducing conditions is that redox sensitive metals such as iron and manganese are mobilized and can coprecipitate with the carbonate cements. These factors, along with the higher temperatures and longer time periods associated with cement formation in the deep-burial environment, result in the production of cements which may have distinctive compositions and morphologies. Compositions that may be distinctive to deep-burial carbonate cements include carbon and oxygen isotope ratios that are often light (Figure 8.16), fluid inclusions reflecting formation in brines, and elevated iron and manganese concentrations in association with low strontium concentrations (e.g., Choquette and James, 1987).

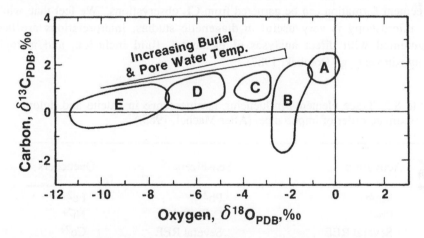

Figure 8.16. A hypothetical trend of changes in the stable isotope composition of carbonate cements in different diagenetic environments. **A**- marine realm; **B**- meteoric realm; **C**- mixing zone; **D**- successively deeper burial for calcite spar; **E**- successively deeper burial for saddle dolomite. **B** through **E** are precipitated in progressively hotter waters. (After Choquette and James, 1987.)

During the 1970's and 1980's cathodoluminescence (CL) microscopy received increasing attention as a method for understanding carbonate cement genesis. The power of this technique stems from its ability to reveal colored bands in carbonate grains that appear homogeneous in transmitted light. The CL-visible bands are the result of chemical inhomogeneities in the cement grains. The banding is believed to be produced by chemical differences among the waters from which the cements formed. Attempts have been made to relate the CL color and hue of cements to Eh-pH conditions at the time of cement formation (e.g., Barnaby and Rimstidt, 1989). Machel (1985) has cautioned that several other factors, such as the availability of iron and manganese, may also influence luminescence variations and that interpretations based solely on variable redox conditions are rarely justified. However, he generally cites literature based on UV-excitation, whereas CL is based on electron-stimulated luminescence. The relation between these two types of luminescence stimulation is not well understood. Considerable effort has also gone into determining the relationship of CL color and hue to the composition of the cement. Most of this research has focused on iron and manganese concentrations and ratios (e.g., Frank et al., 1982; Hemming et al., 1989). Machel (1985) again has raised a cautionary note, stressing in an extensive review that many other cations also may influence CL properties (Table 8.2). Consequently it is not surprising to find a considerable divergence of opinion in the literature as to how much information about cement composition and the chemistry of the waters at the time of cement formation can be garnered from CL observations. We feel that, while CL microscopy is very useful in diagenetic studies, interpretations are best augmented with direct analyses of elemental, fluid inclusion, and isotopic compositions (e.g., Dorobek, 1987).

Table 8.2. Trace elements influencing luminescence in calcite and dolomite in approximate order of importance. (After Machel, 1985.)

Activators	Sensitizers	Quenchers
Mn^{2+}	Pb^{2+}	Fe^{2+}
Pb^{2+}	Ce^{2+}/Ce^{4+}	Ni^{2+}
Several REE	Several REE	Co^{2+}
(particularly Eu^{3+}, Tb^{3+})		
Cu^{2+}		
Zn^{2+}		
Ag^+		
Bi^+		
$Mg^{2+}(?)$		

A major outgrowth of CL microscopy has been the development of the concept of "cement stratigraphy" (e.g., Meyers, 1974, 1978). The basic idea is that cements can be correlated on the basis of their CL properties and patterns over wide areas. Kaufman et al. (1988), for example, have applied this approach to the Burlington-Keokuk Formation, where they were able to map cement stratigraphies that appear to be extensive for over 100,000 km^2. The "cement stratigraphy" technique is not, however, without its critics. Searl (1988) stresses that the chemistry of calcite cements reflects primarily the geochemical environment at the time of precipitation, and that similar environments are likely to reoccur in both space and time. It is also difficult to envision how pore water chemistry could be maintained at close to constant compositions over the large areas considered in many of the studies using "cement stratigraphy." Later in this chapter, we discuss in detail a regional "cement stratigraphy" model.

The major variations in temperature that occur with depth in the deep-burial environment have been invoked to drive fluid circulation and carbonate cementation (e.g., Mackenzie and Bricker, 1971). Models based on this concept face the general difficulties discussed in Chapter 7 for deriving cements from waters flowing over all but very short distances. In addition, temperature variations occur primarily with depth, and are, therefore, also associated with pressure variations. Calcite has the rather peculiar property that its solubility increases with increasing pressure, but decreases with increasing temperature (see Chapter 1). As mentioned previously, Nagy and Morse (1990) have calculated that, for typical geothermal and pressure gradients, the influence of temperature and pressure on calcite solubility approximately cancel each other over a wide depth range. Consequently, it is difficult to obtain highly supersaturated waters simply by having them migrate to different depths. Thus, compositional variations in subsurface water have a much larger influence on calcite solubility in the deep-burial environment than do typical temperature-pressure gradients. It is quite possible that mixing of dissimilar waters in the subsurface may play a major role in deep-burial diagenesis, just as it does in near surface diagenesis.

A complex history of cementation and dissolution events may occur in carbonates buried in the subsurface. To some extent this history is a result of reactions involving the interplay between organic matter degradation in the subsurface and carbonate mineral solubility. The decomposition of organic matter at temperatures greater than 80°C produces significant amounts of organic acids that may be very effective at dissolving carbonates (Meshri, 1986). Also additions of CO_2 to subsurface waters whose pHs are buffered by organic acid anions generally produce waters undersaturated with respect to calcite, causing calcite dissolution and porosity enhancement (Lundegard and Land, 1989). However, pH buffering of subsurface waters by organic acid anions, like acetate, is a complex function of several variables, such as temperature and initial acetic acid concentrations. Thus under some subsurface conditions, addition of CO_2 to a water buffered by reactions involving organic acid species may produce a fluid that is oversaturated with respect to calcite, causing calcite precipitation and porosity destruction (Surdam and Crossey, 1985). Furthermore, at temperatures between 80°C and 150°C, thermochemical sulfate reduction may begin, whereby sulfate is reduced to the sulfur species of H_2S, HS^-, S^{2-} and elemental sulfur, and organic compounds are oxidized to solid bitumen and inorganic carbon species (e.g., Heydari and Moore, 1989; Machel, 1989). It appears that the thermochemical reduction process may involve dissolved and solid sulfate substrates. Whatever the case, this overall process involves the consumption of H^+ and potential subsequent increase in the saturation state of subsurface waters with respect to calcite. Some subsurface calcite and saddle dolomite cements apparently have formed from this thermochemical sulfate reduction process (Machel, 1987, 1989; Heydari and Moore, 1989). Thus in the subsurface, carbonate cementation and dissolution events are strongly linked to organic diagenetic processes, hydrocarbon maturation, and the location and timing of CO_2 production. Consequently, these events can vary considerably in time and space.

Examples of "models" of long-term diagenesis

Although somewhat artificial, we will separate diagenetic processes affecting carbonates on a long-time scale into those occurring in the present oceanic setting and those involving carbonate sediments that are part of the continental setting. In most cases, the discussion revolves around processes of the burial diagenetic environment. In this environment, original seawater composition has been substantially modified, or the seawater has been replaced or mixed with subsurface fluids. The various plate tectonic regimes of oceanic intraplate, craton, and passive and convergent margins (see Chapter 7) are discussed in the context of long-term diagenetic processes.

The Present Oceanic Setting

In this section, we will discuss several examples of the burial diagenesis of carbonates in the environment of the present oceanic sedimentary column. The drilling results of the Deep Sea Drilling Program (DSDP) and the Ocean Drilling Program (ODP) form the basis of the discussion. Because the long-term diagenetic history of a carbonate can be a composite picture of a myriad of complex processes, discussion of the burial diagenesis of calcareous ooze beneath the present seafloor includes consideration of its later history as a chalk.

Burial diagenesis and subsequent alteration of chalk

The Overall Diagenetic Pathway

The majority of chalks begin their history in the deep sea. In the modern oceans, calcareous nannofossil and foraminiferal oozes, and mixtures of these biogenic components, are the precursor materials for formation of chalks and ultimately limestones. These precursor oozes are generally deposited on topographic highs above the CCD in the plate tectonic environment of the mid-ocean ridge and oceanic intra-plate (see Chapter 7). When deposited in this environment, these sediments undergo slow burial to depths of 0.5 to 1 kilometer under normal oceanic gradients of pressure and temperature. However, the basal portions of calcareous oozes may be affected by heating near mid-ocean ridges and by hydrothermal solutions emanating from underlying basalts.

Calcareous nannofossil sediments are classified as ooze, chalk, and limestone depending on their degree of induration which is soft, stiff, and hard, respectively. The diagenetic transformation of calcareous ooze to chalk and subsequently to limestone during burial in the pelagic realm has been well documented by the drilling results of the Deep Sea Drilling Program (e.g., Schlanger and Douglas, 1974; Garrison, 1981). This transformation is another example of the metastability of biogenic calcite and its propensity to alter continuously during diagenesis because of its initial compositional impurity (see Chapter 7). Calcareous oozes are dominantly mixtures of the skeletal components of 0.25-1 μm-size coccoliths and 50-100 μm-size foraminifera. Some oozes are rich in sand-size pteropods, but these biogenic components are aragonite in composition, and their susceptibility to alteration to calcite is well known. The fine-grained, pelagic, biogenic calcites, however, are composed of trace-element-rich and structurally disordered crystals that are unstable thermodynamically relative to well-ordered low magnesian calcite (see Chapter 3). Thus, these phases, like aragonite, are more soluble than stable calcite and have a greater "diagenetic potential" (Schlanger and Douglas, 1974) to alter. In other words, the Gibbs free energy of the pelagic calcite-seawater system at deposition is not a minimum. One reaction that leads to a lowering of the free energy of the system is that involving the "purification" of calcite:

$$Ca_{0.9985}Sr_{0.0015}CO_3 + 0.0015Ca^{2+} \rightarrow CaCO_3 + 0.0015Sr^{2+} \qquad (8.5)$$

using the Sr/Ca atomic ratio observed for foram-nanno ooze at DSDP site 280 (Gieskes, 1983). Strontium is released to the pore water during the recrystallization of calcite, where it may accumulate in aqueous solution or precipitate as celestite (Baker and Bloomer, 1988).

Rapid increases in the interstitial water concentrations of dissolved strontium with increasing burial depth of deep-sea carbonate sediments have been interpreted as evidence of the recrystallization reaction (Baker et al., 1982; Elderfield et al., 1982; Gieskes, 1983). Figure 8.17 shows an example of interstitial-water profiles of dissolved alkaline-earth species from a carbonate nanno-fossil ooze from the Ontong Java Plateau (DSDP site 288; 5°58'S, 161°50'E). At this site calcium and magnesium concentrations are linearly correlated, and their gradients are governed by chemical reactions deep in the sediment column.

The depth distribution of the Sr/Ca ratios of the recrystallized calcites in these sediments was calculated from the Sr^{2+}/Ca^{2+} concentrations in the pore waters and appropriate values of the strontium distribution coefficient as a function of temperature from 5-25°C, where

$$\left(\frac{Sr}{Ca}\right)_{solid} = K_{Sr}\left(\frac{Sr^{2+}}{Ca^{2+}}\right)_{pore\ water} \tag{8.6}$$

(Gieskes, 1983; see Chapter 3). The good agreement between the calculated values and those observed (Figure 8.17), particularly below a few hundred meters sub-bottom depth, implies that considerable recrystallization has occurred in the upper sediment column, and that it is nearly complete in the chalks below 400 m (Gieskes, 1983). This relationship between observed and calculated Sr/Ca ratios has been demonstrated for chalk sequences at a number of DSDP sites (Baker et al., 1982). Thus, the burial diagenesis of calcitic carbonate ooze begins early in the history of the sediment at relatively low temperatures. With continuous burial, chemical changes in trace element and isotopic compositions of the chalk become more substantial. Soluble carbonate components are dissolved, partly because of increased overburden pressure, and a more pure calcite reprecipitates, a process known as "solution-compaction" (Scholle, 1977).

Aside from Sr, the original biogenic components of calcareous oozes also contain variable and minor amounts of other elements, like Mg, Na and S, all of which affect the solubility of the biogenic phases. Magnesium is especially important because it substitutes directly for calcium in the ooze calcite, reaching concentration levels of several thousand ppm. The various species and genera of foraminifera have characteristic magnesium concentrations that affect their solubilities. It would be anticipated that the original $MgCO_3$ content of a calcareous ooze would be determined by the relative amounts of foraminifera, coccoliths and discoasters found in the sediment. During recrystallization of this biogenic calcite, there is normally a slight increase in the Mg content of the phase. This enrichment in Mg is due to the fact that the stable calcite is not pure but contains a few mole % $MgCO_3$ in solid solution (see Chapter 3), whereas coccoliths and pelagic foraminifera contain less than 1% and generally less than 2% $MgCO_3$, respectively. However, some benthic forams found in oozes are more enriched in Mg than stable calcite and their recrystallization should result in loss of Mg. Whatever the case, the compositional impurity of biogenic ooze calcite represents a driving force for dissolution-precipitation reactions leading to a lower Gibbs free energy of the ooze-seawater system.

Larger crystals grow at the expense of smaller ones in the burial of a foraminiferal-nannofossil ooze. In particular, very small calcite crystals that comprise disaggregated coccoliths and the walls of foraminifera represent the most

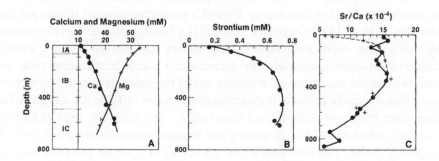

Figure 8.17. Interstitial water chemistry of sediments at DSDP site 288. A. Ca^{2+} and Mg^{2+} concentration-depth profiles. B. Sr^{2+} concentration-depth profile. C. Sr/Ca ratio in carbonates; black dots, observed; pluses calculated from interstitial water data and distribution coefficients. Lithology: 1A, pyrite-bearing, foram-nanno ooze; 1B, bioturbated nanno ooze; 1C, nanno-foram chalks and oozes. (After Gieskes, 1983.)

important phases that undergo dissolution and act as donors of $CaCO_3$ for overgrowths on discoasters and remaining larger coccoliths. This process also lowers the Gibbs free energy of the biogenic calcite-seawater system. In such a system the surface free energy component of the total free energy of the calcite must be taken into account (see Chapter 2). In essence the dissolution-precipitation process is partly dependent on the large difference in surface area between the population of very small biogenic grains and the population of larger ones. The dispersed system of a mixture of very small calcite crystals and larger ones has a greater solubility than that of the condensed system of larger calcite biogenic clasts and overgrowth and pore-filling calcite cements (Schlanger and Douglas, 1974). Thus, both compositional differences between precursor calcite oozes and chalk and limestone diagenetic products and grain-size differences act as driving forces for the dissolution-precipitation process.

Schlanger and Douglas (1974) developed a diagenetic model for the pelagic ooze-chalk-limestone transition that accounts for the observed reduction in porosity and foraminiferal content with increasing age and depth of burial. The model (Table 8.3) is based on the argument that surface energy changes are the principal diagenetic mechanism acting as a driving force for the conversion process. They coined the term diagenetic potential as "the length of the diagenetic pathway left for the original dispersed foraminiferal-nannoplankton assemblage to traverse before it reaches the very low free-energy level of a crystalline mosaic" (Schlanger and Douglas, 1974, p. 133). Diagenetic potential is a function of a number of variables including water depth, sedimentation rate, surface seawater temperature, surface calcareous plankton productivity, the foraminifer + coccolith: discoaster ratio, the ratio between the maximum and minimum size of biogenic calcite and the predation rate. The diagenetic potential is diagramatically shown in Table 8.3 and related to diagenetic realms of shallow- and deep-burial. The residence time of calcitic biogenic carbonate and its petrography and physical properties are given for each diagenetic realm. This table outlines the life history of a pelagic limestone from its beginning as a dispersed assemblage of planktonic organisms in the euphotic zone through its aggregation on the sea floor and its lithification by cementation and compaction. Throughout its history the diagenetic potential decreases as the sediment undergoes chemical and grain-size stabilization and matures. The rock approaches a final state of minimum free energy at which diagenetic change is not probable or is vanishingly slow.

Table 8.3. Diagenetic shallow- and deep-burial realms for pelagic ooze-chalk-limestone transition. (After Schlanger and Douglas, 1974.)

Depth	Realm	Residence Time	Petrography	Porosity (ϕ)% Velocity (V_c)km s⁻¹	Diagenetic potential 0 ⟵⟶ ∞
0-200 m (surface water)	I Initial production	Weeks	Highly dispersed calcite-sea water system; 10-10² forams m⁻³, 10⁴-10⁶ nannoplankton m⁻³		
200 m to sea floor	II Settling	Days to weeks for forams; months to years for coccoliths depending on pelletization	Pelletized coccoliths, ratio of broken to whole nannoplankton increases downward, ratio of living to empty foram tests decreasing during settling		
3000-5000 m (see Diagenetic Potential)	III Deposition	Inversely proportional to sedimentation rate and dissolution rate	"Honeycombed" structure. Large foram test supported by chains of coccolith discs. This surface is actually part of Realm IV.	$\phi \cong 80\%+$ $V_c \cong 1.45\text{-}1.50$ km s⁻¹	← slope at 3000 m
0-1 m (sub-bottom)	IV Bioturbation	50,000 years (at 20 m/10⁶ years sedimentation rate)	Remoulded "honeycomb", slight compaction, burrowing, destruction by ingestion and solution	$\phi \cong 75\text{-}80\%$ $V_c \cong 1.45\text{-}1.6$ km s⁻¹	← slope at 5000 m
1-200 m (sub-bottom)	V Shallow-burial	10x10⁶ years (at 20 m/10⁶ years sedimentation rate)	Ooze affected by gravitational compaction, establishment of firm grain contacts; dissolution of fossils and initiation of overgrowths	$\phi \cong 75\text{-}60\%$ $V_c \cong 1.6\text{-}1.8$ km s⁻¹	
200-1000 m (sub-bottom)	VI Deep-burial	Up to $\cong$ 120x10⁶ years (by then either subducted or uplifted)	Chalk with strong development of interstitial cement and overgrowths; transition down to limestone with dissolution of forams, pervasion by cement and overgrowths - grain interpenetration, welding and "ameboid mosaics"	$\phi \cong 60\%$ down to 35-40% $V_c \cong 1.8$ increasing to 3.3 km s⁻¹	
1-10 km (sub-surface)	VII Metamorphic	10⁶-10⁷ years	Recrystalization trending to "pavement mosaic" of completely interlocking crystals	$\phi \cong 40\%$ down to <5% $V_c \cong 3+$ up to 6 km s⁻¹	

V_c = compressional velocity

Influence of Heat Flow

Heat flow affects the rate of alteration of pelagic ooze to chalk; the higher the temperature, the greater the degree of cementation of the sediment. This effect on the ooze/chalk transition has been convincingly demonstrated by Wetzel (1989) in a study of compositionally similar pelagic carbonate sediments at DSDP sites 504 and 505 located south of the Costa Rica Rift zone.

The thermal conditions at site 504 are very different from those at site 505 (Table 8.4). Physical property measurements at both sites reveal systematic downhole changes and differences between the trends for the two sites. For example, Figure 8.18 shows the porosity-depth curve for the sediments at sites 504 and 505. It is evident that the porosity decreases more rapidly in the higher temperature site 504 calcareous sediments than in the lower temperature site 505 sediments. As would be anticipated from the porosity trends, the compressibility of the sediments decreases systematically with increasing depth. At a similar depth below the sea floor, site 505 sediments are more compressible than those at site 504. All the sediments are over-consolidated, a result of diagenetic cementation.

Table 8.4 Thermal conditions at DSDP sites 504 and 505, south of the Costa Rica Rift zone, eastern Pacific Ocean. (After Wetzel, 1989.)

Condition	Site 504	Site 505
Heat flow (mWm^{-2})	200	50
Temperature gradient $(^{o}C\ 100\ m^{-1})$	25-36	2-3
Basement temperature (^{o}C)	60-90	15-20

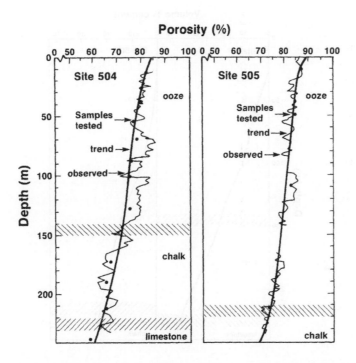

Figure 8.18. Porosity-depth relations at DSDP sites 504 and 505. Thin line, porosity determinations on ship-board; black dots, porosity determinations from samples tested in laboratory; dark thick line, best fit to the data. Notice porosity gradient is steeper at the high heat flow site 504 than at the "cooler" site 505. (After Wetzel, 1989.)

Wetzel (1989) calculated the volume of calcite cement at the two DSDP sites. The results of these calculations are shown in Figure 8.19. It can be seen that the calculated rate of increase of cement volume with increasing depth at the high heat flow area of site 504 is greater than that of the low heat flow site 505.

These differences in the physical properties of pelagic calcareous sediments at DSDP sites 504 and 505 are a result of the temperature difference between the two sites. A more rapid decrease in diagenetic potential is favored by increasing sediment temperature. The rates of the diagenetic solution-precipitation reactions are increased because of the higher temperature (Baker et al., 1980), and the accompanying increased concentrations of ions involved in cementation (Mottl et

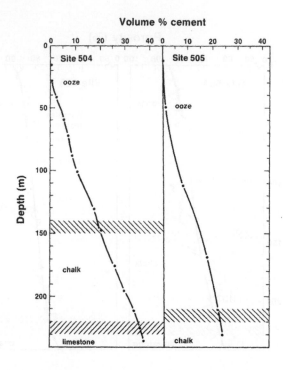

Figure 8.19. Calculated cement volumes for DSDP sites 504 and 505 calcareous sediments. In the high heat flow area of site 504, cementation starts at a shallower depth and a larger cement volume is observed than at the "cooler" site 505. (After Wetzel, 1989.)

al., 1983). Also, pore water flow may be enhanced because of the lower viscosity and density of interstitial fluids, leading to greater reduction of pore volume (Einsell, 1982). This latter effect is not as important as the chemical processes, but all processes lead to more rapid reduction of porosity and higher overconsolidation in heated sediments than in sediments undergoing diagenesis in the more normal temperature gradient of deep-sea sediments of 20-30°C km^{-1}.

Influence of Original Composition

Many chalks undergoing burial diagenesis in the present oceanic realm and those exposed on land were originally pelagic foram-nannofossil calcite oozes. However, calcareous oozes deposited in the periplatform environment are compositionally more complex. These oozes represent transitional carbonate deposits found between carbonate banks and the deep sea (Schlager and James,

1978). Thus, processes leading to their chemical stabilization resemble those occurring in both the shallow-water and the deep-sea environment.

The periplatform oozes are a complex mixture of bank-derived high-Sr aragonite and high-magnesian calcite, plus pelagic components of low-magnesian foraminiferal and coccolith calcite with or without pteropod Sr-aragonite. An early diagenetic event occurring in these sediments is the formation of seafloor or near seafloor marine cements of calcite containing several mole % $MgCO_3$ (e.g., Malone et al., 1989). The mole % $MgCO_3$ in the cement depends on temperature and carbonate saturation state of the seawater from which it is precipitated, with temperature perhaps the most important variable (see Chapters 3 and 5; Videtich, 1985; Burton and Walter, 1987). In periplatform deposits of the Tongue of the Ocean, Bahamas, the calcite cements contain 3.5-5 mole % Mg (Schlager and James, 1978), a composition similar to that calculated for deposits at a similar depth in the Maldives Archipelago, Indian Ocean (Malone et al., 1989).

Further processes leading to the chemical stabilization of these periplatform metastable carbonates and chalk formation have been documented by Malone et al. (1989) for carbonate sediments found at ODP site 716, Maldives Archipelago, Indian Ocean. This site, one of only a few such periplatform sites drilled by the ODP, represents a reasonably continuous recovery of sediments deposited at a nearly constant sedimentation rate.

With increasing burial depth to 260 meters below the seafloor (mbsf), both Sr and Na are lost from the carbonate solids. Figure 8.20 is a plot illustrating the generalized covariance of Sr and Na; shallow carbonate sediments are characterized by high concentrations of Sr and Na, whereas lower concentrations of these elements are found in sediments at deeper depths. The monotonic decrease in Na, and especially Sr, of the carbonate solids is good evidence of burial diagenetic processes. Furthermore, celestite nodules have been found in these sediments at depths as shallow as about 112 mbsf. The formation of celestite requires elevated Sr^{2+} concentrations in the pore waters to induce saturation and precipitation of this phase from the water. At this site, at a depth of approximately 100 mbsf, dissolved Sr^{2+} concentrations reach a maximum of 500 µm (Swart and Burns, quoted in Malone et al., 1989). These Sr^{2+} concentrations require the recrystallization of $CaCO_3$ and release of Sr^{2+} to the interstitial waters during burial diagenesis (Equation 8.5). The decrease in the Sr content of the sediments is probably due to the dissolution of aragonite and some biogenic calcite and reprecipitation of low-magnesian calcite (Malone et al., 1989). During this process Na is also lost from the carbonate solids. Experimental evidence indicates that the Na in carbonate phases probably occupies interstitial positions in the crystal lattice and is limited by the number of crystal defects (Busenberg and Plummer, 1985). It is likely that the number of defects in the original Na-rich carbonate phases at site 716 decreases with increasing burial diagenesis, accounting for the progressive loss of Na from the carbonate solids with increasing burial depth. Thus, once more we see an example of the free energy drive for the diagenetic system leading to purification of the original carbonate solid.

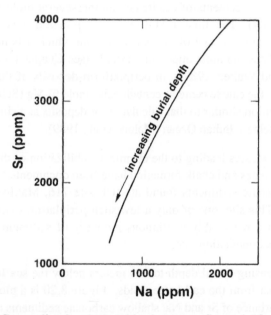

Figure 8.20. Generalized relationship between Sr and Na in CaCO$_3$ for periplatform carbonate sediments at ODP site 716, Maldives Archipelago, Indian Ocean. In general, Sr and Na concentrations in the carbonate phase decrease with increasing burial depth at this site. (After Malone et al., 1990.)

Malone et al. (1989) emphasized the rapidity of stabilization reactions involving periplatform metastable carbonates at site 716. Within 1.1 million years, high-magnesian calcite disappears from the sediment, probably by conversion to low-magnesian calcite. Most of the aragonite has been converted to low-magnesian calcite by 2.5 million years, and is completely gone from the sediment by 5.5 million years. After stabilization of most of the aragonite to calcite, celestite appears at about 3.4 million years. Chalk in discrete layers and as nodules becomes abundant below a depth of 100 m or approximately 3 million years.

It is likely that the diagenesis of periplatform carbonate oozes containing a metastable assemblage of shoal-water and pelagic biogenic debris may lead to the formation of chalk at a faster rate than that of deep-sea calcareous oozes. This is because textural changes in the latter deposits depend on small free energy differences between compositionally impure and very fine-grained biogenic calcite and more pure coarser-grained calcite. The periplatform oozes containing an abundance of aragonite and high-magnesian calcite have an additional driving force to produce the chalk texture: the dissolution of these phases and the reprecipitation of the chemical components as low-magnesian calcite. Their diagenetic potential is greater than that of calcite ooze. This additional diagenetic mechanism may lead to more rapid stabilization and production of the chalk texture.

Progressive Diagenesis of Chalk

The DSDP/ODP investigations of pelagic carbonates have provided substantial information concerning the burial diagenesis of these sediments. In summary, the diagenesis of chalk begins in chemically-modified pore waters and involves a dewatering phase in which as much as 50% of the depositional porosity is lost (Schlanger and Douglas, 1974; Scholle, 1977). As overburden increases, grain breakage occurs, grains are reoriented and intergranular porosity is reduced. At depths of a few hundred meters (Czerniakowski et al., 1984), the sediment has undergone sufficient consolidation to increase the effective stress to a level where pressure solution at grain contacts can occur. Pressure solution results in the liberation of carbonate which reprecipitates as overgrowths on nannofossils and as crystalline mosaics in intragranular and intergranular pore space (Matter et al., 1975). Eventually the chalk may be converted into a limestone exhibiting pervasive grain interpenetration and welding and a mosaic of interstitial and overgrowth calcite cement (Fischer et al., 1967). Observations of DSDP/ODP sediments have documented this general pattern in the progressive diagenesis of ooze/chalk/limestone in the present deep-sea sediment environment. However, some chalks because of tectonic deformation or because of deposition in water depths of only a few hundred meters are now exposed on land or are available for study by drilling on the continents. These sediments provide further insight into chalk diagenesis. In particular, they illustrate that stabilization of calcite chalk is a much slower process than that involving sediment composed of aragonite and magnesian calcite.

In a study of Miocene chalks of Jamaica, Land (1979b) discerned two chalk populations. Pelagic chalk accumulated on the slope of a carbonate bank in late Miocene time in an area that is now part of the north Jamaican coast. Because of deformation between late Miocene and middle Pleistocene time, the bank edge was uplifted, exposing some of the chalk deposits. The lower part of the carbonate slope was down-faulted into the Cayman Trench, and chalks from this part of the slope have never been subaerially exposed. These chalks exhibit little chemical alteration, whereas the subaerially-exposed chalks have undergone varying degrees of stabilization in the vadose meteoric environment (Figure 8.21).

The marine chalks, which have never been subaerially exposed, have high Sr concentrations and slightly positive $\delta^{13}C$ and $\delta^{18}O$ values, chemical signatures indicative of the marine environment. The subaerially-exposed chalks show varying degrees of depletion of Sr, ^{18}O, and ^{13}C relative to their marine precursors and a negative correlation between Sr concentration and $\delta^{18}O$ (Figure 8.21). These chemical signatures are compatible with alteration in the vadose meteoric environment.

Of further interest is the observation that the trend in Sr concentration vs $\delta^{18}O$ observed for Jamaica chalks altered in the vadose environment is parallel to that for Shatsky Rise limestones (Figure 8.21). These limestones were recovered during DSDP drilling and have never been subaerially exposed. The Shatsky Rise limestones show evidence of interaction with the basalt basement. The depletion toward the bottom of the sequence in ^{18}O and Sr is probably not due completely to replacement at elevated temperatures, but also an effect of thermally driven, Sr- and ^{18}O-depleted waters entering the base of the sediment pile from below (Land, 1986). Land (1979b, 1986) pointed out that on the basis of Sr and $\delta^{18}O$ chemical signatures that the burial diagenesis of chalks may be indistinguishable from meteoric diagenesis of this sediment type.

Obviously there are major differences between the ^{13}C content of Jamaican chalks altered in the vadose meteoric environment and that of Shatsky Rise limestones. However, as Land (1979b) pointed out, if chemical stabilization of the Jamaican chalks had occurred in the phreatic meteoric environment with little influence from soil CO_2, Sr and ^{18}O depletions would be observed with little change in $\delta^{13}C$. The limestones formed would be more homogeneous and virtually indistinguishable geochemically from Shatsky Rise sediments.

The degree of openness of the diagenetic system is also important. In closed system burial diagenesis of chalks, the isotopic composition of diagenetic carbonates may be rock-dominated (high rock:water volume ratio). In such systems, the $\delta^{13}C$ and $\delta^{18}O$ values of replacement carbonates may strongly reflect the values of the original dissolved carbonate precursor minerals. For the Cretaceous Austin Chalk, Czerniakowski et al. (1984) observed relatively uniform

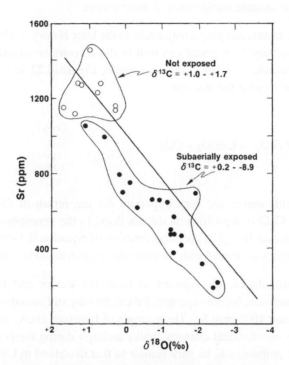

Figure 8.21. Strontium, $\delta^{13}C$, and $\delta^{18}O$ composition of Miocene chalk, northern Jamaica. Subaerially exposed chalks are low in Sr, and depleted in ^{18}O and ^{13}C relative to chalks that have never been exposed to subaerial conditions. The line is the Sr-$\delta^{18}O$ trend for Shatsky Rise limestones recovered by the DSDP program (Matter et al., 1975). Notice the similarity between the Sr-$\delta^{18}O$ trends for the Shatsky Rise limestones, which have undergone burial diagenesis, and that for the Jamaican chalks, which have been subjected to meteoric processes. (After Land, 1986.)

oxygen and carbon isotopic compositions of chalk micrite, intergranular cement, and fracture-filling cement to a depth of 3000 m. There was no strong correlation observed between isotopic composition and burial depth, and the isotopic compositions of diagenetic carbonates were very close to those of the estimated starting composition of Cretaceous marine cement. Thus, in this closed marine burial diagenetic system, it appears that diagenetic reactions were rock-dominated and controlled by the isotopic composition of dissolving carbonates during lithification. If the system were more open, and a volume of subsurface water moved through the sediment relatively rapidly, diagenetic phases might reflect more closely the isotopic composition of these waters.

Tectonic events can play a major role in the later history of chalk diagenesis. Subduction and very deep burial can lead to the conversion of calcite to a calc-silicate with release of CO_2. As an example, Figure 8.22 illustrates the P-T conditions necessary for the reaction

$$CaCO_3 + SiO_2 \rightarrow CaSiO_3 + CO_2 \qquad\qquad (8.7)$$

Reactions of this nature are very important for the return of CO_2, originally sequestered in $CaCO_3$ deposition on the sea floor, to the atmosphere-hydrosphere system (see Chapter 10). The chemical reaction of equation 8.7 can also occur in the contact aureoles associated with the intrusion of igneous bodies into limestone.

On uplift, chalks are exposed to meteoric waters and their chemical stabilization continues; before exposure, the calcite may still contain on the order of 1500 ppm Mg and 1000 ppm Sr. The example of Jamaican chalks is an illustration of some of the geochemical changes chalks undergo during meteoric diagenesis. The diagenetic pathway can be very similar to that described in Chapter 7 for the meteoric diagenesis of the Eocene to Miocene age calcitic Gambier Limestone. Figure 8.23 shows some data from a hydrogeochemical study of the Cretaceous chalk freshwater aquifer in England (Edmunds et al., 1987). In the unsaturated zone waters, stoichiometric dissolution of calcite is occurring. As the water moves down gradient, there is a progressive increase in its Mg and Sr concentrations, and an approach of the $\delta^{13}C$ values of the water to that of the chalk calcite. These trends indicate that the calcite in the chalk aquifer is dissolving incongruently, resulting in precipitation of a compositionally more pure calcite. Mineral saturation calculations show that the waters are very near saturation with respect to pure calcite throughout the aquifer. As the waters move from the unconfined to confined realms of the reservoir, the influence of soil-gas CO_2 depleted in ^{13}C on the $\delta^{13}C$ water values disappears. The calcite-water reactions become rock-dominated and the water is enriched in ^{13}C. These results demonstrate once more the adjustment of a marine sedimentary carbonate-water system with high Gibbs free energy to a lower energy state. The drive is the incongruent dissolution of impure calcite, with loss of Mg and Sr and other impurities, and precipitation of a more pure, and probably coarser-grained and structurally more ordered, calcite phase.

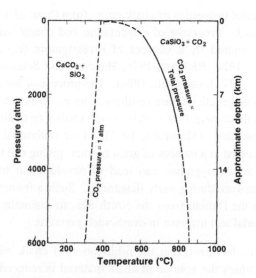

Figure 8.22. Equilibrium curves for the reaction $CaCO_3 + SiO_2 \rightarrow CaSiO_3 + SiO_2$ at a CO_2 pressure of 1 atm and at a CO_2 pressure equal to total pressure on the system. At P-T values between the two CO_2 pressure extremes, calcite and quartz are stable. Heavy solid line is experimentally determined and dashed lines are extrapolated or theoretical. (After Barth, 1962.)

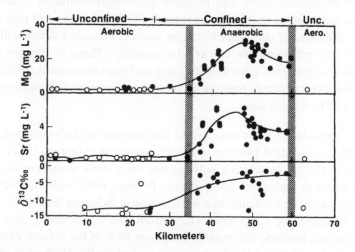

Figure 8.23. Variation of Mg^{2+}, Sr^{2+} and $\delta^{13}C$ along the flow lines of groundwater in the chalks of the Berkshire syncline, U.K. (After Edmunds et al., 1989.)

The mechanical properties of chalk range from those of a stiff soil to those akin to a strong rock. Processes of cementation and consolidation in carbonate rocks have been studied by a number of investigators (e.g., Ebhardt, 1968; Neugebauer, 1973, 1974; Bathurst, 1975; Hancock and Scholle, 1975; Mimran, 1975; Shinn et al., 1977; Jones et al., 1984). The process of hardening of chalk is certainly in part a diagenetic feature resulting from cementation of the rock. It is this early diagenetic cement that enables some chalks to retain a high primary porosity. "Spot welding" (Mapstone, 1975), where individual calcite grains in chalk are spot-cemented at a number of grain contacts giving rise to a meniscus-like cement between adjacent grains, can lead to development of an initial rigid intergranular framework during early diagenesis. Such a framework may enable some chalks, like the Danian from the North Sea, to maintain a relatively high porosity during burial and increase in overburden pressure.

It is unlikely, however, that the lithification of chalk will go on without consolidation, in which the volume of chalk material is reduced in response to a load on the chalk. Consolidation can lead to a reduction in porosity up to about 40%, and an increase in the effective stress (Jones et al., 1984). The increased effective stress is required to instigate the process of pressure solution. Pressure solution provides Ca^{2+} and HCO_3^- for early precipitation of calcite cement in the chalk. However, the inherently low permeability of chalk would inhibit the processes of consolidation and pressure solution/cementation unless some permeable pathways are opened up to permit the dissipation of excess pore pressure created by the filling of pore space by calcite cement. Pressure solution will cease if the permeable pathways are blocked by cement. Thus, it appears that the development of fractures, open stylolites and microstylolitic seams (Ekdale et al., 1988) is necessary to permit pressure solution to continue and lead to large rates of Ca^{2+} and HCO_3^- mobilization.

The formation of joints and elevated pore pressure in chalks may increase the rate of $CaCO_3$ migration by three orders of magnitude (Mimran, 1977). During deformation $CaCO_3$ in solution will migrate from chalk sequences under high effective stress to regions of lower stress. In these latter regions, $CaCO_3$ may precipitate, leading to formation of a well-cemented, indurated limestone. Mimran (1977) calculated for Cretaceous chalk from Dorset, England, that, during deformation and meteoric diagenesis, more than 90% of the volume of the chalk was lost, or about 2.2 g of $CaCO_3$ for every cm^3 of chalk. He further calculated that to dissolve 90% of the $CaCO_3$ in a 200-m-thick succession of chalk would require about 175×10^3 liters of water, and at a flow rate of meteoric water of 10 cm y^{-1}, about 17.6 million years. This dissolution of $CaCO_3$ over this time period represents carbonate for calcite cement that can be precipitated in sediments elsewhere. It is possible that, to overcome the inherent mass transfer problems involving $CaCO_3$ described in Chapter 7 (p. 309), high effective tectonic stresses and high pore fluid pressures are necessary to mobilize sufficient $CaCO_3$ to promote extensive allochthonous cementation in limestones.

Dolomite formation

Organogenic dolomite

Dolomite rhombs have been observed on smear slides of sediments from several DSDP sites, and the occurrence of dolomite in these sediments has been documented quantitatively by Lumsden (1988). Deep-marine dolomite averages about 1 wt % in all sampled DSDP sediments throughout post-Jurassic time. The dolomite is nonstoichiometric, averaging 56 mole % $CaCO_3$ (Figure 8.24), and has a crystal size and appearance similar to that of supratidal dolomite. Lumsden (1988) concludes that most is an early chemical precipitate from seawater. He estimated that about 10% is detrital, but he advocates that more criteria are needed to distinguish dolomite formed in cold, deep marine waters from that formed in supratidal deposits before this estimate can be substantiated.

The occurrence of greater than 1 wt % diagenetic dolomite appears to be restricted to organic-rich sediments from hemipelagic, continental-margin, DSDP sites (Baker and Burns, 1985; Compton, 1988). Compton (1988) coined the term "organogenic" for such dolomite, because it contains significant amounts of carbon derived from bacterial degradation of organic matter, as well as from dissolution of calcite. Organogenic dolomites are not restricted to DSDP sediments but have been recognized in the Miocene Monterey Formation of California (e.g., Garrison et al., 1984; Burns and Baker, 1987), in the Kimmeridge Clay of Dorset, England (Irwin, 1980), the Messinian Tripoli Formation of the Central Sicilian Basin (Kelts and McKenzie, 1984), and in the Miocene Pungo River Formation of North Carolina (Allen, 1985). Organogenic dolomites generally are Fe-rich, with Fe concentrations varying from a few hundred ppm to 18 mole %, slightly enriched in Ca and Mn, and depleted in ^{13}C, with $\delta^{13}C$ values in the range of -15 to -20 ‰ (e.g., Spotts and Silverman, 1966; Pisciotto and Mahoney, 1981; Suess and von Huene, 1988).

To understand the process of formation of these organogenic dolomites, it is necessary to look at the chemistry of the pore waters in which they form. Figure

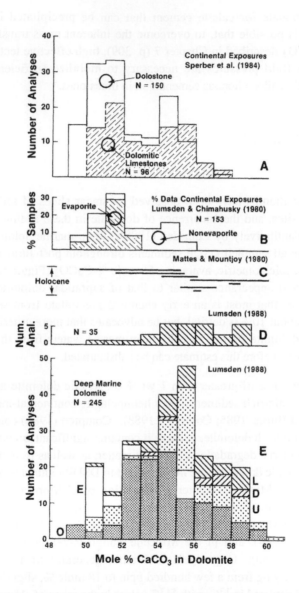

Figure 8.24. Summary of composition of nonstoichiometric dolomite. A. Data for 246 samples from continental exposures of dolostones and dolomitic limestones. B. Data for 153 samples from continental exposures of dolomites related and not related to evaporite sequences. C. Ranges of Holocene dolomite compositions. D. Data for 35 samples of deep-marine dolomite from DSDP cores. E. Histogram of 245 samples of deep-marine dolomite: E evaporite-associated, O organic origin, L samples of Lumsden (1988) above, D detrital, and U uncertain. (After Lumsden, 1988.)

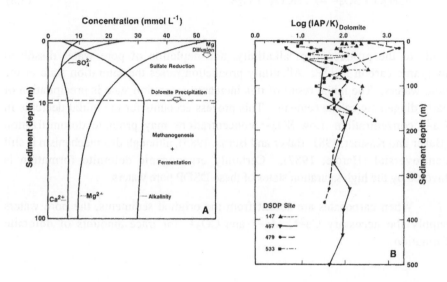

Figure 8.25. Some chemical and diagenetic properties of organic-rich marine sediments as a function of depth based on DSDP interstitial water profiles. A. Schematic gradients of SO_4^{2-}, total alkalinity, Ca^{2+} and Mg^{2+} in pore waters, and zones of sulfate reduction, methanogenesis and fermentation. Magnesium diffuses into the sediment and organogenic dolomite forms at depth. B. Logarithm of calculated saturation states of interstitial waters with respect to dolomite. Dolomite saturation=0. All these pore waters are oversaturated with respect to dolomite. (After Compton, 1988.)

8.25A is a schematic diagram showing the general trends in concentration gradients of Ca^{2+}, Mg^{2+}, SO_4^{2-}, and total alkalinity for DSDP interstitial water profiles from organic-rich marine sediments. Figure 8.25B illustrates the calculated saturation state with respect to dolomite of interstitial waters from several DSDP sites. Relative to seawater, the pore waters of these sediments show a reduction in Mg^{2+} and SO_4^{2-} and an initial decrease then increase in Ca^{2+} concentrations, and an increase in total alkalinity with increasing burial depth. The interstitial waters are oversaturated with respect to dolomite, reaching maximum values at 1 to tens of meters below the sediment-seawater interface, the depths of the minimum in the SO_4^{2-}:alkalinity ratio. This depth is determined partly by the rate of organic matter accumulation, the rate of sulfate reduction, and the sedimentation rate (Baker and Burns, 1985; Burns and Baker, 1987; Compton, 1988).

The bacteriogenic reduction of sulfate owing to

$$2CH_2O + SO_4^{2-} \rightarrow 2HCO_3^- + H_2S \qquad\qquad (8.8)$$

leads to the production of alkalinity, and depletion of pore water dissolved inorganic carbon in ^{13}C. Alkalinity production raises the saturation state of the pore waters. The initial result of this increased saturation state is precipitation of early diagenetic calcite cements. This process accounts for the initial decrease in Ca^{2+} concentration. Low SO_4^{2-} concentrations may promote dolomitization (Baker and Kastner, 1981; Baker and Burns, 1985), although this mechanism is still controversial (Hardie, 1987). Certainly organogenic dolomite formation is favored by the high saturation states of these DSDP pore waters.

When carbonates are absent from the original sediments, the pore waters supply the necessary Ca^{2+}, Mg^{2+} and CO_3^{2-} for trace amounts of dolomite formation:

$$Ca^{2+} + Mg^{2+} + 4HCO_3^- \rightarrow CaMg(CO_3)_2 + 2CO_2 + 2H_2O \qquad\qquad (8.9)$$

If $CaCO_3$ is present, dolomitization may occur by the following mechanism

$$CaCO_3 + Mg^{2+} + 2HCO_3^- \rightarrow CaMg(CO_3)_2 + CO_2 + H_2O, \qquad\qquad (8.10)$$

and if the sediment is very rich in a carbonate mineral precursor,

$$2CaCO_3 + Mg^{2+} \rightarrow CaMg(CO_3)_2 + Ca^{2+} \qquad\qquad (8.11)$$

In the first case, Ca^{2+} and some CO_3^{2-} is supplied by dissolution of $CaCO_3$ and Mg^{2+}, with additional CO_3^{2-}, is derived from the pore water. In the second case, most of the components of the diagenetic dolomite come from precursor carbonate (Isaacs, 1984; Baker and Burns, 1985; Compton and Siever, 1984, 1986; Burns and Baker, 1987; Compton, 1988).

The profile of Mg^{2+} in Figure 8.25 indicates downward diffusion of this constituent into the sediments. Mass balance calculations show that sufficient Mg^{2+} can diffuse into the sediments to account for the mass of organogenic dolomite formed in DSDP sediments (Baker and Burns, 1985; Compton and Siever, 1986). In areas of slow sedimentation rates, the diffusive flux of Mg^{2+} is high, and the pore waters have long residence times. Dolomites form under these conditions in the zone of sulfate reduction, are depleted in ^{13}C, and have low trace element contents. With more rapid sedimentation rates, shallowly-buried sediments have shorter residence times, and dolomites with depleted ^{13}C formed in the sulfate-reduction zone pass quickly into the underlying zone of methanogenesis. In this zone the DIC is enriched in ^{13}C because of the overall reaction

$$2CH_2O \rightarrow CO_2 + CH_4 \tag{8.12}$$

which leads to ^{13}C-depleted CH_4 and ^{13}C-enriched CO_2. Dolomites formed under these conditions have $\delta^{13}C$ values from plus a few parts per thousand to about $+8\%o$ (e.g., Suess and von Huene, 1988), or even higher. Some Monterey Formation dolomites have $\delta^{13}C$ values exceeding 20 ‰. These dolomites may also have higher trace metal concentrations, because of higher sedimentation rates owing to a greater flux of siliciclastic sediments. These sediments provide trace elements for incorporation in the dolomites (Burns and Baker, 1987).

At this point, we should also mention some further implications of the impact of methanogenesis on diagenetic carbonates. Authigenic carbonates derived from DIC produced from the oxidation of methane according to

$$CH_4 + SO_4^{2-} \rightarrow HCO_3^- + HS^- + H_2O \tag{8.13}$$

are known from a number of environmental settings (e.g., Friedman, 1988; Raiswell, 1988; Suess and Whiticar, 1989). However, along convergent plate boundaries, this process may account for substantial amounts of carbonate mineral cements, concretions, and chimneys that commonly are found at subduction vents. Venting of methane-rich fluids is a particularly important process of active continental margins undergoing lateral compression (Suess and Whiticar, 1989). Hydrocarbon seeps and submarine aquifers also discharge methane-rich fluids along passive margins (Paull et al., 1984; Brooks et al., 1987; Howland et al., 1987), but on a global scale the process may be more important at convergent margins, because tectonically-induced fluid movement necessary to carry methane from depth is important in these areas. Suess and Whiticar (1989) calculate that as much as 30% of the total CO_2 in pore fluids of sediments of the Oregon subduction zone may come from methane oxidation. The metabolic CO_2 derived from oxidation of methane is exceedingly depleted in ^{13}C, within the range of $\delta^{13}C$=-35 to -66 ‰ PDB. Diagenetic carbonate derived from such a DIC pool in pore waters also would be substantially depleted in ^{13}C.

The $\delta^{18}O$ values of organogenic dolomites are variable, ranging from -8 to +6‰ PDB. These values can reflect varying temperatures of formation, equilibration with diagenetic fluids and conversion of early, poorly-ordered, ^{18}O-enriched dolomite to ordered, ^{18}O-depleted dolomite by dissolution-precipitation reactions during burial diagenesis (Land, 1985; "wet recrystallization" of Hardie, 1987). There are still problems to be addressed concerning organogenic dolomite formation, but the overall process for its origin in DSDP sediments seems to be well-documented.

Kohout Flow and Dolomitization

Before closing this section on long-term processes of diagenesis below the present seafloor, it is worth commenting on a process of convective circulation of waters through carbonate platforms that probably has important diagenetic implications. Warm saline waters have been observed discharging along the west coast of the Florida Platform (Kohout, 1965; Kohout, 1967; Kohout et al., 1977; Fanning et al., 1981). A model for the origin of these waters has been developed by Kohout and co-workers. At a depth of 500-1000 m, the aquifer is in hydrologic contact with the ocean. Cool seawater from these depths penetrates the carbonate sediments of the Floridan Platform from the east, west and perhaps, the south. As the water moves toward the center of the platform, it is geothermally heated from its initial temperature of 5-10°C to 40°C. The heated water is less dense and therefore buoyant and moves upward through channels and fractures in the overlying sediments that confine the aquifer. The water emerges as warm springs along the west coast of southern Florida.

Fanning et al. (1981) sampled three of these spring waters and analytically determined their compositions. The discharging seawater is depleted in magnesium by 2.7 mmole kg^{-1} and calcium is enriched by 3.6 mmole kg^{-1} relative to normal seawater with the same chlorinity. These authors hypothesize that these chemical changes are consistent with the process of dolomitization of the platform limestones. If this idea is correct, then about 141 m^3 of dolomite can be produced yearly in association with one spring. The obvious question is, how many of these springs exist around the Floridan Platform, and, for that matter, coastal limestone platforms around the world, and what are the seawater fluxes associated with these vents? If the fluxes are quantitatively important, this mechanism could be an important mode of dolomite formation, and a sink of seawater magnesium. Fanning et al. (1981) calculate that 1 million vents with chemistries like those of the Floridan springs could account for removal of 40% of the riverine input of magnesium. This estimate is certainly too high for today's oceans where Mg removal at mid-ocean ridges is very important, but it does provide a feeling for the magnitude of the process, which may have varied throughout geologic time.

"Convection" dolomites may not be restricted to carbonate platforms, because the hydrological conditions favorable for seawater circulation also may exist in mid-ocean atolls, and perhaps, the deep ocean floor. Aharon et al. (1987) have described such a process of dolomitization in atoll carbonates at Niue Island in the South Pacific.

The Present Continental Setting

General considerations In the deep sea, carbonates generally are not subjected to very high pressures and temperatures. An exception is the basal sections of some limestone sequences that have been in contact with heated hydrothermal solutions. In the shallow continental crustal setting, the environmental conditions of carbonate diagenesis are more diverse. Cratonic carbonate sequences may undergo only shallow burial to depths rarely exceeding 3 to 4 km and a maximum pressure and temperature of 1 kb and 100°C, respectively. At subduction zones, passive margins, and in areas of continent-continent collisional tectonics, carbonates may be buried to depths exceeding 10 km where diagenetic pressures and temperatures can reach nearly 3 kb and 250°C (Figure 8.26).

Sedimentary carbonates exposed or available to observation by drilling in the present continental setting (cratonic, passive, and convergent plate margins) may have undergone a complex geologic history. In many cases, these rocks have been subjected to a range of P and T conditions and fluids of different compositions. Their diagenetic history is complex, and quantitative interpretation of this history is fraught with many difficulties. Thus, diagenetic studies of ancient limestones of the continental setting require an integrative approach, a synthesis of stratigraphic, petrographic, and geochemical data with analysis of the burial history of the basin containing the sedimentary carbonate sequence. Models need to be developed to

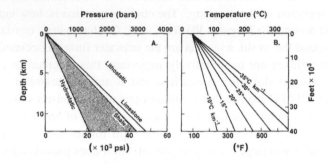

Figure 8.26. Pressure and temperature variations in the burial diagenetic realm. A. Static pressure variations; most pore fluid pressures plot in the stippled area. B. Depth-temperature plots assuming different geothermal gradients; gradients in the shallow crust are generally in the range of 20° to 30° km⁻¹. (After Choquette and James, 1987.)

constrain the thermal history of the basin, fluid migration paths and compositions, and mineral-water reaction pathways in time and space. This type of modeling is in its infancy today. There is some expectation, however, that such an integrated approach may have as some of its output the prediction of porosities of sedimentary carbonates for different depths and locations in a sedimentary basin, and the timing of porosity development. We will explore two contrasting situations of diagenesis in the present continental setting. The first involves the extensive meteoric cementation of cratonic and cratonic margin carbonate sediments; the second, the deep burial diagenesis of carbonates now part of a present-day continental margin. In this latter example, the rocks have been subjected to orogenesis and uplift. In both cases, the rock record is our principal source of information. Original marine pore waters are gone.

"Cement stratigraphy"   Many ancient limestones contain little porosity (Choquette and Pray, 1970), and much of the porosity has been destroyed by cementation with sparry calcite. There are a number of studies dealing with sediment lithification and porosity development in the literature that have employed a variety of petrographic (plane-light and cathodoluminescence) and geochemical techniques (Fe, Mn, Mg, Sr abundances; C, O and Sr isotopes; and fluid inclusion studies) to investigate regionally extensive calcite cementation and dolomitization events (e.g., Meyers, 1974, 1978; Meyers and Lohmann, 1985; Moore, 1985, 1989; Moore and Druckman, 1981; Heckel, 1983; Grover and Read, 1983; Gregg, 1985; Dorobek, 1987; Goldstein, 1988; Niemann and Read, 1988; Kaufman et al., 1988).

Integrated cement stratigraphy studies were pioneered by Meyers (1974). We emphasize here his work (summarized in Meyers and Lohmann, 1985) on the Mississippian carbonate grainstones of southwestern New Mexico, U.S.A., as a case history of calcite cement stratigraphy and porosity destruction in an environment interpreted as that of a regional meteoric diagenetic system. Many of the basic geochemical and geohydrological principles pertinent to meteoric diagenetic systems were discussed in Chapter 7.

The Mississippian limestones of Kinderhookian to Chesterian age (Lower to Upper Mississippian) form a classic transgressive-regressive shoaling upward sequence in southern New Mexico. The bulk of these limestones were deposited in subtidal environments as echinoderm-bryozoan grainstones. Their diagenetic history is relatively simple, because they were not subjected to repeated diagenetic events owing to migration of fresh water lenses associated with shoreline migrations.

The major unit studied by Meyers and co-workers is the Lake Valley Formation of Osagian age, which is a relatively thick (<100-200 m) regressive deposit of crinoidal-lime grainstone. The Lake Valley grainstones were composed originally of high magnesian calcite crinoid grains containing about 10 mole % $MgCO_3$. Throughout the several tens of thousands of km^2 of the area studied by Meyers, the crinoid magnesian calcite has been altered to low-magnesian calcite, and the crinoidal limestones have been cemented by syntaxial, inclusion-free calcite on the echinoderms. This cement makes up 85% of the total intergranular cement.

The inclusion-free syntaxial cements contain four compositional zones, distinguishable on the basis of cathodoluminescence, that are volumetrically important and can be correlated and mapped regionally. It is likely that these zones are time-stratigraphic, and each zone represents an unique set of environmental conditions of formation. The zones designated 1, 2, 3, and 5 are shown in Figure 8.27A, and the proportions of the major cements making up the total volume of intergranular cement in the diagram of Figure 8.27B. The earliest cement zones, 1, 2, and 3, were formed in the regional shallow meteoric environment and are pre-basal Pennsylvanian in age, whereas zone 5 cements are believed to represent later post-Mississippian, and probably pre-Permian, burial cementation processes. Age and overburden information led Meyers (1978, 1980) to conclude that zones 1, 2, and 3 cements were precipitated at a temperature of less than about 30°C, whereas zone 5 cements were formed at a temperature of less than about 60°C. This conclusion was based on an assumed geothermal gradient of 30°C km^{-1}, a normal continental gradient.

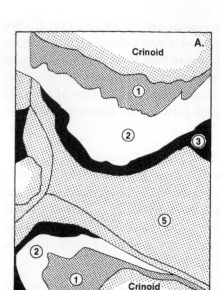
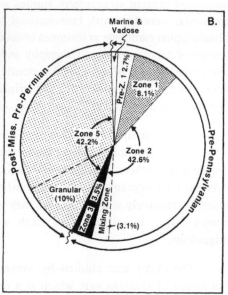

Figure 8.27. Luminescent zoned calcite cements in the Mississippian Lake Valley Formation. A. Schematic diagram illustrating paragenesis of the zonal luminescent cements. B. Proportions of zone 1, 2, 3 and 5 cements making up total intergranular cement. (After Meyers, 1980; as modified by Moore, 1989.)

The trace element compositions of the four major cement zones are shown in Figure 8.28 and Table 8.5. It is evident that each zone is chemically distinct. Zone 1 is rich in Mg and poor in Mn and Fe; Zone 2 is rich in Mn and has intermediate concentrations of Mg and Fe; Zone 3 is poor in all three trace elements; and Zone 5 is rich in Fe and Mg and has intermediate concentrations of Mn. These trace element differences account for the luminescence differences among the cement zones. The distinct chemical signature of each zone supports the petrographic evidence that each zone records a regionally extensive cementation event in a distinct meteoric water chemistry. If the zones were lateral equivalents of each other rather than separate time-stratigraphic events, one would anticipate considerable overlap between the zonal cement chemistries.

To further understand the origin of these calcite cement zones, it is necessary to determine the carbon and oxygen isotopic chemistry of each zone. The ranges of

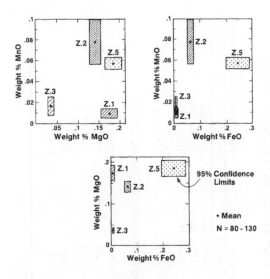

Figure 8.28.   Trace element compositions of zone 1, 2, 3, and 5 cements in Mississippian crinoidal grainstones. Rectangles encompass 95% confidence limits. (After Meyers and Lohmann, 1985.)

isotopic data for the grains and cements of these Mississippian grainstones are shown in Figure 8.29, and means and standard deviations are given in Table 8.5. Once again it can be seen that the various zonal syntaxial calcite cements have distinct isotopic signatures when viewed in carbon-oxygen space. The pre-Pennsylvanian zone 1, 2, and 3 cements exhibit a progressive depletion in ^{13}C and ^{18}O with decreasing age. The post-Mississippian cements are distinctly depleted in ^{18}O relative to cements of the three other zones.

The isotopic trends of the marine components of crinoids, micrite, and marine cements differ markedly one from the other but converge at a composition of roughly $\delta^{13}C = +4.0‰$ and $\delta^{18}O = -1.5‰$. Meyers and Lohmann (1985) conclude that these values represent the initial composition of Mississippian marine carbonate. Similar values for this geologic period were obtained by Brand (1982). Using the average trace element chemistry of zone 1, 2, 3, and 5 calcite cements and appropriate homogeneous distribution coefficients (Chapter 3) at 25°C, Meyers and Lohmann (1985) calculated the average cation activity ratios for water compositions that led to precipitation of these cements (Table 8.5). These

Table 8.5 Geochemical characteristics of calcite cements. (After Meyers and Lohmann, 1985.)

	PRE-PENNSYLVANIAN CEMENTS			POST-MISS. PRE-PERMIAN CEMENT
	ZONE 1	ZONE 2	ZONE 3	ZONE 5
Wt. % MgO	0.17 ± 0.02	$0.14 \pm .01$	$0.06 \pm .008$	$0.19 \pm .02$
Wt. % MnO	$0.01 \pm .004$	$0.08 \pm .02$	$0.02 \pm .009$	$0.06 \pm .006$
Wt. % FeO	$0.007 \pm .002$	$0.07 \pm .01$	$0.01 \pm .003$	$0.24 \pm .05$
$\delta^{13}C$ PDB ‰	$+3.7 \pm .2$	$+2.4 \pm .4$	$-0.8 \pm .4$	$+2.7 \pm .5$
$\delta^{18}O$ PDB ‰	$-1.3 \pm .2$	$-2.8 \pm .9$	$-3.7 \pm .2$	-7.4 ± 1.0

ESTIMATES OF WATER CHEMISTRY DURING PRECIPITATION OF CEMENT ZONES

	ZONE 1*	ZONE 2*	ZONE 3*	ZONE 5**
Mg^{2+}/Ca^{2+}	0.16	0.13	0.06	0.12
Mn^{2+}/Ca^{2+}	3.3×10^{-5}	2.6×10^{-4}	5.7×10^{-5}	5.2×10^{-5}
Fe^{2+}/Ca^{2+}	1.0×10^{-4}	1.0×10^{-3}	1.5×10^{-4}	3.5×10^{-3}
$\delta^{13}C$ PDB ‰	$+0.9$	-0.4	-3.6	-0.8
$\delta^{18}O$ SMOW ‰	$+0.6$	-0.9	-1.8	-1.8

Mississippian $\delta^{13}C$ HCO_3^- = +2 ‰ PDB; $\delta^{18}O$ = +0.5 ‰ SMOW

*assuming 25°C
**assuming 45°C

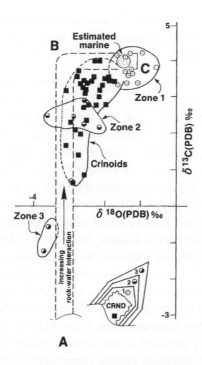

Figure 8.29. Carbon-oxygen isotopic composition of meteoric phreatic calcite cements in Mississippian crinoidal grainstones. Cements formed in ^{13}C-depleted, soil-gas CO_2-charged meteoric waters proximal to recharge areas have compositions near A. Cement compositions arising from dissolution-precipitation reactions along groundwater flow lines evolve toward B. In the distal portion of the groundwater system, cement compositions will evolve toward C and converge with the estimated composition of marine carbonates. The result of this overall process is the cement zoning seen in the schematic diagram of the crinoid grain (crnd) and cements shown in the lower right. See Chapter 7, Figure 7.42, for discussion of this type of J-shaped diagram. (After Meyers and Lohmann, 1985.)

calculations show that the precipitational waters for all these cements were low in Mg^{2+}, and thus seawater was an insignificant component of these pre-Pennsylvanian meteoric waters. It is likely that the small amount of Mg^{2+} in these waters came from the incongruent dissolution of the high magnesian calcite crinoids. Leutloff (1979) estimated that on average, about 25% of the original magnesium content of the crinoids was lost during diagenesis. The calculated activity ratios for Mn^{2+} and Fe^{2+} are within the range of modern groundwaters and are significantly greater than those found for normal oxygenated seawater. Consequently, element ratios for these dissolved species also point to little seawater influence on these pre-Pennsylvanian precipitational groundwaters. Similar conclusions can be reached for Zone 5 meteoric waters, but the calculations are less rigorous because of poor constraints on the geothermal gradient during formation of these pre-Permian cements.

A plot of the carbon-oxygen isotopic evolution of zones 1, 2, and 3 meteoric phreatic calcite cements is revealing (see also Chapter 7, Figure 7.42). Figure 8.29 shows the progressive evolution of the regional meteoric water system. The isotopic composition of Zone 1 cements is near that of the estimated values of Mississippian marine carbonate. This indicates a low water:rock ratio, slow flushing of the meteoric system, and strong buffering of the isotopic signature of the calcite cements by the marine, metastable, high magnesian calcite allochems undergoing mineral stabilization. These events may occur in the distal portion of the meteoric phreatic lens. Zones 2 and 3 cements plot along the meteoric water line but are distinctly different in isotopic composition. Zone 3 cements probably precipitated from ^{13}C-deficient groundwaters enriched in soil-gas CO_2 proximal to a recharge zone. Zone 2 cements are intermediate in carbon-oxygen signature and indicate the progressively decreasing importance of the meteoric water recharge carbon signature on cement composition. These cements probably formed farther down meteoric water flow lines than Zone 3 cements. An individual crinoid grain, because of its changing position in the meteoric water system, could display a cement paragenesis of changing cement composition outward from the crinoid grain surface (Figure 8.29).

Meyers and Lohmann (1985) developed an integrated model of the petrography and trace element and isotopic characteristics of Mississippian zone 1, 2 ,and 3 calcite cements (Figure 8.30). All these zones are interpreted in light of a regional meteoric water system. Zone 1 nonluminescent cements formed during a sealevel drop shortly after deposition of the Lake Valley crinoidal grainstones in an unconfined aquifer system. Metastable magnesian calcite minerals were abundant and acted as a cement source because of their high solubility and relatively rapid dissolution kinetics. Brightly luminescent Zone 2 cements formed during a sealevel stillstand in a meteoric aquifer that was shut off from extensive vadose recharge. Grain-to-grain pressure solution became an important cement source as the grainstones were buried to a depth of a few hundred meters. During a subsequent sealevel rise, a second stage of vadose recharge occurred. Carbon-13 deficient, nonluminescent Zone 3 cements developed in a region of the aquifer strongly influenced by ^{13}C-deficient soil-gas CO_2. At this time, metastable carbonate minerals were rare, and the system was dominated by water alteration reactions (see Chapter 7).

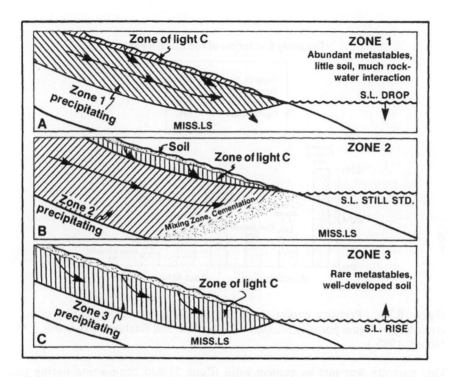

Figure 8.30. A model for the trace element and isotope characteristics of calcite cement zones 1, 2 and 3 in Mississippian Lake Valley grainstones. Zones 1, 2 and 3 calcites formed at different times during different stages in the evolution of the rock-groundwater system. (After Meyers and Lohmann, 1985.)

The wide range of $\delta^{18}O$ values for post-Mississippian Zone 5 cements reflects, in part, precipitation of these cements at elevated temperatures. The trace element contents indicate little seawater influence on water chemistry. The $\delta^{13}C$ values are variable but intermediate and suggest intraformational or extraformational cement carbon sources owing to pressure solution in conjunction with stylolitization during progressive burial of the formation to a depth of about 1000 m (Meyers and Lohmann, 1985; Moore, 1989).

The porosity evolution of the Lake Valley crinoidal grainstones is shown in Figure 8.31. Of the original porosity of 40%, 60% was lost because of cementation and 40% owing to compaction. At the time of burial, after the extensive cementation events of zones 1, 2 and 3, only 45% of the original porosity was left.

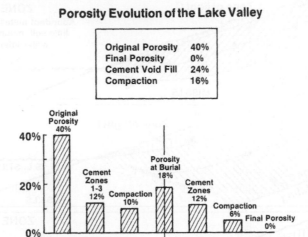

Figure 8.31.   Porosity evolution of the Mississippian Lake Valley crinoidal grainstones. Initial porosity is assumed to be 40% and final porosity is 0%. (After Moore, 1989.)

This porosity was lost by cement infill (Zone 5) and compaction during post-Mississippian time. Thus all porosity in this unit was lost before hydrocarbon maturation within Mississippian rocks could have occurred (Moore, 1989). This statement is not true of all carbonate units that have undergone early meteoric cementation. The Jurassic Smackover Formation of the U.S. Gulf of Mexico, at the time of burial into the subsurface, had a porosity of 0-50%. This porosity was unevenly distributed throughout the unit. The distribution was largely a response to the irregular network of calcite cement derived from stabilization of aragonite oöids in the regional meteoric water system. In the case of the Smackover Formation, it had perhaps as much as 20% porosity still intact at the time of hydrocarbon maturation and migration (Moore, 1989).

Other cement stratigraphic studies of carbonate sequences have revealed diagenetic features like those of the Mississippian Lake Valley Formation. These investigations include the Jurassic Smackover Formation of the U.S. Gulf of Mexico (Wagner and Matthews, 1982; Moore, 1989), the Siluro-Devonian Helderberg Group of the Central Appalachians (Dorobek, 1987), the Mississippian Burlington-Keokuk Formation of Illinois and Missouri (Kaufmann et al., 1988), the Mississippian Newman Limestone of Kentucky (Niemann and Read, 1988), and the Pennsylvanian Holder Formation of New Mexico (Goldstein, 1988). Not all these studies are unequivocal in their conclusions (e.g., the Jurassic Smackover Formation), and in some cases the diagenetic history of the unit is very complex (e.g., the Siluro-Devonian Helderberg Group).

Carbonate sediments composed of metastable mineral assemblages may be buried quickly or isolated from meteoric waters before substantial burial. These sediments enter the subsurface without initial stabilization of highly reactive minerals like aragonite and high-magnesian calcites, and the less reactive, but still unstable, impure low magnesian calcites. Burial-driven stabilization of these rocks could result in diagenetic features significantly different from those of meteoric diagenesis. In some cases, however, it may be difficult to distinguish between these features, particularly in a rock that has witnessed a complex burial history and more than one diagenetic event. In the following section, we take a closer look at burial stabilization and multi-diagenetic events in carbonate rocks.

Burial and multi-diagenetic stabilization

Some General Comments

There appears to be considerable controversy in the literature concerning the relative importance of early surficial diagenetic processes versus burial diagenetic processes in determining the chemical, mineralogical, and fabric evolution of carbonates. As Longman (1980) stated, "much (and perhaps most) cementation and formation of secondary porosity (except fractures) in carbonates occurs at relatively shallow depths in one of four major diagenetic environments: the vadose zone, meteoric phreatic zone, mixing zone, and marine phreatic zone" (p. 461). Certainly these processes are of considerable importance, as documented in innumerable studies, some of which have been discussed in this book. Scholle and Halley (1985), however, articulated another viewpoint: "..., we would argue here that, particularly for rapidly subsiding sedimentary basins, most cementation and much leaching take place in moderately deep to very deep subsurface settings (mesogenetic environments)" (p. 309). It is likely that the truth is somewhere in between; however, any decision on the relative importance of the two realms of diagenesis is biased currently by the four decades of investigation that have gone into studies of early diagenetic processes in Holocene and Pleistocene sediments. No such foundation of well-studied examples of burial diagenesis exists, and theoretical modeling of such processes is crude and in its infancy. The observations (Moore, 1989) that the carbonate geologic record is "replete with examples of total porosity destruction by early marine and meteoric phreatic cementation" (p. 278) and that "the volume and importance of these compaction-related cements (as emphasized by Scholle and Halley, 1985) do not seem to be supported by actual observations of workers dealing with subsurface diagenetic studies" (p. 277) may be, in part, an artifact of the diagenetic data base.

Moore (1989) appears to take somewhat of a middle position with regard to porosity development, arguing that early diagenetic processes of dissolution, cementation, and dolomitization are the major factors governing the ultimate distribution of <u>economic</u> porosity and permeability in the subsurface. Burial diagenetic processes, such as compaction and cementation, are responsible for reduction in porosity volume and quality during the burial history of a carbonate rock. Moore further points out that hydrocarbon migration and reservoir filling usually occur before the end of the burial diagenetic history of a rock. Thus at the reservoir site, these processes are virtually terminated, whereas in inter-reservoir rocks, compaction, cementation, and other burial diagenetic processes may continue.

Part of the controversy concerning early (and relatively shallow burial) versus later (moderate to deep burial) diagenesis revolves around the interpretation of porosity data from carbonate sediments and rocks. Modern carbonate mudstones generally have porosities of around 60 to 70% (Ginsburg, 1956), with very surficial deposits having even higher porosities. Porosity values for grainstones range from 40 to 50% (Halley and Harris, 1979; Enos and Sawatsky, 1981; Halley and Evans, 1983). These initial porosities in both shallow-water carbonate sediments and deep-water chalks change with increasing age and/or increasing burial depth of the sediment. With increasing burial depth, both oozes and shallow-marine carbonates exhibit consistent and usually smooth patterns of porosity decrease with increasing burial depth of the sediments (Figure 8.32) (Scholle and Halley, 1985). These porosity depth profiles are certainly a result of compaction and diagenesis. The question is how much of the pattern is due to shallow diagenetic processes as opposed to processes occurring at deeper depths? Halley and Schmoker (1983) and Scholle and Halley (1985) concluded that these trends were mainly the result of chemical compaction (the dual process of pressure solution and cementation) during burial in a semi-closed system.

From studies of porosity change during mineralogical stabilization of oölites in Florida and the Bahamas (Figure 8.33), several authors (Halley and Harris, 1979; Beach, 1982; Halley and Evans, 1983; Scholle and Halley, 1985) have argued that even though these sediments have undergone progressive meteoric vadose and phreatic diagenesis, they still retain an average high porosity of 45% (Figure 8.33). The range of porosity values increases with increasing intensity of diagenesis, but the mean value of modern oöid sands is very near that of the Pleistocene Miami

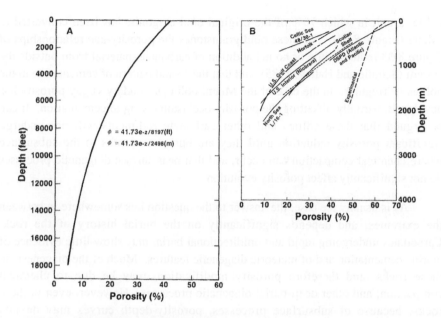

Figure 8.32. Porosity-depth plots of (A) shallow-marine carbonates and (B) deep-sea oozes. (After Scholle and Halley, 1985.)

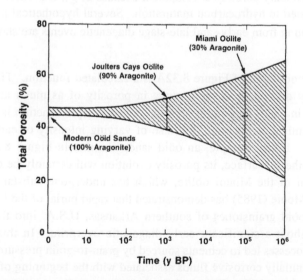

Figure 8.33. Porosity-age relationships for oölitic grainstones. Notice that although the average porosity of these calcareous sands does not change significantly for a million years, the older oölites exhibit a spread in porosity because of dissolution-precipitation events. (After Scholle and Halley, 1985.)

oölite, a unit in which 70% of the original oöid aragonite has been converted to calcite (Figure 8.33). For these oöid grainstones the porosity-age relationships of Figure 8.33 indicate little or no net addition of carbonate material from outside the system (Scholle and Halley, 1985), and that the spatial pattern of cementation in the oölites is irregular. In the case of the Miami oölite, secondary vuggy porosity has developed, partially offsetting the porosity occlusion owing to cementation. It can be argued that these oölites and other carbonate sediments will not undergo significant porosity reduction until they are buried to depths in the subsurface where chemical compaction can occur, and that near-surface diagenetic processes do not significantly affect porosity evolution.

As is usually the case, the answer to the question lies somewhere in between the extremes, and depends significantly on the burial history of the rock. Carbonates undergoing rapid and unidirectional burial may show little evidence of marine cementation and of meteoric diagenetic features. Much of the diagenesis in these rocks, and therefore porosity modification, may be due to chemical compaction, and other deep-burial diagenetic processes. However, even in these rocks, because of subsurface processes, porosity-depth curves may deviate substantially from the smooth relationships shown in Figure 8.32. These deviations, for example, are due to overpressuring in the sedimentary sequence and late-stage secondary porosity development because of generation of organic acids and CO_2 related to hydrocarbon maturation. Several hypothetical porosity-depth curves stemming from early- and late-stage diagenetic events are shown in Figure 8.34.

The smooth plot of Figure 8.32A is a calculated function. The actual data points at any one depth cover a range in porosity of as much as 30%. This observation, in itself, may imply that the porosity-depth function is not a simple result of chemical compaction, but also of varying intensity of early diagenetic overprinting. If for example, an oöid sand, like that in Figure 8.33, is buried quickly into the subsurface, its porosity evolution will certainly be different than that of burial of the Miami oölite, which has undergone substantial meteoric diagenesis. Moore (1985) has demonstrated that rapid burial of the Upper Jurassic Smackover oöid grainstones of southern Arkansas, U.S.A. into the subsurface occurred without a significant early diagenetic overprint. In this unit, burial diagenetic processes led to cements sourced by grain-to-grain pressure solution and possibly by initially corrosive fluids associated with the beginning of hydrocarbon maturation and expulsion of waters from shales.

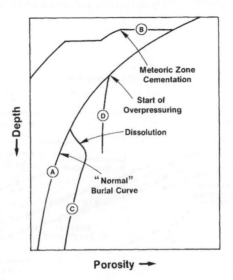

Figure 8.34. Hypothetical porosity-depth curves for various diagenetic situations. A. "Normal" burial curve of chalk sequences. B. A typical curve stemming from porosity changes brought about by dissolution and cementation processes during meteoric zone diagenesis. C. A scenario for late stage porosity development brought about by a dissolution event related to hydrocarbon maturation and destruction. D. Porosity preservation resulting from conditions of overpressuring. (After Choquette and James, 1987.)

Another means of assessing the relative role of early and late diagenetic processes in the evolution of a carbonate sequence is to consider the relationship between porosity and thermal maturity. In Figure 8.35 the porosity-depth relations and the temperature-time history of carbonates from various basins are portrayed in a porosity-time-temperature index (TTI) plot. TTI is a measure of the thermal maturity of a rock and is dependent on both the age of the rock and the amount of time the rock has spent at various burial temperatures. Higher values of TTI indicate greater thermal maturity. The plot of Figure 8.35 shows that there is no single porosity-thermal maturity function for all carbonate units; however, for a carbonate unit from a single sedimentary basin, there is a strong log-linear functional relationship. Details of this kind of plot are given in Schmoker (1984). One interpretation is that the subparallel plots of Figure 8.35 reflect chemical compaction, and that differences between basins are indicative of differences in the initial fabric and texture of the carbonate unit because of different depositional processes and the intensity of early diagenetic processes of cementation, dissolution and dolomitization.

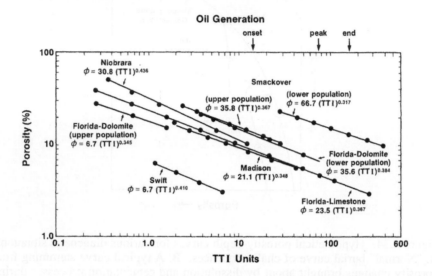

Figure 8.35. Porosity versus time-temperature index of maturity (TTI) for various carbonate rock units. The TTI values are based on Lopatin's (1971) method. The onset and end of oil generation (the "oil generation window") in terms of TTI units are from Waples (1980). The log linear equations of best fit between porosity and TTI data are also shown. ϕ = porosity. (After Schmoker, 1984.)

Subsurface diagenesis at moderate (several hundred meters) and deep burial depths (several thousand meters) has been demonstrated for several carbonate units (e.g., the Lower Cretaceous Stuart City Trend of south central Texas, U.S.A., Prezbindowski, 1985; the Lower Cretaceous Pearsall and Glen Rose formations of south Texas, U.S.A., Woronick and Land, 1985; the Upper Jurassic Smackover Formation of Arkansas, U.S.A., Moore, 1985; and the Siluro-Devonian Helderberg Group of the central Appalachians, U.S.A., Dorobek, 1987). As mentioned previously, diagenesis starts early in the history of a carbonate rock and continues throughout its evolution to a lower free energy state. It is likely that the intensity of diagenesis is not uniform in time and space and depends significantly on the thermal and fluid evolutionary history of a basin. Many older carbonate rocks have seen multiple stages of deformation and diagenesis. To unravel their diagenetic histories is a difficult task, particularly because we know so little about the processes leading to the formation of late diagenetic features. In the following section, we describe the petrography, geochemistry, and origin of burial diagenetic facies in the Siluro-Helderberg Group of the central Appalachians, U.S.A. (Dorobek, 1987). We chose this unit as a case study because it well illustrates the necessity of an integrated approach to carbonate diagenesis in which all available basin-wide information, including petrographic, geochemical, basinal stratigraphy and thermal history, are used to construct a diagenetic model (Moore, 1985).

Diagenesis of the Siluro-Devonian Helderberg Group

The Siluro-Devonian Helderberg Group is exposed in northeast-trending fold structures throughout the Valley and Ridge province of the central Appalachians. Dorobek (1987), using a variety of petrographic and geochemical techniques, coupled with a well-defined stratigraphic framework and geologic history, attempted to unravel the complex diagenetic history of the Helderberg Group skeletal limestones and mudstone buildups. Within the area studied by Dorobek (Virginia, West Virginia, and Maryland, U.S.A.), the Helderberg Group (0-150 m thick) generally overlies the Silurian Tonoloway Limestone and is overlain by Middle Devonian clastics that are the basal units of the Acadian clastic wedge of the central Appalachians. A schematic stratigraphic cross-section of units composing the Helderberg Group is shown in Figure 8.36.

Helderberg sediments in the central Appalachians were deposited on a gently-eastward-sloping carbonate ramp during a tectonically quiet period between the Taconic (Middle to Late Ordovician) and Acadian (Middle Devonian) orogenies. The ramp was bordered on the east by the low-relief remnants of an accretionary prism formed during the Taconic orogeny (Read, 1980). During Helderberg time, this prism was not a source of significant siliciclastic detritus, and Helderberg carbonate deposition took place in an open marine environment that had developed across the whole central Appalachian basin. During the Acadian orogeny, the accretionary prism was reactivated and siliciclastic sediments prograded westward, quickly burying the Helderberg carbonates and intercalated sands (Dorobek and Read, 1986).

A plot illustrating the burial history of Lower Ordovician through Upper Devonian strata in the central Appalachians is shown in Figure 8.37. To construct this plot, Dorobek (1987) assumed an average geothermal gradient of 30°C km^{-1}, and a surface temperature of 20°C. During Helderberg deposition, the central Appalachian basin was subsiding at a rate of about 10 m my^{-1}, but underwent more rapid burial during deposition of the thick Middle Devonian clastic wedge, a result

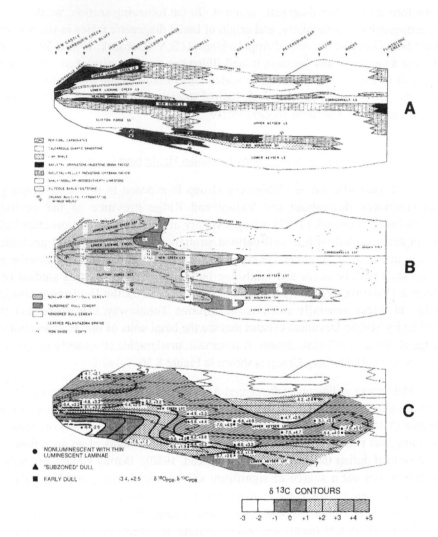

Figure 8.36. Cross-sections of Helderberg Group rocks in the central Appalachian basin, U.S.A. All sections oriented southwest to northeast. A. Stratigraphic units and nomenclature. B. Regional distribution of cathodoluminescent calcite cement zones; nonluminescent cements coincide with distribution of updip sandstones and skeletal limestones. The distribution of leached pelmatozoan grains (L), and iron oxide coatings on these grains (Fe) are also shown. C. Variation in stable isotopic composition of early calcite cements; the increase in $\delta^{13}C$ values basinward and upsection is interpreted as representing original flow paths of meteoric groundwaters. (After Dorobek, 1987.)

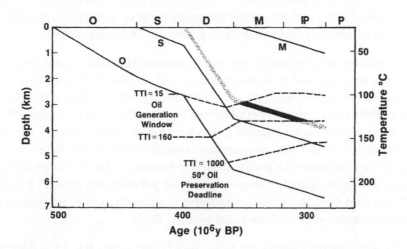

Figure 8.37. Burial history plot for Helderberg Group rocks and other strata. Helderberg Group is shaded, whereas the base of Ordovician strata, base of Silurian strata and base of Mississippian strata are shown as thin lines marked O, S and M. Dashed lines are time-temperature indices of maturity (TTI; see Waples, 1980). The oil generation window represents the necessary conditions of time and temperature for potential oil generation. The 50° oil preservation deadline is the upper TTI limit for occurrence of oil with API gravity of 50°. (Modified from Dorobek, 1987.)

of the Acadian orogeny. The maximum depth of burial of the Helderberg carbonates was about 4 km at an ambient burial temperature of approximately 150°C. The timing of generation and destruction of hydrocarbons for various source beds is also shown in Figure 8.37. Of interest is the fact that the Helderberg Group sediments and older strata were heated for a time sufficiently long to generate hydrocarbons.

Synsedimentary marine cements are the earliest diagenetic feature recognized by Dorobek (1987). These calcite cements predate all other cements. Three major cement fabrics are recognized: bladed or relict rhombohedral, inclusion-rich syntaxial rim cement, and microcrystalline hardground cement. The bladed or relict rhombohedral cements were interpreted as marine because they are interlayered with marine internal sediment and predate equant calcite cement of shallow- to deep-burial origin. The original mineralogy of these cements is equivocal; they could have been originally low-magnesian or high-magnesian calcites. The inclusion-rich syntaxial rim cements were most likely marine high-magnesian calcites because they have abundant microdolomite inclusions which define rhombohedral terminations.   The microcrystalline hardground cements were deposited in the mud-mound and deep ramp facies of the Helderberg, and have all the characteristics of sea-floor cements (James and Choquette, 1983).

Following development of the marine cements, Helderberg grainstones and mudstones were exposed to meteoric waters, and diagenetic features developed typical of this realm of diagenesis. Pelmatozoan (orginally attached echinoderms) grains in Helderberg carbonates exhibit three major end-member diagenetic fabrics produced in meteoric water aquifers: grains with well-developed original stereome (calcareous tissue in the mesodermal endoskeleton of an echinoderm), neomorphosed inclusion-rich grains, and leached grains. Table 8.7 shows the trace element and stable isotopic composition of the pelmatozoan grains. Dorobek (1987) has used these characteristics and other considerations as constraints on the chemical evolution of the shallow-burial meteoric fluids.

In southwestern Virginia, the Helderberg Group is only 0 to 25 m thick and consists primarily of sandstone units that are bounded by disconformities (Dorobek and Read, 1986). Carbonate grains are extensively leached or replaced by iron oxide cement. Dorobek argues that these thin sands are the preserved parts of the Helderberg aquifers that were closest to the recharge area. Thus, their waters would be the most aggressive to carbonate minerals and the most oxidizing; hence the extensive leaching and iron mineral cementation observed in this region of the Helderberg Group rocks.

Neomorphism of both original aragonite molluscan and original magnesian calcite pelmatozoan grains is extensive, but only select pelmatozoan grains have their stereome preserved or are leached. From these and other observations, Dorobek (1987) concluded that the groundwaters in which these sediments underwent meteoric diagenesis were near saturation with aragonite and most magnesian calcite compositions. Under these conditions, neomorphism proceeded by thin-film inversion (dissolution-reprecipitation along a narrow front) of high-magnesian calcite and aragonite to low magnesian calcite. Pelmatozoan grains with original stereomes preserved were probably armored by iron oxide precipitate in the up-dip, recharge region of the meteoric aquifers prior to leaching or neomorphism. Continuous dissolution of the grains probably occurred in the oxidizing pore fluids of the updip portions of the aquifers during shallow burial.

Helderberg carbonates are extensively cemented by clear equant calcite cement that postdates marine cement and fills most of the original pores. These cements are compositionally zoned as shown by cathodoluminescence and trace element analyses. These precipitates include nonluminescent, bright and dull cements. The luminescent cement is the earliest generation of cement followed by bright, then dull cement. The geochemical characteristics of these cements are given in Table 8.7.

Nonluminescent cement is low in Mn and Fe and intermediate in Mg content compared to bright and dull cements. Dull cements contain the highest Mg and Fe concentrations. Bright cement is high in manganese.

Primary two-phase inclusions are only found in the ferroan-rich dull cements (Table 8.7). The fluids contained in these inclusions, as determined from eutectic and melting (ice) temperatures, are dilute solutions (2.9 wt % NaCl) to complex (>23 wt % NaCl) brines. Homogenization temperatures indicate maximum temperatures of fluid entrapment of 90° to 200°C.

Table 8.7. Summary of geochemical data for cements and pelmatozoan grains (from Dorobek, 1987).

			NONLUMINESCENT	BRIGHT	DULL	PELMATOZOANS
PERCENT TOTAL CEMENT		N	66	84	122	
		RANGE	0 - 60%	0 - 53%	23 - 100%	
		MEAN	28%	13.8%	76.2%	
TRACE ELEMENTS	Mn	N	69	42	743	215
		RANGE	0 - 403	31 - 3408	0 - 1053	0 - 798
		MEAN	20	1164	293	106
	Fe	RANGE	0 - 700	0 - 1143	0 - 42,500	0 - 21,145
		MEAN	150	376	1261	1417
	Mg	RANGE	0 - 2111	0 - 3492	0 - 10,699	0 - 9288
		MEAN	1146	803	1801	2106

		LATE DULL CEMENT	QUARTZ CEMENT
PRIMARY TWO-PHASE FLUID INCLUSIONS	N	20	3
	T_E	-45.2 TO -24.0°C	-25.5 TO -23.9°C
	T_{M-ICE}	-28.8 TO -1.7°C	-5.0 TO -1.4°C
	T_{HL-V}	46.2 TO 160.3°C	139.6 TO 175.0°C
	T_{HP-C}	128.1 TO 201.3°C	181.6 TO 220.0°C
	WEIGHT% NaCl	2.9 TO >23	2.4 TO 7.9

			NONLUMINES-CENT WITH THIN LUMINESCENT LAMINAE	SUBZONED DULL	EARLY DULL	PELMATO-ZOAN GRAINS	DIFFERENCE BETWEEN PELMATOZO-ANS AND EARLIEST CEMENT	LATE DULL
STABLE ISOTOPES	$\delta^{18}O$	N	15	7	5	22	23	65
		RANGE	-7.6 TO -4.7	-6.8 TO -4.7	-7.0 TO -4.7	-8.7 TO -4.4	-4.0 TO +2.0	-12.0 TO -4.7
		MEAN	-5.8	-6.0	-5.4	-6.0	-0.3	-8.7
		σ	0.8	0.7	0.9	1.2	1.2	1.8
	$\delta^{13}C$	N	15	7	5	22	23	65
		RANGE	-2.9 TO +4.2	+0.6 TO +4.6	+2.6 TO +4.7	+0.4 TO +4.7	-2.5 TO +3.3	-1.0 TO +5.8
		MEAN	+2.6	+2.4	+3.5	+3.1	+0.2	+2.8
		σ	1.7	1.7	1.0	1.3	1.1	1.8

N = number of analyses, T_E = eutectic temperature, T_{M-ICE} = melting temperature of ice; T_{HL-V} = homogenization temperature of liquid and vapor phases, T_{HP-C} = 400 bar hydrostatic pressure-corrected trapping temperature, σ = standard deviation. Stable isotopes reported relative to PDB standard.

The regional distribution of the cathodoluminescent cement zones and variation in their stable isotopic composition (Table 8.7) are shown superimposed on the stratigraphic cross-section of Figure 8.36A in Figures 8.36B and 8.36C. The nonluminescent cement makes up 0 to 60% of the equant calcite cement and decreases in abundance basinward. Bright cement comprises 0 to 53% of the equant calcite cement and postdates nonluminescent cement in updip sections of the Helderberg carbonates, or occurs immediately downdip of the last occurrence of nonluminescent cement. Dull cement comprises 23 to 100% of the total void filling calcite cement, and occurs in all lithofacies that have coarse-grained calcite cement as a final product of calcite cementation. In general, the nonluminescent and early dull cements become progressively enriched in ^{13}C basinward and upsection. The late dull ferroan cements show no stratigraphic variation in isotopic composition.

Other diagenetic phases have been identified in the Helderberg carbonates. These include silica replacing calcite, quartz overgrowths on detrital quartz grains, late diagenetic saddle dolomite (coarse-grained, iron-rich with curved faces, and cleavage planes) and rare fluorite.

A summary of the timing and origin of the calcite cements and related diagenetic events, as proposed by Dorobek (1987), is given in Figure 8.38. It can be seen from this diagram that the Helderberg Group rocks have gone through a complex history of diagenesis, including marine, shallow-burial meteoric, and deep-burial diagenesis. Dorobek (1987) discusses these diagenetic processes in detail. A few of his conclusions are emphasized below primarily to demonstrate how an overall diagenetic model for a complexly altered unit can be derived from a variety of geochemical, stratigraphic and burial history information. Many of the early diagenetic features observed in the Helderberg Group rocks were discussed in Chapters 6 (marine cements) and 7 (meteoric diagenesis).

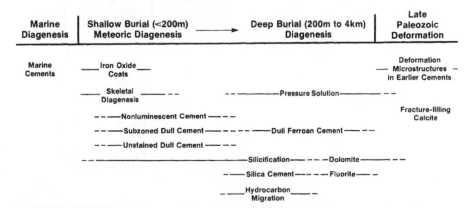

Figure 8.38. Summary of diagenetic events affecting Helderberg Group rocks in the central Appalachians. Hydrocarbon migration occurred during deep burial and development of quartz overgrowths in the oil-generative window of Figure 8.37, and at a temperature of about 180 to 220ºC. (After Dorobek, 1987.)

Skeletal diagenesis, development of iron oxide coatings, and precipitation of nonluminescent, bright and early non-ferroan-rich dull cements occurred during shallow-burial meteoric diagenesis. During Helderberg deposition, the meteoric aquifers were unconfined, and freshwater wedges did not extend far offshore. After deposition of overlying (Middle Devonian) impermeable shales, Helderberg aquifers were confined, and meteoric waters extended at least 150 km offshore. The similarity in stable isotopic composition between early calcite cements and pelmatozoan grains (Table 8.7) indicates that the carbonate for the early cements came from dissolution of the pelmatozoan grains in the waters of these paleoaquifers. Nonluminescent cements were precipitated in the proximal oxidizing waters of the recharge areas of these aquifers along the eastern flank of the Appalachian basin. The early dull cements precipitated from reducing meteoric waters under shallow burial conditions (<300 m) in the more distal portions of the Helderberg aquifers, whereas the Mn-rich bright cements formed in transitional redox zones in the paleoaquifers. The enrichment of these early cements in ^{13}C downdip and upsection is indicative of progressive water-rock interaction along groundwater flow paths.

In contrast to these early calcite cements, late, ferroan, ^{16}O-enriched dull cement probably formed at temperatures of 90 to 200°C during progressive burial of the Helderberg units from 300 to 4000 m. The subsurface fluids responsible for these cements were dilute to saline Na-Ca-Cl waters with stable isotopic compositions similar to those from modern oil fields.

Silica, derived principally from dissolution of sponge spicules, replaced calcite during all stages of diagenesis, whereas quartz overgrowths on detrital quartz grains postdate early calcite cements. These quartz overgrowths have primary hydrocarbon inclusions defining crystal faces. These inclusions are evidence for the migration of oil during the time of this diagenetic event. Fluid inclusion studies indicate that the quartz overgrowth cements were formed in dilute Na-Cl brines with temperatures of 180 to 220°C. The saddle dolomite and fluorite are very late diagenetic phases, as has been observed in burial diagenetic studies from other areas (e.g., Prezbindowski, 1985), and probably formed just prior to and during the Alleghenian orogeny.

This case history of burial diagenesis of the Helderberg Group is just one scenario for the diagenetic processes affecting carbonates as they are progressively buried in the subsurface. Because of the inherent complex nature of the processes affecting carbonates during burial, the different burial histories of sedimentary basins, and the fact that many ancient basins have seen several cycles of subsidence and uplift, it is likely that burial and multidiagenetic stabilization studies will need to be basin specific.

Concluding remarks

In terms of our knowledge of the diagenesis of sedimentary carbonates, the subsurface still remains an unknown. Observational data on the composition of carbonate solids and fluids in the subsurface are not plentiful on a global scale and are biased by studies from petroleum producing areas, particularly the Gulf Coast, U.S.A. An understanding of the processes of the chemical evolution and migration of subsurface fluids requires more data and the development of multidisciplinary analytical and numerical models. These models need to incorporate thermodynamic and kinetic factors of carbonate mineral-organic matter-aqueous solution reactions within a framework that includes the hydrodynamics and geologic evolution of a basin. Furthermore, it is necessary to recognize that carbonates are not a system onto themselves but interact strongly with other lithologies in the subsurface, sourcing cements in these lithologies or acting as sinks of chemical components derived from them. These chemical components may travel hundreds to thousands of kilometers, or only several microns, in the subsurface. The subsurface environment of diagenesis is an exceedingly variable region of P, T, and solution compositions. Because of this complexity, we are probably in for a number of surprises in the future.

In the last two chapters of this book, the carbonic acid system and sedimentary carbonates are viewed in the context of global environmental change, past, present, and future. Much of our information concerning change in Earth's surface environment comes from studies of the carbonate system. Chemical, mineralogical, and isotopic data obtained from carbonate rocks are used to interpret the origin and evolution of Earth's surface environment. It must be kept in mind that interpretation of some of these data may be equivocal, because it is difficult to unravel primary chemical signatures related to environment of deposition from those due to diagenesis. Thus, discussion in this chapter and the preceding two chapters is germane not only to considerations of the nature and extent of carbonate diagenesis, fluid migration and evolution, and porosity modification in the subsurface but also to the broader picture of the evolution of the planet.

Chapter 9

THE CURRENT CARBON CYCLE
AND HUMAN IMPACT

Introduction

During much of this book, we have been concerned with carbonate mineral-water reactions in the modern environment and during early diagenesis. In this chapter and in Chapter 10, we will consider carbonate chemistry, and carbonate sediments and sedimentary rocks in the context of the evolution of Earth's surface environment. Trends in the mineralogy, chemistry isotopic composition, and carbonate rock volume and lithic type with geologic age are an important source of information concerning evolutionary and cyclic changes in the hydrosphere, lithosphere, atmosphere, and biosphere of the Earth during geologic time. Knowledge of the present-day carbon cycle has proven to be an important component in interpretation of the recent increase in atmospheric CO_2, because of the burning of fossil fuels, land use activities, and future climatic consequences of this environmental change. A recent volume, *The Carbon Cycle and Atmospheric CO_2: Natural Variations Archean to Present* (Sundquist and Broecker, 1985), reflects the strong interest of the scientific community in the carbon cycle both on a human time perspective and on geologic time scales. Because of the intense research effort involving studies of the carbon cycle and atmospheric CO_2, knowledge of this important cycle is increasing rapidly, and even as we write this book our concepts are changed and modified.

To begin the discussion, we will present briefly a view of the modern carbon cycle, with emphasis on processes, fluxes, reservoirs, and the "CO_2 problem". In Chapter 4 we introduced this "problem"; here it is developed further. We will then investigate the rock cycle and the sedimentary cycles of those elements most intimately involved with carbon. Weathering processes and source minerals, basalt-seawater reactions, and present-day sinks and oceanic balances of Ca, Mg, and C will be emphasized. The modern cycles of organic carbon, phosphorus, nitrogen, sulfur, and strontium are presented, and in Chapter 10 linked to those of Ca, Mg, and inorganic C. In conclusion in Chapter 10, aspects of the historical geochemistry of the carbon cycle are discussed, and tied to the evolution of Earth's surface environment.

Any discussion of the historical geochemistry of an element or compound, or a complex of species is necessarily speculative. Much is known, but much remains to be found out. Our work draws on the efforts of innumerable scientists, starting with Arrhenius (1896) and Chamberlain (1898), and continuing with the authors referenced in this chapter and Chapter 10.

Modern biogeochemical cycle of carbon

A Model for the Cycle of Carbon

It seems appropriate to begin our investigation of global historical aspects of carbon and carbonate sediment cycling with a look at the modern biogeochemical cycle of carbon. This approach will give the reader some feel for the magnitudes of reservoir carbon contents, fluxes, and processes associated with carbon cycling. Furthermore, one can see qualitatively from examination of the cycle how changes in fluxes during geologic time can influence atmospheric CO_2 content. Atmospheric methane, carbon monoxide, and carbon dioxide, organic matter (commonly expressed by geochemists as CH_2O), and inorganic carbonate minerals are the chief constituents in the carbon cycle. The fact that organic matter contains reduced carbon as C^0, whereas carbonate minerals contain oxidized carbon as C^{4+}, is especially important to our later discussion of the meaning of the geologic age distribution of carbonate sediment carbon isotopic values. Figure 9.1 is a detailed interpretation of the modern biogeochemical cycle of carbon, and represents an updated version of the original cycle of Garrels et al. (1975). A less complex cycle is presented later in this book. Other representations of the carbon cycle are available in the literature (see references); of particular interest is the recent compendium by Sundquist (1985).

From Figure 9.1, it can be seen that the major form of carbon in the atmosphere is $CO_2(g)$, constituting over 99% of atmospheric carbon. Carbon dioxide makes up 0.035% by volume of atmospheric gases, or 350 μatm = 350 ppmv. The atmosphere has a mass of CO_2 that is only 2% of the mass of total inorganic carbon in the ocean, and both of these carbon masses are small compared to the mass of carbon tied up in sediments and sedimentary rocks. Therefore, small changes in carbon masses in the ocean and sediment reservoirs can substantially alter the CO_2 concentration of the atmosphere. Furthermore, there is presently 3 to 4 times more carbon stored on land in living plants and humus than resides in the atmosphere. A decrease in the size of the terrestrial organic carbon reservoir of only 0.1% y^{-1} would be equivalent to an increase in the annual respiration and decay carbon flux to the atmosphere of nearly 4%. If this carbon were stored in the atmosphere, atmospheric CO_2 would increase by 0.4%, or about 1 ppmv y^{-1}. The

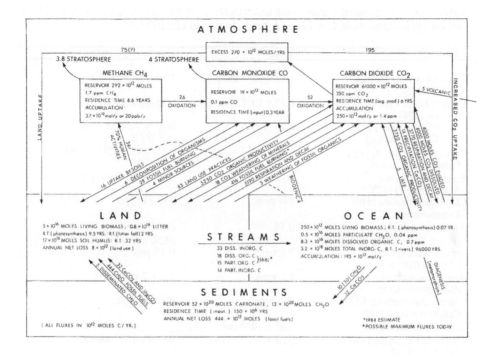

Figure 9.1. The modern biogeochemical cycle for carbon. One mole C = 12 g.

important point is that small changes in the mass of the ocean, land, and sediment carbon reservoirs can greatly affect atmospheric CO_2 contents. The rate of CO_2 change depends principally on the time scale for carbon changes in the large carbon reservoirs and on the rates and direction of the feedback mechanisms that operate between the atmospheric carbon reservoir and other reservoirs (see e.g., Lerman, 1979).

Now let us take a brief tour through the carbon cycle, emphasizing some, but not all, processes and fluxes. It should be kept in mind that the fluxes shown in Figure 9.1 are not agreed on by all authors but are reasonable estimates to give a balanced cycle. Table 9.1 shows some estimated values for the global carbon cycle to give the reader a feeling for the uncertainties involved.

Table 9.1.   Estimated values and uncertainties of key parameters in the global carbon cycle. One mole C = 12 g. (After Solomon et al., 1985.)

Parameter	Value	Uncertainty Range
Total Fossil Fuel Release (1860-1982)	14.2×10^{15} mole C	$12.5\text{-}15.8 \times 10^{15}$ mole C
Current Fossil Fuel Release (1982)	0.42×10^{15} mole C	$0.38\text{-}0.46 \times 10^{15}$ mole C
Annual Increase in Fossil Fuel CO_2 Emissions	0.00 (1982)	Highly variable in past and future
Recoverable Conventional Fossil Fuel Resources	0.33×10^{18} mole C	$0.31\text{-}0.35 \times 10^{18}$ mole C
Total Carbon in the Atmosphere (1982)	60×10^{15} mole C	$59\text{-}61 \times 10^{15}$ mole C
CO_2 Concentration in the Troposphere (1982)	341 ppm	339-343 ppm
Airborne Fraction (Fossil Fuels, 1958-1982)	0.58	$(0.54\text{-}0.63)$[a] $(0.50\text{-}0.80)$[b]
Annual Atmospheric CO_2 Oscillation Range	7 ppm	0-15 ppm
Annual Increase in Atmospheric CO_2	1.5 ppm	1.3-1.7 ppm
Historic Atmospheric CO_2 Concentration (1800)	280 ppm	260-285 ppm
Surface Area of Ocean	3.61×10^{14} m^2	
Carbon in Ocean Surface Layer (0-75 m)	52.5×10^{15} mole C	$49\text{-}60 \times 10^{15}$ mole C
Carbon in Intermediate and Deep Ocean	3.2×10^{18} mole C	$2.7\text{-}3.5 \times 10^{18}$ mole C
Carbon in Ocean Sediments	10^{22} mole C	?
Marine Particulate Carbon Flux	0.33×10^{15} mole C	$0.17\text{-}0.5 \times 10^{15}$ mole C
Gross Annual Ocean CO_2 Uptake	8.9×10^{15} mole C	$8.8\text{-}9 \times 10^{15}$ mole C
Net Ocean CO_2 Uptake[c]	0.2×10^{15} mole C	$0.12\text{-}0.28 \times 10^{15}$ mole C
River Export of Carbon to Oceans	0.06×10^{15} mole C	$0.03\text{-}0.11 \times 10^{15}$ mole C
Historic Terrestrial Biomass	75×10^{15} mole C	$58\text{-}92 \times 10^{15}$ mole C
Contemporary Biomass	47×10^{15} mole C	$35\text{-}55 \times 10^{15}$ mole C
Historic Soil Carbon	0.14×10^{18} mole C	$0.14\text{-}0.17 \times 10^{18}$ mole C
Contemporary Soil Carbon	0.13×10^{18} mole C	$0.1\text{-}0.15 \times 10^{18}$ mole C
Net Carbon Flux From Land Biosphere Since 1800	10×10^{15} mole C	$7.5\text{-}15 \times 10^{15}$ mole C
Gross Annual Terrestrial Plant CO_2 Uptake	10×10^{15} mole C	$7.5\text{-}10 \times 10^{15}$ mole C
Net Primary Production	5×10^{15} mole C	$3.8\text{-}5.2 \times 10^{15}$ mole C
Annual Tropical Forest Area Conversion (1970-1980)	0.004	0.003-0.006
Annual Net Carbon Flux From Land Conversion (1970-1980)	0.11×10^{15} mole C	$0.0\text{-}0.22 \times 10^{15}$ mole C

[a] Based on atmospheric and fossil fuel emissions record; [b] based on combined carbon cycle uncertainties; [c] prior to extensive fossil fuel burning and deforestation activities, the oceans were a net source of CO_2 to the atmosphere (Smith and Mackenzie, 1987; Wollast and Mackenzie, 1989).

Methane and Carbon Monoxide Fluxes

First consider methane, which in general, is produced at the Earth's surface principally by decomposition of organic matter under reducing conditions, whether in the rumens of cows or in swamps and rice paddies. The present total flux of methane to the atmosphere is a contested value (cf. Ehhalt, 1979; Senum and Gaffney, 1985; Cicerone and Oremland, 1988); source and sink terms are given in Table 9.2. Measurements of atmospheric CH_4 concentrations in ice and air during the last 200 years show that the gas is presently accumulating in the atmosphere at a rate of 3.7×10^{12} moles CH_4, or 0.02 ppmv, per year (Figure 9.2). The average global methane concentration has risen by 11% from 1978 to 1987 (Blake and Rowland, 1988).

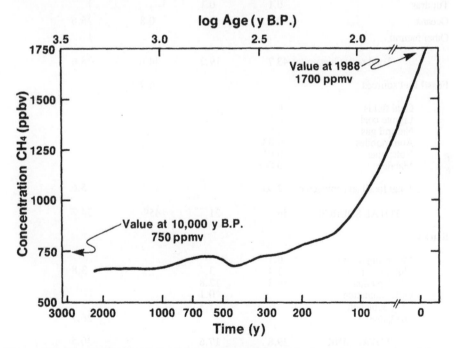

Figure 9.2. Generalized trend of the concentration of methane in air extracted from dated ice cores. (After Cicerone and Oremland, 1988.)

Table 9.2. Global methane sources and sinks; fluxes in units of 10^{12} moles C y^{-1}.

Sources	Ehhalt (1979)	Senum and Gaffney (1985)	Cicerone and Oremland (1988)	[a]Figure 9.1 values
Biomass burning			4.6	5.4
Enteric fermentation				
Animals	9.4	9.4	}	}
Termites		3.1	}10	}5.6
Humans		0.3	}	}
Rice paddy fields	17.5	3.4	9.2	7.6
Swamps and marshes	15.3	1.9	}	3.4
Freshwater sediments	0.8	0.4	}10	}
Tundras	0.1	0.1	}	}
Oceans	0.6	0.6	0.8	}6.6
Other natural				}
Total recent biogenic	43.7	19.2	34.6	28.6
Fossil fuel sources			6.7	
Coal fields	0.9			
Lignite coal	0.2			
Natural gas	0.9			
Automobiles	0.03			
Volcanoes	0.01			
Mantle	0.006			
Total fossil and abiogenic	2.05			5.6
TOTAL SOURCE	46	21	45[b]	34.2
Sinks				
Micro-organisms	0.1			
Flux into stratosphere	3.4	3.4		3.8
OH depletion	36.3	12.8		
O('D) depletion		0.1		
NO$_3$ depletion		1.3		
Oxidation				26.7
TOTAL SINK	39.8	17.6		30.5

[a]Adapted from Wollast and Mackenzie (1989); [b]includes landfill emissions and methane hydrate destabilization.

The total flux of CH_4 to the atmosphere today is about 34×10^{12} moles y^{-1}; about 20% of this flux results from enteric fermentation processes in animals, termites, and humans. Of the remaining flux, about 60% comes from the process of methanogenesis in sediments of various types--primarily muds of rice paddy fields, swamps, and marshes. The overall bacterial reactions in sediments are the disproportionation of organic matter to CH_4 and CO_2, that is:

$$2CH_2O \rightarrow CO_2 + CH_4 \tag{9.1}$$

and the reduction of CO_2 to CH_4 according to:

$$CO_2 + 4H_2 \rightarrow CH_4 + 2H_2O \tag{9.2}$$

The main sink for methane in the atmosphere is oxidation by OH radical to carbon monoxide, which in turn is oxidized to carbon dioxide according to the reaction scheme first proposed by Levy (1971) and McConnell et al. (1971):

$$CH_4 + OH \rightarrow CH_3 + H_2O \tag{9.3}$$

$$CH_3 + O_2 \rightarrow CH_3O_2 \tag{9.4}$$

$$CH_3O_2 + NO \rightarrow CH_3O + NO_2 \tag{9.5}$$

$$CH_3O + O_2 \rightarrow CH_2O + HO_2 \tag{9.6}$$

$$\begin{aligned} CH_2O + OH &\rightarrow CHO + H_2O \\ CH_2O + h\nu &\rightarrow H_2 + CO \\ CH_2O + h\nu &\rightarrow HCO + H \end{aligned} \tag{9.7}$$

$$HCO + O_2 \rightarrow CO + HO_2 \tag{9.8}$$

$$CO + OH \rightarrow CO_2 + H \tag{9.9}$$

OH is eventually regenerated by:

$$HO_2 + NO \rightarrow NO_2 + OH \tag{9.10}$$

Reaction of CH_4 with OH accounts for the destruction of 76% of the methane flux to the atmosphere. Other sinks of tropospheric methane are shown in Table 9.2.

The main source of carbon monoxide, aside from the fossil fuel burning flux, is from oxidation of CH_4 (Table 9.3). The remainder comes from decomposition of organic matter in soils, and bacteria and algae in the ocean that actively generate CO from respiration processes according to:

$$2CH_2O + O_2 \rightarrow 2CO + 2H_2O \tag{9.11}$$

The release of carbon monoxide from fossil fuel burning, principally from transportation sources, is currently about 28×10^{12} moles C y^{-1}. The residence

Table 9.3.   Global carbon monoxide sources and sinks; fluxes in units of 10^{12} moles C y^{-1}. (After Seiler, 1974.)

Sources

Anthropogenic (mainly fossil fuel burning)	28
Decomposition of organic matter in soils	6
Ocean	4
Bush and forest fires, burning of agricultural waste	2
Oxidation of hydrocarbons	2
Total	42
Oxidation of methane	26[a]

Sinks

Soil uptake	16
Flux into stratosphere	4
Total	20
Oxidation by OH	52[b]

[a]From Table 9.2; [b]Figure 9.1

time of CO in the atmosphere is short, reflecting its rapid conversion to CO_2 by reaction (9.9). Thus most of the CH_4 and CO that enters the troposphere ends up as CO_2. This CO_2 is photosynthesized in the natural carbon cycle to organic material on land and less importantly in the ocean, which gives rise by decomposition to more CH_4 and CO. The average composition of the plant material synthesized in the ocean and on land is, respectively, $(CH_2O)_{106}(NH_3)_{16} H_3PO_4$ and $(CH_2O)_{800}(NH_3)_{10} H_3PO_4$. Land plants have a much higher C/N ratio than do marine plants and a somewhat lower N/P ratio. Thus, burning of a unit weight of land-derived organic matter (e.g., coal) releases relatively more carbon than burning of an equivalent amount of natural gas, an important point to keep in mind when considering the effect of the burning of fossil fuels on the release of CO_2 to the atmosphere.

CO_2 Fluxes

To the extreme right of Figure 9.1 is the major gaseous reservoir of carbon in the atmosphere--CO_2. The residence time (τ) of about 6 years means that the gas should be reasonably well-mixed in the atmosphere, but there is a small gradient in mean concentration from the northern hemisphere to the southern. The residence time is calculated with respect to total organic productivity (although the system is not at steady state and τ changes with time):

$$\tau_{CO_2} = \frac{\text{moles of } CO_2 \text{ in atm}}{\text{total organic productivity flux}} = \frac{61,000 \times 10^{12} \text{moles}}{10,000 \times 10^{12} \text{moles y}^{-1}} = 6 \text{ years}$$

(9.12)

This residence time is two-thirds that for living terrestrial plants calculated with respect to photosynthesis and only about 1.5% of that calculated with respect to total inorganic carbon in the ocean and the total evasion flux of 7800×10^{12} moles C y^{-1} (Figure 9.1). Thus, atmospheric CO_2 concentration is very delicately balanced; any significant change in the ratio of photosynthesis to respiration plus decay could change CO_2 levels several-fold in 100 years. Interestingly this time scale of response of atmospheric CO_2 to potential changes in rates of natural processes is on the order of the time passed since humankind has been burning fossil fuels in large quantities, and atmospheric CO_2 concentrations have been rising.

Notice that CO_2 fluxes involving weathering, diagenesis and carbonate precipitation are considerably smaller than fluxes involving the biosphere, ocean exchange, or fossil fuel burning and land use practices. It is the former fluxes, however, and changes in their magnitudes that help maintain the balance of CO_2 in the atmosphere on a geologic time scale. As a matter of comparison, it appears that the total degassing rate of "mantle" carbon, principally as CO_2, is on the order of 1 to 8 x 10^{12} moles C y^{-1} (cf. Holland, 1978; Arthur et al., 1985; Des Marais, 1985). This value includes metamorphic (diagenetic) and magmatic fluxes from oceanic arcs, ridges, and off-ridge, submarine volcanic areas.

Carbon dioxide fluxes related to land use practices and fossil fuel burning will be discussed in detail later. Now let us turn to the large reservoirs of carbon in the land, ocean, and sediments. It is the interactions between these reservoirs and the atmosphere that to a large extent govern atmospheric carbon levels on a long-time scale. For a start, a comparison of the reservoir masses of carbon in the atmosphere with those of land, ocean, and sediments is illuminating. These masses, sediment-C/ocean-C/land-C/ atmospheric-C, are in the ratio 106,600/53/4/1. The carbon mass of sediments is huge and is dominated by inorganic C in carbonate minerals; the ratio of oxidized C in carbonates to reduced C in organic matter is about 4.3. The residence time of C with respect to input in the sediment reservoir is 150 million years, orders of magnitude greater than the residence times of dissolved inorganic carbon in the ocean or living or dead organic matter on land.

The terrestrial living biomass is about 200 times larger than the marine biomass, although it turns over at a rate that is 140 times slower than the oceanic rate. Notice that for both the land and ocean reservoirs the dead organic carbon masses are greater than the living. The former represent, if oxidized, large sources of CO_2 for the atmosphere.

Exchange between the land and ocean is mainly via streams, whereas that between sediments and the ocean and land is by sedimentation and diagenesis, and by uplift, respectively. Some particulate carbon is transported by winds from the continents to the ocean to fall on the sea surface, on the order of 2 x 10^{10} moles C y^{-1}. The DIC of rivers is primarily as HCO_3^- (the average pH of river waters is about 6.6), and today this flux and the total riverine organic carbon flux to the oceans are about equal. Today's organic carbon flux is probably higher than in pre-human times because of deforestation and cultivation on land. Mackenzie (1981) estimated that this flux may be double its pre-humankind value. It is also the case that the accumulation of organic matter on the sea floor today may be greater than in pre-industrial times because of increased stream input of organic matter, and because of eutrophication of marine coastal areas (Mackenzie, 1981; Peterson and

Melillo, 1985). The accumulation of excess carbon on the sea floor acts as a minor sink today for fossil fuel and land use carbon. The magnitude of this sink is controversial (Table 9.1) but may represent as much as 10% of the yearly carbon storage in the atmosphere. The accumulation of 30×10^{12} moles C as organic matter yearly would represent 4.5×10^{12} moles N and 0.3×10^{12} moles P at a C:N:P ratio of 106:16:1.

Inorganic carbon sedimenting and accumulating on the sea floor is mainly as $CaCO_3$; little dolomite is forming today, and the average Mg content of modern calcareous sediments is about 0.7%, representing a flux of $MgCO_3$ to the sea floor of 3×10^{11} moles y^{-1}. The Ca^{2+} for precipitation comes both from rivers and reactions at the sea floor between basalts and sea water, discussed later in this chapter. The overall reaction is:

$$Ca^{2+} + 2HCO_3^- \rightarrow CaCO_3 + CO_2 + H_2O, \qquad (9.13)$$

In this reaction, the CO_2 is thought to return to the ocean-atmosphere system. It is possible, however, that $CaCO_3$ precipitation is intimately tied to organic metabolism (Smith, 1985), and carbon dioxide evasion from the ocean prior to fossil fuel and deforestation activities of humankind was due to net organic oxidation (net heterotrophy; Smith and Mackenzie, 1987). The carbon released as CO_2 in reaction 9.13 is incorporated initially in organic matter and becomes tied to the turnover time of the organic carbon reservoir. During later diagenesis-metamorphism some of the $CaCO_3$ is converted to dolomite, and some is decarbonated to a silicate mineral plus CO_2, returning CO_2 to the atmosphere. This latter reaction is partially responsible for regulating the CO_2 content of the atmosphere on a geologic time scale of 10^6 years or greater.

Uplift returns carbon from the moderately and deeply buried sediments to the land. Both carbonate rocks and fossil organic carbon are exposed to weathering processes. For every mole of organic carbon buried on the sea floor and not respired-decayed by processes like:

$$CH_2O + O_2 \rightarrow CO_2 + H_2O, \qquad (9.14)$$

a mole of O_2 accumulates in the ocean-atmosphere system. In a completely steady-state system, each mole of CH_2O must be oxidized by O_2 on uplift or O_2 can accumulate in the atmosphere on a time scale of 10's to 100's of millions of years.

Thus, we see a link between the cycles of O_2 and CO_2, to be discussed in more detail later in Chapter 10.

Note once more that the land, ocean, and sediment reservoirs are not at steady state because of man's activities. It is also the case because we are dealing with such small differences in values between large reservoir sizes and fluxes, that the natural system may not be balanced today and may not have been in steady state prior to humans coming on the scene. Thus, we have a real problem when we try to assess the baseline for the carbon cycle against which inputs from human activities can be evaluated.

Finally a word with respect to feedback mechanisms. With some idea of processes, reservoirs, and fluxes in the carbon cycle, it is convenient at this stage to introduce the concept of feedbacks in the system. A feedback mechanism is a process, or series of processes, that enhance or reduce a perturbation to a system; that is, there are positive and negative feedback mechanisms. Feedbacks are very important in regulating the composition of a system, and maintaining a system near a steady-state condition, and, therefore, can often be considered as buffers.

For example, in the carbon cycle consider the balance between terrestrial photosynthesis and respiration-decay. If the respiration and decay flux to the atmosphere were doubled (perhaps by a temperature increase) from about 5200 x 10^{12} to 10,400 x 10^{12} moles y^{-1}, and photosynthesis remained constant, the CO_2 content of the atmosphere would be doubled in about 12 years. If the reverse occurred, and photosynthesis were doubled, while respiration and decay remained constant, the CO_2 content of the atmosphere would be halved in about the same time. An effective and rapid feedback mechanism is necessary to prevent such excursions, although they have occurred in the geologic past. On a short time scale (hundreds of years or less), the feedbacks involve the ocean and terrestrial biota. As was shown in Chapter 4, an increase in atmospheric CO_2 leads to an increase in the uptake of CO_2 in the ocean. Also, an initial increase in atmospheric CO_2 could lead to fertilization of those terrestrial plants which are not nutrient limited, provided there is sufficient water, removal of CO_2, and growth of the terrestrial biosphere. Thus, both of the aforementioned processes are feedback mechanisms that can operate in a positive or negative sense. An increased rate of photosynthesis would deplete atmospheric CO_2, which would in turn decrease photosynthesis and increase the oceanic evasion rate of CO_2, leading to a rise in atmospheric CO_2 content. More will be said later about feedback mechanisms in the carbon system.

Human impact on carbon fluxes

The Fossil Fuel and Land Use Fluxes

The burning of fossil fuels in stationary and transportation sources results in the emission of a variety of chemical substances into the atmosphere, including C, N, S, and HC gases, and particulates containing potentially toxic trace metals. Some of these substances remain in the atmosphere and circulate globally; others are deposited as dry fallout or in rain on the land and oceans and enter ecosystems of local, regional, or global extent. Most of these substances are elements or compounds that have circulated freely throughout Earth's surface environment during geologic time; that is, they represent additions to natural biogeochemical cycles that existed prior to man's activities. Thus, the potential environmental problems arising because of the burning of fossil fuels depend fundamentally on the fact that rates of transfer of chemical substances from one environment to another have been increased by this activity. Coal currently found below ground will be exposed at the Earth's surface because of uplift and erosion and be oxidized, resulting in release of gases to the atmosphere. Mining of coal and its burning, however, result in increased fluxes of C, N, and S to the atmosphere. These increased fluxes can be substantial enough that the natural system can no longer deal with them, and the chemical substances accumulate in the environment, giving rise to a potentially damaging situation.

Certain of the gases and particulates released to the atmosphere by the burning of fossil fuels can accumulate there. Carbon dioxide is the classic example of such a gas. Its effect on the temperature of Earth's surface environment and consequent climatic change have been known for nearly a century (Arrhenius, 1896; Chamberlain, 1898).

Radiation from the sun is in part reradiated as long wavelength (infrared) radiation from the Earth's surface and is absorbed by small amounts of water vapor, carbon dioxide, ozone, and other compounds in the atmosphere (Table 9.4). The ability of these components to intercept infrared radiation is shown in Figure 9.3. The upper boundary of the stippled area is the emission of the ocean's surface, whereas the lower boundary is the radiation measured at the distance of satellites. The difference is the net energy absorbed and trapped in the lower atmosphere. On a global average annual basis, this trapped infrared radiation is equal to 153 watts per square meter of the Earth's surface.

The reradiation of the absorbed energy leads to an observed warming of Earth's surface environment of about 35°C; that is, an average surface temperature of 15°C rather than -20°C. This warming has been termed the "greenhouse effect." Enhancement of this "effect" because of changes in the absorption characteristics of the atmosphere owing to the activities of humankind has received much attention during the last two decades. In particular, the "CO_2 problem"--the relationship between fossil fuel burning (and other anthropogenic activities), increasing levels of atmospheric CO_2, and climate--is of particular concern to the scientific community, government, and the public. Other minor atmospheric gas constituents produced by anthropogenic activities also can affect climate (Table 9.4).

Table 9.4.   Some examples of surface temperature changes owing to atmospheric constituents other than CO_2.

Constituent	Mixing Ratio Change (ppb)		Surface Temperature Change (°C)
	From	To	
Nitrous oxide (N_2O)	300	600	0.3-0.4
Methane (CH_4)	1500	3000	0.3
CFC-11 ($CFCl_3$)	0	1	0.15
CFC-12 (CF_2Cl_2)	0	1	0.13
CFC-22 (CF_2HCl)	0	1	0.04
Carbon tetrachloride (CCl_4)	0	1	0.14
Carbon tetrafluoride (CF_4)	0	1	0.07
Methyl chloride (CH_3Cl_3)	0	1	0.013
Methylene chloride (CH_2Cl_2)	0	1	0.05
Chloroform ($CHCl_3$)	0	1	0.1
Methyl chloroform (CH_3CCl_3)	0	1	0.02
Ethylene (C_2H_4)	0.2	0.4	0.01
Sulfur dioxide (SO_2)	2	4	0.02
Ammonia (NH_3)	6	12	0.09
Tropospheric ozone (O_3)	F(Lat,ht)	2 F(Lat, ht)	0.9
Stratospheric water vapor (H_2O)	3000	6000	0.6

Adapted from National Research Council (1983).

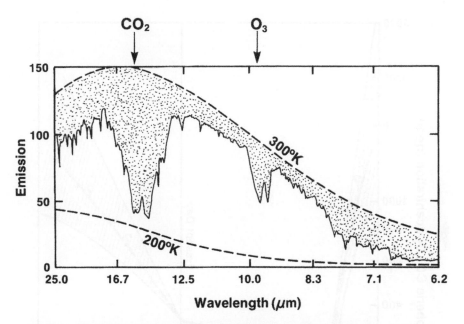

Figure 9.3. Emission spectra from the infrared interferometer spectrometer on board the Nimbus 3 satellite for the tropical Pacific Ocean under clear-day conditions. The dashed lines represent outgoing long-wave, blackbody radiation at the temperature indicated. CO_2 absorbs in the spectral window of 13-17 μm, O_3 between 9-10 μm, and H_2O in the entire spectral domain. Emission is measured in arbitrary units. (After Ramanathan, 1988.)

Figure 9.4 shows one interpretation of the atmospheric CO_2 history of Earth over a time period of 100 million years. The geologic record of atmospheric CO_2 will be addressed in detail in Chapter 10; of interest here is the possibility that humankind activities of fossil fuel burning and land use practices could lead to future atmospheric CO_2 levels rivaling those of the past. Furthermore the future time scale of atmospheric CO_2 change may be shorter than any period of CO_2 change the Earth has experienced in 100 million years.

The impact of humankind on the carbon cycle, evident in Figure 9.1, stems mainly from the release of CO_2 to the atmosphere by fossil fuel burning and likely land use practices, such as deforestation and cultivation. Fossil fuel burning resulted in 1984 in the release of 444×10^{12} moles C to the atmosphere, or about 5% of the total CO_2 released by natural processes of respiration and decay. Because of the small size of the atmospheric CO_2 reservoir, this release can have a substantial effect on the CO_2 concentration of the atmosphere.

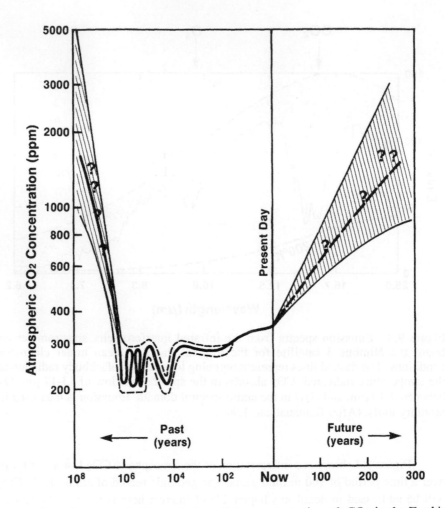

Figure 9.4. One scenario for evolution of the concentration of CO_2 in the Earth's atmosphere over the past 100 million years compared with potential future change. (After Gammon et al., 1985.)

The cumulative production of CO_2 from fossil fuel use and cement manufacturing (less than 2% of the total) is shown in Figure 9.5. This production should have resulted in an increased atmospheric CO_2 concentration since the late 1800's of about 90 ppmv. The observed cumulative increase for the period 1958-1984, however, coupled with a value of 290 ppmv in 1860 (e.g., Baes et al., 1976), indicates that only about one-half of this production has remained in the atmosphere; the remainder has been stored in the oceans and other sinks. The parallelism of the cumulative production curve and the observed increase in atmospheric CO_2 concentration curve for the Mauna Loa Observatory data was one of the first hints that the increase in atmospheric CO_2 recorded at the Observatory was due to fossil fuel burning.

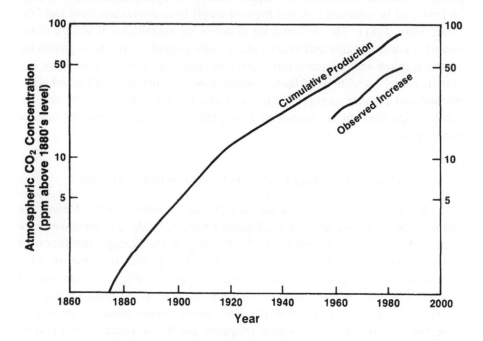

Figure 9.5. Cumulative production of CO_2 from fossil fuel burning and cement manufacturing, and cumulative observed increases in atmospheric CO_2 from 1958-1984 at the Mauna Loa Observatory, Hawaii, U.S.A. Parallelism of curves suggests that atmospheric increase is due to fossil fuel burning. (After Baes et al., 1976.)

Conventional wisdom dictates that about 56% of all the fossil fuel CO_2 released has remained airborne (Table 9.1); the rest has been taken up by other reservoirs, primarily the oceans. As shown in Figure 9.1, if land use practices (primarily deforestation and cultivation) have led to release of CO_2 to the atmosphere, then it is likely that some anthropogenic carbon may have been stored on land, perhaps in unperturbed forest areas in trees, litter, and soil humus, or the carbon cycle is not balanced. Current thought is that there has been a net loss of carbon from the land reservoir during the past 200 years. The magnitude of this loss depends primarily on the relative magnitudes of the carbon fluxes owing to land use and storage, and streams (organic C). In particular, if the land use flux is high, then the compensatory increased fluxes (over ancient times) related to the sinks of CO_2 solution in ocean water, storage in the atmosphere, CO_2 reaction with shallow-water carbonate minerals, organic carbon burial in sediments, and storage on land must be enhanced over that required simply to accommodate fossil fuel CO_2 (Mackenzie, 1981). The problems are to define the magnitudes of today's fluxes related to potential major and minor sinks of anthropogenic CO_2, and to predict the course of future anthropogenic fluxes and sink strengths. It is this knowledge that permits model calculations of future concentrations of atmospheric CO_2 and hence thermal and climatic change. In the next chapter, we will discuss further the relationships between the long-term geological cycle of carbon and its modern counterpart.

Observed Atmospheric CO_2 Concentration Increase

Increases in observed atmospheric CO_2 concentrations for the Mauna Loa and other observatories are shown in Figures 9.6 and 9.7. Many of these data have been collected and analyzed by C. D. Keeling at the Scripps Institution of Oceanography. For the period 1958-1984, the global increase in CO_2 concentration in the atmosphere was about 29 ppmv, or 9%. A regular annual oscillation in concentration can be seen in the monthly averages, related principally to the withdrawal of CO_2 from the atmosphere during summer seasons of important photosynthesis in northern temperate and boreal zones. It can also be seen that the South Pole data trend is similar to that of Mauna Loa, showing that the response of the atmosphere to the annual cycle of photosynthesis and respiration in the northern hemisphere, and the input of fossil fuel and land use CO_2, are a global phenomenon. The damping of the annual cycles of CO_2 in the southern hemisphere reflects the fact that the sources and sinks controlling this cyclicity are mainly in the northern hemisphere. The time lag between the seasonal variations at Mauna Loa and those at the South Pole may reflect the isolation of the latter area from the lower latitudes by circumpolar winds.

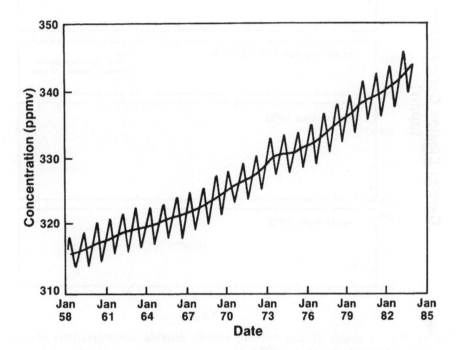

Figure 9.6.   Mean monthly concentrations of atmospheric CO_2 at Mauna Loa Observatory. The yearly oscillation is explained mainly by the annual cycle of photosynthesis and respiration of plants in the northern hemisphere. (After Sundquist, 1985.)

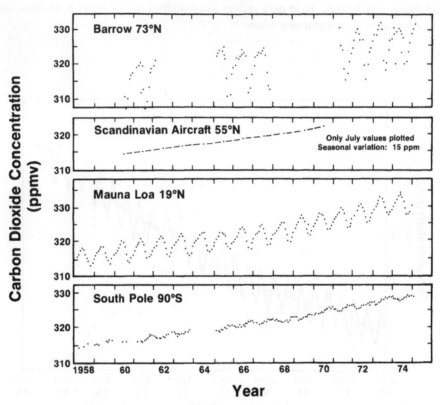

Figure 9.7.   Growth of atmospheric carbon dioxide concentrations at four locations. (After Machta et al., 1977.)

C. D. Keeling (in US/USSR Workshop, 1981) has shown that the exponential trend in mean monthly concentrations of atmospheric CO_2 at Mauna Loa and the South Pole from 1958-1980 are consistent with a series of calculated yearly atmospheric CO_2 concentrations based on assuming that 56% of the annual fossil fuel CO_2 production has remained airborne over that time period. This calculation provides good evidence that, at least for the past 25 years, fossil fuel CO_2 has probably been the most important flux of CO_2 to the atmosphere from human activities. This statement is in accord with recent analyses of the importance and change in emissions of CO_2 from the terrestrial biosphere during the past 200 years. Figure 9.8 shows the atmospheric CO_2 increase since the middle 1700's as indicated from measurements on air trapped in old glacial ice and the Mauna Loa data. Figure 9.9 compares the annual carbon flux to the atmosphere because of fossil fuel burning with that resulting from land use activities of deforestation and cultivation (Houghton et al., 1983). Land use fluxes are based on model calculations whereas fossil fuel fluxes are estimated from data on the world's consumption of fossil fuels and their carbon content. It can be seen from this figure that the annual carbon release from the terrestrial biosphere may have been greater than that from fossil fuels until the early 1900's. Care must be taken in interpretation of this figure in that the land use flux is not a net flux to the atmosphere and is based on model calculations.

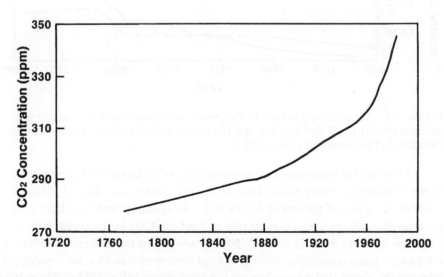

Figure 9.8. Atmospheric CO_2 increase in the past 200 years as indicated by measurements on air trapped in old ice from Siple Station, Antarctica, and by the annual mean values from Mauna Loa Observatory. The line is a spline fit through all data . (After Siegenthaler and Oeschger, 1987.)

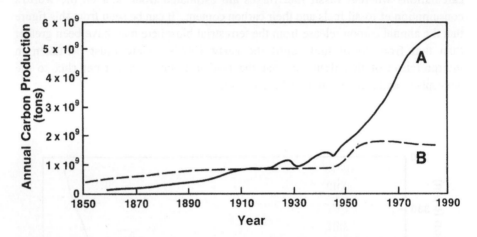

Figure 9.9.   Annual carbon flux to the atmosphere in units of 10^9 metric tons C because of (A) fossil fuel burning and (B) deforestation and cultivation activities. (Modified after Houghton et al., 1983.)

Quantitative estimates of the emission rates of fossil fuel and non-fossil CO_2 to the atmosphere, using as one basis of control the ice core data of Figure 9.8, confirm the *general* pattern of Figure 9.9. Siegenthaler and Oeschger (1987) estimate that the net cumulative release of land use CO_2 to the atmosphere until 1980 was 7.5 to 12.5 x 10^{15} moles C; 50%, however, was released prior to 1900. In the past 30 years their calculations give a net release of 0 to 75 x 10^{12} moles y^{-1}, whereas the fossil fuel release has ranged from about 160 x 10^{12} to 444 x 10^{12} moles y^{-1} from 1954 to 1984. Thus it appears that land use activities were more important sources of CO_2 release to the atmosphere during the 19th and the early part of the 20th century than fossil fuel emissions. A controversy today still exists concerning the exact value of the net biospheric CO_2 emission (Detwiler and Hall, 1988), and more importantly, how those emissions may change in the future with potentially increased deforestation and cultivation activities, particularly in developing nations of the tropics and subtropics.

Future Atmospheric CO_2 Concentration Trends

The future rate of change of CO_2 depends significantly on future rates of industrial CO_2 emissions, and therefore, on future energy development. Land use practices also will play a role. If fossil fuel energy consumption in the next 20 years remaines the same as that in the previous two decades, the atmospheric CO_2 concentration will be about 394 ppmv in the year 2000.

Predictions of atmospheric CO_2 change depend not only on choices concerning future energy sources but also on the rate of world population growth, and the ultimate maximum energy consumption per capita. These estimates are difficult to make, but various scenarios have been developed to provide some limits on future CO_2 concentrations.

Demographers predict a world population of about 6 billion in the year 2000 and a saturation level of 12 billion at the beginning of the 22nd century (Figure 9.10). For limiting values of per capita energy consumption of 10 and 20 kilowatts,

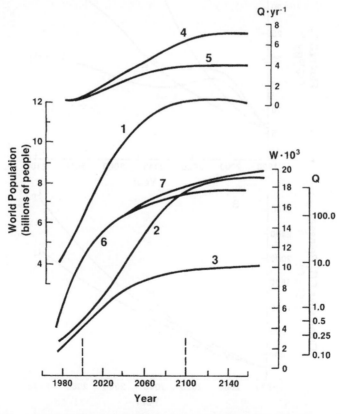

Figure 9.10. World population and energy production estimates. 1 = world population. 2,3 = per capita energy consumption for limiting values of 20 kw and 10 kw, respectively. 4,5 = annual energy consumption, and 6,7 = total heat consumption for above limiting values. $1Q = 2.5 \times 10^{17}$ kcal $= 36 \times 10^9$ tons coal. (After US/USSR Workshop, 1981.)

ultimate energy consumption for the world's population would be about 4 to 7 quads per year, respectively, equivalent to 144-252 x 10⁹ tons of coal per year. At the higher rate, it would take only 30 years to use up all the world's supply of fossil fuels eventually recoverable (Hubbert, 1970).

Carbon dioxide emissions to the atmosphere depend on the future course of energy consumption and very likely on forest management. The former has been investigated thoroughly, the latter hardly at all. Figure 9.11 illustrates

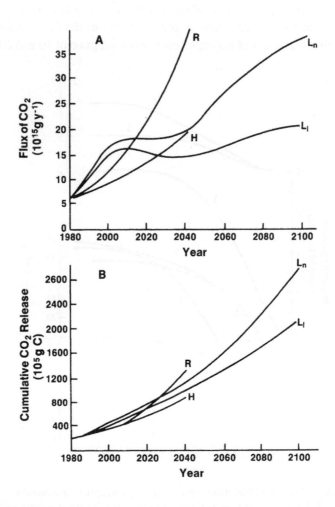

Figure 9.11.   Rates of industrial CO_2 emissions to the atmosphere (A) and cumulative CO_2 release (B) for various future energy scenarios. (After US/USSR Workshop, 1981.)

future rates of industrial CO_2 emissions to the atmosphere, and the cumulative release for various energy scenarios. The results of the Russian scientists Legasov and Kuz'min (1981, curves L_n and L_1 in Figure 9.11) are predicated on the assumption that nuclear power will play a significant role in supplying the world's energy needs between the years 2000 and 2100. Rotty (1979, curve R, Figure 9.11) and Hafele (1979, curve H, Figure 9.11) place less emphasis on nuclear power, at least until the mid-21st century. Furthermore, Hafele assumes that the present energy disparity between various regions of the globe will remain the same, whereas Rotty and Legasov and Kuz'min predict a significant increase in energy use by developing countries. Because the various assumptions balance one another, all predictions of cumulative industrial CO_2 emissions to the atmosphere are remarkably similar at least until the mid-21st century.

The various scenarios of energy development and resultant CO_2 emissions, assuming a constant airborne fraction of 0.56 as obtained from CO_2 monitoring data, lead to the prediction of an atmospheric CO_2 concentration in the year 2000 of 394 ± 9 ppmv. Scenarios of atmospheric CO_2 changes beyond the year 2000 require calculations based on models describing the exchange of carbon between its various global reservoirs, particularly the atmosphere, ocean and terrestrial biota. Calculated future atmospheric CO_2 concentrations are shown in Figure 9.12. The X's denote the observed trend for the period 1958-1988, the shaded area, the range in estimates of future concentrations. The range depends on the various scenarios for future energy development and on assumptions concerning the role of the terrestrial biota as a source or sink of CO_2.

The maximum rate of change of CO_2 concentrations can be predicted for the case where future energy use heavily favored fossil fuels, and the land biota were a significant source of atmospheric CO_2. Minimum rates of change would be obtained for opposite conditions. It appears that a doubling of atmospheric CO_2 concentration over its late 1800's value of 290 ppmv may occur between the years 2030 and 2050. This increase in CO_2 in the atmosphere can lead to global climatic effects that have a wide range of ecologic consequences.

Consequences of Increased Atmospheric CO_2 Levels

Table 9.5 lists some of the climatic and ecologic consequences resulting from increasing atmospheric CO_2 concentrations. The obvious consequence is that of a temperature increase. The question of whether or not this temperature increase

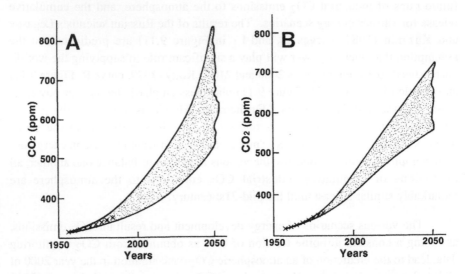

Figure 9.12. Calculated CO_2 concentrations in the atmosphere assuming various scenarios of future energy development. Shaded area indicates range of estimates and points to a doubling of atmospheric CO_2 concentration over its late 1800's value of 290 ppmv between the years 2030 and 2050. A. High range energy scenerio. B. Low range energy scenerio. (After US/USSR Workshop, 1981.)

Table 9.5. Consequences of increasing atmospheric CO_2.

- INCREASING TEMPERATURE/WARMING OF GLOBAL CLIMATE

- CHANGE IN PRECIPITATION PATTERNS

- GLACIAL MELTING AND RISE OF SEA LEVEL

- MELTING OF SEA ICE

- CHANGE IN DISTRIBUTION OF ARID AND SEMI-ARID LANDS AND
  AGRICULTURAL LANDS

- CHANGE IN DISTRIBUTION OF MICROBIAL AND INSECT PESTS

- DECREASED BIOLOGICAL AND ECONOMIC PRODUCTION OF ALL
  CLIMATE ZONES EXCEPT THE WETTEST

- CHANGE IN WATER BALANCE OF WHOLE STATES AND REGIONS

- OCEANOGRAPHIC RESPONSES (e.g., changes in pH and total dissolved carbon,
  air-sea exchange of CO_2, ocean currents, and sea level)

- SOCIO-ECONOMIC CONSEQUENCES OF ALL OF THE ABOVE

---

already has been observed is presently a matter of some debate. Variations in the mean annual air temperature over the past 100 years (Figure 9.13) have been large enough that the recent warming trend since 1950 is difficult to ascribe to increasing CO_2 levels.

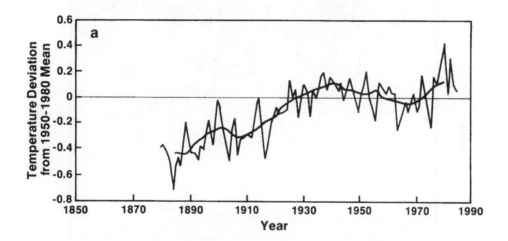

Figure 9.13. Global temperature change. The spikey curve is the annual mean temperature; the smooth curve is the five-year running mean. Since the mid-1960's, there has been a trend of increasing temperature of about 0.08°C/10 years. (After Hansen and Lebedeff, 1987.)

Future temperature changes owing to a doubling of the CO_2 concentration, for example, must be calculated on a computer using sophisticated climate models. A pictorial representation of the results of one such calculation is shown in Figure 9.14. Manabe's calculation of the global distribution of temperature increase because of a doubling of atmospheric CO_2 concentration indicates that a doubling would result in a 2 to 3°C increase in average surface temperature. If other greenhouse gases are included in the modeling calculations, the additional range of potential temperature increase is 1.5 to 5.5°C. Latitudinal temperature changes may be more important than simply the average change, with polar regions attaining future temperatures nearly 8°C warmer than those of today.

are only just observed is presently a matter of some debate.   Variations in the mean annual air temperature over the past 100 years (Figure 9.13) have been such that the recent warming trend since 1950 is difficult to ascribe to increasing

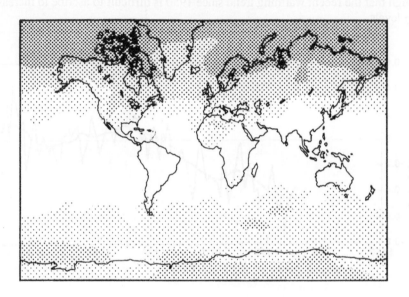

Figure 9.14.   Manabe's calculation of global distribution of temperature increase owing to a doubling of atmospheric CO_2 concentration. Estimated ranges of temperature increase (°C): unstippled region <2.3° to 3.4°; lightly stippled region 3.4° to 6.8°; darkly stippled region >6.8°. (After Olsen, 1982.)

This warming of the world's climate could conceivably alter precipitation patterns. Table 9.6 gives estimated values of the mean surface air temperature change for different years because of increasing atmospheric CO_2 concentration, as well as of the latitudinal distribution of temperature and precipitation changes owing to a doubling of CO_2 levels. The latter estimated changes are derived from both general atmospheric circulation models and paleoclimatic data. It can be seen that the mid-latitudes may experience the least change in precipitation whereas low- and high-latitude regions may become substantially wetter. Such a latitudinal change in precipitation can lead to a change in the distribution of semi-arid and arid lands and agricultural lands, as well as a change in the water balance of whole states and regions. Furthermore, precipitation changes coupled with temperature changes can affect the distribution of microbial and insect pests, and could lead to decreased biological and economic production of all regions but the wettest.

Table 9.6. Estimated temperature and precipitation changes for a doubling of atmospheric CO_2 concentrations.

Year	ESTIMATED CHANGE IN THE GLOBAL MEAN SURFACE AIR TEMPERATURE (°C)
2000	1 - 2
2025	2 - 3
2050	3 - 4

NORTH LATITUDE	0-10°	10-20°	20-30°	30-40°	40-50°	50-60°	60-70°	70-80°	80-90°
General atmospheric circulation model (Manabe and Wetherald,1980), °C	1.7	2.0	2.5	3.1	2.8	4.3	5.2	6.8	7.6
Paleoclimatic data	1.2	1.4	1.6	2.4	4.1	4.6	6.8	7.9	9.5

NORTH LATITUDE	10-20°	20-30°	30-40°	40-50°	50-60°	60 70°	70-80°
General atmospheric circulation model (Manabe and Wetherald, 1980), cm y^{-1}	10	14	3	-1	10	9	13
Paleoclimatic data	12	12	2	2	8	9	13

After US/USSR Workshop, 1981.

Figure 9.15 is another attempt to provide a scenario of possible moisture changes stemming from a warmer Earth. Notice that the pattern of warmer and drier than now regions for the northern hemisphere is more complex than the latitudinally-zoned precipitation results derived from the general circulation model and paleoclimatic data (Table 9.6). These differences reflect the difficulties inherent in making projections of future changes in precipitation patterns caused by the "greenhouse effect."

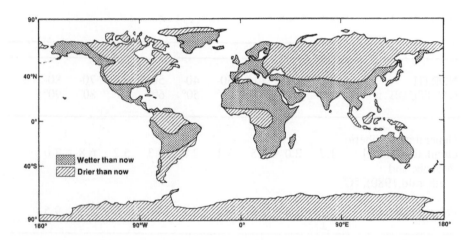

Figure 9.15. A scenario of possible soil moisture patterns on a warmer Earth in terms of regions where it may be wetter or drier than now in summer, based on climate model experiments and reconstructions of past climate. (After Kellogg, 1987.)

The predicted pronounced latitudinal change in surface air temperature resulting from a doubling of atmospheric CO_2 could lead to melting of sea ice and eventually to glacial melting and a rise in sea level. Figure 9.16 illustrates observed sea level changes over time. The calculated range of sea level variation because of thermal expansion of the ocean owing to temperature change appears to be insufficient to account for the observed sea level rise. It is possible that there already has been some melting of continental ice sheets (Gornitz et al., 1982), as well as a decline in the occurrence and distribution of sea ice. These conclusions are strongly debated today.

Figure 9.17 shows possible changes in sea level because of a greenhouse warming resulting from accumulation of CO_2 and other trace gases in the

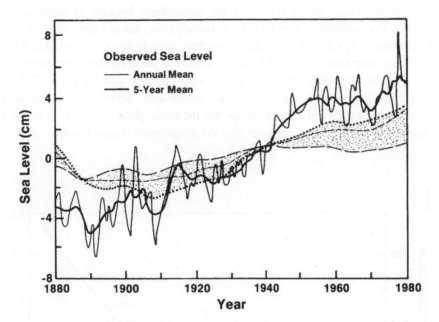

Figure 9.16. Observed sea level changes over time and calculated range (shaded area) of sea level variation owing to thermal expansion of ocean water. Dotted line = ΔT_{eq} of 5.6°C; Short-dashed line= ΔT_{eq} of 2.8°C; Long-dashed line = ΔT_{eq} of 1.4°C. (After Gornitz et al., 1982.)

atmosphere. It is obvious that if these sea level changes were to take place, many areas of the world would face potential flooding conditions. For example, some Pacific atoll nations would be nearly drowned. The Bangladesh coastal area could be inundated, island and continental beaches and cities jeopardized. Furthermore, if the maximum sea level rise projection of Figure 9.17 were to develop, it is possible that the vertical accretion rate of reef ecosystems could not keep pace, leading to drowning of these systems (Buddemeier and Smith, 1988). Melting of all glacial ice would raise sea level about 60 meters. The increased uptake of CO_2 by the ocean will also have a major impact on the chemistry of seawater and deposition of carbonate minerals in sediments. This topic will be discussed further in Chapter 10 and was introduced in Chapter 4.

Before closing this section on the consequences of a potential change in atmospheric CO_2 and trace gas concentrations, a word of caution is necessary. The release of CO_2 to the atmosphere by human activities is indeed a great "Geophysical Experiment" (Revelle and Suess, 1957), and as such the consequences are difficult to assess, and are debatable. Some scientists feel that the temperature effect and the climate change in general have been overrated (e.g., Idso, 1982). Whatever the case, once CO_2 enters the atmosphere because of humankind's activities, the removal of that CO_2 to "pristine" levels takes a long time. Figure 9.18 illustrates that if the CO_2 in the atmosphere were to rise to levels seven times its 1800's concentration, it would take thousands of years for the concentration to return to near original levels, if the ocean were the only sink for CO_2. There is need for further research, and perhaps most importantly, development of plans for responses to the potential climate change that the entire globe may face. It seems to us that this is the time to prepare, not after the temperature change has taken place.

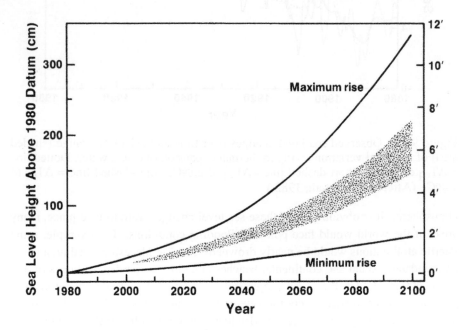

Figure 9.17.   Various scenarios for sea level rise owing to temperature increase resulting from increased greenhouse gases. Stippled area shows probable range of sea level rise. (After Hoffman et al., 1983.)

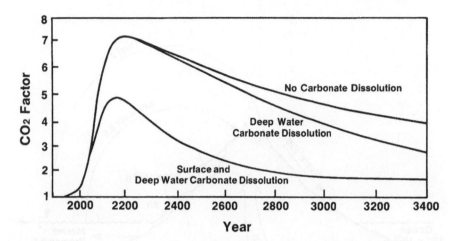

Figure 9.18. Graph illustrating that once CO_2 enters the atmosphere because of human activities, the return to original conditions takes a long time, regardless of whether or not $CaCO_3$ minerals are dissolved. (After Bacastrow and Keeling, 1979.)

The oceanic system

Sources of Calcium, Magnesium, and Carbon for Modern Oceans

In the previous section the present-day cycle of carbon was discussed in some detail. We can now turn our attention to the sources of carbon to the ocean, and of calcium and magnesium, the major elements (other than oxygen and hydrogen) with which carbon interacts. Because carbon dioxide is the major acid gas involved in both carbonate and silicate mineral weathering reactions at the Earth's surface, it is informative to consider the sources of other elements as well.

Figure 9.19 is a schematic diagram illustrating the major agents, or transport paths, by which materials are transported to the ocean. Rivers dominate the input, and we will begin our discussion by considering river input and its sources.

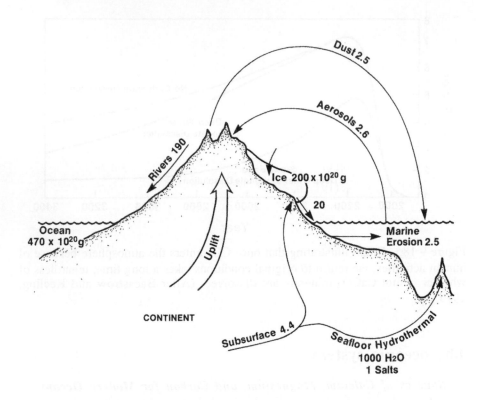

Figure 9.19.   Major agents responsible for transfer of dissolved and particulate materials to and from the oceans. Fluxes in units of 10^{14} g y^{-1}.

River input The principal source of carbon in rivers is from the uptake of atmospheric CO_2 by plant production and the subsequent oxidation of organic matter in soils. Calcium and magnesium come from the weathering of rocks. Nearly 90% of the total material entering the ocean annually is transported by streams as dissolved and suspended materials. Until recently, Livingstone's (1963) pioneering compendium of river-water analyses formed the basis for calculation of the chemical composition of average river water. Meybeck and Martin (Meybeck, 1976, 1977, 1979a,b;   Martin and Meybeck, 1978, 1979) recently extended Livingstone's work, and that of Alekin and Brazhnikova (1960, 1962,1968; Alekin, 1978), and arrived at an improved estimate of the average composition of river waters, as well as the chemical composition of the particulate load of rivers. In the following discussion, we rely heavily on the work of Meybeck and Martin.

To evaluate the net riverine influx of dissolved species to the ocean, the river load has to be corrected for sea salts transported via the atmosphere from the ocean to the continents and rained out mainly in coastal precipitation. Table 9.7 shows the average concentration of selected dissolved and particulate elements in rivers from Martin and Meybeck (1979), and the corresponding net fluxes corrected for sea-salt cycling from Martin and Meybeck (1979). The corrections of fluxes for cyclic salts and pollution are still debatable estimates (e.g., Holland, 1978; Maynard, 1981), and affect mainly the evaluation of the net flux of Na^+ by perhaps as much as 20%.

Table 9.7. Elemental content and fluxes of world major rivers (adapted from Martin and Meybeck, 1979).

	Concentration		Net Flux		
	Dissolved 10^{-3} g L^{-1}	Particulate 10^{-3} g g^{-1}	Dissolved[a] 10^{12} moles y^{-1}	Particulate[b] 10^{12} moles y^{-1}	Total 10^{12} moles y^{-1}
Na	5.1	7.1	5.7	4.8	10.5
Mg	3.8	13.5	5.4	8.7	14.1
Al	0.05	94.0	0.07	53.9	54
Si	5.42	285	7.3	157.8	165
K	1.35	20.0	1.3	7.9	9.2
Ca	14.6	21.5	12.4	8.3	20.7
Fe	0.04	48.0	0.03	13.3	13.3
C[c]	10.4	10.8	32.4	13.9	46.3
S	2.7	---	3.2	----	3.2
Cl	3.2	---	3.4	----	3.4

[a]Based on total river discharge of 3.74×10^{16} L y^{-1} and corrected for oceanic salts and pollution
[b]Based on total river particulate load of 15.5×10^{15} g y^{-1}
[c]Inorganic carbon values

It can be seen in Table 9.7 that the particulate load constitutes by far the most important contribution (88%) of total river discharge of materials to the ocean. The amount carried as solids should be increased by bed load transport, which usually is considered to be about 10% of the total suspended load (Blatt et al., 1980). The mean chemical composition of river suspended matter closely approximates that of average shale (Table 9.8). This resemblance is expected because suspended solids in rivers are derived mainly from shales. Sedimentary rocks constitute about 66% of the rocks exposed at the Earth's surface; fine-grained rocks, like shales, comprise at least 65% of the sedimentary rock mass. Thus, roughly 50% of surface erosion products come from shaly rocks.

Table 9.8. Comparison of the average chemical composition of river suspended material, shales and soils.

	River suspended material[a] (wt. %)	Shales[b]	Soils[c]
SiO_2	61.0	61.9	70.7
Al_2O_3	17.8	16.9	13.5
Fe_2O_3[d]	6.86	7.20	5.44
MgO	2.20	2.40	1.04
CaO	2.94	1.49	1.92
Na_2O	0.95	1.07	0.85
K_2O	2.40	3.70	1.64

[a]Calculated from the data of Meybeck (1977)
[b]After Clarke (1924)
[c]After Vinogradov (1959)
[d]Total iron as Fe_2O_3

Table 9.9. Mineralogical composition of river suspended load and shales (wt.%). (After Wollast and Mackenzie, 1983.)

	Quartz	Plagioclase	K-feldspar	Kaolinite	Illite	Montmorillonite	Chlorite
Amazon	15	2	2	27	31	20	5
Mississippi	30	6	4	8	13	38	
Shales	31	<----------4.5-------->		8	15	36	2

The mineralogy of the suspended matter carried by rivers is not well documented. There are numerous analyses either of the clay fraction or of sands carried by rivers, but only a few total quantitative analyses are reported in the literature. As examples, the average mineralogical composition of two large river systems, the Amazon and the Mississippi, are presented in Table 9.9. This table also includes the mean mineralogical composition of shales for comparison with river suspended sediments. The overall average of 300 samples of shales analyzed by Shaw and Weaver (1965) is 30.8% quartz, 4.5% feldspar, 60.9% clay minerals, and

7% of other non-silicate materials. According to Weaver (1967), 60% of the clay fraction in recent shales is represented by montmorillonite, 25% by illite, 12% by kaolinite, and 3% by chlorite.

One means of assessing sources of dissolved constituents transported to the oceans is to consider the chemical composition of average river water. The average composition of river waters (corrected for pollutional inputs (Meybeck, 1979a) and charge balance) entering the ocean in $\mu eq\ L^{-1}$ or $\mu mole\ L^{-1}$ (SiO_2) is:

SiO_2	Ca^{2+}	Mg^{2+}	Na^+	K^+	Cl^-	SO_4^{2-}	HCO_3^-	Total
173	671	281	223	33	166	175	867	2589

Meybeck calculates that about 12% of the total mass of dissolved constituents delivered to the oceans today is due to human activities. The elements most affected are Cl, S, and Ca.

Dissolved inorganic carbon is about 34% of the salinity of average river water. Because the source of dissolved C in river waters is intimately linked to the origin of other constituents, it is necessary to examine the sources of all the major dissolved components. The following balance is modeled after the works of Garrels and Mackenzie (1971a), Holland (1978), and Meybeck (1979a). Details of some of the following calculations are given in these references and in the more recent works of Wollast and Mackenzie (1983) and Meybeck (1987).

The first step in accounting for the dissolved species in river water is subtraction of the amounts of dissolved materials in precipitation. This step is particularly important in balancing NaCl, the most important component of sea aerosol washed out of the atmosphere by precipitation. After the precipitation component of river water is removed, the remainder represents dissolved constituents derived from rocks, and in the case of HCO_3^-, in part from the atmosphere:

SiO_2	Ca^{2+}	Mg^{2+}	Na^+	K^+	Cl^-	SO_4^{2-}	HCO_3^-	Total
173	671	281	223	33	166	175	867	2589
	-3	-16	-64	-1	-76	-8		-168
173	668	265	159	32	90	167	867	2421

Sufficient Na^+ is now subtracted to remove all the Cl^-. This step accounts for Cl^- present in rocks as halite (NaCl), as NaCl in fluid inclusions, and as NaCl in concentrated aqueous films surrounding mineral grains in rocks. The remainder is:

SiO_2	Ca^{2+}	Mg^{2+}	Na^+	K^+	Cl^-	SO_4^{2-}	HCO_3^-	Total	NaCl (μmole) Removed
173	668	265	159	32	90	167	867	2421	
			-90		-90			-180	
173	668	265	69	32	0	167	867	2241	90

Sulfate is principally derived from the weathering of $CaSO_4$ minerals (gypsum and anhydrite) in sedimentary rocks. Some sulfate in rivers, however, comes from the weathering of magnesium sulfate salts in sedimentary rocks and from oxidation of sulfides (primarily FeS_2, pyrite) in sedimentary and crystalline rocks. The latter process also liberates small amounts of the cations Ca^{2+}, Mg^{2+}, Na^+, and K^+ by reactions like:

Step 1. Oxidation of pyrite

$$2FeS_2 + {}^{15}/_2 O_2 + 4H_2O \rightarrow Fe_2O_3 + 4SO_4^{2-} + 8H^+ \qquad (9.15)$$
pyrite   hematite

Step 2. Weathering of rock minerals by "H_2SO_4"

$$2CaCO_3 + H_2SO_4 \rightarrow 2Ca^{2+} + 2HCO_3^- + SO_4^{2-} \qquad (9.16)$$
calcite

and

$$2KAlSi_3O_8 + H_2SO_4 + 4H_2O \rightarrow Al_2Si_4O_{10}(OH)_2$$

K-feldspar   montmorillonite-like
                                        mineral

$$+ 2K^+ + SO_4^{2-} + 2H_4SiO_4 \qquad (9.17)$$

Notice that in step 2, the weathering of an aluminosilicate mineral by the aggressive attack of "H_2SO_4" waters may result in the release of dissolved silica, as well as cations. The ratio of silica to cations released depends on the composition of the primary mineral weathered and its alteration product. The weathering of a carbonate mineral gives cations and bicarbonate.

About 95% of the sulfur as SO_4^{2-} in river waters from rock weathering is derived from sedimentary rocks; the remainder is from crystalline rocks (Meybeck, 1979a). Of the total sulfur in river waters derived from rock weathering, about 30% comes from the weathering of sulfides in sedimentary and crystalline rocks, 25% from sediments, and 5% from crystallines (Holland, 1978). Using these proportions, the Na/K and Ca/Mg ratios of cations accompanying sulfate oxidation in rocks (Meybeck, 1979a), and the assumptions that all the Na^+ and K^+, 15% of the Ca^{2+}, and 50% of the Mg^{2+} are derived from silicate mineral weathering reactions giving an *average* molar ratio of SiO_2 to HCO_3^- of 1:1 (Garrels and Mackenzie, 1971a; Holland, 1978), the following balance is obtained:

SiO_2	Ca^{2+}	Mg^{2+}	Na^+	K^+	Cl^-	SO_4^{2-}	HCO_3^-	Total	Reactants Removed
173	668	265	69	32	0	167	867	2241	
-14	-41	-22	-5	-3		-46	-25	-156	
159	627	243	64	29	0	121	842	2085	14 silicate
									25 carbonate
									12 FeS_2

This is a tentative balance because of a number of poorly known estimates, but it does illustrate the link between sulfide mineral weathering and resultant acid attack on carbonate and silicate minerals.

The remaining sulfate comes from calcium and magnesium sulfates in sedimentary rocks in the proportion of $3Ca^{2+}$ to $1Mg^{2+}$; thus, the remainder is:

SiO$_2$	Ca^{2+}	Mg^{2+}	Na$^+$	K$^+$	Cl$^-$	SO$_4^{2-}$	HCO$_3^-$	Total	(Ca,Mg)SO$_4$ Removed
159	627	243	64	29	0	121	842	2085	
	-91	-30				-121		-242	
159	536	213	64	29	0	0	842	1843	60

The constituents left are derived from the weathering of silicate and carbonate rocks by CO$_2$-charged soil and ground waters. From the data of Garrels and Mackenzie (1971a), Holland (1978), and Meybeck (1979a), it is estimated that the weathering of carbonate rocks supplies 85% of the remaining Ca^{2+} and 50% of the remaining Mg^{2+} according to a reaction of the type:

$$CaCO_3 + CO_2 + H_2O \rightarrow Ca^{2+} + 2HCO_3^- \qquad (9.18)$$

A similar reaction holds for the magnesium component of carbonate rocks. The balance now is:

SiO$_2$	Ca^{2+}	Mg^{2+}	Na$^+$	K$^+$	Cl$^-$	SO$_4^{2-}$	HCO$_3^-$	Total	Carbonate Removed
159	536	213	64	29	0	0	842	1843	
	-456	-107					-563	-1126	
159	80	106	64	29	0	0	279	717	282

The molar ratio of Ca^{2+} to Mg^{2+} in the world's rivers calculated as carbonate removed is about 4:1. This ratio is slightly greater than that obtained for North America of 3:1 (Garrels and Mackenzie, 1971a), and that given by Meybeck (1979a) of 3:1 for the world. It is the same ratio, however, calculated by Holland (1978) for the world's rivers and found in the average limestone composition of Clarke (1924). It is slightly less than the average ratio in continental carbonate rocks, 5:1 (Ronov, 1980), the most important source of dissolved constituents in rivers. A Ca:Mg ratio of 4:1 indicates that the ratio of limestone to dolomite weathered is about 3:1 for the world, which is certainly a reasonable estimate. The total amount of HCO$_3^-$ now removed in carbonate minerals is 588 μeq L^{-1}; about one half of this amount comes from the atmosphere, the remainder from carbonate rocks.

Now we must attempt to determine the source of the rest of the HCO_3^- in river waters. The HCO_3^- is derived from two major sources: atmospheric CO_2 that has interacted directly with silicate minerals or has been incorporated into plants by photosynthesis, later to be released by plant decay or root respiration; and CO_2 produced by oxidation of fossil carbon already present in rocks (Garrels and Mackenzie, 1971a, 1972; Holland, 1978).

The budget can be balanced by consideration of a variety of silicate reactions producing different ratios of cations: HCO_3^- : SiO_2. The simplest approach was that of Garrels and Mackenzie (1971a), in which they considered the reactions of feldspars with CO_2-charged water to produce kaolinite and a montmorillonite-type mineral:

$$2NaAlSi_3O_8 + 2CO_2 + 11H_2O \rightarrow Al_2Si_2O_5(OH)_4$$

albite   kaolinite

$$+ 2Na^+ + 2HCO_3^- + 4H_4SiO_4 \tag{9.19}$$

and

$$2NaAlSi_3O_8 + 2CO_2 + 6H_2O \rightarrow Al_2Si_4O_{10}(OH)_2$$
albite   "montmorillonite"

$$+ 2Na^+ + 2HCO_3^- + 2H_4SiO_4 \tag{9.20}$$

(similar reactions can be written for K-feldspar, $KAlSi_3O_8$).

Notice that in these reactions the ratio of SiO_2 to HCO_3^- varies from 1:1 to 2:1. The weathering of montmorillonite to kaolinite also produces one molecule of SiO_2 for each molecule of HCO_3^-. However, weathering of the anorthite ($CaAl_2Si_2O_8$) component of feldspars to kaolinite produces no dissolved silica:

$$CaAl_2Si_2O_8 + 2CO_2 + 3H_2O \rightarrow Al_2Si_2O_5(OH)_4 + Ca^{2+} + 2HCO_3^- \tag{9.21}$$
anorthite   kaolinite

and weathering of chlorite to kaolinite produces little SiO_2 relative to HCO_3^-:

$$Mg_5Al_2Si_3O_{10}(OH)_8 + 10CO_2 + 5H_2O \rightarrow Al_2Si_2O_5(OH)_4$$

chlorite   kaolinite

$$+ 5Mg^{2+} + 10HCO_3^- + H_4SiO_4 \qquad\qquad (9.22)$$

Because it is likely that most of the Na^+ and K^+ in river waters derived from silicates comes from weathering reactions like those involving feldspars, and most of the Ca^{2+} and Mg^{2+} stems from reactions like those above for anorthite and chlorite, we will use the stoichiometry of these reactions to complete the balance. First, we remove the remaining Ca^{2+} and Mg^{2+} according to reactions (9.21) and (9.22):

SiO_2	Ca^{2+}	Mg^{2+}	Na^+	K^+	Cl^-	SO_4^{2-}	HCO_3^-	Total	Silicate Removed
159	80	106	64	29	0	0	279	717	
-11	-80	-106					-186	-383	
148	0	0	64	29	0	0	93	334	11

A balance for the remaining materials can be obtained by utilizing about 75% of the remaining silica in the chemical weathering of feldspars to kaolinite, and the rest in reaction of feldspars to montmorillonite:

Feldspars to Kaolinite

SiO_2	Ca^{2+}	Mg^{2+}	Na^+	K^+	Cl^-	SO_4^{2-}	HCO_3^-	Total	Feldspars Removed
148	0	0	64	29	0	0	93	334	
-111			<----56--->				-56	-223	
37	0	0	<---37--->		0	0	37	111	56

Feldspars to Montmorillonite

SiO2	Ca2+	Mg2+	Na+ K+	Cl-	SO4 2-	HCO3-	Total	Feldspars Removed
37	0	0	<---37---> 0	0		37	111	
-37			<--37--->			-37	-111	
0	0	0	<--- 0 ---> 0	0		0	0	37

The ratio of silica used for the two weathering reactions of 3:1 is that required by the stoichiometry of the reactions; interestingly, it is also the ratio of dissolved silica loads derived from tropical and temperature regions (3.2:1; Meybeck, 1979a). In general, tropical weathering of silicate minerals produces kaolinite as the principal alteration product, whereas temperate weathering results in formation of montmorillonite (e.g., Loughnan, 1969).

Although a balance has been obtained that seems consistent with present ideas, it is not necessarily a true representation. The balance, however, gives a rough approximation of the relative contributions of different rock types to chemical weathering and sources of dissolved constituents in river water. If we compute the weight percentages of rock types dissolved from the moles of minerals weathered, the dissolved load of the world's rivers transported to the oceans is derived from the following sources: 7% from beds of halite and salt disseminated in rocks, 10% from gypsum and anhydrite deposits and sulfate salts disseminated in rocks, 38% from limestones and dolomites, and 45% from weathering of one silicate to another.

The composition of the average plagioclase feldspar weathered is $Na_{0.65}Ca_{0.35}Al_{1.35}Si_{2.65}O_8$, andesine, and is similar to the composition of plagioclase feldspar found in granodiorite, a reasonable average igneous rock composition supplying much of the Na^+ and Ca^{2+} in rivers derived from silicate weathering. The molar ratio of $Na^+:Ca^{2+}$ in rivers derived from weathering of plagioclase feldspar is essentially that found in the average plagioclase feldspar. Of the HCO_3^- in river water, 56% stems from the atmosphere, 35% from carbonate minerals, and 9% from the oxidative weathering of fossil organic matter. Reactions involving silicate minerals account for about 30% of river HCO_3^-.

This balance accords to silicate mineral weathering a greater importance than that given originally by Meybeck (1979a), reflecting our opinion that 10-20% of the Ca^{2+} and 40-60% of the Mg^{2+} in river waters is derived from silicate weathering. Meybeck assigned a trivial portion of Ca^{2+} and Mg^{2+} in rivers to silicate weathering; thus, he arrived at a ratio of silicate to carbonate weathered much lower than our estimate. Our value is more in accord with Holland's (1978) balance for the world's rivers. Inclusion of the pollution component of river waters in the balance would modify slightly the calculated ratios of rock types weathered.

Meybeck, in a recent (1987) comprehensive article on the global sources of dissolved constituents in world's rivers, arrived at a balance more similar to the one presented here. He presented a compilation of various authors' estimates of the sources, given here in Table 9.10. There is seemingly good agreement between the works of Holland (1978), Wollast and Mackenzie (1983; this book), Berner et al. (1983), and Meybeck's latest work (1987). Table 9.10 provides the reader with some idea of the history of source estimates for the dissolved materials of rivers and their range of values.

In conclusion, it should be noted that the dissolved silica and cations brought to the ocean by streams must be removed, if the ocean is a steady-state reservoir. Also, the anions Cl^- and SO_4^{2-} must enter into chemical reactions for the same reason. Bicarbonate should be removed by reactions resulting in return of CO_2 to the atmosphere. During the past two decades controversy has raged over the steady-state ocean concept, the mechanisms and fluxes responsible for element removal from the ocean, and the time scale for removal of these elements and return of CO_2 to the atmosphere, i.e., early or late diagenesis or metamorphism.

Submarine basalt-seawater reaction fluxes In early studies of geochemical cycles and sources and sinks of materials for the ocean, the rivers were considered the main purveyors of dissolved constituents and solids to the ocean. Sedimentation of biogenic or inorganic solids was thought to be the major removal mechanism for most elements. With the advent of the concept of plate tectonics, material transport into and out of the mantle has added a new dimension to our ideas concerning element cycling within the exogenic cycle of atmosphere-ocean-sediments (continental crust). At mid-ocean ridges, plate motion is divergent and new rock is added to the lithosphere; at trenches and subduction zones the oceanic crust and part of the mantle are returned to the Earth's interior. Both boundaries of lithospheric plates can be regions of significant mass transfer, and of interaction between the exogenic cyclic system of uplift, erosion, deposition, burial, and uplift, and the endogenic system involving the mantle. At mid-ocean ridges and off-axis regions, submarine basalts react with sea water at low to moderately high temperatures, perhaps as high as 380°C. At subduction zones, sediment can be returned to the mantle, oceanic crust and mantle may be accreted to continents as ophiolites, and material added to the crust by island arc volcanism.

Table 9.10. Sources of river dissolved load carried to the oceans (in percent of total amount derived from weathering). (After Meybeck, 1987.)

	SiO$_2$	Ca^{2+}	Mg^{2+}	Na$^+$	K$^+$	Cl$^-$	SO$_4^{2-}$ pyr SO$_4$ (1)	SO$_4^{2-}$ SO$_4$ (2)	HCO$_3^-$	Σ$^+$
MEYBECK MODEL (1987)										
atmosphere									67	
silicates	92.5	26	48	46	95	0	40	18		35
carbonates	0	67	42	0	0	0	0	0	33	51
evaporites	0	7	10	54	5	100	0	42		14
amorphous silica	7.5									
WORLD RIVERS (Garrels, 1976)										
silicates		10	50	34						
carbonates		80	50	0						
evaporites		(10)	0	66						
WORLD RIVERS (Holland, 1978; based on Livingstone, 1963)										
atmosphere									65	
silicates	100	18	57	48	83	0				
carbonates	0	68	37	0	0	0	50	50		14
evaporites	0	14	≤4	52	17	100			35	86
WORLD RIVERS (Meybeck, 1979a)										
atmosphere									56	
silicates	100	5	8	44	100	0	28	0	0	14
carbonates	0	82	81	0	0	0	0		44	68
evaporites	0	13	11	56	0	100		72	0	18
WORLD RIVERS (Berner, Lasaga and Garrels, 1983; based on Meybeck, 1979a)										
atmosphere										
silicates		20	60							
carbonates		72	40							
evaporites		8	0							
WORLD RIVERS (Wollast and Mackenzie, 1983, p. 48; based on Meybeck, 1979a)										
atmosphere									56	
silicates	100	18	48	43	100					45
carbonates	0	68	40						35	38
evaporites	0	13.5	11	57			28	72		17
fossil org. matter									9	
AMAZON BASIN (Stallard, 1980, measured budget)										
atmosphere									70.5	
silicates	100	29	59	61	100	0		36	0	43
carbonates	0	62	41	0	0	0		0	29.5	45
evaporites	0	9	0	39	0	100		64	0	12
WORLD RIVERS (Berner and Berner, 1987)										
atmosphere							40		(4)	65
silicates	100	20	60	34	94					
carbonates		71	40				20			35
evaporites		9		66$^{(3)}$	6		100$^{(4)}$		40	

Σ$^+$ = sum of cations
(1) SO$_4^{2-}$ --- resulting from pyrite and organic sulfur oxidation.
(2) SO$_4^{2-}$ --- resulting from sulfate mineral dissolution.
(3) Including NaCl from shales and thermal springs.
(4) Natural biogenic emissions to atmosphere delivered to land by rain (31%), plus volcanism (9%).

The reaction of submarine basalts with sea water can occur over a range of temperatures, and not by a single well-defined process. Certainly the process investigated in most detail has been that involving the extensive hydrothermal circulation of sea water through mid-ocean ridge systems. Numerous authors have described serpentinites and greenstones collected from the sea-floor, resulting from the reaction of seawater with mid-ocean ridge basalts. Typical mineral assemblages found in hydrothermally altered basalts include zeolites, smectite, and mixed-layer chlorite-smectite. In laboratory experiments (e.g., Bischoff and Dickson, 1975; Bischoff and Seyfried, 1978; Mottl and Holland, 1978) and from theoretical calculations involving basalt-seawater reactions at elevated temperatures (Wolery, 1978), information has been provided concerning the factors that govern these hydrothermal basalt-seawater reactions and their mineral assemblages.

One set of schematic major equations describing the reactions between seawater and basalt is (Drever et al., 1987):

$$4Ca\,Al_2Si_2O_8 + 6Mg^{2+} + 4H_2O \rightarrow$$
$$\text{anorthite}$$

$$Mg_6Si_8O_{20}(OH)_4 + 4Al_2O_3 + 4Ca^{2+} + 4H^+ \quad\quad (9.23)$$
$$\text{talc} \quad\quad\quad\quad \text{corundum}$$

$$11Fe_2SiO_4 + 2Mg^{2+} + 2SO_4^{2-} + 2H_2O \rightarrow$$
$$\text{fayalite}$$

$$Mg_2Si_3O_6(OH)_4 + FeS_2 + 8SiO_2 + 7Fe_3O_4$$
$$(9.24)$$
$$\text{sepiolite} \quad\quad \text{pyrite} \quad \text{quartz} \quad \text{hematite}$$

$$CaAl_2Si_2O_8 + 2Na^+ + 4SiO_2 \rightarrow Ca^{2+} + 2NaAlSi_3O_8 \quad\quad (9.25)$$
$$\text{anorthite} \quad\quad\quad \text{quartz} \quad\quad\quad \text{albite}$$

$$Na^+ + KAlSi_2O_8 \rightarrow K^+ + NaAlSi_3O_8 \quad\quad (9.26)$$
$$\text{K-spar} \quad\quad\quad \text{albite}$$

Calcium released from these reactions may precipitate as $CaCO_2$ in veins and interstitial areas of the basalt or be released from the hydrothermal system via hot springs or diffuse flow. More will be said about this calcium release in Chapter 10 with regard to the apparent excess of calcium in the sedimentary rock column. Notice that in the above reactions, and for the precipitation reaction of $CaCO_3$ (equation 9.13), protons are released, new alteration minerals formed, and some dissolved constituents are removed from seawater. Consequently, the mid-ocean ridges have been considered important sites for "reverse weathering" (Garrels, 1965; Mackenzie and Garrels, 1966a,b) by several authors (Edmond and Von Damm, 1983; Drever et al., 1987). This concept will be developed in detail later.

Table 9.11 is a summary mainly based on Wolery and Sleep's work (1987) of estimates of fluxes between basalt and seawater. Flux estimates include both those owing to hydrothermal reactions at axial ("high" temperature) and off-axial ("low" temperature) portions of ridge systems, as well as estimates associated with alteration processes between submarine basalts and ambient ($\sim 2°C$) temperature seawater ("weathering"). Little is known about mass transfer today associated with mid-plate volcanic activity or with back-arc basin spreading centers. Arthur et al. (1985) concluded that, because mid-plate volcanism in the Cretaceous was so extensive in the Pacific basin, this volcanism on the Pacific plate alone could have amounted to as much as 20% of the estimated Cretaceous CO_2 emission from mid-ocean ridge and oceanic arc volcanism. Investigations of the chemistry of mid-plate volcanism today at Loihi Seamount, Hawaii, and of back-arc volcanism in the Marianas are currently underway and should shed more light on the importance of these tectonic provinces in the balance of carbon and other elements.

To conclude this section, a few comments on carbon fluxes associated with volcanism and diagenesis-metamorphism are appropriate. A review of the literature (Garrels and Mackenzie, 1971a; 1972; Javoy et al., 1982; 1983; Berner et al., 1983; Lasaga et al., 1985; Des Marais, 1985) shows that these fluxes are difficult to assess, and thus, poorly known. Table 9.12 lists some estimates from the recent literature. The important point is that carbon in mantle basalt is vented to the ocean-atmosphere system at mid-ocean ridges and arcs. This CO_2 takes part in weathering reactions and is converted to HCO_3^-. As might be expected, if mantle degassing is on the order of 1 to 8×10^{12} moles C y^{-1} (Des Marais, 1985), it would take a time equal to the age of the Earth to fill all the carbon reservoirs of the exogenic cycle--atmosphere, land, ocean, and sediments. In Figure 9.1 the mantle flux from ridges and arcs is given as 5×10^{12} moles C y^{-1}, and a similar value is assigned to the carbon flux to Earth's surface because of late diagenetic and metamorphic reactions.

Table 9.11. Fluxes involving submarine basalt-seawater reactions (units of 10^{12} moles y^{-1}). (After Wolery and Sleep, 1988.)

Component	Low-temp alteration	Off-axis hydrothermal	Axial hydrothermal	Total hydrothermal	Net flux	Ref
Na	-0.35 to-0.04	0.48	?	?	0.13 to 44	(1)
	0.009			-1.96		(2)
	1.39			-10.39		(3)
				0 to -3.83		(4)
			-8.6 to 1.9			(8)
	-0.47	0.17				(9)
K	-0.51 to -0.10	0.33	0.21	0.54	0.03 to 0.44	(1)
	-0.31			0.61		(2)
	-1.54			0		(3)
				<0.26		(4)
	0 to -0.17					(5)
				1.28		(6)
			1.9 to 2.3			(8)
	-0.26	-0.56	0.13 to 1.26	-0.43 to 0.69	-0.69 to 0.44	(9)
Ca	0.25 to 1.25	1.20	0.60	1.80	2.05 to 3.05	(1)
	0.06			0.38		(2)
	4.68			15.58		(3)
				2.58 to 7.05		(4)
	0.15 to 0.70			3.80		(5)
			2.05 to 4.30			(6)
	0.20 to 1.25			1.63 to 3.38		(7)
			0.24 to 1.5			(8)
	0.32	1.18	0.32 to 3.25	1.50 to 4.43	1.83 to 4.75	(9)
Mg	0.29 to 1.89	-1.23	-1.27	-1.50	-2.21 to 0.61	(1)
	0.05			-1.73		(2)
	4.98			-8.39		(3)
	0.41 to 0.58			-4.52		(5)
	0.04 to 0.45			-2.71		(7)
			-7.5			(8)
	1.21	0.46	-7.79 to -4.17	-7.33 to -3.71	-6.13 to -2.5	(9)
SiO_2	0.08 to 0.47	0.37	0.52	0.89	0.97 to 1.36	(1)
	0.87			3.27		(2)
	4.83			6.67		(3)
	0 to 0.63					(5)
			3.10			(6)
	0.50 to 1.33				0.50 to 1.50	(7)
			2.2 to 2.8			(8)
	1.88	0.71	0.31 to 3.10	1.02 to 3.81	2.96 to 5.71	(9)

1) Wolery and Sleep, 1988; 2) Maynard, 1976; 3) Hart, 1973; 4) Mottl, 1976; 5) Seyfried, 1976; 6) Edmond et al., 1979; 7) Wolery and Sleep, 1976; 8) Von Damm, 1983; 9) Thompson, 1983a,b

Table 9.12. Some estimates of CO_2 fluxes from metamorphism (diagenesis) and mantle basalt (units of 10^{12} moles C y^{-1}).

Source	CO_2 Flux	Reference
Mid-ocean ridge	0.7 - 7.5	(a)
	20	(b)
	4	(c)
Oceanic arc	0.3 - 0.75	(a)
	3.9	(c)
Magmatic-metamorphic	5.8	(d)
Metamorphic-diagenetic	5	(e)
Volcanic degassing	7.5	(f)
	7.2	(g)

(a) Des Marais, 1985; (b) Javoy et al., 1983; (c) Arthur et al., 1985; (d) Berner et al., 1983; (e) Mackenzie, 1981; (f) Holland, 1978; (g) Javoy et al., 1982.

Sediment pore water fluxes Diagenetic processes occurring in marine sediments may significantly affect the composition of interstitial waters and lead to concentration gradients responsible for exchange of dissolved species with the overlying water column. Any evaluation of fluxes associated with diagenetic processes may be limited to a significant degree by artifacts produced by temperature and pressure changes during pore water sampling procedures, by lack of knowledge of concentration gradients near the sediment-water interface, and by uncertainties in the values of the effective diffusion coefficients used to calculate fluxes. It is important to know these fluxes, however, because they may represent a significant source or sink of dissolved materials for the ocean. Furthermore, evaluation of these fluxes can provide an idea of the diagenetic reactions obtaining in sediments.

Details of the approaches used to calculate these fluxes are given in many references, including Lerman (1979) and Berner (1980). Methods used to calculate global fluxes and estimates obtained are found in articles by the authors referred to in Table 9.13, a summary of fluxes for exchange of dissolved constituents across the sediment-water interface. Notice that burial of pore water removes constituents from the oceans and may be a major removal process for Na^+. The Ca^{2+}, Mg^{2+}, and HCO_3^- fluxes are more fully discussed in a later section.

Atmospheric transport of elements to the oceans We will conclude our considerations of transport agents and fluxes of materials to the ocean with a brief look at airborne transport of solids. Wind stress on the land surface results in advection of soil into the atmosphere; the dust later falls on the sea surface as particulate matter in rain or as dry fallout. Buat-Menard and Chesselet (1979) evaluated fluxes to the sea surface of various elements in particulates for the North Atlantic open ocean and found for Al a flux of 5×10^{-6} g cm^2 y^{-1}. For a mean Al content in soils of about 7%, a total flux of solid is obtained of 7×10^{-5} g cm^2 y^{-1}. If this flux is representative of all the ocean surface area (360×10^{16} cm^2), the amount of solid transferred from the continents to the ocean via the atmosphere is estimated as 2.5×10^{14} g y^{-1}. This value is within the range of 0.5 to 5×10^{14} g y^{-1} reported earlier by Mackenzie and Wollast (1977), and similar to that of Buat-Menard (1983).

The input from the atmosphere is small compared to the suspended load of rivers of 155×10^{14} g y^{-1}. As was pointed out by Buat-Menard and Chesselet (1979), however, the relative importance of this source of particulates increases for

Table 9.13. Some estimates of exchange of dissolved constituents across the sediment-water interface. Fluxes in units of 10^{12} moles y^{-1}.

	Maynard (1976)	Manheim (1976)	Sayles (1979)	Wollast and Mackenzie (1983)	Drever et al. (1987)	Burial[a]
Na^+	+0.04	-0.61	+4.4	+1.1	+2.5	-1.23
K^+	-0.01	-0.72	-2	-0.64	-0.9	-0.025
Mg^{2+}	-0.03	-2.5	-5.3	-3.7	-2.3	-0.14
Ca^{2+}	-0.02	+0.75	+6.3	+0.7	+3.6	-0.03
H_4SiO_4[b]				+6.6		
HCO_3^-					+4.7	

[a] Mean of Wollast and Mackenzie (1983) and Drever et al. estimates (1987).
[b] Heath (1974) gives a range of 7 to 14×10^{12} moles H_4SiO_4 per year; Wollast (1974) obtained 6.8×10^{12} moles Si y^{-1}.
+ denotes input to seawater; - denotes removal from seawater

the open ocean, where the contribution of river materials is reduced substantially. For example, the occurrence of quartz in deep-sea sediments of central parts of the oceans (e.g., Windom, 1976) has been related to quartz-bearing dust transported from arid regions of the continents by prevailing winds. Similarities in the distributions of feldspar and quartz in these sediments, and the mineralogy of dust samples collected over the open ocean support the conclusion that alkali feldspars also have an eolian origin in certain regions of the ocean.

The major element content and mineralogy of air-borne particles reflect closely those of continental soils and shales, although atmospheric particulates also include materials of oceanic origin (Delaney et al., 1967), and show considerable enrichments in some trace metals (Buat-Menard and Chesselet, 1979). The average composition of shales and soils (Table 9.8) was chosen to represent the properties of dust transported from the continents to the ocean. Fluxes of elements in atmospheric transport to the ocean are given in Table 9.14.

Table 9.14. Estimates of fluxes of elements to the sea surface via atmospheric transport from land. A total dust flux of 2.5×10^{14} g y^{-1} is assumed.

	Atmospheric Dust[a] (wt. %)	Element Flux (10^{12} moles y^{-1})
SiO_2	66.3	2.8
Al_2O_3	15.2	0.75
Fe_2O_3	6.3	0.20
MgO	1.7	0.11
CaO	1.7	0.08
Na_2O	0.96	0.08
K_2O	2.7	0.07
CO_2[b]	2.6	0.02

[a] average of shale and soil composition from Table 9.8.
[b] after Clarke (1924).

Mass Balance of Ca, Mg, and C in Present Oceans

There are several different ways to approach the cycles of Ca, Mg, C, and other elements in the oceanic system depending on the choice of boundary conditions. For example, discussion can be restricted to the water column bounded by continents, atmosphere, and the sediment-water interface. In this case, the various fluxes between the sea-floor and the water column must be defined. Another approach is to include the marine sediments in the oceanic system and to evaluate from the sedimentary record the element fluxes and uplift rates of marine sediments. Furthermore, the water column could be subdivided into a photic zone, intermediate waters, and deep waters; and the sedimentary column into an early diagenetic layer and deeply buried sediments where slow rates of diagenetic reactions obtain. In fact, the literature is confused on the subject of element cycle models because of the lack of definition of boundary conditions and fluxes used in these models.

Various fluxes and processes considered in the following discussion are represented schematically in Figure 9.20, adapted from Wollast and Mackenzie (1983). Steady-state conditions require that the mass balance for each element is fulfilled for the entire system and also separately for the water and sedimentary columns. To write these mass balances, we must consider the global rate of the reactions occurring in the two subsystems. Therefore, we define R and D as the annual amount of a given element transferred from the solid phase to the aqueous phase or vice versa. The fluxes are taken as positive if there is a net input to seawater, and negative if there is a net output. To maintain the concentration of an element constant in seawater, the net flux resulting from $R_d + H_d + L_d + P_d$ in Figure 9.20 must be balanced by precipitation or dissolution reactions in the water column:

$$R_d + H_d + L_d + P_d = R \qquad (9.27)$$

with R > 0 for precipitation and < 0 for dissolution. R can be obtained for any element for which there is sufficient information. In Table 9.15 we summarize the various fluxes of Ca, Mg and C given previously and give the computed value of R. The net fluxes for the three elements are all positive, and if the ocean is at steady state, these fluxes must correspond to removal rates of the elements from seawater by physico-chemical and biological processes.

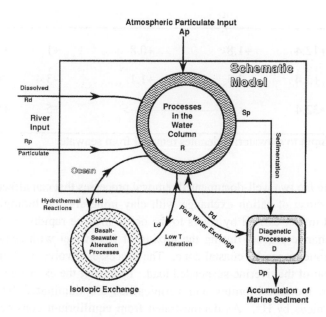

Figure 9.20.   Schematic diagram of fluxes and processes evaluated for the global cycle of an element. R_d, R_p, S_p, D_p, H_d, L_d, and P_d are fluxes related to riverine dissolved and particulate matter transport, oceanic sedimentation, and accumulation, basalt-seawater hydrothermal and low temperature alteration reactions, and pore water exchange, respectively: d refers to dissolved flux, p to particulate, and R and D are annual amounts of an element transferred between the solid and the aqueous phase.

Table 9.15. Our best estimate of the mass balance for dissolved Ca^{2+}, Mg^{2+}, and HCO_3^- in the oceanic water column in units of 10^{12} moles y^{-1}.

	River input	Hydrothermal reactions	Low temperature alteration	Pore water flux	Net balance (R)
Ca^{2+}	+12.4	+1.8	+0.8	+1	+16
Mg^{2+}	+5.4	-2.5	+1.1	-3.4	+0.6
HCO_3^-	+32.4	<------------5------------>		+5	+42.4

+ denotes input to seawater; - denotes removal from seawater

There are two well-documented removal processes that can affect these three elements; those of cation exchange with clay minerals and biological uptake. Suspended matter carried by rivers to the ocean may be rapidly involved in the cation exchange process during the mixing of freshwater with seawater in the estuarine zone or nearby coastal zone. The processes involve predominantly the clay fraction of the riverine suspended load. Typically, the exchange positions in river-borne montmorillonites and vermiculites are dominated by Ca^{2+}, and degraded micas by H^+. As demonstrated from equilibrium considerations and experimental evidence, these cations are replaced by major cations present in abundance in seawater, mainly Na^+, Mg^{2+}, and K^+. Two estimates of global fluxes involving exchange reactions are given in Table 9.16.

The process of removal of calcium by marine organisms in the water column is well known. Production of calcium carbonate by water column biological processes may be estimated from primary productivity and from the mean chemical composition of plankton. After death of the organisms and removal of the organic protective layer, the skeletons may undergo dissolution if they encounter water undersaturated with respect to their mineral composition. Active dissolution of calcium carbonate occurs mainly near the sediment-water interface in deep waters that are undersaturated with respect to both calcite and aragonite (see Chapter 4). Thus, calcium is regenerated from calcareous skeletons and, finally, only a small fraction of the initial production of these materials accumulates in sediments. An extended discussion of these processes can be found, for example, in the detailed review by Berger (1976), and in Chapter 4.

Table 9.16. Estimates of fluxes involving exchange reactions between seawater and solids and biological and chemical removal of some elements from seawater. Units of 10^{12} moles y^{-1}.

	Exchange		Biological and Chemical Precipitates	
	A	B	A	B
Na$^+$	-1.87	-1.53		
K$^+$	-0.12	-0.20		
Mg^{2+}	-0.29	-0.32	-0.29	-0.26
Ca^{2+}	+1.18	+0.96	-15	-12.5
SO$_4$$^{2-}$			-0.6	
HCO$_3$$^-$		+0.42	-24.2	
H$_4$SiO$_4$^{0}			-13.4	

A - Wollast and Mackenzie (1983); B - Drever et al. (1987)
+ denotes input to seawater; - denotes removal from seawater.

SiO$_2$ and CaCO$_3$ constitute the bulk of the skeletal material precipitated by marine organisms; however, impurities also are incorporated into the skeletons of organisms and removed from the water column. Calcite precipitated by marine organisms contains between 0 and 30 mole percent magnesium carbonate. Magnesian calcite skeletal materials, however, principally accumulate in shallow waters. Shallow-water calcareous sediments contain about 5% MgCO$_3$ (Chave, 1954a,b), whereas calcareous constituents of deep-water oozes contain about 0.5% MgCO$_3$ (Drever, 1974). Assuming that 20% and 80% of the Ca deposited annually are removed in shallow-water and deep-water calcareous sediments, respectively, Wollast and Mackenzie (1983) arrived at a biogenic Mg removal of 0.3 x 10^{12} moles y^{-1}. Two estimates of the fluxes involving biological removal of elements from seawater are given in Table 9.16.

The cation exchange and biological removal fluxes can be used to constrain a net mass balance for the water column. In Figure 9.21 we have constructed balances for Ca, Mg, and inorganic C in the modern oceanic system. Within the uncertainties of the individual fluxes for these elements, it appears that the ocean today may be near steady state with respect to these elements. Precipitation of Ca^{2+} and Mg^{2+} as carbonate minerals in the oceans today requires $2(16 \times 10^{12} + 0.3 \times 10^{12}) = 32.6 \times 10^{12}$ moles HCO_3^- y^{-1}. One half of this carbon accumulates on the sea floor in carbonate sediments; the remaining half (16.3×10^{12} moles y^{-1}) enters the ocean as CO_2 (see reaction 9.14). Because of the pH and high buffer capacity of seawater, much of this CO_2 is retained in the ocean. This carbon most likely initially enters into reactions involving synthesis of marine organic matter (Smith, 1985). At some time in its history, this organic carbon is oxidized to CO_2 for weathering of carbonate minerals on land. If this were not the case, the atmosphere would be depleted of CO_2 because of weathering of carbonate minerals in less than 5000 years [(atmospheric $CO_2 = 61,000 \times 10^{12}$ moles) / (CO_2 flux to weather carbonates $\approx 12 \times 10^{12}$ moles y^{-1}) ≈ 5000 years]. The weathering and oxidation of the organic carbon may occur on land after uplift and exposure of the sedimentary carbon.

Figure 9.21.   Schematic diagrams illustrating balance of dissolved Ca, Mg, and HCO_3 in oceanic system. Fluxes are in units of 10^{12} moles y^{-1}.

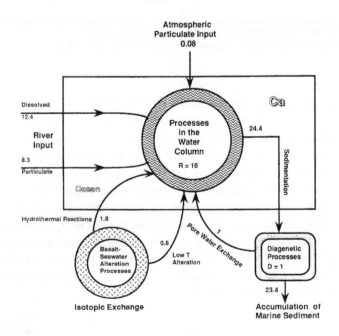

Figure 9.21 continued.

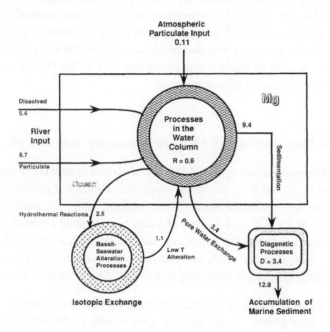

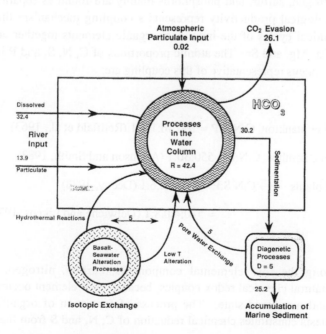

In Figure 9.21 all of the carbon eventually used in weathering of minerals by CO_2-charged soil water is shown as entering the atmosphere. The difference between the flux of CO_2 owing to precipitation of carbonate minerals in the ocean and the total CO_2 released from the ocean is that CO_2 used to weather silicate minerals on land, and agrees with the calculations of riverine source materials made earlier in this chapter, in which it was shown that 30% of the HCO_3^- in river water comes from weathering of silicate minerals.

Oceanic Mass Balance of Elements Interactive with Ca, Mg, and C

The biogeochemical cycling behavior of carbon is intimately tied to that of the elements N, P, and S by biological processes. Biological productivity on land and in marine and terrestrial waters forms organic materials made of six major elements -- C, H, O, N, P, and S -- and of a dozen or so minor elements necessary to the maintenance of organic structures and physiological functions of living organisms. Inorganic carbonate skeletons of marine organisms contain Ca, Mg, and Sr, and their behavior is intimately tied to that of the "organic" elements.

The major elements of organic matter are present in different proportions in aquatic and land plants. In the surface environment of the Earth, the elements carbon, nitrogen, sulfur, and phosphorus mainly are found as separate chemical forms. Biological productivity represents a coupling mechanism that joins the biogeochemical cycles of the individual organic elements together and with the elements Ca, Mg, and Sr. The atomic proportions of C, N, S, and P in terrestrial and aquatic plants representative of this coupling are:

marine plankton C:N:S:P = 106:16:1.7:1 (Redfield et al., 1963)

marine benthos C:N:P = 550:30:1 (Atkinson and Smith, 1983)

land plants   C:N:S:P = 882:9:0.6:1 (Deevy, 1973)

              = 510:4:0.8:1 (Delwiche and Likens, 1977)

Among the four elemental components, carbon, nitrogen, and sulfur represent natural chemical redox couples, because each element occurs both in an oxidized and a reduced state. The process of formation of organic matter by photosynthesis constitutes chemical reduction of C, N, and S from their inorganic forms. The oxidized inorganic forms of the four elements are such commonly occurring species as carbon dioxide (CO_2), dissolved carbonate (HCO_3^- and CO_3^{2-}), nitrate (NO_3^-) and nitrogen oxides, and sulfate (SO_4^{2-}). The reduced inorganic forms of the biologically important elements -- methane (CH_4), hydrogen sulfide (H_2S), ammonia (NH_3), and phosphine (PH_3)-- are thermodynamically unstable in the presence of oxygen, and they become oxidized sooner or later.

The net photosynthetic reactions that produce organic matter with the stoichiometric proportions of C, N, S, and P as given above can be written for the mean composition of marine plankton and land plants in a form that balances oxidized inorganic reactants with organic matter and free oxygen as the reaction products. The two reactions are:

In water:

$$106CO_2 + 16HNO_3 + 2H_2SO_4 + H_3PO_4 + 120\ H_2O$$

$$\rightarrow\ C_{106}H_{263}O_{110}N_{16}S_2P + 141O_2 \tag{9.28}$$

On land:

$$882CO_2 + 9HNO_3 + H_2SO_4 + H_3PO_4 + 890H_2O$$

$$\rightarrow\ C_{882}H_{1794}O_{886}N_9SP + 901.5O_2 \tag{9.29}$$

In the marine photosynthesis reaction, for convenience, the proportion of sulfur was raised from 1.7 to 2, and for land photosynthesis the proportion of S was rounded up to 1. The proportions of hydrogen and oxygen in organic matter are nearly H:O = 2:1; this is the reason for the abbreviated chemical notation of organic matter usually written as CH_2O.

In an approach similar to that used to obtain the balances of Figure 9.21, the present-day oceanic mass balances for the elements C (organic matter), N, P, S, and Sr were obtained and are shown schematically in Figures 9.22 and 9.23. These figures provide some tentative estimates of present-day element fluxes involved with the production of organic and inorganic carbon in the ocean. For these element cycles the ocean reservoir, except for sulfur, appears to be steady-state, with fluxes in equalling fluxes out. Because virtually no sulfur is deposited today as oxidized sulfur in evaporite deposits, sulfur as sulfate may be accumulating in seawater. The deposition of evaporites through geologic time is a secular variable controlled by tectonics. Tectonics provides the necessary paleogeographic settings for sulfate precipitation: sabkhas, lagoons, or deep basins. Once sulfate precipitation starts, it may proceed at a rate of as much as 70 mg S cm^{-2} y^{-1} (Holser, 1979), far faster than pyrite formation. If we assume a C/S ratio of 106/1.7 for marine plankton, then the particulate organic carbon flux from the oceans into sediments would imply an organic sulfur flux of 0.2 x 10^{12} moles S y^{-1}.

Figure 9.22.   Oceanic cycles of organic carbon, nitrogen, and phosphorus. DOC = dissolved organic carbon; POC = particulate organic carbon; NPP = net primary production; DN = dissolved nitrogen; PN = particulate nitrogen; DP = dissolved phosphorus, PP = particulate phosphorus. C and N fluxes are in units of 10^{12} moles C and N y^{-1}; P fluxes are in units of 10^{10} moles P y^{-1}.

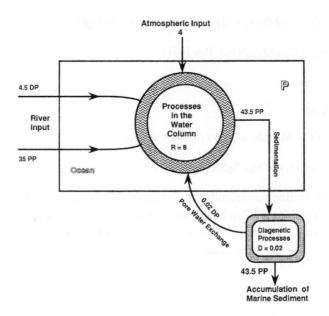

Figure 9.22 continued.

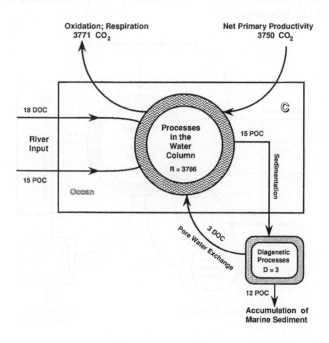

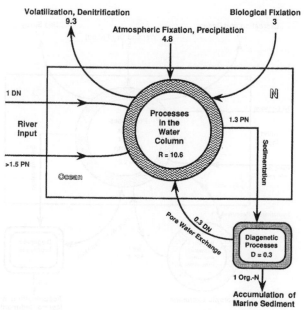

Figure 9.23.   Oceanic cycles of strontium and sulfur. Sr fluxes are in units of 10^9 moles y^{-1}; S fluxes are in units of 10^{12} moles y^{-1}. The imbalance for S suggests storage of SO_4^{2-} in seawater.

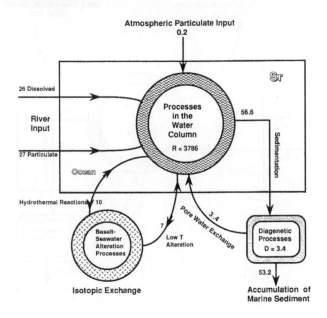

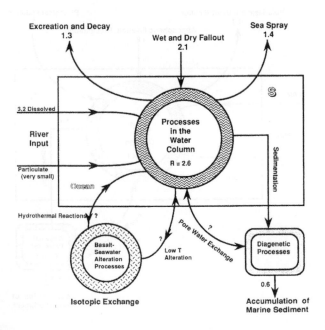

The particulate organic carbon flux includes both that involved with skeletal material and that representing degradation products of soft-bodied planktonic and benthic organisms. The organic matter accumulating on the sea floor has a C/N/P ratio of 250/20/1, representing a mixture of marine and terrestrial organic matter and differential recycling during early diagenesis of N and P relative to C. During early diagenesis, aerobic and anaerobic processes result in decomposition of organic matter and lead to release of dissolved species of C, N, and P to sediment pore waters, and dissolution or precipitation of carbonate minerals (see Chapter 6).

Concluding remarks

Although during the last two decades considerable research has gone into investigating the global carbon cycle, much remains to be done. This heightened interest has arisen because of society's concern over the "greenhouse effect". Today's global carbon cycle is difficult to balance if land use practices result in significant quantities of CO_2 release to the atmosphere. It is possible if this release is real that the oceanic sink of fossil fuel CO_2 has been underestimated and other sinks, such as dissolution of shoal-water benthic carbonates, CO_2 fertilization of terrestrial productivity, and enhanced sequestering of organic carbon in the oceanic system, play an important combined role in balancing the modern carbon cycle.

The magnitudes of the source and sink fluxes of atmospheric methane are poorly known. This gas is accumulating in the atmosphere owing to releases from human activities, but future concentrations are difficult to assess because of lack of knowledge concerning methane cycling. As a greenhouse gas, one molecule of CH_4 in the atmosphere is 20 times more effective in increasing temperature than one molecule of CO_2 (Blake and Rowland, 1988).

Inputs and outputs of calcium, magnesium, and carbon can be balanced for the modern ocean. This balance necessitates reverse weathering reactions in the ocean system in which CO_2 consumed in weathering on land is released to the ocean-atmosphere system during the formation of minerals in the ocean. If this were not the case, regardless of juvenile emissions of CO_2 to the ocean-atmosphere system, atmospheric CO_2 concentrations could be reduced to vanishingly small values in less than 5000 years.

Studies of the modern global biogeochemical cycle of carbon form one basis for understanding the geologic history of atmospheric CO_2. In turn, these investigations of the history of carbon dioxide in the atmosphere provide knowledge that can be used to interpret the future of atmospheric CO_2 levels and the impact of anthropogenic activities on the carbon cycle. In the next chapter, we focus our attention on the chemical evolution of the Earth's surface environment, as gleaned particularly from knowledge of the carbon cycle and carbonate sediments.

Chapter 10

SEDIMENTARY CARBONATES IN THE EVOLUTION OF EARTH'S SURFACE ENVIRONMENT

Introduction

In the last chapter we discussed the modern biogeochemical cycle of carbon, and presented sources and sinks of Ca, Mg, and C in modern oceans. In this chapter, we explore in some detail historical aspects of the carbon cycle with emphasis on sedimentary carbonates. Such an exploration requires synthesis of data and speculation on evolution of the Earth's surface environment. The further back in geologic time we explore, the greater the degree of speculation, because sedimentary rock mass per unit time becomes progressively less and more altered by diagenetic and metamorphic events with increasing geologic age.

An evaluation of the evolutionary and cyclic properties of carbonate rocks, and other lithologies, is necessary for both the sedimentary geologist and the geochemist. For the sedimentary geologist, knowledge of the secular or cyclic variation of the initial composition of marine carbonate precipitates with geologic age is prerequisite for interpretation of pathways of diagenesis. For the geochemist, knowledge of the trend in dolostone/limestone ratios with geologic age may help in constraining models of atmosphere-ocean evolution. Thus, this chapter relies heavily on trends in sedimentary rock properties with geologic age as the basis for understanding the long-term geochemical cycling behavior of carbonates.

In recent years innumerable publications have dealt with the natural carbon cycle and its alteration by human activities. Some summary works of interest in this chapter are *Atmospheric Carbon Dioxide and the Global Carbon Cycle* (ed. Trabalka, 1985), *The Carbon Cycle and Atmospheric CO_2: Natural Variations, Archean to Present* (eds. Sundquist and Broecker, 1985), *Chemical Cycles in the Evolution of the Earth* (eds. Gregor, Garrels, Mackenzie, and Maynard, 1988), *History of the Earth's Atmosphere* (Budyko, Ronov, and Yanshin, 1985), and *The Chemical Evolution of the Atmosphere and Oceans* (Holland, 1984). The interested reader is referred to these volumes for further discussion of material presented here.

Sedimentary rock mass-age distribution

The atmosphere, ocean, and biosphere leave their record in sedimentary rocks. It is likely that this record reflects both secular and cyclic evolutionary processes. The cyclic processes involve chemical mass transfer of materials in and out of global reservoirs like the atmosphere, ocean, and sedimentary rocks. If inputs and outputs of these reservoirs are nearly balanced so that over long periods of geologic time the mass and composition of the reservoirs remain constant, a quasi-steady state is maintained. Hand-in-hand with this cyclicity go changes in Earth's surface environment reflecting secular evolution of the planet, an aging process.

Figure 10.1 is a generalized model of the rock cycle. Both the sediments and the continental crust are recycled; the former by processes of weathering, erosion, transportation, deposition, burial, uplift, and re-erosion; and the latter by resetting

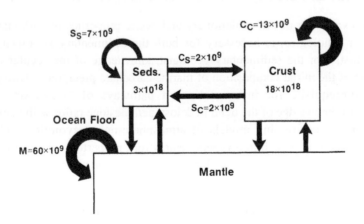

Figure 10.1. A generalized diagram for the steady-state rock cycle. Sediments, S, and continental crystalline crust, C, masses are in units of metric tons. S_S, C_S, S_C, C_C, and M are fluxes in units of 10^9 tons y^{-1} due to erosion of sediments, metamorphism, erosion of crystalline rocks, recycling of crystalline rocks (resetting of ages during tectogenesis), and cycling of oceanic crust, respectively. Total sedimentation rate is 9×10^9 tons y^{-1}. (After Gregor, 1988.)

of ages when the deeper portions of the crystalline crust undergo tectogenesis. The exchange between sediments and crust involves erosion of crystalline rocks to form sediments and conversion of sediments by metamorphic processes to crystalline rocks. There are two-way fluxes between the mantle and the crystalline and sedimentary crusts, and a recycling of the mantle itself by the plate tectonic processes of ridge volcanism and subduction of ocean floor. If the fluxes of materials between mantle and crust are not balanced, then their degree of imbalance represents potential secular change in the compositional evolution of the crust-ocean-atmosphere system. We will return later in this chapter to consideration of cyclic versus secular processes. Now the mass-age distribution of sedimentary rocks and its role in setting the stage for modern ideas of sedimentary cycling will be discussed.

Sedimentary rocks are formed by depositional processes involving principally the agents of water and wind and are destroyed when eroded or transformed chemically into other kinds of rocks like metagneiss. The sedimentary rock mass today, as estimated from geochemical mass balance methods, is 30,000 x 10^{20}g (Garrels and Mackenzie, 1971a; Li, 1972; Veizer, 1988), with 86% of the mass lying within the continental and shelf region of the globe, and 14% a part of the deep ocean floor (Ronov et al., 1980). This rock mass estimate includes the classic lithologies of sandstone, shale and carbonate, as well as their metamorphic equivalents of quartzite, slate, phyllite, low-grade schist and marble. The current mass is that preserved, not the total mass deposited throughout geologic time. Total sedimentary deposition over the last 3.5 billion of years of Earth history has been about 130,000 x 10^{20}g.

Similar to a human population, rock masses can be assigned birth rates and death rates, and they can be subdivided into age groups. Figure 10.2 shows the observed distribution of the sedimentary rock population for Phanerozoic and later Precambrian time. The stippled area of the diagram is based on direct estimates of stratigraphic data for the Phanerozoic and later Precambrian rock mass (Ronov, 1968; Gregor, 1970; 1985; Veizer and Jansen, 1985). For the Phanerozoic mass of 20,000 x 10^{20}g, approximately 30% is in the deep sea and part of the continental margin, and 70% is under land.

The observed Phanerozoic rock mass distribution appears to consist of two mass maxima separated by a minimum in the mass of rock preserved per unit time. Gregor (1985) has demonstrated that the mass-age distribution for Carboniferous and younger sediments has a log linear relationship such that:

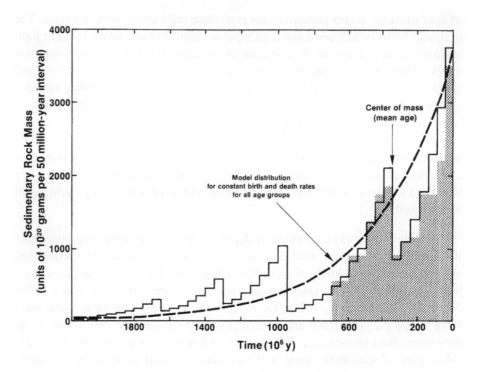

Figure 10.2.   Observed and model sedimentary rock mass-age distributions.
Observed mass distribution is represented by stippled area. Dashed curve is a
model calculation for constant deposition and destruction rates of sediments of all
age groups, whereas the histogram represents the results of a model calculation in
which variations in these rates are permitted. The mean age of the sedimentary
rock mass is about 350 million years. (After Garrels et al., 1976.)

$$\log S = 10.01 - 0.24t \tag{10.1}$$

where S is survival rate, defined as mass of a System divided by duration of a
Period, and t is duration of a Period (Figure 10.3). Earlier mass-age data
suggested a similar exponential function for Devonian and older Phanerozoic
sediments (Gregor, 1970; Garrels and Mackenzie, 1971b), but later work (Gregor,
1985) has not confirmed this relation (Figure 10.3).

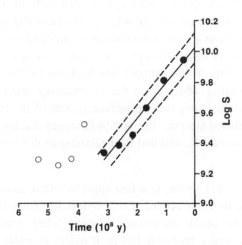

Figure 10.3. Mass-age distribution of the Phanerozoic Systems plotted as the logarithm of the survival rate (S, mass of System divided by duration of Period) versus age. (After Gregor, 1985.)

Two model calculations of mass-age relationships are shown in Figure 10.2; the first (dashed line) allows for constant birth and death rates of sediments for all age groups, and the second (histogram) permits variations in the vulnerability of the sediments of different ages to destruction by erosion and metamorphism, while maintaining sediment deposition and destruction rates and total mass constant. Details of calculations of this nature are given in Garrels and Mackenzie, 1971a, b; Li, 1972; Garrels et al., 1976; and Veizer, 1988.

Because of lack of mass and age data for sediments older than 700 ma, it is possible to obtain only an idea of the mass distribution prior to this time from model calculations and some tentative estimates of Precambrian mass. The model ages of maxima in Precambrian sedimentary mass were chosen to correlate with the ages (as derived from radiometric dating) of world-wide events of granite formation and metamorphism (Gastil, 1960; Engle and Engle, 1970). These events were presumably driven by episodic renewals of plate movement and mountain building throughout geologic time. It can be seen from Figure 10.2 that a model can be constructed to fit the Phanerozoic rock mass distribution. The maxima and minima in the model calculations prior to 700 ma, however, are entirely speculative.

It appears that the present-day age distribution of the population of sedimentary rocks is consistent with the assumption of a nearly constant total rate of deposition and destruction of sediments throughout much of geologic time and is in accord with the concept of an average steady-state cycling of sedimentary mass during Phanerozoic time. This argument may hold true for the past 2 billion years of Earth history, during which time the sedimentary mass has been largely recycled, that is cannibalistic, and the surface system of the Earth, the exogenic system, virtually a closed system. Veizer (1988) argues that the sedimentary cycle today is 90 ±5% cannibalistic and that the beginning of this system was 2.5×10^9 years ago.

As a corollary to the above, to a first approximation, steady-state cycling of mass can be shown to be in accord with a model in which the composition of materials reaching the ocean via streams, glaciers, wind, groundwater flow and submarine basalt-seawater reactions has been nearly constant over Phanerozoic time, suggesting that there has been little evolutionary change in seawater composition during the Phanerozoic (Garrels and Mackenzie, 1972, 1974; Mackenzie, 1975). Furthermore, the average rate of erosion deduced from the model is about 75×10^{14} g y^{-1}, about one third of the present rate of continental denudation (Garrels and Mackenzie, 1971a,b). This lower rate is about that expected for Phanerozoic time, if human activities have greatly accelerated erosional rates (Judson, 1968) and if continental area today is about 20% greater than the average for Phanerozoic time (Sloss and Speed, 1974).

The hypothesis that the present mass-age distribution of sedimentary rocks represents cyclic processes leading to better preservation of the younger part of the sedimentary record has acted as a stimulus to further research in sedimentary and geochemical cycling. For example, the observed changes with geologic age in proportions of lithological types and the chemistry and mineralogy of sediments were interpreted in terms of cyclic processes (Garrels and Mackenzie, 1971a, 1972; Garrels et al., 1976). Phanerozoic secular trends in the carbon isotopic composition of carbonates and the sulfur isotopic composition of evaporitic sulfates were interpreted as representing transfers of carbon and sulfur between global sedimentary reservoirs containing minerals of these elements (Garrels and Perry, 1974; Veizer et al., 1980). A recent book, *Chemical Cycles in the Evolution of the Earth* (Gregor et al., 1988), deals thoroughly with cyclic processes involving chemical mass transfer between global reservoirs. In a well-documented chapter of this book, Jan Veizer discusses the evolving exogenic cycle and integrates concepts concerning secular and cyclic evolution of the rock record.

Because sedimentary carbonates represent primarily chemical and biochemical precipitates from seawater, and because they make up 20% of the common sedimentary rock record, these rock types have been particularly good sources of chemical and mineralogical data for interpretation of the secular and cyclic evolution of the Earth's surface environment. This carbonate rock record as a function of geological age is now explored as are age trends in other rock types and sediment properties. With this information as background material, we can then discuss what these relationships tell us about the history of carbonates and the exogenic system throughout geologic time.

Secular trends in sedimentary rock properties

Lithologic Types

The lithologic composition of sedimentary rocks and the relative percentages of sedimentary and volcanic rocks are shown in Figure 10.4. These data compiled by Ronov and colleagues (Ronov, 1964; Budyko et al., 1985), represent the relative percentages of rock types for *continental regions*. It should be remembered that with increasing geologic age, the total sedimentary rock mass diminishes and a given volume percentage of rock 3 billion years ago represents much less mass than that for rocks 200 million years old. Figure 10.5 illustrates our tentative interpretation of the Phanerozoic distribution of carbonate rocks versus sandstones plus shales. The distribution of Phanerozoic System sedimentary masses with geologic age was obtained from Gregor's (1985) estimates. The mass of carbonate rock in each system was calculated from the estimates of CO_2 found in Phanerozoic carbonates (see Figure 10.27; Ronov et al., 1980; Hay, 1985), and Given and Wilkinson's (1987) calculation of dolomite abundances for all major compilations to date of data on the composition (Mg/Ca or $MgCO_3/CaCO_3$ ratios) of Phanerozoic carbonate rock samples. Table 10.1 shows the results of these calculations. Both Figures 10.4 and 10.5 should be viewed cautiously because they must be considered tentative representations. Some general trends, however, in lithologic rock types agreed on by most investigators are evident.

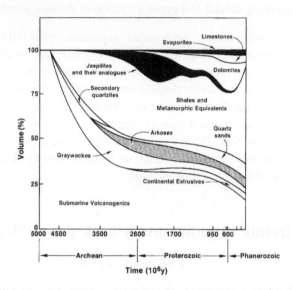

Figure 10.4.   Volume percent of sedimentary rocks as a function of age.
Extrapolation beyond about 3 billion years is hypothetical. (After Ronov, 1964.)

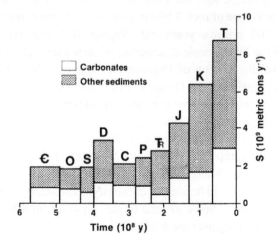

Figure 10.5. Mass-age distribution of carbonate rocks and other sedimentary rock
types plotted as survival rate (S) versus age. Total rock mass data from Gregor
(1985) and estimates of carbonate rock mass from Table 10.1.

Table 10.1. Phanerozoic carbonate mass distribution.

Period	Duration (10^6 years)	Total Carbonate (10^16 tons)	Total Dolomite (10^16 tons)	Total Calcite (10^16 tons)	Calcite/Dolomite Ratio	Total Dolomite (10^8 tons y^{-1})	Log S Dolomite (10^8 tons y^{-1})	Total Calcite (10^8 tons y^{-1})	Log S Calcite (10^8 tons y^{-1})	Total Carbonate (10^8 tons y^{-1})	Log S Carbonate (10^8 tons y^{-1})
Tertiary	66	19.42	2.92	16.50	5.65	4.42	8.65	25.0	9.40	29.42	9.47
Cretaceous	66	10.48	3.88	6.60	1.70	5.88	8.77	10.0	9.00	15.88	9.20
Jurassic	53	6.76	3.99	2.77	0.69	7.53	8.88	5.23	8.72	12.76	9.11
Triassic	50	2.40	1.63	0.77	0.47	3.26	8.51	1.54	8.19	4.80	8.68
Permian	45	4.44	0.31	4.13	13.32	0.69	7.84	9.18	8.96	9.87	8.99
Carboniferous	65	6.44	2.68	3.76	1.40	4.12	8.62	5.78	8.76	9.90	9.00
Devonian	55	5.79	2.76	3.03	1.10	5.02	8.70	5.51	8.74	10.53	9.02
Silurian	35	1.71	0.68	1.03	1.51	1.94	8.29	2.94	8.47	4.88	8.69
Ordovician	55	3.21	1.28	1.93	1.51	2.33	8.37	3.51	8.55	5.84	8.77
Cambrian	80	5.88	0.12	5.76	48.0	0.15	7.18	7.20	8.86	7.35	8.87
Total	570	66.53	20.25	46.28	2.29						

Carbonate rocks are relatively rare in the Archean, and those preserved are predominantly limestones (Veizer, 1973). It should be recalled, however, that the outcrops of very old Archean rocks are few and thus may not be representative of the original sediment compositions deposited. During the early Proterozoic the abundance of carbonates increases markedly and for most of this Era the preserved carbonate rock mass is dominated by epicontinental lagoonal to sublittoral early, and perhaps primary, dolostones (Veizer, 1973). In the Phanerozoic, carbonates constitute about 30% of the total sedimentary mass of sandstone, shale, and limestone plus dolomite. The relative percentages of limestone and dolostone preserved in the rock record younger than about 600 million years may be cyclically distributed, with greatest dolostone abundance during the middle Paleozoic and Cretaceous (Given and Wilkinson, 1987). In a later section this conclusion and its validity will be discussed in more detail. Throughout most of the Phanerozoic until Jurassic time, littoral and neritic skeletal limestones and their dolomitized equivalents prevail. The appearance of abundant carbonate-secreting planktonic organisms in the Jurassic has led to an increasing importance of the deep-sea relative to shallow-water areas as a site of carbonate deposition with decreasing age from the Cretaceous until today.

Other highlights of the lithology-age distribution of Figure 10.4 are:

1)   A marked increase in the abundance of submarine volcanogenic rocks and immature sandstones (graywackes) with increasing age.

2)   A significant percentage of arkoses in the early and middle Proterozoic, and an increase in the importance of mature sandstones (quartz-rich) with decreasing age. Red beds are significant rock types of Proterozoic and younger age deposits.

3)   A significant "bulge" in the relative abundance of jaspilites, chert-iron associations, in the early Proterozoic.

4)   Lack of evaporitic sulfate (gypsum, anhydrite) and salt (halite, sylvite) deposits in sedimentary rocks older than 800 million years.

These trends in lithologic features of the sedimentary rock mass can be a consequence of both evolution of the surface environment of the planet and recycling and post-depositional processes. It has been argued (e.g., Mackenzie, 1975; Veizer, 1988) that both secular and cyclic evolutionary processes have played a role in generating the lithology-age distribution we see today. For the past 1.5-2.0 billion years, the Earth has been in a near present-day steady state, and the temporal distribution of rock types since then has been controlled primarily by recycling. The cycling rates appear to be related to plate tectonic processes. More will be said about this hypothesis later.

Chemistry and Mineralogy

As might be anticipated, the trends observed for lithologic types versus age are reflected in trends in the chemistry and mineralogy of carbonates and other sedimentary rocks. Discussion of some of these latter trends, emphasizing first the overall trends for the past 3.8 billion years, then those of the Phanerozoic, are presented below.

Long-term trends Perhaps the most remarkable set of long-term secular trends in the chemical composition of sedimentary rocks has been compiled by Veizer (Veizer and Jansen, 1979; Veizer 1988), and is shown in Figure 10.6. This figure is somewhat generalized and the sharp inflections at about 2 billion years are not well-defined, although Veizer (1988) argues that the apparent changes at around this time in sedimentary rock properties may have occurred over a time interval of less than 600 million years. It should be kept in mind that this is the duration of the whole Phanerozoic Eon!

All these chemical trends, however, are consistent with the concept that the source areas of most Archean sedimentary rocks were more mafic in composition than the source areas of late Archean and younger rocks. The abundance of Archean submarine volcanogenic rocks and immature sandstones of mafic composition shown in the Ronov diagram of Figure 10.4 supports this interpretation. The age trends in chemical composition show a relatively rapid transition around the Archean-Proterozoic boundary. The difference in the composition between Archean and later sediments is also seen in their minor and trace element contents. Figure 10.7 shows histograms of manganese and iron distributions in Archean and Phanerozoic carbonates. Both elements are enriched in Archean limestones and dolostones, reflecting the possibility of a more mafic source at this time, and very likely a lower oxidation state of the Archean ocean (Holland, 1984; Veizer, 1988). Veizer and Jansen (1979) interpret this late Archean-Proterozoic transition as reflecting a primary feature of the sedimentary lithosphere -- an initial quasi-steady state sedimentary system involving recycling of mafic-rich materials followed by a more felsic-rich system recycling at a different rate.

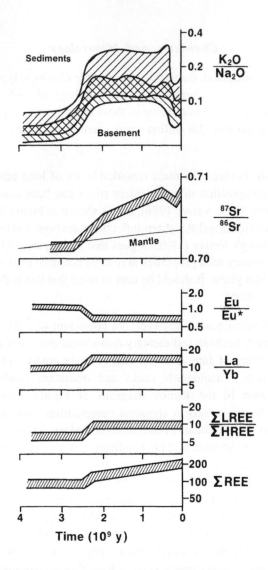

Figure 10.6. Observed secular trends in the chemical composition of sedimentary
rocks. Eu/Eu* = ratio of observed normalized Eu concentration to that predicted
(Eu*) from extrapolation between REE Gd and Sm. REE = rare earth element;
LREE = light REE; HREE = heavy REE. Data from several authors as presented
in Veizer (1988).

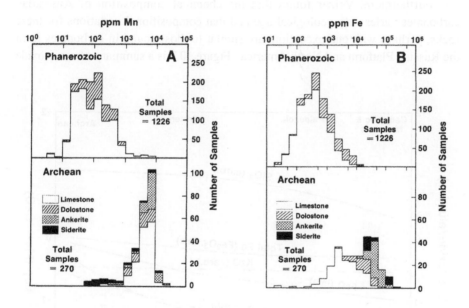

Figure 10.7. Histograms illustrating the differences in the Mn (A) and Fe (B) distributions in Archean and Phanerozoic carbonates. Data from several authors as presented in Veizer (1985).

A number of authors have studied secular variations in the chemistry and mineralogy of sedimentary carbonate rocks (Daly, 1909; Chilingar, 1956; Vinogradov and Ronov, 1956; Ronov and Migdisov, 1970, 1971; Veizer and Garrett, 1978; Veizer, 1978; Schmoker et al., 1985; Given and Wilkinson, 1987). Most of these investigations focused on the Ca/Mg ratio or dolomite/calcite ratio of these rocks. Veizer, however, investigated secular variations in the major and minor elements of carbonates, primarily from Australia, covering a 2.8 billion year time interval from Archean to Recent. Chemical and mineralogical analyses of these rocks revealed that (1) K, Rb, and Na are found principally in illite and authigenic potash feldspars with some Na bound in carbonate phases, (2) Ca, Mg,

Fe, and Mn are present mainly in the lattice positions of calcite and dolomite, (3) Si is found in limestones as quartz and in dolostones as chert, and (4) Ti and P are predominantly associated with the coarser fractions of the insoluble residue; TiO_2 possibly as rutile, ilmenite, brookite, or anatase and phosphorus probably as apatite.

Furthermore, Veizer found that the chemical composition of Australian carbonates varies with geological age and that compositional variations for these rocks, perhaps with one exception, are similar to those found in carbonates from the Russian Platform and North America. Figure 10.8 is a summary of some oxide

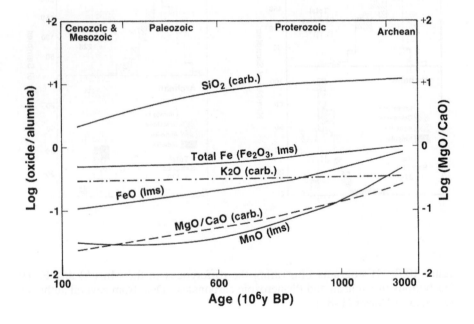

Figure 10.8. Trends in the chemical composition of Australian carbonate rocks plotted as a function of oxide ratio versus geologic age. lms. = limestone, carb. = limestone + dolostone. (Data from Veizer, 1978; Veizer and Garrett, 1978.)

ratio versus geologic age trends for Australian sedimentary carbonates. The K_2O/Al_2O_3, K/Rb, total Fe/Al_2O_3, Fe^{2+}/Al_2O_3, MnO/Al_2O_3, MnO/total Fe, MgO/CaO, and SiO_2/Al_2O_3 ratios of these rocks decrease from the Precambrian to the Cenozoic. In carbonates from North America and the Russian Platform, the K_2O/Al_2O_3 and K/Rb ratios diverge in the Precambrian from the Australian trend, decreasing with increasing geologic age. Veizer (1978) concluded that "primary" phenomena are responsible for these observed secular compositional trends in sedimentary carbonates, although secondary processes, diagenesis and recycling, may play some role. In particular, he argued that the trends are compatible with lower P_{O_2} and higher P_{CO_2} in the atmosphere-hydrosphere system, higher rates of volcanic degassing, or a greater degree of submarine leaching of volcanogenic sediments during the early stages in Earth's surface evolution. The "primary" versus "secondary" arguments for these and other trends will be discussed later in more detail.

Long-term trends in the stable isotopic composition of sedimentary materials are shown in Figures 10.9-10.11. In general, oxygen isotope ratios in carbonates are thought to represent diagenesis or equilibration with formation waters (Veizer and Hoefs, 1976). However, consistent variation of $\delta^{18}O$ of carbonates and cherts with geologic age has been demonstrated (Knauth and Lowe, 1978; Perry et al., 1978; Popp et al., 1986; Veizer et al., 1986). These variations are still quite controversial and may reflect temperature changes in Earth's surface environment or changes in the $\delta^{18}O$ of seawater. If temperature were the principal cause of the trend exhibited by the upper line in Figure 10.9, the 3.3 billion year old Onverwacht Group cherts would have been deposited in an environment with a temperature of about 70°C!

The long-term secular trends in $\delta^{13}C$ of carbonates and organic matter and $\delta^{34}S$ of evaporites are well documented and their variations, particularly through Phanerozoic time, have been used to argue for global environmental changes about a quasi-steady state mean. Because modern ideas of the historical aspects of the geochemical cycle of carbon are so intimately tied to hypotheses concerning the meaning of secular trends in these isotopes, they are considered in more detail.

Figure 10.12 illustrates the carbon and sulfur isotopic compositions of a variety of materials. For both carbon and sulfur, there is an important fractionation that obtains when organic processes are involved. Organic material is depleted in ^{13}C, and sulfide produced from bacterial reduction of sulfate is depleted in ^{34}S. In the exogenic carbon cycle, there are two principal reservoirs of carbon: the oxidized inorganic carbon reservoir, which is mostly carbonate

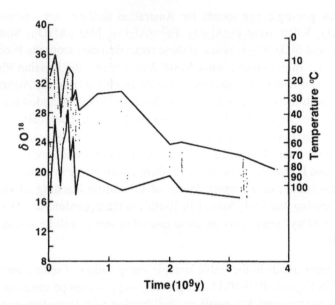

Figure 10.9. Range of oxygen isotope variations of cherts with time. The dots are individual analyses and the upper and lower curves connect, respectively, the highest and lowest $\delta^{18}O$ values for cherts at a given time. The $\delta^{18}O$ values in cherts can represent the $^{18}O/^{16}O$ ratio and temperature of the waters in which they formed and diagenetic effects. (Modified from Knauth and Lowe, 1978 and Holland, 1984.)

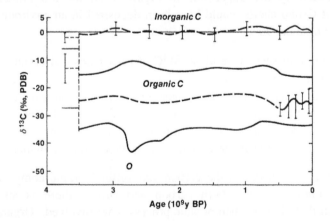

Figure 10.10. Trends in the carbon isotopic composition of carbonate carbon and organic carbon through geologic time. The isotopic fractionation between the two forms of carbon over geologic time averages about 26‰. (Modified from Schidlowski, 1987, 1988; Holser et al., 1988.)

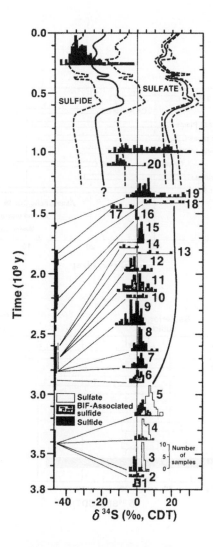

Figure 10.11. Isotopic composition of sedimentary sulfide and sulfate through geologic time in terms of histograms of δ34S values for units of various ages. The Phanerozoic portion of the sulfate record is the best documented. Prior to 1.2 billion years ago, the δ34S values of sulfate are restricted primarily to sedimentary barite (BaSO4). The later Proterozoic and Phanerozoic record of δ34S of sulfide is a construct based on a 40‰ difference between sedimentary sulfide and sulfate. Numbers refer to Precambrian sedimentary units. These are described in Holser et al. (1988). (After Holser et al., 1988.)

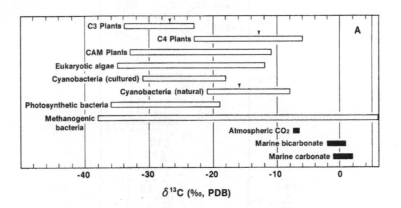

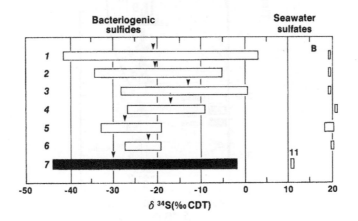

Figure 10.12. A. Carbon isotopic composition of major groups of higher plants and autotrophic microorganisms compared with oxidized carbon (CO_2, HCO_3^-, CO_3^{2-}). The triangles are mean values. (After Holser et al., 1988; Schidlowski, 1988). B. Sulfur isotopic composition of bacteriogenic sulfide in modern marine anaerobic sediments (open bars, 1-6) and in the Permian Kupferschiefer (black bar, 7) compared with oxidized sulfate of modern (1-6) and Permian seawater (+11‰). The black triangles are mean values. (After Holser et al., 1988.)

sediments containing 52 x 10^{20} moles of carbon (a trivial amount of oxidized carbon is stored in seawater, 0.033 x 10^{20} moles, and the atmosphere, 0.00061 x 10^{20} moles), and the reduced organic matter reservoir containing 13 x 10^{20} moles of carbon. The reduced carbon reservoir preferentially sequesters ^{12}C owing to biological fractionation processes. Thus, as Broecker (1970) so succinctly pointed out, if total exogenic carbon has remained constant throughout geological time, the $\delta^{13}C$ values of marine carbonates are functions of the ratio of organic carbon to carbonate carbon in the sedimentary carbon reservoirs. If the size of one reservoir grows, the other must shrink, and the mean $^{13}C/^{12}C$ ratios of each reservoir must change.

For the exogenic sulfur cycle similar relationships exist. Secular variations in the sulfur isotopic composition of evaporitic sulfate minerals are thought to reflect changes in the sulfur isotopic composition of seawater. Changes in the $\delta^{34}S$ of seawater, in turn, reflect variations in the masses of the two principal sedimentary reservoirs of sulfur: the oxidized inorganic sulfate reservoir and the reduced sedimentary sulfide reservoir. The reduced sulfur reservoir, because of bacterially-mediated sulfate reduction reactions in sediments, preferentially accumulates the lighter (^{32}S) isotope of sulfur. The oxidized sulfate reservoir contains about 2.4 x 10^{20} moles of sulfur, mainly in $CaSO_4$ minerals (about 17% of the total oxidized sulfur is in the oceans), whereas the reduced sulfur reservoir contains 2.90 x 10^{20} moles of sulfur, principally in pyrite (FeS_2).

Thus, in an analogous fashion to carbon, if total exogenic sulfur has remained constant, then the $\delta^{34}S$ values of sulfate in seawater, and consequently in evaporite minerals, are functions of the ratio of oxidized sulfur to reduced sulfur in the sedimentary reservoirs of this element. If the size of the sulfate reservoir increases, then the sulfide reservoir shrinks, and the mean $^{34}S/^{32}S$ ratios of each reservoir changes.

Referring back to Figure 10.10, we see that the $\delta^{13}C$ of carbonates averages around zero (-0.4‰) with fluctuations of ±1.5‰ for *all* of geological time. The bulk of organic carbon values fall in the range $\delta^{13}C$ = -27 ±7‰ (Schidlowski, 1982). For the oldest part of the sedimentary record, represented by the 3.8 billion year old Isua Supracrustal Belt series of West Greenland, the $\delta^{13}C$ values are probably the result of amphibolite-grade metamorphism of these sediments (Schidlowski, 1982, 1988). The constancy of $\delta^{13}C$ values of marine carbonates indicates that seawater dissolved inorganic carbon isotopic composition has changed little through geologic time. Marine carbonate carbon isotopic values reflect very closely seawater dissolved inorganic carbon values. Also, the spread in carbon isotopic values of organic carbon is similar to that of recent organic matter, indicating that autotrophic pathways of biological carbon fixation have occurred for all time represented by the sedimentary rock record (Holser et al., 1988; Schidlowski, 1988).

The extreme uniformity of the $\delta^{13}C$ record is remarkable. It argues for an early beginning of the modern carbon cycle and relative stability of the cycle throughout geologic time. For a two-reservoir system of organic carbon and inorganic carbon, and a $\delta^{13}C$ of primordial mantle carbon of -5‰, the isotopic mass balance is:

$$\delta^{13}C_{primitive\ carbon} = \frac{F_0}{F_0 + F_c}(\delta^{13}C_{organic}) + (1 - \frac{F_0}{F_0 + F_c})\delta^{13}C_{carbonate}$$

$$-5 = \frac{F_0}{F_0 + F_c}(-27) + (-0.4 + 0.4\frac{F_0}{F_0 + F_c})$$

$$\frac{F_0}{F_0 + F_c} \approx 0.2,$$

where F_0 and F_c represent, respectively, organic carbon and inorganic carbonate fluxes to the sedimentary reservoirs of these elements. Hence the ratio of fluxes of organic carbon to inorganic carbon has been approximately 1:4, as has the ratio of the reservoir masses of these carbon species, for at least 3.5 billion years of Earth history (Holser et al., 1988). The small variations in the $\delta^{13}C$ of carbonates will be discussed in more detail later when considering the extensive record of Phanerozoic $\delta^{13}C$ carbonate values.

In a general sense, for much of late Precambrian and Phanerozoic time the distribution of $\delta^{34}S$ in evaporitic sulfate minerals is an approximate mirror image of the $\delta^{13}C$ distribution of carbonates. The sulfur isotopic record for much of the Precambrian, however, is less complete than that for the $\delta^{13}C$ of carbonates, because major beds of evaporitic sulfate are absent from rocks older than 1.4 billion years. Thus, the $\delta^{34}S$ record in sedimentary materials before this time comes from sulfide minerals and bedded barites which, although still controversial, are thought to be sedimentary chemical precipitates or early diagenetic products (Heinrichs and Reimer, 1977; Dunlop et al., 1978). Two major questions involved with the sulfur cycle and the $\delta^{34}S$ distribution of Figure 10.11 are the times of emergence of abundant oceanic sulfate and of dissimilatory sulfate reduction involving heterotrophic bacteria like *Desulfulvibrio* sp. and production of ^{34}S-depleted sedimentary pyrite. Schidlowski (in Holser et al., 1988) concludes that the oldest δ^{34} patterns of Figure 10.11 that might be interpreted as biogenic and represent bacterial sulfate reduction according to

$$2CH_2O + SO_4^{2-} \rightarrow H_2S + 2HCO_3^-$$ (10.2)

are those from approximately 2.7 billion year old rocks (numbers 9, 11, and 12 of Figure 10.11). Prior to this time, Archean $\delta^{34}S$ patterns show small ranges in values, indicating small fractionations, and are skewed toward positive values, patterns not indicative of bacterial sulfate reduction. Thus, bacterial sulfate reducers gained control of the sulfur cycle early in Earth history and led to partitioning of primordial sulfur with a value of $\delta^{34}S \approx 0$ (Nielsen, 1978) into a bacterially-processed light sulfide reservoir with a $\delta^{34}S \approx -30$ to -13‰ and a heavy residual reservoir of sulfate with a $\delta^{34}S \approx 11$-20‰. The average fractionation is about 30-40‰ between the two sulfur species (Holser et al., 1988). As with carbon, the average ratio of the mass of the reduced sulfide reservoir to that of the oxidized sulfate reservoir can be calculated and is found to be close to 1.2, with slightly more sulfur bound up as pyrite in the reduced sulfur reservoir.

It is difficult to say from the $\delta^{34}S$ record of sedimentary sulfur-bearing minerals when the oceans accumulated significant oxidized sulfur. It is likely that the activity of photosynthetic green and purple sulfur-oxidizing bacteria produced the first free sulfate in the environment by reactions like

$$H_2S + 2CO_2 + 2H_2O \xrightarrow{h\nu} 2CH_2O + 2H^+ + SO_4^{2-}$$ (10.3)

Thus, the appearance of free sulfate does not require the advent of free oxygen in the Archean environment. Certainly sufficient free sulfate had appeared in the hydrosphere prior to development of the pathway of dissimilatory sulfate reduction. Schidlowski (1979) argues that the small fractionations observed between sulfide and sulfate $\delta^{34}S$ values of pre-2.7 billion year rocks (Figure 10.11) are consistent with the hypothesis that the oxidation of sulfide to sulfate by photosynthetic bacteria preceded the bacterial pathway of dissimilatory sulfate reduction and may have been responsible for early free dissolved sulfate concentrations in the hydrosphere.

Now that we have some feel for the longer term changes in properties of
sedimentary rocks with geologic age, we can look more carefully at the
Phanerozoic record for which there are more data.

Phanerozoic trends   In the 1970's, Vail and colleagues (e.g., Vail et al., 1977)
using seismic stratigraphic data to determine the extent of onlap of coastal deposits
in marine sequences derived a relative sea level curve for the Phanerozoic. The
latest representation of this curve is shown in Figure 10.13, and compared with that

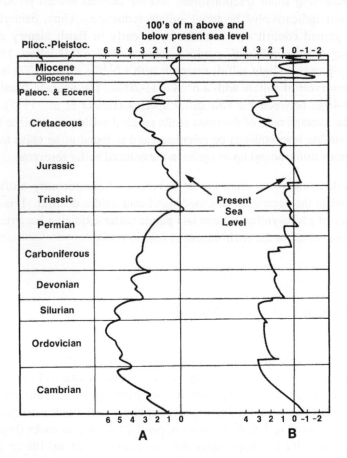

Figure 10.13. Two interpretations of the Phanerozoic sea level curve. A. Hallam,
1984. B. Vail et al., 1977.

of Hallam (1984) based on paleogeographic reconstructions and a hypsometric curve. The sharp regressions in the Vail et al. curve are not real, and the Hallam curve better represents these intervals. Of particular interest is the apparent long-term cyclic pattern of sea level change, superimposed on which are higher frequency changes. The relative changes in sea level have been calibrated by use of an absolute datum (Sleep, 1976; Vail et al., 1977; Turcotte and Burke, 1978) to show the elevation of sea level relative to present sea level. There is still considerable debate concerning the exact elevation of paleo-sea level through Phanerozoic time. For example, the Hallam curve has the Phanerozoic maximum sea level in the late Ordovician, whereas the Vail et al. curve shows a late Cretaceous maximum. Nevertheless the general trends of both curves are reasonably consistent.

The Phanerozoic began at a time of relatively low sea level. Sea level rose to a maximum in late Cambrian-early Ordovician time only to return to a level like that of today in the late Carboniferous. This period of relatively low sea level lasted about 100 million years and was followed by a period of rising sea level that culminated in the major Cretaceous transgression, and then by a general lowering of sea level to that of today. These long-term changes in sea level are certainly caused by global plate tectonic events (e.g., Worsley et al., 1984). There are apparent correlations between the generalized sea level curve and properties of the Phanerozoic sedimentary rock mass. Interpretation of these correlations, and in some cases their actual statistical validity, are controversial. Table 10.2 gives some of these correlations and their possible significance. These relationships and others are discussed below.

In the exogenic cycle, ^{87}Sr and ^{86}Sr are stable isotopes, although initial variations in the ratio of these isotopes stem from the radioactivity of rubidium in igneous rocks. Because continental crust has a high Rb/Sr ratio, it generates more radiogenic strontium than oceanic crust. As shown in Figure 10.14, a schematic diagram of the behavior of strontium in the exogenic cycle, the $^{87}Sr/^{86}Sr$ ratio of seawater reflects the relative proportions of fluids derived from these two rock sources. Sedimentary precipitates from seawater, in turn, record the $^{87}Sr/^{86}Sr$ ratio of the seawater from which they formed.

Figure 10.15 illustrates variations in the $^{87}Sr/^{86}Sr$ ratio of marine precipitates in detail for the Phanerozoic. The generalized Precambrian data are shown in the insert. The data for the Precambrian are still quite scant and fluctuations like those observed for the Phanerozoic may have occurred during the former Eon. Causes of the $^{87}Sr/^{86}Sr$ variations are as yet uncertain. Veizer (1989) has succinctly

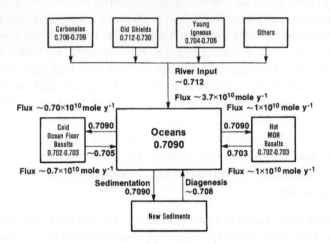

Figure 10.14. The geochemical cycle of strontium. The values within the boxes are the mean $^{87}Sr/^{86}Sr$ ratios of the contained strontium. Estimates of total Sr flux for various flows are also shown. (After Holland, 1984.)

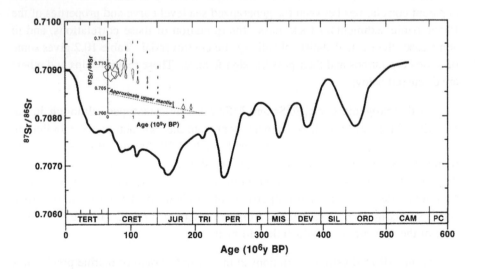

Figure 10.15. Generalized trends in the $^{87}Sr/^{86}Sr$ ratio of sedimentary precipitates. The Phanerozoic trend is from Burke et al. (1982), and that for the past 3.3 billion years (insert) from Veizer (1976).

summarized the causal factors influencing strontium isotopes in seawater. He emphasizes that $^{87}Sr/^{86}Sr$ fluctuations in marine precipitates need to be interpreted in light of the time scale of stratigraphic resolution. A sharp, short-term fluctuation may have a different cause than a monotonic or irregular, long-term change.

Holland (1984) has argued that the present oceans are roughly at steady state with respect to strontium. This steady state condition is shown in Figure 10.14. Sources of strontium for the ocean are river water and fluids resulting from low and high temperature basalt-seawater reactions. It is likely that the strontium isotopic composition of seawater, and hence carbonate precipitates, is controlled to a significant extent by the magnitude of the river plus groundwater input of Sr to the ocean, and their mean $^{87}Sr/^{86}Sr$ ratio, relative to the exchange fluxes involving the reaction of seawater with oceanic basalts. Other factors, however, can play a role, and interpretation of the $^{87}Sr/^{86}Sr$ curve for marine precipitates awaits further research.

The generalized secular variation of the stable isotopes of carbon and sulfur in Phanerozoic rocks is now well established (Claypool et al., 1980; Veizer et al., 1980; Holser et al., 1988). Data are now being obtained that represent a higher resolution of the stratigraphic column (e.g., Holser, 1984). These secular variations for $\delta^{13}C$ in carbonates and $\delta^{34}S$ in evaporitic sulfates (Figure 10.16) are correlated with the generalized sea level curve of Figure 10.13 (Table 10.2), and the secular variations of the two isotopes are roughly a mirror image of each other.

Table 10.2. Factor analyses of the relations among the age curves of strontium isotopes in carbonate, carbon isotopes in carbonate, sulfur isotopes in sulfate, and sea level for the Phanerozoic (from Holser et al., 1988).

	Factor 1	Factor 2	Factor 3	Communality
$^{87}Sr/^{86}Sr$(carb)	0.47	0.57*	0.32	0.65
$\delta^{13}C_{carb}$	-0.83*	-0.54*	0.06	0.98
$\delta^{34}S_{sulfate}$	0.92*	0.15	0.17	0.90
Sea-level stands	0.16	0.76*	0.03	0.60
Percentage of variation	81.9	15.3	2.9	
Factor interpretation	Organic redox coupling	Endogenic tectonic phenomena	Nonrepresentative data	

*significant at the 0.95 level

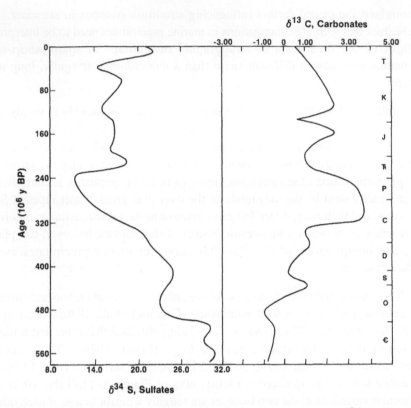

Figure 10.16. Generalized antithetical relationship between $\delta^{34}S$ of marine evaporite sulfate (after Saltzman et al., 1982, and Lindh, 1983) and $\delta^{13}C$ of marine whole rock or fossil calcite (after Lindh et al., 1981). Curves redrawn as in Holser et al. (1988).

The generalized pattern of secular variations in $\delta^{13}C$ shows an enrichment of ^{13}C in late Proterozoic and Permian carbonates, and a depletion in early Paleozoic rocks relative to the intermediate $\delta^{13}C$ values of the Mesozoic and Cenozoic (Veizer et al., 1980). As mentioned previously, variations of this nature can be interpreted as an indication that the $\delta^{13}C$ of dissolved inorganic carbon in seawater has varied during geologic time (Broecker, 1970). These variations in the isotopic composition of seawater bicarbonate can be attributed to variations in the ratio of the depositional fluxes of organic carbon to carbonate carbon to the sea floor, and thus, eventually to variations in the relative sizes of the sedimentary reservoirs of carbon (Garrels and Perry, 1974; Fischer and Arthur, 1977; Veizer et al., 1980).

The generalized pattern of secular variations in $\delta^{34}S$ is antithetical to that for $\delta^{13}C$: there is a depletion of ^{34}S in late Proterozoic and Permian sulfates and an enrichment of ^{34}S in early Paleozoic rocks relative to the intermediate $\delta^{34}S$ values of the Tertiary. These variations in $\delta^{34}S$, reflecting variations in the isotopic composition of seawater sulfate, may be attributed to a combination of variations in the ratio of the depositional fluxes of sulfate sulfur to sulfide sulfur to the sea floor, and variations in the flux and isotopic composition of continentally-derived sulfur reaching the oceans. Ultimately, the relative sizes of the sedimentary reservoirs of sulfur must change in response to the above processes. The factors affecting the $\delta^{34}S$ sulfate values are still somewhat a matter of controversy; the various arguments are presented in Nielsen (1965), Holser and Kaplan (1966), Rees (1970), Holland (1973), Garrels and Perry (1974), Claypool et al. (1980), Veizer et al. (1980), Holser (1984), and Holser et al. (1988).

There are several other properties of Phanerozoic sedimentary rocks that appear to bear some relationship to the generalized sea level curve of Figure 10.13, and some that do not. Although still somewhat controversial (e.g., Bates and Brand, 1990), the textures of oöids appear to vary during Phanerozoic time. Sorby (1879) first pointed out the petrographic differences between ancient and modern oöids: ancient oöids commonly exhibit relict textures of a calcitic origin, whereas modern oöids are dominantly made of aragonite (see Chapter 5). Sandberg (1975) reinforced these observations by study of the textures of some Phanerozoic oöids and a survey of the literature on oöid textures. His approach, and that of others who followed, was to employ the petrographic criteria, among others, of Sorby: i.e., if the microtexture of the oöid is preserved, then the oöid originally had a calcite mineralogy; if the oöid exhibits textural disruption, its original mineralogy was aragonite. The textures of originally aragonitic fossils are usually used as checks to deduce the original mineralogy of the oöids. Sandberg (1975) observed that oöids of inferred calcitic composition are dominant in rocks older than Jurassic.

Following this classical work, Sandberg (1983, 1985) and several other investigators (Mackenzie and Pigott, 1981; Wilkinson et al., 1985; Bates and Brand, 1990) attempted to quantify further this relationship. Figure 10.17 is a schematic diagram representing a synthesis of the inferred mineralogy of oöids during the Phanerozoic. This diagram is highly tentative, and more data are needed to document the trends. It appears, however, that although originally aragonitic oöids are found throughout the Phanerozoic, an oscillatory trend in the relative percentage of calcite versus aragonite oöid mineralogy may be superimposed

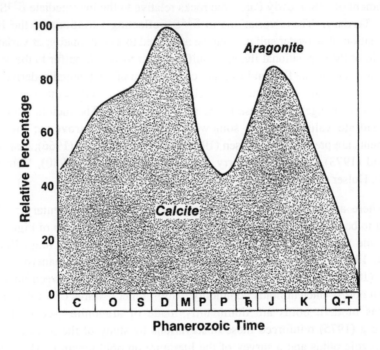

Figure 10.17. Trend in relative percentage of originally calcite and aragonite oöid compositions during the Phanerozoic. (After Mackenzie and Agegian, 1989.)

on a long-term evolutionary increase in oöids with an inferred original aragonitic mineralogy. Although the correlation is not strong, the two major maxima in the sea level curve appear to correlate with times when calcite oöids were important seawater precipitates. Wilkinson et al. (1985) found that the best correlation between various data sets representing global eustasy and oöid mineralogy is that of inferred mineralogy with percentage of continental freeboard. Sandberg (1983) further concluded that the cyclic trend in oöid mineralogy correlates with cyclic trends observed for the inferred mineralogy of occurrences of carbonate cements. Van Houten and Bhattacharyya (1982; and later Wilkinson et al., 1985) showed that the distribution of Phanerozoic ironstones (hematite and chamosite oölitic deposits) also exhibits a definite cyclicity in abundance during Phanerozoic time (Figure 10.18). This cyclicity, as well as that for the occurrences of oölites (Figure 10.18),

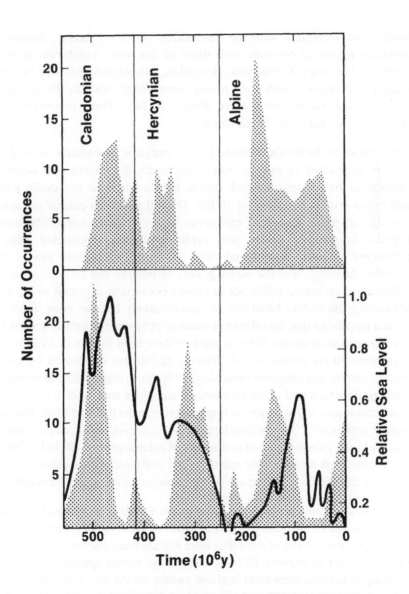

Figure 10.18. Number of occurrences of Phanerozoic ironstones (upper diagram, data from Van Houten and Bhattacharyya, 1982) and oölitic limestones (lower diagram, data from Wilkinson et al., 1985) as a function of geologic age. The relative sea level curve is that of Hallam (1984). Minima in occurrences appear to correlate with times of sea level withdrawal from the continents and major cycles of orogenesis (Caledonian, Hercynian, and Alpine).

appears to be correlated with the generalized sea level curve. Minima in occurrences appear to correlate with times of sea level withdrawal from the continents. The major Caledonian, Hercynian, and Alpine cycles of mountain building and orogenesis also seem to bear some relationship to the occurrence distributions and the sea level curve (Figure 10.18). These relationships are discussed in more detail later in this chapter.

In contrast to the trends observed for the inorganic precipitates of oöids and ironstones, mineralogy vs. geologic age trends for fossilized calcareous organisms are present in the Phanerozoic rock record, but do not show any cyclic pattern related to sea level change (Figure 10.19). Overall there is a general increase in the number of major groups of calcareous organisms, i.e., coccolithophorids, pteropods, hermatypic corals, and coralline algae. Organisms with all mineralogies are well represented over the past 100 million years. It is noteworthy, however, that the major groups of pelagic and benthic organisms contributing to carbonate sediments in today's ocean first appeared in the fossil record during the middle Mesozoic and progressively became more abundant. There is a possibility that the relative abundance of specific organism groups may have followed cyclic trends. What is most evident from Figure 10.19 is a long-term increase in the production of biogenic carbonates dominated by calcite, magnesian calcite, and aragonite mineralogies. Because organo-detrital carbonates are such an important part of the Phanerozoic carbonate rock record, this increase in metastable mineralogies played an important role in the pathway of diagenesis of carbonate sediments. The preponderance of low magnesian calcite skeletal organisms in the Paleozoic led to production of calcitic organo-detrital sediments whose original bulk chemical and mineralogical composition was closer to that of their altered and lithified counterparts of Mesozoic, and particularly Cenozoic, age.

More data are now probably available on the variation of sediment properties with geologic age for the past 100 million years of Earth history than for most of prior geologic time. Some of these data have already been discussed. A few more trends are shown in Figures 10.20-10.23. These trends appear to bear some relationship to the long-term trend in global eustasy for the past 100 million years. Based on paleobotanical data and inferences about the relationships of the $\delta^{18}O$ content of benthic and planktonic foraminifera and paleotemperature, the Tertiary Period can be interpreted as a time of a general unsteady decrease in temperature. This decrease culminated in the fluctuating temperatures of the Pleistocene ice age. The overall pattern of temperature decline in the Tertiary tracks the general withdrawal of the sea from the continents during this Period. Some of the short-term variations in Tertiary temperatures as inferred from the $\delta^{18}O$ chronology

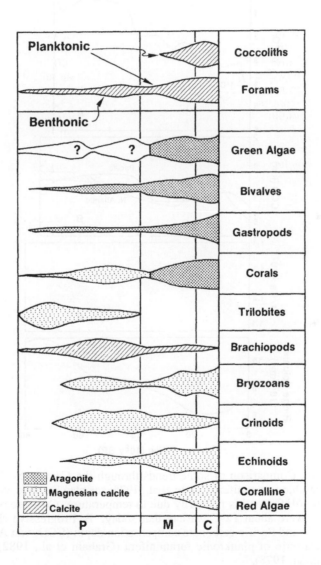

Figure 10.19. Mineralogical evolution of benthic and planktonic organisms during the Phanerozoic based on summaries of Milliken and Pigott (1977) and Wilkinson (1979).

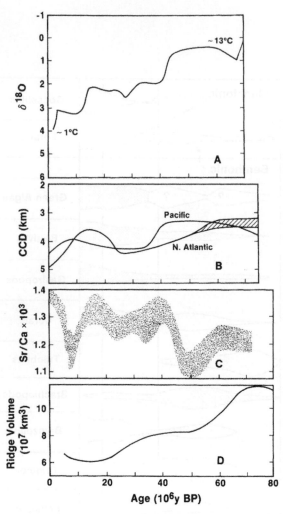

Figure 10.20. Comparison of some trends through the Cenozoic. A. The $\delta^{18}O$
content of benthic foraminifera (Savin et al., 1975; see also Prentice and Matthews,
1988). If the $\delta^{18}O$ trend is primarily due to temperature, Cretaceous deep water
temperatures were about 12°C warmer than today. B. Progressive change of the
North Atlantic and Pacific carbonate compensation depth (CCD; van Andel, 1975).
C. The Sr/Ca ratio of planktonic foraminifera (Graham et al., 1982). D. Ridge
volume (Pitman, 1978).

may also correlate with short-term features of the global eustatic curve. For example, a number of the major sharp withdrawals of the sea from the continents in Tertiary time at about 10, 30, 40, and 60 million years ago coincide very closely with cooling of oceanic bottom waters.

These changes of temperature and sea level for the past 100 million years of Earth history are accompanied by trends in other sediment properties. The Sr/Ca ratio of sedimentary materials and the Sr content of pelagic carbonates decrease irregularly back into the Cretaceous. The Na/Ca ratio of foraminifera declines precipitously with increasing age, and the $^{143}Nd/^{144}Nd$ ratio of sedimentary materials exhibits an irregular increase back to about 50 million years ago, then declines precipitously. Caution must be taken in interpretation of the Sr/Ca and Na/Ca trends. Both of these overall decreases with geologic age may represent progressive diagenesis of carbonate materials and loss of Sr and Na from the solid phase. Marine carbonates as they undergo progressive burial in the deep sea have been shown to lose Sr with increasing burial depth and age (Manghanani et al., 1980). During this recrystallization process, Na may also be lost, as has been observed in carbonates undergoing progressive diagenesis in meteoric water (Chapter 7). The neodynium trend is particularly intriguing because this trend may reflect the relative importance through time of seawater cycling through mid-ocean ridges.

In contrast to the Sr content of pelagic carbonates, their Mg content appears to increase moderately and irregularly with increasing age. This trend would not seem to be related to a diagenetic process, because it would be anticipated that pelagic carbonates would progressively lose Mg, if the trend were due to recrystallization of the pelagic carbonate to a more stable form. The trend, however, may reflect changes in the magnesium content or temperature of the ocean.

It is possible that the dolomite content of deep-sea sediments bears some correlation to the sea level curve (Figure 10.23) and to the apparent change in the Mg/Ca of pelagic carbonates. It can be seen from comparison of Figures 10.23 and 10.21 that the general decrease in the abundance of dolomite in pelagic sediments from early Cretaceous to Paleogene time corresponds to the general decrease in the Mg content of pelagic carbonates seen over this time interval. Both the peak in abundance of dolomite in deep-sea sediments and the moderate decline in the magnesium content of pelagic carbonates appear to precede the maximum transgression of the Cretaceous sea.

The final trends discussed in this section are the Phanerozoic volume distributions of NaCl and CaSO$_4$ in evaporite rocks and the Mg/Ca ratio of carbonates. Marine evaporites owe their existence to an unlikely combination of tectonic, paleogeographic, and sea level conditions. Seawater bodies must be restricted to some degree, but also must exchange with the open ocean to permit large volumes of seawater to enter these restricted basins and evaporate. Environmental settings of evaporite deposition may occur on cratons or in rifted basins. Figure 10.24 illustrates that because evaporite deposition requires an unusual combination of circumstances, a "geological accident" (Holser, 1984), the intensity of deposition has varied significantly during geologic time. This conclusion implies that the volume preserved of NaCl and CaSO$_4$ per unit of time in the Phanerozoic reasonably reflects the volume deposited, and there has been minor differential recycling of these components of the sedimentary rock mass relative to other components (Garrels and Mackenzie, 1971a). Because the oceans are important reservoirs for these components, these large variations in the rate of NaCl and CaSO$_4$ output from the ocean to evaporites may imply changes in the salinity of seawater.

For several decades it has been assumed that the Mg/Ca ratio of carbonate rocks increases with increasing Phanerozoic rock age. The magnesium to calcium weight ratios in carbonate rocks from the Russian Platform and North America are shown in Figure 10.25. The trend of the ratio represents a general, but erratic, decline in the calcium content and increase in the magnesium content of these rocks with increasing age (see Vinogradov and Ronov, 1956a,b; Chilingar, 1956). The magnesium content is relatively constant in these carbonates for about 100 million years, then increases gradually. The magnesium content of North American and Russian Platform continental carbonate rocks appears to increase at a geologic age that is very close to, if not the same as, the age of the beginning of the general increase in the Mg content of pelagic limestones (100 million years before present). Thus, the increase in magnesium content of carbonate rocks with increasing age appears to be a global phenomenon, and to a first approximation, not lithofacies related.

Voluminous research on the "dolomite problem" (Chapters 7 and 8) has shown that the reasons for the high magnesium content of carbonates are diverse and complex. Some dolomitic rocks are primary precipitates; others were deposited as CaCO$_3$ and then converted entirely or partially to dolomite before deposition of a succeeding layer; still others were dolomitized by migrating underground waters tens or hundreds of millions of years after deposition. However, the observation that the magnesium content of pelagic limestones that have not been uplifted, or

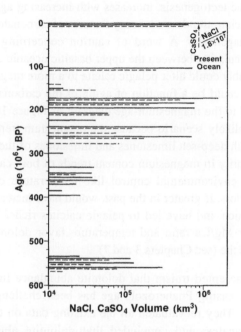

Figure 10.24. Phanerozoic distribution of NaCl (dark line) and CaSO₄ (dashed line) in marine evaporites. (After Holser, 1984.)

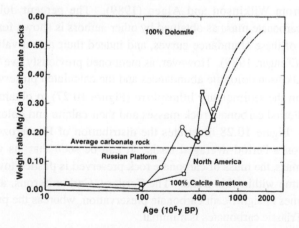

Figure 10.25. Magnesium to calcium weight ratios in Russian Platform and North American carbonate rocks as a function of age. (Data from Vinogradov and Ronov, 1956a,b, and Chilingar, 1956; figure modified from Garrels and Mackenzie, 1971.)

otherwise undergone tectogenesis, increases with increasing age suggests that the
magnesium and calcium trends are primary features of carbonate rocks and not due
to recycling or diagenesis. A word of caution concerning this statement is
needed. Waters circulating between the upper basaltic oceanic crust and overlying
sediments conceivably could alter pelagic calcite to a more magnesium-rich phase,
and this alteration could be a function of aging of the carbonate sediment. This
process might lead to the magnesium-age diagram of Figure 10.21; however, we
consider this an unlikely scenario because the magnesium trend for the past 150
million years in both deep-sea limestones and limestones of the continents is very
similar. The similarity in magnesium content trends of both carbonate lithofacies
suggests a global environmental control like temperature or Mg/Ca ratio of
seawater. Both factors, if greater in the past, would have increased the probability
of dolomite formation and have led to pelagic calcites richer in Mg than today.
Increased seawater Mg/Ca ratio and temperature favor dolomite formation and
magnesium-rich calcite (see Chapters 3 and 7).

Recently the accepted truism that dolomite abundance increases relative to
limestone with increasing Phanerozoic age has been challenged by Given and
Wilkinson (1987). They reevaluated all the existing data on the composition of
Phanerozoic carbonates, and concluded that dolomite abundances do vary
significantly throughout the Phanerozoic but may not increase systematically with
age. Figure 10.26 is a summary of the global data set for Phanerozoic dolostone
abundances from Wilkinson and Algeo (1989). The percent dolomite in the
Phanerozoic carbonate mass as obtained by other authors is shown for comparison.
The meaning of these abundance curves, and indeed their actual validity, are still
controversial (Zenger, 1989). However, as mentioned previously, we have used the
Given and Wilkinson dolomite abundances and the calculated preservation of CO_2
as carbonate in the sedimentary lithosphere (Figure 10.27) to obtain estimates of
Phanerozoic Period carbonate rock masses and their calcite and dolomite contents
(Table 10.1). Figure 10.28 illustrates the distribution of Phanerozoic carbonate
rock masses on a Period-averaged basis. It can be seen that, as with the total
sedimentary mass, the mass of carbonate rock preserved is pushed toward the front
of geologic time within this trend. The Tertiary, Carboniferous, and Cambrian
Periods are times of significant carbonate preservation, whereas the preservation of
Silurian and Triassic carbonates is minimal.

The Period-averaged mass ratio of calcite to dolomite (Figure 10.29) is
relatively high for Cambrian, Permian, and Tertiary System rocks, whereas this
ratio is low for Ordovician through Carboniferous age sediments and rises in value
from the Triassic through the Recent. The generalized sea level curve of Vail et al.

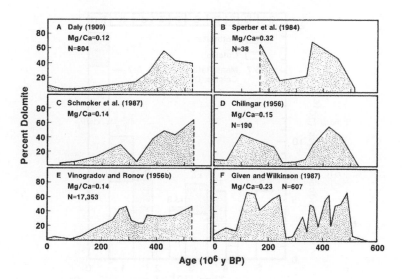

Figure 10.26. Estimates of percent dolomite in Phanerozoic cratonic carbonate rocks as a function of age. Mg/Ca = average ratio. (After Wilkinson and Algeo, 1989.)

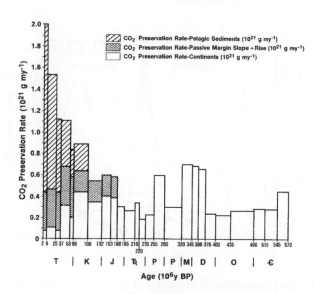

Figure 10.27. The preservation ("survival") rate of Phanerozoic sedimentary carbonate expressed as 10^{21} g CO_2 per million years as a function of age. The sedimentary carbonate mass has been divided into the pelagic, passive margin slope plus rise and continental realms. (Modified from Hay, 1985.)

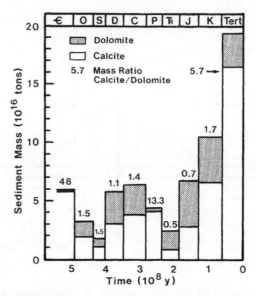

Figure 10.28. The Phanerozoic sedimentary carbonate mass distribution as a function of geologic age. Period masses of calcite and dolomite and the Period mass ratios of calcite/dolomite are also shown.

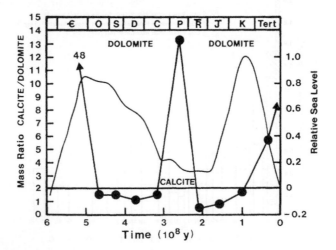

Figure 10.29. The Phanerozoic distribution of the mass ratio of calcite/dolomite in sedimentary carbonates as a function of age. The relative sea level curve is that of Vail et al. (1977). In general, times of low sea level seem to be times of high calcite/dolomite ratios in sedimentary carbonates.

(1977), and also that of Hallam (1984), appears to correlate crudely with the calcite/dolomite ratio through Phanerozoic time. The Phanerozoic starts out with sea level rising, and Cambrian carbonate strata are enriched in calcite. For much of the Paleozoic when sea level was high or declining from its Ordovician maximum, the calcite/dolomite ratio remains around 1-1.5, increasing sharply in the Permian to about 13. As global sea level rises towards the Cretaceous maximum sea level transgression, the calcite/dolomite ratio remains low, but trends towards the higher ratios of the Tertiary and Quaternary as sea level falls, and dolomite becomes less and less abundant in the sedimentary record. Mackenzie and Agegian (1986) and Given and Wilkinson (1987) were the first to suggest this possible cyclicity in the calcite/dolomite ratio during the Phanerozoic, and Lumsden (1985) observed a secular decrease in dolomite abundance in deep marine sediments from the Cretaceous to Recent, corresponding to the general fall of sea level during this time interval. These cycles in calcite/dolomite ratios correspond crudely to Fisher's (1984) two Phanerozoic super cycles and Mackenzie and Pigott's (1981) oscillatory and submergent modes.

In Figure 10.30 the survival rate of the total sedimentary mass for the different Phanerozoic systems is plotted and compared with survival rates for the total carbonate and dolomite mass distribution. The difference between the two latter survival rates for each system is the mass of limestone surviving per unit of time. Equation 10.1 is the log linear relationship for the total sedimentary mass, and implies a 130 million year half-life for the post-Devonian mass, and for a constant sediment mass with a constant probability of destruction, a mean sedimentation rate since post-Devonian time of about 100×10^{14} g y^{-1}. The modern global erosional flux is 200×10^{14} g y^{-1}, of which about 15% is particulate and dissolved carbonate. Although the data are less reliable for the survival rate of Phanerozoic carbonate sediments than for the total sedimentary mass, a best log linear fit to the post-Permian preserved mass of carbonate rocks is

$$\log S_{carb.} = 9.54 - 0.35t \qquad (10.4)$$

and corresponds to a half-life for the post-Permian carbonate mass of 86 million years, and a mean sedimentation rate of these sediments of about 35×10^{14} g carbonate per year; the present-day carbonate flux is 30×10^{14} g y^{-1}. The difference in half-lives between the total sedimentary mass, which is principally sandstone and shale, and the carbonate mass may reflect more rapid recycling of the

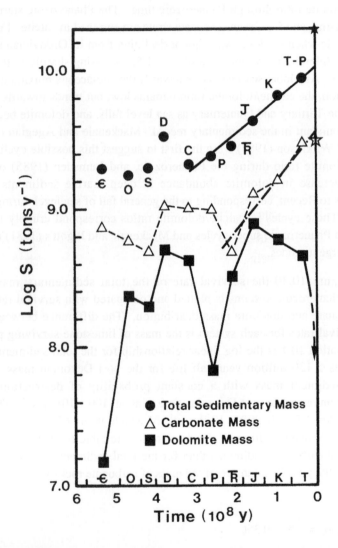

Figure 10.30. Phanerozoic sedimentary rock mass-age relationships expressed as the logarithm of the survival rate in tons y^{-1} versus time. The straight lines are best fits to the total mass data (solid line) and to the carbonate mass data (dash-dot line) for particular intervals of Phanerozoic time. The difference between the logarithm of S for the carbonate mass and that of the dolomite mass is the survival rate of the calcite mass. Filled star is present total riverine flux to the oceans, whereas open star is carbonate flux.

carbonate mass at a rate about 1.5 times the total mass. This is not an unlikely situation. With the advent of abundant carbonate-secreting organisms in the Jurassic, the site of carbonate deposition shifted significantly from shallow-water areas to the deep-sea. Thus, it is possible that at that time the probability of destruction of carbonate sediments by subduction and subsequent metamorphism increased, leading to more rapid recycling of the carbonate mass. Veizer (1988) has shown that the rate of recycling of oceanic crust exceeds that of the continental basement by a factor of 30. Also, Southam and Hay (1981) using a half-life of 100 million years for pelagic sediment, estimated that as much as 50% of all sedimentary rock formed by weathering of igneous rock may have been lost by subduction during the past 4.5 billion years.

In conclusion, it appears, as originally suggested by Garrels and Mackenzie (1971a), that the carbonate component of the sedimentary rock mass may have a cycling rate different than that of the total sedimentary mass. These authors argued that the differential recycling rates of the different components of the sedimentary lithosphere were related to their resistance to chemical weathering and transport. Evaporites are the most easily soluble; limestones are next; followed by dolostones; and shales and sandstones are the most inert. Although resistance to weathering may play some role in the selective destruction of sedimentary rocks, it is likely that differences in the recycling rates of different tectonic regimes in which sediments are deposited are more important. Table 10.3 is a summary of recycling rates and half-lives for major global tectonic realms. It is evident from the table that carbonates deposited on the oceanic crust may recycle at a rate about 6 times faster than those which are part of platform environments. Gradual, and on-going, transposition of significant carbonate deposition from shallow-water environments to the deep sea would increase the rate of destruction of the global carbonate mass relative to the total sedimentary mass from Jurassic time on, giving rise to the smaller half-life of post-Permian carbonates as compared to the total sedimentary mass. The carbonate mass distribution and calcite/dolomite ratios will be discussed further in the following section.

Carbon cycling modeling

Introduction and Development of Global Model

In the last two decades, a great deal of progress has been made in the modeling of the carbon cycle and of parameters related to the formation of carbonate rocks. These models attempt to show quantitatively the interrelated mass relationships

Table 10.3. Summary of recycling rates and half-lives for major global tectonic realms. (After Veizer, 1988.)

Tectonic Realms	Recycling Rate (b^{10}) in units of 10^{-10} y^{-1}		Theoretical Half-L (τ_{50}) x 10^6 y
	Steady-State Since 4500 x 10^6 y Ago	"Preferred" or Logistic Growth Model	
Active margin basins	223.0 ± 51.0	223.0 ± 51.0	27
Oceanic intraplate basins	126.0	126.0	51
Oceanic crust	110.0 ± 5.0	110.0 ± 5.0	59
Passive margin basins	88.0 ± 8.0	88.0 ± 5.0	75
Immature orogenic belts	85.0 ± 35.0	85.0 ± 35.0	78
Mature orogenic belts (roots)	19.3	18.4 ± 0.5	355
Platforms	19.0	18.6 ± 0.6	361
Continental basement:			
Probability of being affected by			
Low-grade metamorphism	10.2	8.5 ± 0.7	673
High-grade metamorphism	7.0	3.7 ± 1.3	987
Recycling into mantle	4.0	2.1 ± 1.0	1728

The "preferred" or logistic growth model assumes an initiation of generation and/or preservation of continental crust ~4000 x 10^6 years ago, fast rate of crustal growth 32 x 10^9 years ago and near steady state since 2000 x 10^6 years ago. Note that the τ_{50}'s for the steady state and logistic growth models are comparable because slower growth and recycling rates in the latter alternative yield internal age distribution patterns similar to those of the steady state.

among the reservoirs that affect the carbon cycle. The major reservoirs of carbon considered are the atmosphere, ocean, sediments, land, and terrestrial and marine biota. Commonly these reservoirs may be subdivided into smaller reservoirs; for example, upper ("mixed layer") and deep ocean subreservoirs or the troposphere and stratosphere subreservoirs of the atmosphere. The choice of reservoirs depends to a certain extent on the mass and the residence time of carbon in the reservoir. For example, if one is interested in the buildup of carbon dioxide in the atmosphere because of the burning of fossil fuels on a time scale of a few hundreds of years it is not necessary to consider the huge reservoir of carbon in sedimentary rocks. The residence time of carbon in this reservoir is on the order of 300 million years and the perturbations of carbon in the atmosphere induced by fossil fuel burning will be little affected by the mass of carbon in this reservoir. A hierarchical series of carbon cycle box models has been constructed varying from a short-term carbon dioxide model, which can be used to simulate the effects of the addition of anthropogenic CO_2 to the ocean-atmosphere system, to models that have time responses on the order of hundreds of thousands to hundreds of millions of years. Sundquist (1985) has shown that the carbon cycle can be represented without sediment interactions for time scales up to a few hundred years, whereas the upper portion of the sediment column must be incorporated in models of carbon cycle behavior over thousands of years. On longer time scales, the major reservoirs of sedimentary rocks must be included. Even when considering carbon cycle models on a millenial time scale, it might be necessary to include the reservoirs and fluxes involved with the long-term sedimentary cycle. The point is that the choice of reservoirs and fluxes in a model of the carbon cycle depends strongly on the time scale of interest.

The masses of most chemical species are far greater in their sedimentary reservoirs than in the ocean. Oceanic residence times for most elements are less than 10 million years, whereas for the same elements, residence times in sediments are characteristically on the order of hundreds of millions of years. The transfers of materials from one sedimentary reservoir to another over long periods of geologic time amount to many times the amount of that material stored in the ocean today. It is also the case that these mass transfers are substantially greater than the probable variations of the oceanic content of these materials through geologic time. Thus, as pointed out previously, the ocean when viewed on a time scale of millions of years is a medium of transfer rather than one of element storage. However, on shorter times, elements may be stored in the ocean. For example, the addition of carbon dioxide to the atmosphere today owing to fossil fuel burning is resulting in the storage of about half of that carbon release in the ocean. Also, it is likely that because large evaporite deposits are not forming today that the oceans may be storing some sodium chloride and calcium sulfate, which will be released later in Earth history when the proper conditions of evaporite deposition are met.

Figure 10.31 is a simplified three-box model illustrating the principles behind most carbon cycle models. The three boxes or reservoirs may be carbon masses in the ocean, atmosphere, and biota. The arrows between the boxes represent processes by which carbon is transferred from one reservoir to another. Associated with these processes are rates of transfer, that is, fluxes of carbon from one reservoir to another. The equations shown in Figure 10.31 represent the material balance equations necessary to satisfy mass conservation in this system of reservoirs and fluxes. If the system is at steady-state, then the rate of change of mass, that is carbon mass in this case, is zero; if not, then the mass of carbon in one or more of these reservoirs is changing and the system is in transient state. Figure 10.32 is a schematic steady-state system showing average relationships of mass transfer between sedimentary reservoirs over the past few hundreds of millions of years. Several of these reservoirs may affect atmospheric carbon dioxide content and the cycling of carbon through the system. This is a rather complicated diagram and it is worth exploring briefly how the diagram, with its various reservoirs and fluxes, was arrived at. In so doing, we will look at the long-term and modern CO_2 biogeochemical cycles and emphasize some aspects briefly discussed in Chapter 9, but brought out in more detail here. The long-term geological cycle of carbon as shown in Figure 10.33 is that for the behavior of carbon dioxide prior to the industrial revolution and prior to extensive land use activities that led to deforestation and increased erosion rates. It is not the representation shown in the multi-million year average cycle of Figure 10.32. In that cycle, the fluxes have been averaged for post-Precambrian time and thus the values of the flux estimates are lower than those of the long-term cycle shown in Figure 10.33. Some important features of Figures 10.33A and 10.33B are:

1. Prior to extensive fossil fuel burning, land use and cement manufacturing activities of human society, the atmosphere contained about 49600×10^{12} moles of CO_2, or 280 ppmv (cf. Barnola et al., 1987; Siegenthaler and Oeschger, 1987). The CO_2 concentration value of the late-1980's of 350 ppmv implies a 25% increase in atmospheric CO_2 since the middle of the 18th century.

2. This increase in atmospheric CO_2 concentration for the past 200 years is primarily a result of the increased flux of carbon gases (CO_2, CO, CH_4) to the atmosphere from the land because of the burning of fossil fuels and deforestation and cultivation practices of society. It is likely that land use fluxes were more important than fossil fuel sources in the 19th and early 20th century, but during much of the latter century fossil fuel CO_2 releases appear to have been greater than

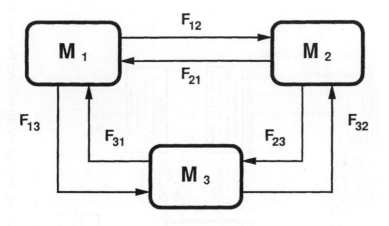

$$\frac{dM_1}{dT} = F_{21} + F_{31} - (F_{13} + F_{12})$$
$$= k_{21}M_2 + k_{31}M_3 - M_1(k_{12} + k_{13})$$

$$\frac{dM_2}{dT} = k_{12}M_1 + k_{32}M_3 - M_2(k_{21} + k_{23})$$

$$\frac{dM_3}{dT} = k_{13}M_1 + k_{23}M_2 - M_3(k_{31} + k_{32})$$

Figure 10.31. Schematic diagram of a three-box (reservoir) model of a closed-system geochemical cycle of a substance (e.g., carbon). The reservoir masses are designated M_1, M_2, and M_3, and the rates of transfer (fluxes) of a substance between boxes are shown as F_{ij}, where i and j = 1, 2, 3, but i≠j. The mass balances for the three reservoirs are given by the three differential equations. k_{ij} are first-order rate constants (units of 1/T) and T is time.

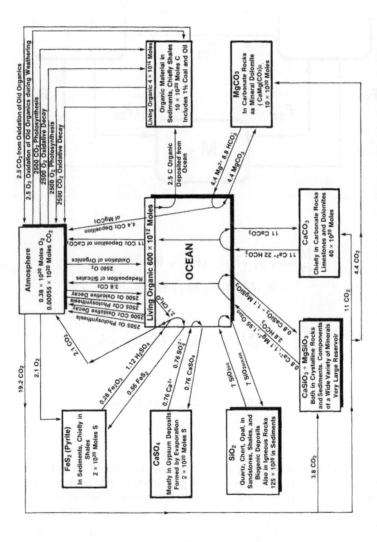

Figure 10.32. Schematic steady state sedimentary rock system suggesting average relations over the past few hundreds of millions of years. Fluxes are for dissolved constituents in units of 10^{12} moles y^{-1}. Not all reservoirs are shown and fluxes of most solid materials are not included. (Modified from Garrels et al., 1975.)

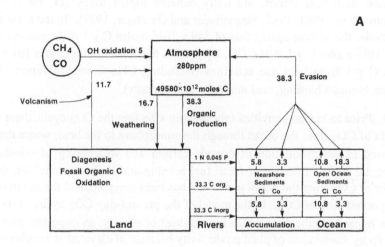

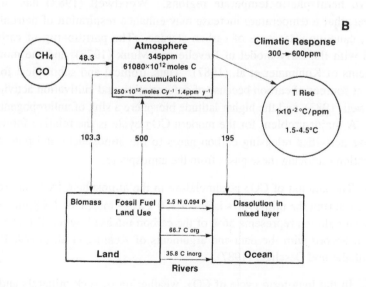

Figure 10.33. Model of the long-term geological cycle of CO_2 (A) and the perturbation to this cycle because of the fossil fuel burning and deforestation activities of humankind (B). c_i and c_o are, respectively, inorganic (inorg.) and organic (org.) carbon fluxes. Fluxes are in units of 10^{12} moles y^{-1}. See text for detailed explanation. (After Wollast and Mackenzie, 1989.)

net biospheric fluxes. The relative magnitudes of the fossil fuel and land use fluxes, and their historical trends, are hotly debated topics today (cf, for instance, Houghton et al., 1983, 1987; Siegenthaler and Oeschger, 1987). In the present-day CO_2 cycle, the anthropogenic flux of 549×10^{12} moles C y^{-1} is the summation of 448×10^{12} moles C y^{-1} (CH_4, CO, CO_2) from fossil fuel burning and 101×10^{12} moles C y^{-1} from land use activities (including CH_4 from cultivation of rice paddies, biomass burning, coal mining, and ruminants).

3. Prior to human activities substantially affecting the CO_2 cycle, there was a net flux of CO_2 from the ocean through the atmosphere to the land, where the CO_2 was used in net production of organic carbon and weathering of minerals in continental rocks. Because of fossil fuel burning and land use activities, the net transfer of CO_2 from the ocean to the land has been reversed, and the ocean is now an important sink of CO_2. In the model of the present-day CO_2 cycle, of the total carbon released to the atmosphere by activities of society, an important portion is taken up by stimulation of plant productivity because of elevated atmospheric CO_2 levels, and increases in forest growth and humus accumulation, particularly in northern hemispheric temperate regions. Woodwell (1983) has suggested, however, that a temperature increase may enhance respiration of detrital organic matter, causing a decrease of carbon storage. The partitioning of carbon is in accord with the early model of Revelle and Munk (1977), and the more recent arguments of Kohlmaier et al. (1987). Thus, tropical and subtropical forests are apparent sources of carbon because of deforestation and cultivation activities (e.g., Woodwell, 1983), and the higher latitude biosphere a sink of anthropogenic carbon gases. A major problem for the modern CO_2 cycle is the relative future role of land use activities releasing carbon gases to the atmosphere and potential CO_2 fertilization removing these gases from the atmosphere.

4. The amount of CO_2 accumulating in the atmosphere from anthropogenic activities during the mid-1980's is about 250×10^{12} moles, or 1.4 ppmv, per year. This accumulation represents 56% of the carbon released by fossil fuel burning, a value in accord with the data and arguments of Keeling et al. (1979, 1980) and Siegenthaler and Oeschger (1987).

5. In the long-term cycle of CO_2, weathering of rock minerals and organic production in the terrestrial realm produced inorganic and organic carbon, respectively, which were transported to the ocean as dissolved inorganic carbon (DIC) and dissolved and particular organic carbon (DOC, POC). Some detrital carbonate, about 1.4×10^{12} moles C y^{-1}, was also transported by rivers. The long-term weathering and organic production fluxes on land were largely supported by fluxes related to CO_2 evasion from the ocean, oxidation of the reduced carbon gases

CH_4 and CO, derived mostly from biogenic sources on land, volcanic and diagenetic processes, and a small flux from weathering by atmospheric O_2 of fossil organic matter in sedimentary rocks exposed at Earth's surface.

In the oceans half of the DIC riverine input accumulated in nearshore (coastal zones, oceanic banks and atolls, shelves) and open ocean (hemipelagic and pelagic) sediments as calcium and magnesium carbonate minerals (calcite, aragonite, magnesian calcite), and half *eventually* returned to the atmosphere owing to precipitation (e.g.):

$$Ca^{2+} + 2HCO_3^- \rightarrow CaCO_3 + CO_2 + H_2O \qquad\qquad (10.5)$$

Of the organic carbon entering the ocean, about 35% accumulated in marine sediments as organic carbon with a molar C/N/P ratio of about 250/20/1 (Mackenzie, 1981); the rest was respired and oxidized to CO_2 and evaded the ocean. The nitrogen and phosphorus riverine fluxes of 1 and 0.045 x 10^{12} moles y^{-1}, respectively, represented the nutrient fluxes necessary to support the organic matter accumulation of 11.7 x 10^{12} moles C y^{-1}.

6. In the present-day CO_2 cycle, the interactions between land and ocean via rivers appear to have been affected substantially by the activities of humankind. Herein lies an important link between the carbon cycle and those of the nutrients N and P. This link is shown in Figure 10.33B, which also demonstrates, when compared to the long-term carbon cycle of Figure 10.33A, the effects of anthropogenic carbon on land-ocean-atmosphere CO_2 coupling.

First, the application of fertilizers to the land surface has led to increased runoff of N and P from the land surface. Discharge of human wastes has augmented this flux. It appears that the total global nutrient flux today is about 2.5 times greater than the long-term geologic flux, and results in excess *accumulation* of organic carbon in the ocean (Meybeck, 1982; Wollast, 1983). Furthermore, because of land use activities, the flux of POC, DOC (Likens et al., 1981), and DIC (Meybeck, 1982) from land via rivers to the ocean has been enhanced.

It is difficult to say what happens to these enhanced carbon fluxes today. In Figure 10.34, two opposing scenarios are presented for the oceans. In one case (Figure 10.34A), most of the excess carbon brought by rivers is respired and oxidized to CO_2, and the excess organic carbon, supported by excess nutrient loading, is accumulated, primarily in nearshore environments. Also, all the

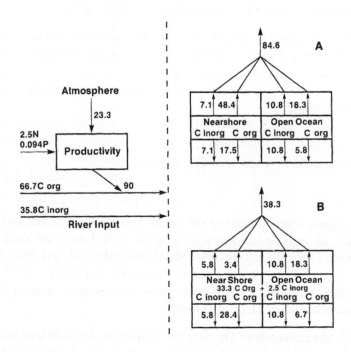

Figure 10.34. Two scenarios of the fate of excess carbon input to the ocean because of human activities. A. Most excess carbon is respired and oxidized to CO_2. B. Excess carbon is stored in the oceanic system and the CO_2 evasion flux is maintained at the geologic rate. See text for discussion. Fluxes are in units of 10^{12} moles C y^{-1}. (After Wollast and Mackenzie, 1989.)

riverine DIC is partitioned equally between deposition of carbonate minerals and CO_2 evasion to the atmosphere.

In the second case (Figure 10.34B), the excess carbon is stored in the ocean, either in the water column or in sediments, and the CO_2 evasion flux is maintained at the geologic rate. In this case, the oceanic reservoir gains 59.2×10^{12} moles C y^{-1}, an increase in storage capacity of 11% of the anthropogenic flux of carbon to the atmosphere. Much of this increase takes place in the nearshore realm, although potential transport of DOC, POC, and DIC to the open ocean would also increase the loading of this part of the ocean.

In the long-term geological cycle of carbon, approximately 6.6×10^{12} moles of carbon per year were deposited as organic carbon. This deposition represents the link between the carbon dioxide cycle and that of oxygen. This is carbon that has escaped oxidation-respiration in the oceanic system and represents a flux of oxygen from the ocean to the atmosphere of 6.6×10^{12} moles of oxygen per year. With an atmospheric oxygen content of 0.38×10^{20} moles, the residence time of oxygen in the atmosphere with respect to this flux would be 5.7 million years. That is, if the long-term situation of Figure 10.33A continued for several millions of years, the oxygen content of the atmosphere would be doubled unless there were sinks for this oxygen. There are two major sinks, the oxidation of fossil organic matter and the oxidation of sulfides, principally iron sulfides in sediments. Figure 10.35A shows the CO_2-O_2 connection. In this figure, the fluxes are average 100 million year fluxes in units of 10^{12} moles per year. Included in the diagram, besides the oxygen mass in the atmosphere and the ocean reservoir, is the mass of carbon in the fossil organic carbon reservoir and the mass of sulfide in the pyrite sedimentary reservoir. The oxygen released from the ocean on a long-time scale is consumed by the oxidation of fossil organic matter and pyrite in the ratio 2:5:1. Notice that production of pyrite in sediments in the ocean during early diagenesis results in release of carbon dioxide to the ocean-atmosphere system, and that oxidation of organic matter in the sedimentary organic carbon reservoir also results in release of carbon dioxide to the atmosphere. We now have involved, besides carbon and oxygen, the elements iron and sulfur in the cycling model. The connection between C-O-S is shown in Figure 10.35B, in which an additional reservoir, calcium sulfate, is introduced. These two figures taken together show how carbon, oxygen, and sulfur are interconnected in the global cycling behavior of these elements. The reactions involved in C-O-S cycling behavior are:

Water column-sediment aerobic oxidation

$$CH_2O + O_2 \rightarrow CO_2 + H_2O \tag{10.6}$$

Anaerobic oxidation

$$7.5CH_2O + 4H_2SO_4 + Fe_2O_3 \rightarrow 2FeS_2 + 7.5CO_2 + 11.5H_2O \tag{10.7}$$

Weathering

$$CH_2O + O_2 \rightarrow CO_2 + H_2O \tag{10.8}$$

$$2FeS_2 + 7.5O_2 + 4H_2O \rightarrow Fe_2O_3 + 4H_2SO_4 \tag{10.9}$$

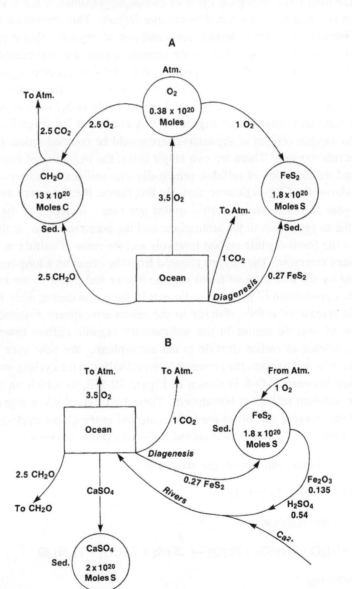

Figure 10.35. Coupling between the carbon, oxygen, and sulfur cycles. A. CO_2-O_2
connection. B. CO_2-O_2-S connection. Fluxes are average Phanerozoic values in
units of 10^{12} moles y^{-1}. See text for discussion.

Deposition and accumulation of organic matter and net oxygen escape:

$$CO_2 + H_2O \rightarrow CH_2O + O_2 \qquad\qquad (10.10)$$

Deposition of pyrite (equation 10.7) and deposition of $CaSO_4$:

$$Ca^{2+} + H_2SO_4 \rightarrow CaSO_4 + 2H^+ \qquad\qquad (10.11)$$

By continuing this line of reasoning and building up the number of sedimentary reservoirs active in the sedimentary recycling of materials, the diagram of Figure 10.32 was constructed. Now that we have some idea how the cycle diagrams are put together, we can explore some cycling models and conclusions derived from these models about the Phanerozoic history of carbon dioxide and carbonate rock cycling.

Glacial-Interglacial Changes of Carbon Dioxide

Various evidence from ice cores and deep sea sediments shows that atmospheric CO_2 concentrations have varied by up to 40% over the past few hundreds of thousands of years. Perhaps the most impressive record of this variation is found in the Vostok Antarctic ice core drilled by the Soviet Antarctic expeditions. In a series of three articles (Barnola et al., 1987; Genthon et al., 1987; Jouzel et al., 1987), a 160,000-year record of atmospheric CO_2 as obtained from samples of air trapped in the glacial ice is presented along with a continuous isotopic temperature record for this period of time. The atmospheric carbon dioxide record is shown in Figure 10.36, and in Figure 10.37 this record is compared with a temperature record for the last 160,000 years as obtained from deuterium data at the Antarctic site. The climatic record correlates remarkably well with the marine $\delta^{18}O$ climatic record back to approximately 110,000 y BP. The marine $\delta^{18}O$ record is thought to represent principally continental ice volume change, which is a parameter of global significance. The ice volume changes produce changes in global sea level, which are also shown in Figure 10.37. Thus, the Vostok temperature record is at least qualitatively of relatively large geographical significance.

The details of the spectral characteristics of the CO_2 and climate records in the Rostov ice core led the above authors to conclude that climatic changes are triggered by insolation changes with the relatively weak orbital forcing being strongly amplified by possibly orbitally-induced CO_2 changes. There is much controversy concerning the exact reasons for the atmospheric CO_2 concentration

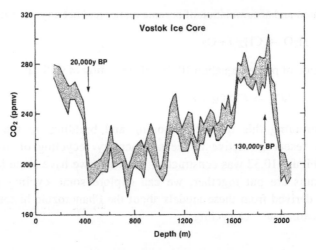

Figure 10.36. The CO_2 content of air bubbles trapped in Antarctic glacial ice at the Vostok station as a function of depth in the ice core. Two ages of the ice are shown. (After Barnola et al., 1987.)

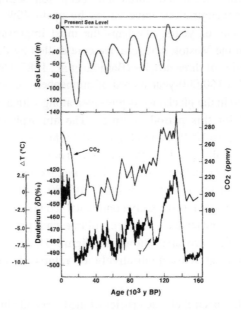

Figure 10.37. Comparison of CO_2 and deuterium trends obtained from the Vostok ice core (Barnola et al., 1987; Jouzel et al., 1987) with a polynomial model of sea level data for the past 140,000 years. (After Pinter and Gardner, 1989.)

variation over the past 160,000 years; however, most investigators agree that the source of atmospheric change must be the ocean, because most of the rapidly exchangeable carbon resides in that body. A wide range of models has been put forth to explain these ice age variations in atmospheric CO_2. The early models (Broecker, 1982, 1983) necessitated a 40% increase in the nutrient content of the whole glacial ocean relative to the modern ocean. Broecker (1982) envisioned an increase in the phosphorus content of the ocean because of oxidation of organic matter exposed on continental shelves during glacial epochs. This additional phosphorus induced higher productivity in the oceans and resulted in more efficient operation of the biological pump in which carbon is driven from the shallow ocean into the deep ocean. More recent models consider the Antarctic as a source of the carbon dioxide change or evaluate the effects of changing productivity in the low latitudes. These models do not necessarily require an external source of nutrients to enhance biological production during glacial stages, but depend on changes in the intensity of ocean circulation. Increasing ocean circulation during climatic cooling is called upon to provide a mechanism to increase the nutrient flux to the photic zone, resulting in increased biological productivity and removal of CO_2 from the atmosphere.

The exact way in which productivity in the ocean affects long-term atmospheric carbon dioxide is a matter of much debate today (e.g., Broecker, 1982, 1983; Boyle, 1986, 1988a,b; Mix, 1989). Indeed, Broecker and Peng (1987; 1989), following the lead of Boyle (1988a), argue that the last glacial-interglacial change in atmospheric CO_2 is due to an alkalinity change in polar surface seawater, and is not necessarily linked to a major change in productivity. However, the argument that marine productivity is increased during glacial stages and decreased during interglacial episodes appears to be supported by measurements of the ^{13}C and cadmium contents of carbonate tests in deep sea cores (e.g., Boyle and Keigwin, 1985; Shackleton, 1985; Shackleton and Pisias, 1985; Berger and Vincent, 1986; Boyle, 1988a, b; Broecker and Denton, 1989; Mix, 1989). Whatever the case, ocean box models and direct assessments of changes in patterns of productivity between the last glacial and modern times lead to the conclusion that some portion of the observed CO_2 change between glacial and interglacial times may be driven by variation in biological productivity.

It is possible to reproduce CO_2 trends observed in the Vostok ice core using the model developed by Garrels et al. (1976) for CO_2 - O_2 coupling. The processes involving carbon and oxygen transfer in this model for the steady-state situation are shown in Figure 10.38. Seven reservoirs are considered: atmospheric CO_2 and O_2, oceanic living biomass and seawater, and the sediment reservoirs of pyrite, organic

matter, and limestone. Fluxes associated with transfer between these reservoirs are
shown in Figure 10.38. The operation of the system is best visualized through an
example. Consider an increase in production of organic matter in the ocean. This
increase leads to removal of CO_2 from the atmosphere and an increased flux of
organic carbon to the sea floor. The excess accumulation of organic carbon on the
sea floor gives rise to increased atmospheric oxygen, which in turn is utilized in
oxidation of organic matter and pyrite exposed on the land. The weathering of old
organics returns CO_2 directly to the atmosphere, whereas pyrite weathering
produces the oxidized substances sulfate and iron oxide. These are transported to
marine sediments where these constituents are involved in oxidation of organic
carbon to produce CO_2, which is returned to the atmosphere. The perturbation
owing to excess organic matter production is relieved by the two negative feedback
mechanisms involving weathering of fossil organics and pyrite. The effect of a
perturbation stemming from a doubling of the rate of marine productivity on

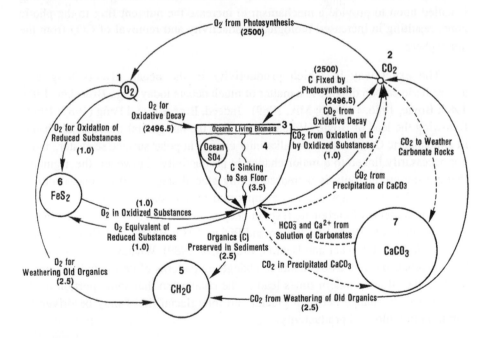

Figure 10.38. Model of CO_2-O_2 geochemical cycling describing interrelationships
on a several thousand to 100 million year time basis. Fluxes resulting from the
various processes shown are in units of 10^{12} moles y^{-1}. (After Garrels et al., 1976.)

several components of the CO_2 - O_2 coupled geochemical cycle is shown in Figure 10.39. Notice, in particular, the decrease in the size of the CO_2 reservoir on a time scale of less than 100,000 years. This substantial change reflects the small size of the atmospheric CO_2 reservoir and indicates that on relatively short time scales, changing marine productivity can result in atmospheric compositional change. Because this model responds on a time scale of glacial-interglacial change to changes in marine productivity, it may be applied to the 160,000-year record of atmospheric CO_2 change obtained from the Vostok ice core.

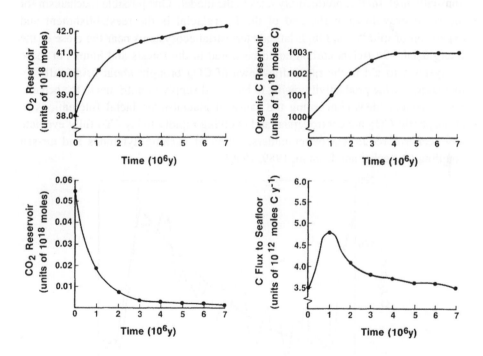

Figure 10.39. Perturbation of the CO_2-O_2 geochemical cycle illustrated in Figure 10.38. In this case, the perturbation was a doubling of the carbon flux related to marine productivity. Changes in reservoir masses and the carbon flux to the seafloor are shown as a function of time. (After Garrels et al., 1976.).

Figure 10.40 shows the results of model calculations of atmospheric CO_2 content in which the dominant mechanism of atmospheric CO_2 change is marine productivity. These calculations were performed by Michael E. Lane III of the University of Hawaii on a Macintosh II computer using the modeling program STELLA. Superimposed on the modeling calculations of Figure 10.40 is the generalized CO_2 trend seen in the Vostok ice core. The trends agree remarkably well, confirming arguments that the oceans are a principal source of atmospheric CO_2 change, and that this change is linked to the production of organic material. The modeling calculations also showed, however, that the rapid drawdown of atmospheric CO_2 near the close of the last Interglacial could not be accounted for simply by increased marine productivity. Such rapid drawdown necessitated unusually high marine productivity rates in the model. One possible mechanism for the rapid drawdown at the end of the Interglacial is the reestablishment and expansion of middle- and high-latitude terrestrial ecosystems near the close of the Interglacial. Sufficient carbon could be stored in the forests and humus of these ecosystems to aid in the rapid drawdown of CO_2 brought about principally by increased marine productivity owing to increased supply of nutrients to the oceanic photic zone. Ideas concerning the cause of glacial-interglacial fluctuations in atmospheric CO_2 are controversial and evolving rapidly today. To fully evaluate any scenario requires a better understanding of oceanic dynamics and air-sea coupling (Broecker and Denton, 1989, 1990).

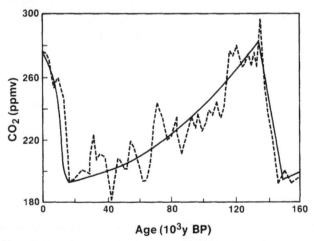

Figure 10.40. Comparison of a model calculation (solid line) for glacial-interglacial changes in atmospheric CO_2 with the observed CO_2 record (dashed line) from the Vostok ice core. The model is that of the CO_2-O_2 geochemical cycle in Figure 10.38.

Long-Term Changes of Atmospheric CO_2

An important area of progress during the last 20 years has been the development of geochemical models that quantitatively interrelate mass relations and fluxes among the large sedimentary reservoirs. Figure 10.32 was one of the first attempts to do so. The early models were based on arguments of a nearly constant total sedimentary mass with a nearly constant average chemical composition for the last 600, and perhaps 1500, million years of sedimentary cycling. Deviations from this chemical constancy, at least for Phanerozoic time, have been small, but these deviations have been responsible for change in Earth's surface chemical environment. The estimated values for the reservoir masses and the fluxes to and from them shown in Figure 10.32 are for a system in steady state. There is no net transfer from one reservoir to another. This initial steady-state condition has been employed in most geochemical models dealing with carbon cycling. Because the geochemical cycle of carbon is tied closely to that of other substances, like oxygen and sulfur, these species may be included in such models. Geochemical models have been constructed for the cycling of carbon, sulfur and oxygen through geologic time (Garrels and Lerman, 1981, 1984; Berner and Raiswell, 1983; Berner, 1987), and its effect on atmospheric oxygen (Kump and Garrels, 1986; Berner, 1989; Berner and Canfield, 1989), and the global carbonate-silicate geochemical cycle and its influence on atmospheric CO_2 (Garrels and Berner, 1983; Berner et al., 1983; Lasaga et al., 1985), to name a few. The latter geochemical model is of importance to an understanding of the geologic history of atmospheric CO_2. A schematic diagram illustrating the processes involved in the geochemical carbonate-silicate cycle is shown in Figure 10.41, and the model used to quantify the processes in Figure 10.42.

The computer model is based on the following processes: weathering on the continents of calcite, dolomite, and calcium- and magnesium-bearing silicate minerals, biogenic precipitation and removal of $CaCO_3$ from the ocean, magnesium removal from the ocean by reactions involving submarine basalts and hydrothermal seawater, metamorphic breakdown of carbonate to silicate plus CO_2, magmatic production of CO_2, and organic matter burial. In order to perturb the steady state of Figure 10.42, it is necessary to obtain rate expressions for all the transfer processes. These rate expressions are derived to reflect changes over the past 100 million years in the following rate-controlling parameters: continental land area and its effect on weathering fluxes; seafloor spreading rate and its effect on metamorphic and magmatic production of CO_2 and the uptake of Mg^{2+} from seawater because of basalt-seawater reaction; progressive conversion of dolomite to calcite through the 100-million year interval and its effect on weathering and

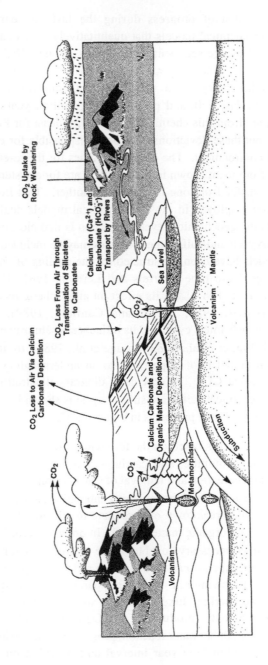

Figure 10.41. Schematic diagram of processes involved in the carbonate-silicate geochemical cycle. (After Berner and Lasaga, 1989.)

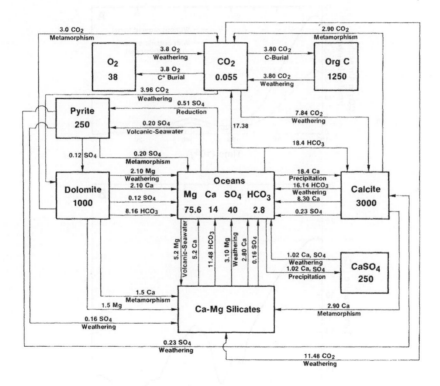

Figure 10.42. A quantitative box model of the carbonate-silicate geochemical cycle. Reservoir masses are in units of 10^{18} moles, and fluxes in units of 10^{18} moles per million years. Comparison with Figure 10.32 gives some idea how flux values and portrayal of the cycle have changed during the last decade and a half. (After Lasaga et al., 1985.)

metamorphic decarbonation rates; atmospheric CO_2 level and its effect on global
temperature and consequently weathering rates and continental runoff; burial rate
of organic carbon and its effect on atmospheric CO_2; and the Ca and HCO_3
concentrations of the ocean and atmospheric PCO_2 levels and their effect on $CaCO_3$
deposition in the ocean. Details of the computer model and calculations are given in
Berner et al. (1983) and Lasaga et al. (1985). Figure 10.43 illustrates two results of
model calculations of the carbonate-silicate geochemical cycle.

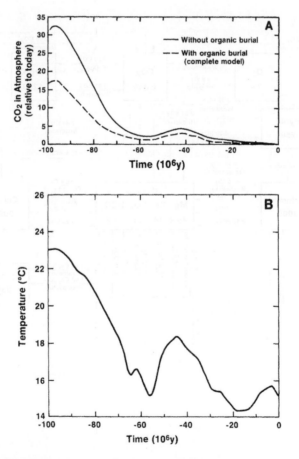

Figure 10.43. Results of model calculations of the carbonate-silicate geochemical
cycle illustrated in Figure 10.42. The corresponding changes in atmospheric CO_2
and temperature during the last 100 million years are evident. Notice how organic
carbon burial may play a strong role as a negative feedback mechanism for a
perturbation in atmospheric CO_2 driven by tectonics. (After Lasaga et al., 1985.)

These results indicate the likelihood of increased atmospheric CO_2 levels about 100 million years ago, declining somewhat irregularly to values more like today over the 100 million year time interval. These increased levels of atmospheric CO_2 in the past produced an enhanced greenhouse effect giving rise to temperatures approximately 100 million years ago about 1.5 times that of today's average surface temperature of 15^oC. The temperature history of the Earth's surface since then has been one of irregular decrease. These results are very sensitive to changes in seafloor spreading rates, and of secondary importance, to the rate of burial of organic carbon (see Figure 10.43A). Thus, plate tectonic processes on a long-time scale can control climate, as shown by this model, and in agreement with the arguments of Hays and Pitman (1973), Fischer and Arthur (1977), and Mackenzie and Pigott (1981). It should be pointed out, however, that other processes can affect atmospheric CO_2 on a long-time scale, and thus, climate. Volk (1989 a,b) argues that biological factors may be important. In late Cretaceous-early Tertiary time, there was a dramatic increase in the diversity of angiosperms. The spread of angiosperm - deciduous ecosystems at this time could have led to increased rates of continental weathering. These increased rates of weathering may have led to lower atmospheric CO_2 levels, and as a result of a decreased greenhouse effect, progressive cooling of the Cenozoic atmosphere.

Another biological factor that may have influenced climate over the past 100 million years is that of the evolution and diversification of planktonic foraminifera and coccolithophores in the late Mesozoic. This evolutionary event resulted in progressive transfer of carbonate deposition from shoal-water areas to the deep sea during the Cenozoic. Thus, there should have been an increase in the transfer of pelagic carbonate to subduction zones. The metamorphic flux of CO_2 to the atmosphere should have increased during the Cenozoic as these carbonates were converted to silicates on subduction slabs with CO_2 release and degassing from island-arc systems and surrounding environments. Volk (1989b) argues that this increased metamorphic CO_2 flux may have been a warming factor influencing Earth's climate. It may be possible that Earth's climate over the past 100 million years has been determined by a balance affected by a number of factors that influenced atmospheric CO_2 concentrations, including plate tectonic processes, and the biological events of evolution of calcareous marine plankton and evolution and spread of angiosperm-deciduous ecosystems on land. It is likely that until this combination of factors resulted in an Earth cooler than at the beginning of the Cenozoic that Milankovitch forcing was not strong enough to induce glaciations. At that time, atmospheric CO_2 on a glacial-intergalcial time scale became regulated principally by ocean-related changes in the flux of CO_2 owing to changes in marine biological production rates.

The burial of organic carbon has an additional effect on atmospheric composition. Burial of organic matter implies an excess of photosynthesis over bacterial respiration. This difference results in accumulation of oxygen in the ocean-atmosphere system. If all other factors remain constant, times of high organic matter burial should have given rise to high atmospheric oxygen levels. Berner and Canfield (1989) recognized the implications of these relationships and derived a model for Phanerozoic atmospheric O_2 levels based on rates of burial and weathering of organic carbon and pyrite sulfur. Their best estimate calculation is shown in Figure 10.44. The maximum in atmospheric oxygen for the Carboniferous and Permian periods might be anticipated, but its magnitude is debatable. Such a high oxygen concentration implies a greatly enhanced probability of forest fires at this time. Evidence for such conflagration has not been documented as yet in the rock record. However, just before the Carboniferous, vascular plants evolved and spread over the continents, providing a new source of organic matter resistant to degradation. This material entered the geochemical cycle of carbon, some was respired, and some was buried in the vast coastal lowland swamps that existed in the Carboniferous and Permian. These two periods were the most important coal-forming intervals in Earth's history. Thus, excess organic carbon burial at this time could have lead to increased atmospheric O_2 levels, as seen in the Berner-Canfield model.

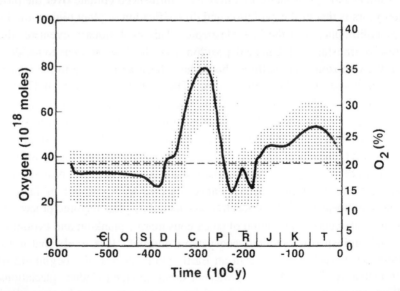

Figure 10.44. Model calculation of atmospheric O_2 concentration during the Phanerozoic. (After Berner and Canfield, 1989.)

Phanerozoic Cycling of Sedimentary Carbonates

The subject of the cycling behavior of sedimentary carbonates was introduced earlier in this chapter. In this section we expand the discussion with particular emphasis on mass-age relationships and compositional trends. Model calculations provide much of the input for the discussion. The mass-age relationships for the Phanerozoic global carbonate reservoir were presented earlier in this chapter. The relationships were developed on a Period-averaged basis, because of concern for the significance of the data on a finer time scale. Wilkinson and Walker (1989), however, have developed mass-age data for sedimentary carbonates using Phanerozoic epochs as time intervals. Furthermore, they distinguished mass-age data for continental, oceanic, and global carbonate reservoirs as shown in Figure 10.45. Mass-age models similar to those used by Garrels and Mackenzie (1971a) were developed for these carbonate reservoirs. The mass-age relations are approximated by exponential decay functions for which two major assumptions are employed. The first is that the rate of destruction of carbonate mass per unit time

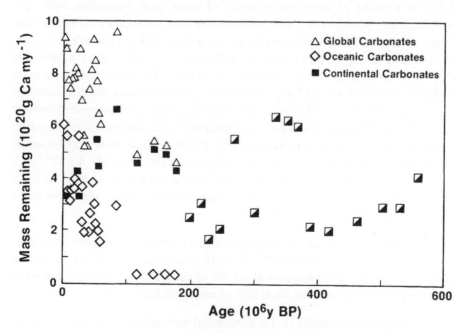

Figure 10.45. Mass-age relationships of sedimentary carbonates. The global mass is shown as open triangles when representing the sum of pelagic and cratonic masses and as half-solid rectangles when entirely continental. Solid rectangles are cratonic masses when the global mass consists of both continental and deep oceanic carbonate. (After Wilkinson and Walker, 1989.)

by erosion or metamorphism is proportional to the mass of rock present. Preferential preservation is not important over the time periods considered. The second assumption is that each mass-age unit cycles at the same rate; that is, there is a decay constant that describes the rate of recycling of the carbonate mass. Models may be constructed in which the mass of total carbonate rock remains constant or grows linearly (constant mass and linear accumulation models of Garrels and Mackenzie, 1971a), or for a continuum of growth possibilities (power law models of Veizer and Jansen, 1979). The models may be constrained to yield the best estimate of reservoir mass or the best estimate of the present rate of cycling (mass-fit and flux-fit approximations of Wilkinson and Walker, 1989).

Figure 10.46 shows model fits to global, continental, and oceanic (pelagic and slope-rise) sedimentary carbonate masses. Recently, Wilkinson and Algeo (1989) have modified slightly these model fits and expressed the preserved sedimentary mass in terms of "extant" carbonate flux. They also attempted to separate slope-rise and pelagic mass-age data and model them separately. Of importance here is the conclusion that continental and oceanic reservoirs of sedimentary carbonates have different patterns of exponential decay of mass with increasing rock age. Wilkinson and Walker (1989) concluded that the total sedimentary carbonate mass comprises about 3500×10^{20} g, expressed as calcium, with a range of estimates of 3170-3850×10^{20}g. We estimate for the Phanerozoic (Table 10.1), a carbonate mass of about 2200×10^{20} g, expressed as calcium. The global carbonate mass cycles at a rate of 8.6×10^{20} g Ca per million years and has a decay constant of 0.0025 ma^{-1}. The present flux of carbonate to the sea floor, according to Wilkinson and Walker, is between 7.7 and 9.5×10^{20} g Ca ma^{-1}. Our estimate of the dissolved calcium and magnesium flux as carbonate to the seafloor today (Chapter 6) corresponds to accumulation rates over a million year time span of 6.4×10^{20} g Ca and 0.07×10^{20} g Mg. If we add the particulate carbonate flux of today of 0.7×10^{20} g ma^{-1}, expressed as Ca, the resultant total carbonate flux is very close to the lower estimate of Wilkinson and Walker.

In contrast to the total sedimentary carbonate mass, oceanic carbonate oozes have a decay constant of 0.02 ma^{-1}, about 10 times that of the global carbonate mass. The pelagic oozes compose about 8% of the global sedimentary carbonate mass but presently account for about 60% of carbonate deposition.

The exponential model fit for continental carbonates, as shown in Figure 10.46, is not good. These rocks dominate the sedimentary carbonate mass older than 100 million years. One can fit the pre-Cretaceous continental carbonate data with a negative exponential function having a decay constant of 0.0025 ma^{-1}, as done by Wilkinson and Walker (1989). However, based on analysis of the mass-age

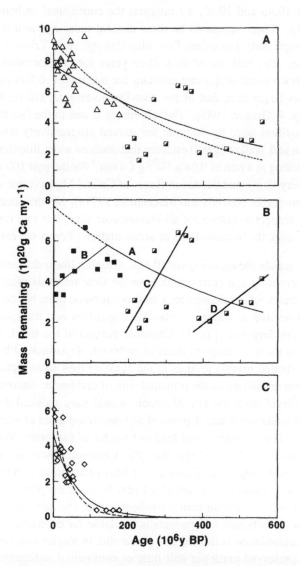

Figure 10.46. Mass-age relationships of sedimentary carbonates and examples of model fits to the mass-age data. (A) Global carbonates fitted using two approaches: the best statistical approximation of total mass ("mass fit," dark line), and the best statistical mass fit constrained to intercept the value of the Pliocene mass remaining ("flux fit," dashed line). (B) Cratonic carbonates fitted using least-squares fit of mass-age data compared with exponential fit. (C) Pelagic carbonates fitted using "mass fit" (solid line) and "flux fit" (dashed line) models. (After Wilkinson and Walker, 1989.)

data of Figure 10.45 and 10.46, we interpret the continental carbonate mass-age data differently. There appear to be three distinct mass-age trends representing Cambrian through early Devonian, Devonian through early Triassic, and Triassic to present time. The half-life of 86 million years for post-Permian sedimentary carbonates gives a value of the rate constant for cycling of 0.008 ma^{-1}, which is about 1.5 times larger than that of the post-Devonian total sedimentary mass of 0.0054 ma^{-1} (p. 4; Gregor, 1985). This difference is due to the fact that the site of carbonate deposition since the Jurassic has moved progressively toward the deep sea. Wilkinson and Walker (1989) estimate that shallow water limestone deposition has been decreasing at a rate of 0.04 x 10^{20} g Ca ma^{-1} for the past 100 million years. Because it is very likely that the cation fluxes of Ca and Mg have not varied greatly during the Phanerozoic (Garrels and Mackenzie, 1971a), and the oceans have been saturated with respect to calcite for all Phanerozoic time, this rate represents, to a first approximation, the increased rate of accumulation of deep-sea carbonate ooze.

The three trends shown in Figure 10.46B exhibit a linear decrease in mass per epoch with decreasing age corresponding to the time intervals mentioned above. The Cenozoic trend, which extends back to about Jurassic time, has been interpreted to represent a decline in cratonic carbonate deposition and transposition of that deposition to the deep sea. For the Cenozoic portion of the trend, there are two possible reasons for this change in locus of carbonate deposition. First, the advent of planktonic shelled marine protists in the Jurassic may have replaced shallow-water calcareous organisms as the principal sink of carbonate. Second, the general decline of sea level since the late Mesozoic would have reduced the geographic extent of shoal-water carbonate deposition and decreased areas of warm carbonate-saturated seas. This change would lead to transfer of carbonate from shallow to deep environments. It is likely that the Cenozoic increase in carbonate accumulation in the deep sea and decrease in rates of accumulation in cratonic and other shoal-water settings are a result of both evolution of the planktonic calcareous organisms and increasing continental freeboard since the late Mesozoic. Only a finite reservoir of carbonate components is available for deposition; if the area of shoal-water accumulation is limited and a new sink in marine calcareous plankton provided, the preserved mass per unit time of continental sedimentary carbonate will decrease.

The other trends of decreasing epoch mass with decreasing age of continental sedimentary carbonates (Figure 10.46B) for Cambrian to early Devonian and Devonian to early Triassic strata correlate very well with the continental freeboard curve (relative elevation of continents with respect to sea level) (Figure 10.47) for the cratonic interior of the United States and southern Canada. The preserved mass

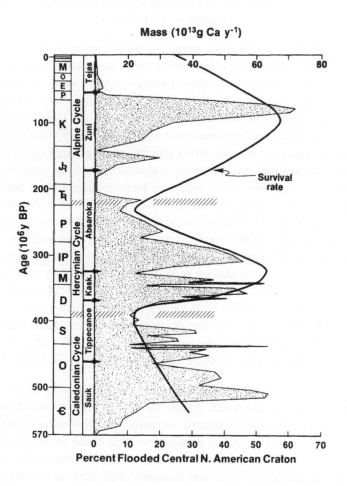

Figure 10.47. Phanerozoic continental freeboard curve (stippled pattern) of the North American craton (Wise, 1974) compared with generalized Epoch-interval survival rate of cratonic carbonates calculated as mass of carbonate rock in a Series divided by duration of Epoch. (Data from Wilkinson and Walker, 1989.)

distribution of continental sedimentary carbonates expressed as Epoch mass preserved per 10^6 years tracks reasonably well the freeboard curve (Figure 10.47). The terminations of the major cycles are very close in time to the end of the Tippecanoe and Absaroka cratonic sequences as defined by Sloss (1963; Figure 10.47). These sequence boundaries appear to be marked by regional, and probably

global, emergences and unconformities. Sloss (1976) has also demonstrated that the
sedimentary mass per unit time of a cratonic sequence decreases with decreasing
sequence age as the end of the sequence is approached.

It is tempting to argue that there are three cycles of decreasing epoch
continental carbonate sedimentary mass with decreasing rock age in the
Phanerozoic. The origin of the youngest was discussed above. The origin of the
two older cycles may be related to one of the assumptions underlying the
construction of mass-age models; that is, the assumption that there is a single decay
constant for the continental carbonate mass that applies equally to any unit mass of
this carbonate reservoir. It seems likely that this assumption of equal jeopardy of
the continental carbonate rock reservoir to destruction may be incorrect. Within
each cycle, the younger units of the cycle may be more susceptible to destruction by
erosion and thus cycle at a faster rate than the older units. It is also possible that
with decreasing freeboard as shown for the Caledonian and Hercynian cycles, there
was increased transport of shoal-water carbonate derived from platform margins to
the deeper ocean, where its susceptibility to destruction by subduction would be
enhanced. Also, it is likely that, in order to maintain a steady state ocean with
respect to Ca^{2+} and DIC, that times of low sea level coincided with enhanced
inorganic precipitation of carbonate in the deep sea. The cycling rates of sediments
in rise and deep ocean tectonic regimes is greater than that in cratonic settings. This
mechanism is not unlike what happened during the last 100 million years of the
Alpine cycle, but in that case carbonate transfer was due to biological evolution.

Finally, this tripartite cyclicity is also seen in the frequency of occurrence of
Phanerozoic ironstones and oölites (Figure 10.18). As sea level withdrew from the
continents and continental freeboard increased, shallow-water areas with the
requisite environmental conditions necessary to form oölite and ironstone deposits
decreased in extent. Thus, as calcium carbonate deposition increased on slopes and
in the deep sea, carbonate oölite and ironstone deposition on shelves and banks
nearly ceased.

Synopsis of the origin and evolution of the hydrosphere-atmosphere-sedimentary lithosphere

Origin of the Hydrosphere

We are now at a stage, based on the data presented in the initial part of this
chapter and some further information, to discuss briefly the origin and evolution of

the hydrosphere-atmosphere-sediment system, the exogenic system. To begin, let us explore the origin of the hydrosphere.

It is not very likely that the total amount of water at the Earth's surface has changed significantly over geologic time. Based on the age of meteorites, the Earth is thought to be 4.6 billion years old. The oldest rocks known date about 3.9 billion years in age, and these rocks although altered by post-depositional processes show signs of having been deposited in an environment containing water. There is no direct evidence for water for the period between 4.6 and 3.9 billion years. Thus, ideas concerning the early history of the hydrosphere are coupled to our thoughts about the origin of the Earth.

The subject of Earth's origin is complex, difficult, and intimately a part of ideas concerning the nature and evolution of the solar nebula and the formation of the planets in general. It is likely that the Earth originated from a slowly rotating disk-shaped cloud (the "solar nebula") of gas and dust of solar composition. There has been much debate concerning whether Earth accreted as a cold or a hot body and whether the planet accreted homogeneously or inhomogeneously (e.g., Anderson, 1989; Atreya et al., 1989). Some degree of inhomogeneity in the chemistry of the accreting material is required; otherwise, H_2O would not exist at the Earth's surface. In the inhomogeneous accretion hypothesis, as the solar nebula cools, the Earth and other planets accrete materials that are in equilibrium with the temperature of the nebula or materials that condensed earlier and escaped accretion. The mean composition of the Earth therefore became less refractory and more volatile as the radius of the planet increased. It is likely that, because of the decay of short-lived radioactive elements and because of the entrapment of much of the gravitational energy of accretion within the planet, Earth started life as a relatively hot body. Differentiation of the Earth into a core, mantle, and crust probably occurred nearly contemporaneously with accretion. The depletion of the very light and volatile elements such as H, He, C, and N in the Earth relative to the Sun or carbonaceous chondrites argues for loss of some primitive atmospheric components of Earth. These components were blown away because of the strong solar wind present early in Earth's history and because of gigantic impacts that blasted materials into space. The present atmosphere and oceans of Earth are probably derived as a late-accreting veneer from impacts of comets and carbonaceous asteroids (e.g., Chyba, 1990) and from concurrent and later degassing of the planet's interior. The outgassing for Earth has been relatively efficient because much of the ^{40}Ar produced by decay of ^{40}K in the planetary interior now resides in the atmosphere.

Thus, at an early stage the Earth did not have liquid water at its surface. Once the Earth's surface had cooled sufficiently, water contained in the materials of accretion and water released at depth and escaping to Earth's surface, instead of being lost to space, could cool and condense to form the initial hydrosphere. A large, cool Earth serves as a better trap for water than a small, hot body, because the lower the temperature the less likely for water vapor to escape, and the larger the Earth the stronger its gravitational attraction for water vapor. Whether most of the degassing took place during core formation or shortly thereafter or whether there has been significant degassing of Earth's interior throughout geologic time is still uncertain. However, it is most likely that the hydrosphere attained its present volume early in Earth's history, and since that time there have been only small losses and gains. Gains would be from continuous degassing of the Earth and perhaps accretion of extraterrestrial particles containing H_2O. The present degassing rate of juvenile water today is only 0.3 km^3 per year. Water loss is by photodissociation by the energy of ultraviolet light to hydrogen and oxygen atoms in the upper atmosphere. Hydrogen is lost to space and the oxygen remains behind. Only about 4.8 x 10^{-4} km^3 of water vapor is presently destroyed each year by photodissociation. This low rate is because the very cold temperatures of the upper atmosphere result in a cold trap at about 15 km in the atmosphere, where most of the water vapor condenses and returns to lower altitudes, escaping photodissociation. The low rate of loss is consistent with the observation that the $\delta^{18}O$ of water vapor is constant with altitude. Since the early formation of the hydrosphere, the amount of water vapor in the atmosphere has been regulated by the temperature of the Earth's surface, hence its radiation balance. Higher temperatures imply higher atmospheric water vapor concentrations, lower temperatures, lower atmospheric levels.

The Early Stages

The gases released from the Earth by impact processes and degassing in its early history, including water vapor, have been called excess volatiles because their masses cannot be accounted for simply by rock weathering. An estimate of the excess volatiles is given in Table 10.4. These volatiles would have formed the early atmosphere of the Earth. If the Earth had an initial crustal temperature of about 600°C, almost all of these compounds, including H_2O, would be in the atmosphere. The sequence of events that occurred as the crust cooled is difficult to construct. Below 100°C all of the H_2O would have condensed, and the acid gases would have reacted with the original igneous crustal minerals to form sediments and an initial hydrosphere, dominated by a salty ocean. One possible pathway that assumes

Table 10.4 Estimate of "excess volatiles" (units of 10^{20} grams). (After Walker, 1977.)

Water	16600
Total C as carbon dioxide	2500
Chlorine	300
Nitrogen	49
Sulfur	44
Boron, bromine, argon, fluorine, etc.	4

reaction rates are slow relative to cooling is that the 600°C atmosphere contained, together with other compounds, H_2O, CO_2, HCl, and SO_2 in the ratio 120:20:4:1 and cooled to the critical temperature of H_2O. If the CO_2 had remained in the atmosphere, it is likely that its pressure of 50 atmospheres would have been sufficient to cause a "runaway greenhouse" like that of Venus, which has an atmospheric CO_2 pressure of greater than 90 atmospheres and a temperature of 450°C (Walker, 1977). The H_2O would have condensed into an early hot ocean. At this stage, the hydrochloric acid would be dissolved in the ocean (~1 mole per liter), but most of the CO_2 would still be in the atmosphere with about 0.5 mole per liter in the ocean water. This early acid hydrosphere would react vigorously with crustal minerals, dissolving out silica and cations and creating a residue that consisted principally of aluminous clay minerals that would form the sediments of the early ocean basins.

This is one of the several possible pathways for the early surface of the Earth. Whatever the actual case, after the Earth's surface had cooled to 100°C, it would have taken only a short time for the remaining acid gases to be consumed in reactions involving igneous rock minerals. The presence of cyanobacteria in the fossil record of rocks older than 3 billion years attests to the fact that the Earth's surface had cooled to temperatures lower than 100°C by this time, and neutralization of much of the original acid volatiles had taken place. However, it is likely that in the Archean, the Earth's atmosphere still contained more CO_2 and higher concentrations of CH_4 than today. If its atmospheric composition had been the same as today, the Earth would have been frozen. The reason is that the Sun 3-4 billion years ago was 20-30 percent fainter than today (Figure 10.48). Higher atmospheric carbon dioxide and methane levels in the past, providing an additional greenhouse effect, is the most likely reason that surface temperatures of the Earth stayed above freezing throughout the Archean.

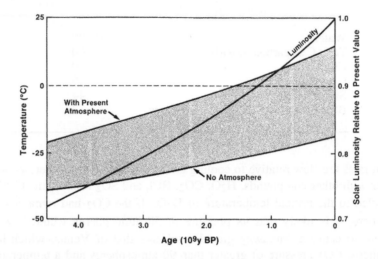

Figure 10.48. Trends in the solar luminosity and temperature of the Earth. The upper temperature curve is calculated for an atmospheric composition like that of today whereas the lower temperature curve is for an airless Earth. The difference between the curves (the shaded region) represents the magnitude of the greenhouse effect. It is likely that atmospheric CO_2 concentrations were higher in the earlier Precambrian and, consequently, temperatures were higher than shown. (After Kasting et al., 1988.)

If most of the degassing of primary volatile substances from the Earth's interior occurred early, the chloride released by reaction of HCl with rock minerals would be found in the oceans and seas or in evaporite deposits, and the oceans would have a salinity and volume comparable to that of today. This conclusion is based on the assumption that there has been no drastic changes in the ratios of volatiles released through geologic time. The overall generalized reaction indicative of the chemistry leading to formation of the early oceans and sediments can be written in the form:

Primary igneous rock minerals + acid volatiles + $H_2O \rightarrow$

sedimentary rocks + oceans + atmosphere   (10.12)

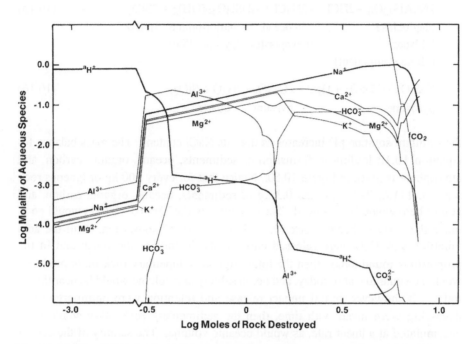

Figure 10.49. Model calculation simulating changes in ocean composition as primary igneous rock minerals react with a solution containing the proportions of "excess volatiles" shown in Table 10.4. Concentrations of dissolved species are shown relative to 1 kilogram of water, and the extent of reaction is measured by the amount of igneous-rock minerals destroyed. Changes in slopes on diagram are due to formation of sedimentary minerals. (After Lafon and Mackenzie, 1974.)

In Figure 10.49 a computer calculation is given, simulating the gigantic acid-base titration that led to the formation of the neutralization products of ocean and sediments. The importance of this calculation is that it demonstrates the essence of equation 10.12; as primary igneous rock minerals react with water and acids, the excess volatiles, by reactions like:

$$CaSiO_3 + CO_2 \rightarrow CaCO_3 + SiO_2 \qquad (10.13)$$
pyroxene   calcite   quartz

$$CaAl_2Si_2O_8 + CO_2 + H_2O \rightarrow CaCO_3 + Al_2Si_2O_5(OH)_4 \qquad (10.14)$$
plagioclase   calcite   kaolinite
feldspar
(anorthite component)

$$2NaAlSi_3O_8 + 2HCl \rightarrow 2NaCl + Al_2Si_4O_{10}(OH)_2 + 2SiO_2 \qquad (10.15)$$

plagioclase　　　　　　　　　in ocean　montmorillonite　quartz
feldspar　　　　　　　　　or evaporites　(pyrophyllite)
(albite component)

$$CaSiO_3 + H_2SO_4 + H_2O \rightarrow CaSO_4 \cdot 2H_2O + SiO_2, \qquad (10.16)$$

pyroxene　　　　　　　　　gypsum　　quartz

the resultant solution pH increases as does its NaCl content. The mass balance for equation 10.12 leading to formation of sediments, oceans, organic carbon, and atmosphere is given in Figure 10.50. Notice that for every 100 kg of igneous rock reacted, 111.6, 79.9, 1.3, and 0.3 kg of sediments, oceans, organic carbon, and atmosphere would be produced. The components $CaCO_3$ and $MgCO_3$ would go to make the early carbonate rock mass. Equation 10.12 shows that if all the acid volatiles and H_2O were released early in the history of the Earth and in the proportions found today, then the total original sedimentary rock mass produced would be equal to that of today, and ocean salinity and volume would be near that of today. If, however, initial impact release and retention were unimportant and degassing were linear with time, then the sedimentary rock mass would have accumulated at a linear rate, as would oceanic volume. The salinity of the oceans would remain nearly the same if the ratios of volatiles degassed did not change with time. The most likely situation is that presented here, namely, that major release occurred early in Earth's history, after which minor amounts of volatiles were degassed episodically or continuously for the remainder of geologic time. The salt content of the oceans based on the constant proportions of volatiles released would depend primarily on the ratio of sodium chloride (NaCl) locked up in evaporites to that dissolved in the oceans. If all the NaCl in the evaporites were added to the oceans today, the salinity would be approximately doubled. This value gives a sense of the maximum salinity the oceans could have attained throughout geologic time.

One component missing from the early Earth's surface was free oxygen, because it would not have been a constituent released from the cooling crust. Early production of oxygen was by photodissociation of water in the Earth's atmosphere as a result of absorption of ultraviolet light. The reaction is $2H_2O + hv \rightarrow O_2 + 2H_2$ in which hv represents the photon of ultraviolet light. The hydrogen produced would escape into space, and the O_2 would react with the early reduced gases by reactions such as $2H_2S + 3O_2 \rightarrow 2SO_2 + 2H_2O$. Oxygen production by photodissociation gave the early reduced atmosphere a start toward present-day conditions, but it was not until the appearance of photosynthetic organisms approximately 3.5 billion years ago that it was possible for much oxygen to be released to Earth's surface environment. Significant accumulation of oxygen in the ancient atmosphere, however, did not begin until about 1.5 billion years ago.

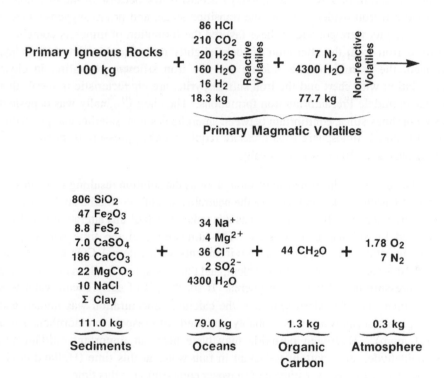

Figure 10.50. Schematic mass balance for the origin of the sedimentary rock mass, oceans, atmosphere, and buried organic carbon. The reaction coefficients are given in units of relative moles. (After Li, 1972.)

The Transitional Stage

The nature of the rock record from the time of the first sedimentary rocks (~3.8 billion years ago) to about 1 to 2 billion years ago suggests that the amount of oxygen in the Earth's atmosphere was significantly lower than today, and that there were continuous chemical trends in the sedimentary rocks formed and, more subtly, in hydrosphere composition. Figure 10.6 illustrates how the chemistry of rocks shifted dramatically during this transitional period. The source rocks of sediments during this time may have been more basaltic than later ones;

sedimentary detritus was formed by the alteration of these rocks in an oxygen-deficient atmosphere and accumulated primarily under anaerobic marine conditions. The chief difference between reactions involving mineral-ocean equilibria at this time and today was the role played by ferrous iron. The concentration of dissolved iron in today's oceans is low because of the insolubility of oxidized iron oxides. During the transition stage, and prior, oxygen-deficient environments were prevalent; these favored the formation of minerals containing ferrous iron from the alteration of rocks slightly more basaltic rich than today. Indeed, the iron carbonate siderite and the iron silicate greenalite, in close association with chert and the iron sulfide pyrite, are characteristic minerals that occur in middle Precambrian iron formations. The chert originally was deposited as amorphous silica; equilibrium between amorphous silica, siderite, and greenalite at 25°C and 1 atmosphere total pressure requires a CO_2 pressure of about $10^{-2.5}$ atmospheres, or 10 times today's value.

The oceans at this time can be thought of as the solution resulting from an acid leach of basaltic rocks, and because the neutralization of the volatile acid gases was not restricted primarily to land areas as it is today, much of this alteration may have occurred by submarine processes. The atmosphere at the time was oxygen deficient; anaerobic depositional environments with internal CO_2 pressures of about $10^{-2.5}$ atmospheres were prevalent, and the atmosphere itself may have had a CO_2 pressure near $10^{-2.5}$ atmospheres. If so, the pH of early ocean water was lower than that of modern seawater, the calcium concentration was higher, and early global ocean water was probably saturated with respect to amorphous silica (~120 ppm). Hydrogen peroxide may have been an important oxidant and formaldehyde, an important reductant in rain water at this time (Holland et al., 1986). Table 10.5 is one estimate of seawater composition at this time.

Garrels and Mackenzie (1971a) envisioned the following scenario simulating the nature of the Earth's surface system during the transitional stage.

"To simulate what might have occurred, it is helpful to imagine emptying the Pacific Basin, throwing in great masses of broken basaltic material, filling it with HCl so that the acid becomes neutralized, and then carbonating the solution by bubbling CO_2 through it. Oxygen would not be permitted into the system. The HCl would leach the rocks, resulting in the release and precipitation of silica and the production of a chloride ocean containing sodium, potassium, calcium, magnesium, aluminum, iron, and reduced sulfur species in the proportions present in the rocks. As complete neutralization was approached, Al could begin to precipitate as hydroxides and then combine with precipitated silica to form cation-deficient aluminosilicates. The aluminosilicates, as the end of the neutralization process was reached, would combine with more silica and with cations to form minerals like chlorite, and ferrous iron would combine with silica and sulfur to make greenalite and pyrite. In the final solution, chlorine would be balanced by Na and Ca in roughly equal proportions with subordinate K and Mg; Al would be quantitatively removed, and silicon would be at saturation with amorphous silica. If this solution were then carbonated, Ca would be removed as calcium carbonate, and the Cl balance would be maintained by abstraction of more Na from the primary rock. The sediments produced in this system would contain chiefly silica, ferrous iron silicates, chloritic minerals, calcium carbonate, calcium-magnesium carbonates, and minor pyrite." (p. 295-296) (see also Garrels and Mackenzie, 1974).

Table 10.5. An example of early seawater composition. (After Garrels and Perry, Jr., 1974.)

$P_{O_2} = 10^{-70.07}$ atm	$Ca^{2+}_{aq} = 10^{-1.95}$ (0.11 m)
$P_{CO_2} = 10^{-1.20}$ atm	$Mg^{2+}_{aq} = 10^{-1.52}$ (0.030 m)
$P_{CH_4} = 10^{-4.39}$ atm	$Fe^{2+}_{aq} = 10^{-4.13}$ (0.000074 m)
$P_{H_2} = 10^{-0.11}$ atm	$K^+_{aq} = 10^{-2}$ (0.01 m)
$P_{NH_3} = 10^{-6.95}$ atm	$Na^+_{aq} = 10^{-0.330}$ (0.47 m)
$P_{H_2S} = 10^{-8.34}$ atm	$Cl^-_{aq} = 10^{-0.26}$ (0.55 m)
$HS^-_{aq} = 10^{-5.41}$ m	$SiO_{2aq} = 10^{-2.7}$ (0.002 m)
$SO^{2-}_{4aq} = 10^{-10.4}$ m	pH = $10^{-6.68}$
$HCO^-_{3aq} = 10^{-2.02}$ (0.0094 m)	

Calculations were made assuming equilibrium at 25°C with $FeSiO_3$, FeS_2, $FeCO_3$, $CaCO_3$; concentrations of ions and values of activity coefficients were assumed to be the same as in the present ocean.

If the HCl added was in excess of the CO_2, the resultant ocean would have a high content of calcium chloride, but the pH would still be near neutrality. If the CO_2 added was in excess of the Cl, Ca would be precipitated as the carbonate until it reached a level approximately that of the oceans today, namely, a few hundred parts per million.

If this newly created ocean were left undisturbed for a few hundred million years, its waters would evaporate and be transported onto the continents in the form of precipitation; streams would transport their loads into it. The sediment created in this ocean would be uplifted and incorporated into the continents. Gradually, the influence of the continental debris would be felt and the pH might shift a little. Iron would be oxidized out of the ferrous silicates to make iron oxides, but the water composition would not vary a great deal.

The primary minerals of igneous rocks are all mildly basic compounds. When they react in excess with acids such as HCl and CO_2, they produce neutral or mildly alkaline solutions plus a set of altered aluminosilicate and carbonate reaction products. It is improbable that ocean water has changed through time from a solution approximately in equilibrium with these reaction products, which are clay minerals and carbonates.

Modern Conditions

It is likely that the hydrosphere-sediment system achieved many of its modern chemical characteristics 1.5 to 2 billion years ago. The chemical and mineralogical compositions and the relative proportions of sedimentary rocks of this age differ little from their Paleozoic counterparts. Calcium sulfate deposits of late Precambrian age testify to the fact that the acid sulfur gases had been neutralized to sulfate by this time. Chemically precipitated ferric oxides in late Precambrian sedimentary rocks indicate available free oxygen, whatever its percentage. The chemistry and mineralogy of middle and late Precambrian shales are similar to those of Paleozoic shales. The carbon isotopic signature of carbonate rocks has been remarkably constant for over 3 billion years, indicating a remarkable stability in size and fluxes related to organic carbon. The sulfur isotopic signature of sulfur phases in rocks indicates that the sulfur cycle involving heterotrophic bacterial reduction of sulfate was in operation 2.7 billion years ago. Thus, it appears that continuous cycling of sediments like those of today has occurred for 1.5 to 2 billion years and that these sediments have controlled hydrospheric, and particularly oceanic, composition.

It was once thought that the saltiness of the modern oceans simply represents the storage of salts derived from rock weathering and transported to the oceans by fluvial processes. With increasing knowledge of the age of the earth, however, it was soon realized that, at today's rate of delivery of salt to the ocean or even at much reduced rates, the total salt content and the mass of individual salts in the oceans could be attained in geologically short time intervals compared to the Earth's age. The total mass of salt in the ocean can be accounted for at today's rates of stream delivery in about 12,000,000 years. The mass of dissolved silica in ocean water can be doubled in only 20,000 years by addition of stream-derived silica; to double sodium would take 70,000,000 years. It then became apparent that the oceans were not simply an accumulator of salts, but as water evaporated from the oceans, along with some salt, the introduced salts must be removed in the form of minerals deposited in sediments. Thus, the concept of the oceans as a chemical system changed from that of a simple accumulator to that of a steady-state system, in which rates of inflow of materials into the oceans equal rates of outflow. The steady-state concept permits influx to vary with time, but it would be matched by nearly simultaneous and equal variation to efflux.

In recent years this steady-state conceptual view of the oceans has been modified somewhat. In particular, it has been found necessary to treat components of ocean water in terms of all their influxes and effluxes, and to be more cognizant of the time scale of application of the steady-state concept. Certainly the recent increase in the carbon dioxide concentration of the atmosphere owing to the burning of fossil fuels may induce a change in the pH and dissolved inorganic carbon concentrations of surface ocean water on a time scale measured in hundreds of years. If fossil fuel burning were to cease, return to the original state of seawater composition could take thousands of years. Ocean water is not in steady state with respect to carbon on these time scales, but on a longer geological time scale it could be. However, even on this longer time scale, oceanic composition has varied because of natural changes in the carbon dioxide level of the atmosphere and because of other factors.

It appears that the best description of modern seawater composition is that of a chemical system in a dynamic quasi-steady-state. Changes in composition may occur over time, but the system always seems to return to a time-averaged, steady-state composition. In other words, since 1.5-2 billion years ago, evolutionary chemical changes in the hydrosphere have been small when viewed against the magnitude of previous change.

Modern seawater chemistry has been characteristic of the past 600 million years of ocean history. Evaporite sediments provide strong evidence that the composition of seawater during the Phanerozoic has not varied a great deal. It seems likely, however, that fluctuations did occur, particularly in the concentrations of Ca^{2+}, Mg^{2+}, and SO_4^{2-}. The isotopic composition of sulfur in seawater SO_4^{2-} as recorded in evaporites has varied dramatically during the past billion years. Although it is difficult to relate these isotopic fluctuations to the Ca^{2+}

and SO_4^{2-} concentrations of seawater, some scientists believe these fluctuations imply changes in these concentrations. Furthermore, the major features of the sulfur isotopic curve for evaporites versus Phanerozoic time is similar to that of the $^{87}Sr/^{86}Sr$ ratio, and perhaps the Sr/Ca ratio, of sedimentary materials during this time interval. Such covariation is consistent with a model in which fluxes related to alteration of seafloor basalts and continental river runoff vary with time, resulting in variation in seawater composition. The time scale of variation may be long-term or abrupt, as seen, for example, in the abrupt shift of the $\delta^{18}O$ and $\delta^{13}C$ record of marine abiotic carbonate cements at the Devonian-Carboniferous transition (Lohmann and Walker, 1989).

With respect to sedimentary carbonates, Ca and Mg fluxes involving river and groundwater discharge, ridge exchange reactions and limestone and dolostone reservoirs probably did not vary by a factor of more than 120% during the Phanerozoic (Wilkinson and Algeo, 1989). Such changes in fluxes would have resulted in variation in the Mg/Ca ratio of seawater of less than a factor of 5. Wilkinson and Algeo (1989) conclude that the rate of marine dolomite formation is the most important process controlling seawater Mg concentration and Mg/Ca ratio. This conclusion is based on their argument that most sedimentary dolomite is a synsedimentary marine precipitate, a conclusion that is still strongly debated (e.g., Hardie, 1987; Zenger, 1989). If dolomite is mainly formed early in the history of a carbonate rock, then an estimate can be made of the amount of river-borne magnesium removed by the processes of dolomitization versus hydrothermal alteration during Phanerozoic time (Figure 10.51). The apparent decreased importance of dolomitization today, in the late Paleozoic-early Mesozoic, and in the Cambrian is reflected in the cyclic nature of the calcite/dolomite ratio of sedimentary carbonates (Figure 10.29).

The chemistry of the atmosphere has certainly changed significantly during the past 1 billion years of Earth's history. Such changes imply changes in the chemistry of the hydrosphere. Oxygen in the atmosphere rose significantly between 2 billion years ago and the beginning of the Phanerozoic Eon, whereas atmospheric carbon dioxide levels probably decreased (Kasting, 1987). This change led in general to a progressively more oxygenated and less acidic hydrosphere. It is likely that the development of higher land plants in the Devonian gave rise to an increase in atmospheric oxygen and a decrease in carbon dioxide. The geochemical record of carbon and oxygen isotopes, iridium, and other metals in Permian/Triassic boundary carbonates in the Carnic Alps of Austria suggests important changes in the carbon cycle. In this case, the isotope shifts and chemical events may be due to causes related to the major Permian regression of the sea

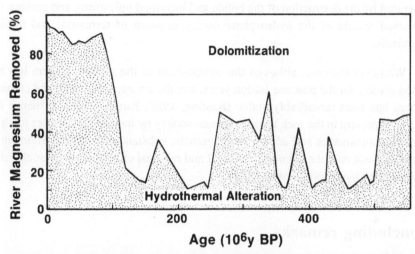

Figure 10.51. Calculated percent of river Mg removed during hydrothermal
alteration and marine dolomitization during the Phanerozoic. (After Wilkinson
and Algeo, 1989.)

(Holser et al., 1989). Air trapped in bubbles in Arctic and Antarctic ice show that
the CO_2 content of the atmosphere during the climax of the last ice age was about
180 ppmv, whereas prior to extensive fossil fuel burning and deforestation
activities of society and during the last great Interglacial of 120,000 years before
present, atmospheric CO_2 levels were about 280 ppmv. These atmospheric changes
in themselves influence the chemistry of the hydrosphere, but coupled with these
changes are other changes in the rock-ocean-atmosphere-biota system that strongly
affect hydrospheric chemistry. For example, although surface waters probably
remained oxygenated during the Cretaceous and Devonian Periods of Earth
history, there is evidence that intermediate and deep ocean waters were more
anoxic than today. These "anoxic events" are characterized by dramatic changes in
Earth's ocean-atmosphere system, including changes in the rates of cyclic transfer
of elements at Earth's surface and atmospheric composition (e.g., Arthur et al.,
1988). The probable impact of a bolide and increased volcanism at the end of the
Cretaceous, although hotly debated topics today, certainly could have caused short-
term changes in the chemistry of Earth's atmosphere-hydrosphere and
bioproductivity (e.g., Zachos et al., 1989). Some speculative results of such an
event would be (1) changes in the Earth's radiant energy balance because of vast
amounts of particulate and gaseous input into the atmosphere, and subsequent
cooling, (2) acid rain stemming from the input of nitrogen gases to the atmosphere

generated by the bolide impact, (3) increased trace metal fluxes to the hydrosphere generated by the destruction of the bolide and increased volcanism, and perhaps (4) increased anoxia of the hydrosphere owing to death of terrestrial and marine organisms.

Whatever the case, although the composition of the modern system has not varied greatly for the past one billion years, and the atmosphere-ocean and climatic system has been remarkably stable (Kasting, 1989; Kump, 1989), evidence for change is present in the rock record. Human society by increasing the rates of input of natural substances and adding new, synthetic substances to the environment is modifying not only the chemistry of local and regional environments, but also that of the global hydrosphere-atmosphere.

Concluding remarks

Much of the chemical, mineralogical, and isotopic data on which our ideas of the evolution of Earth's surface environment are based come from analysis of sedimentary carbonates. Unfortunately because of the increased susceptibility of the sedimentary carbonate mass to alteration and destruction by the processes of weathering, subduction, and metamorphism, much original information is lost from the rock record with increasing age of the carbonate mass. We are left with a biased record of the preserved sedimentary carbonate rock mass. The Precambrian Eon, which includes more than 85% of Earth history, contains only about 25% of the mass of sedimentary carbonates in existence!

It appears that the major control on carbonate rock composition through geologic time is exerted by processes within the depositional and early diagenetic environments, with secondary processes of diagenesis playing a subordinate, but important, role. Therefore many of the compositional trends of carbonates with geologic age indicate evolutionary and cyclic changes in the composition of solutions from which carbonate minerals formed. These solutions include seawater and its early diagenetic derivatives and meteoric water, and mixtures of the two waters. In turn, these changes in solution composition are linked to changes in atmospheric composition. Although still somewhat controversial, long-term trends in the chemistry of sedimentary carbonates with geologic age point to lower atmospheric P_{O_2} and probably a greater degree of submarine alteration of volcanic materials earlier in Earth history. At this time, seawater was a more reduced fluid and ferrous iron and manganous manganese were more concentrated in the hydrosphere than today.

For much of the later Precambrian and the Phanerozoic the system was more like today. Cyclic changes in the composition of sedimentary carbonates dominate the record, with less evidence of evolutionary changes. It should be remembered, however, that many of these cyclic compositional changes are poorly documented and controversial. If real, these changes represent variation in the chemistry of the hydrosphere-atmosphere system. If these changes were principally driven by plate tectonic mechanisms, Figure 10.52 summarizes changes in environmental conditions and sedimentary precipitate composition during the Phanerozoic, based on data and arguments of this chapter.

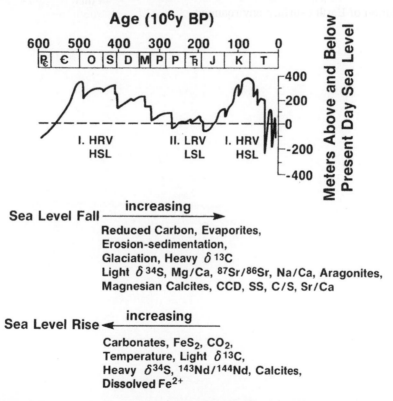

Figure 10.52. Generalized sea level curve for the Phanerozoic (Vail et al., 1977) identifying times of submergence, high sea level (HSL), and high ridge volume (HRV) and times of emergence, low sea level (LSL), and low ridge volume (LRV). Predicted, long-term changes in the isotopic and elemental composition and the deposition of sedimentary materials are summarized for periods of rising and falling sea level. CCD = carbonate compensation depth; SS = saturation state. (After Mackenzie and Agegian, 1989.)

    The interpretation of the secular compositional variation found in sedimentary carbonate rocks is not an easy task. Besides knowledge of the data base involved, the carbonate petrologist or geochemist must be familiar with the chemical principles developed in the first third of this book, the carbonate processes and products of the middle third, and the behavior of the biogeochemical cycle of carbon and interactive elements, as developed in this chapter and Chapter 9. The scientific study of sedimentary carbonates has come a long way; its future lies in integration of a variety of subdisciplines that are commonly separated (e.g., petrology verses geochemistry) and endowed with a great number of specialists. The need for such an integration is evident in our discussion in this chapter of the evolution of Earth's surface environment.

EPILOGUE

Introduction

We would like to conclude this book in a rather unusual manner with a summary overview of what we feel are the major accomplishments in the field of sedimentary carbonate geochemistry, where the field stands at the time this book was written, and our opinion of future directions for the field. In the more than three years it took us to write this book, we have come to realize that this field of scientific endeavor has a rich and unique character reflecting not only its relationships with other areas of basic science, but also those with the petroleum industry and public concern about human impact on the environment. It is hard to think of many other fields that contain such a diversity of activities as monitoring the chemistry of the atmosphere in the Antarctic ice cores, chipping rocks in the Alps, sampling the deep ocean with submersible vehicles, diving on tropical reefs, carrying out complex laboratory experiments and constructing global computer models.

The road traveled

Although important studies of the mineralogy and chemical composition of carbonate sediments and sedimentary rocks have been conducted for over 100 years, like many other areas of geochemistry, the "modern" era did not begin in earnest until the 1950's. This early research was largely spurred on by the growth of new analytical techniques, such as stable isotope mass spectrometry, and a growing interest by the petroleum industry in geochemical approaches to hydrocarbon exploration. During this period, the central importance of carbonate geochemistry to studies in sedimentary petrology, sediment diagenesis and marine chemistry was clearly established. Many of the "fundamental" questions in the field were first asked at this time and initial attempts were made at quantitative answers.

The 1960's can be viewed as the "golden age" of sedimentary carbonate geochemistry. This "golden age" was in no small part due to the "gold" which started to flow rapidly and relatively easily into university research. There was an explosive growth in the number of investigators devoting their careers to the study of sedimentary carbonates. Much of the work in the first part of this decade was exploratory or observational in nature, and there was plenty of virgin territory to explore. Volumes of data were produced on the compositions of carbonate rocks, biogenic carbonates and natural waters in carbonate units. By the middle of the decade, interests shifted to application of experimental techniques, chemical thermodynamics and models of processes to explain the growing body of data.

However, while the geochemists were having a great time with all of their new data and theories, carbonate petrologists and sedimentologists were starting to get suspicious. In spite of all their equipment and increasingly complex jargon, the geochemists seemed to be unable to come up with explanations for many of the most basic problems involved with the behavior of sedimentary carbonates (e.g., Why isn't dolomite forming in large amounts in modern sediments?). Indeed, amazingly complex arguments were presented to explain the apparently obvious (e.g., carbonate cements form in shallow marine water because it is supersaturated and carbonates dissolve in the deep ocean because it is undersaturated!). The situation was compounded by the growing debates among the carbonate geochemists over the interpretation of geochemical data and the validity of competing, and frequently loosely founded, theories.

It is hard to say why the study of sedimentary carbonate geochemistry started the 1960's with such vigor and ended the decade in such a chaotic manner. The tumultuous nature of the times, sudden infusion of previously undreamed of levels of funding, and rapid addition of new investigators to the field certainly were all part of it. However, it also seems probable that frustration with the unexpected complexities of the chemical behavior of carbonates and failure to arrive at solutions of basic problems also contributed substantially to the chaotic state of the field as the 1960's ended. These were the wild adolescent years of the field.

As the 1970's began, the field of sedimentary carbonate geochemistry started to evolve into distinct subfields and continued to evolve increasingly away from its roots in sedimentary petrology and sedimentology. Carbonate geochemists now had questions of their own to answer. These questions often claimed priority over those posed by carbonate petrologists or sedimentologists, and a schism between these groups formed. Probably the largest subgroup of carbonate geochemists was created in the rapidly growing oceanographic institutions. The study of the chemistry of the oceans and atmosphere was a focus of their research. The pressing questions in this area were not primarily concerned with carbonate rocks, shoal-water carbonates, or the behavior of carbonate sediments more than a few centimeters below the sediment-water interface. Their primary interest in carbonate sediments, other than those of recent origin, was for paleo-environmental reconstruction.

A second major subgroup that evolved during this time was concerned with understanding the chemical behavior of carbonates in a more basic manner than had previously been attempted. These investigators were motivated by the realization that simple approaches to carbonate geochemistry, based largely on observational data, were inadequate to explain the complex behavior of these minerals. A new body of thermodynamic data was needed to determine solubilities of impure minerals in complex solutions, and the influences of temperature and pressure. It was also clear that reaction kinetics, surface chemistry and a host of other chemical aspects of carbonate mineral behavior were not going to be understood through an equilibrium thermodynamic approach. The field was increasingly populated by scientists with generally stronger backgrounds in chemistry and weaker backgrounds in geology.

The interaction of subsurface waters with lithified sedimentary carbonates became a major focus of research in the 1970's. Efforts were made not only to sample and analyze these waters and the associated carbonates, but also to construct increasingly complex models of their interactions. These models were primarily aimed at establishing how a carbonate-aqueous system evolves through time and incorporated hydrologic considerations.

The 1980's saw the maturing of the field of carbonate geochemistry, with many of the approaches initiated during the late 1960's and 1970's continuing to gain in sophistication. In many areas, the study of carbonate geochemistry was pushed close to the limits existing in the field of pure chemistry. Major advances were made in understanding the chemistry of complex electrolyte solutions, behavior of ion partition coefficients, carbonate mineral surface chemistry and reaction kinetics, and the fine-scale structure of sedimentary carbonate minerals. These advances also continued to demonstrate the complexities of sedimentary carbonates and their interaction with natural waters. A major result of these advances in understanding carbonate chemistry was to bring into serious question some of the most cherished ideas regarding the interpretation of compositional data and the utility of equilibrium thermodynamics in understanding carbonate behavior. There was a growing uneasy feeling that the more we learned about carbonate chemistry the more difficult it was becoming to apply it to the "real world".

A major change was also occurring during this period in the field of carbonate sedimentary petrology where there was an increasing awareness of the need to apply more sophisticated approaches to the study and interpretation of carbonate rocks. The utility of using a wide range of geochemical data, as well as the rapidly advancing concepts about the behavior of carbonate minerals, to interpret the formation and evolution of these rocks was clearly demonstrated. As the decade came to a close, there were clear signs that the fields of sedimentary carbonate geochemistry and petrology were starting, albeit uneasily, to grow closer with renewed appreciation that they are to a large extent inseparable.

A major increase in attention to the "CO_2 problem" during this decade also stimulated considerable activity. Much of this activity was devoted to modeling current systems, such as those described in Chapter 9. While a major portion of this activity focused on the current environment and human impact, the "CO_2 problem" also sparked interest in trying to understand long-term processes, such as the role of sedimentary carbonates in the evolution of the Earth's atmosphere.

The state of the art

We feel that it is worthwhile to attempt to view the current status of carbonate geochemistry in light of the fundamental problems that were largely defined 20 to 30 years ago. Although one frequently encounters the attitude that so much as been done in the field of carbonate mineral geochemistry relative to other mineral systems that all the major problems have been solved, it is our opinion that this assertion is far from the truth. On the contrary, many of the most important problems are largely unresolved in spite of this mammoth effort.

It is reasonable to expect that our understanding of carbonate sediments which are currently forming would be the best, because we can directly observe their formation and the waters associated with them. Determination of the factors controlling carbonate accumulation in pelagic sediments has been a long standing problem that would appear readily amenable to solution. By the mid-1970's it was well established that a straight-forward thermodynamic explanation was inadequate and that complex reaction kinetics were important. More recently, it has been shown that processes occurring near the sediment-water interface may considerably alter the relationship between the overlying water chemistry and carbonate dissolution in deep-sea sediments. A general quantitative model for these interfacial processes awaits further observational and modeling efforts. Consequently, while it has been possible to gain an insight into the processes and general mechanisms that control carbonate deposition in pelagic sediments, it is not possible, at present, to predict if and to what extent carbonates will be deposited over much of the ocean floor.

Many of the early basic questions regarding carbonate geochemistry were asked with shoal-water carbonate sediments in mind. In spite of a large body of research, relatively little progress has been made in resolving several of the major problems. The major mineralogic component in some of these sediments is aragonite needle mud. The question of a biogenic or abiotic origin of these muds still remains largely unresolved. Whitings represent a possible source of muds in some areas. However, even the most studied area of these micrite deposits in the Bahamas remains largely enigmatic as to the origin of the micrites. Oöids are also commonly encountered in shoal-water carbonate sediments. An abiotic process is generally believed to be responsible for their formation, but the several hypotheses presented for their formation are not without limitations. Carbonate cements are common in reefs, but surprisingly scarce in shoal-water carbonate sediments. The factors controlling cement distribution and mineralogy remain unknown except at the most general descriptive level. Carbonate geochemists have been particularly frustrated in their efforts to understand the factors that control the composition of high magnesian calcite cements. The rare occurrence of dolomite in modern sediments also presents several major unresolved problems -- both as to why dolomite is not found more commonly, and why dolomite is found in certain environments. Although its formation has been observed to be associated with high salinity and organic-rich environments, a comprehensive geochemical explanation for its preferential formation in these environments remains an elusive problem.

It appears that studies of the meteoric diagenesis of Holocene and Pleistocene carbonates have provided reasonable first approximation models of the complex dissolution, precipitation, and replacement reactions that take place in metastable carbonate mineral-groundwater systems. The applicability of these studies to more ancient rocks, where starting carbonate mineralogies and water chemistries may have been different, is difficult. Quantitative models incorporating reaction kinetics and present-day or inferred ancient groundwater flow regimes are in their infancy state. It is still difficult in some cases to conclude unequivocally that a carbonate rock has been altered in the burial diagenetic realm as opposed to the meteoric setting.

The role of burial processes in carbonate rock diagenesis remains an open question. Is the scarcity of burial cements observed in most well-documented studies of burial diagenesis a general truism? Or have our thoughts been prejudiced by the fact that most studies to date are from the northern Gulf of Mexico, U.S.A.? Fluid migration and mixing processes in the subsurface are poorly understood. How do these processes affect the solubility, reaction kinetics, and mass transfer involving carbonate minerals in the subsurface? Chemical data for subsurface water compositions, chemical and mineralogic data for the composition of subsurface cements and replacement products, and basic petrographic microscopic data are reasonably abundant in the literature and increasing rapidly. However, we are still far from developing quantitative models of burial diagenetic and multi-diagenetic processes. These processes can involve strong linkages between hydrocarbon maturation and migration, porosity destruction and development, and geochemical and hydrologic parameters at depth in the subsurface. Furthermore, much of the basic thermodynamic and kinetic information for carbonate minerals in the subsurface environment of increased P and T and changing fluid compositions is lacking.

The chemistry, mineralogy and isotopic composition of sedimentary carbonates are grossly known. Trends in the values of these parameters with geologic age are apparent in the rock record, but the meaning of many of these trends is controversial. Does the general enrichment of sedimentary carbonates in ^{16}O with increasing age reflect a temperature trend or a change in the $\delta^{18}O$ of seawater? Is the age distribution of the calcite/dolomite ratio in Phanerozoic carbonates cyclic, or does this ratio progressively increase with increasing rock age? Many fundamental questions of this nature remain unanswered, and thus our knowledge of the evolution of Earth's surface environment is riddled with holes.

Where do we go from here? This question is addressed in the following section.

Ever onward

Many of the most basic questions regarding the behavior of sedimentary carbonates that were posed thirty or more years ago remain substantially unresolved and are likely to remain a central focus of research for many years to come. The continuing evolution of related fields, such as chemistry, development of increasingly sophisticated analytic instruments and the ever increasing speed of computers will certainly influence the approaches that carbonate geochemists will use to attack these problems. However, we feel that in many instances it is probably necessary to question seriously long held concepts and attempt to rethink the problems. In this area, as in other areas of science, it is essential to question constantly "dogma", if new ideas are to flourish and major progress is to be achieved.

It is probable that environmental concerns and an increasingly "global" perspective in Earth sciences will stimulate and influence research directions in the study of sedimentary carbonates. The field will almost certainly evolve in a more interdisciplinary direction with increasing input from studies in other fields such as physical oceanography, climatic modeling and tectonophysics. If the general research pattern in Earth sciences is followed, a team approach to many basic problems, perhaps involving basic chemists, geochemists, petrologists, hydrologists, structural geologists and biologists, is likely to become much more common.

We conclude our crystal ball gazing with a list of research topics or approaches that we feel are likely to be of importance in the future. After reading this book, it should be obvious that this list is far from exhaustive. Our list follows the general organization of this book and is submitted in a terse format without details or justification.

Basic Chemistry

1) More information is needed on carbonate reaction kinetics near equilibrium in complex and high ionic strength solutions, and the influences of temperature and pressure on kinetics to near metamorphic conditions.

2) The carbonic acid system constants can probably be improved by close to an order of magnitude.

3) Interactions of carbonate minerals with organic matter and the influences of these interactions require more investigation.

4) Interactions of thin water films with carbonates need further investigation.

5) Use of TEM to study very slow reactions under realistic conditions may provide new breakthroughs.

6) Methods should be developed to determine more accurately surface free energies and the influences of adsorbates on these free energies.

7) Methods need to be developed to measure and determine the influences of factors such as strain, crystal geometry, dislocation density, low concentrations of coprecipitates, etc. on subtle solubility and reactivity differences of carbonate minerals.

8) A more sophisticated chemical understanding needs to be obtained of the problem of why elevated ionic strength seems to influence so many important reactions such as the aragonite to calcite transformation and dolomite formation.

Modern Marine Sediments and the Ocean

1) The saturation state of deep seawater with respect to calcite and aragonite on a large scale needs to be determined within an order of magnitude more accuracy than it is now known.

2) Work needs to be continued on sediment-water interfacial processes that control carbonate accumulation in deep-sea sediments.

3) Transport of shoal-water carbonates to the deep ocean and to the slope, and their dissolution require more extensive study.

4) Early carbonate diagenesis and cementation of reefs need to be quantitatively and mechanistically examined.

5) The diagenesis of carbonates in terrigenous sediments requires further research and its role in global cycles assessed.

6) High magnesian calcite cements, oöids and whitings all necessitate innovative new approaches to explain their formation and distribution.

7) Fluxes of dissolved components into and out of shoal-water carbonates should be quantified.

Diagenesis

1) There is a need to investigate more examples of the stabilization of calcitic carbonates in the meteoric environment.

2) Criteria for differentiating limestones that have undergone diagenesis in the meteoric environment from those that have been altered in the deep subsurface should be developed and extended.

3) There is a need to utilize modern instrumentation (XRD, IR, Raman, stable and radiogenic isotope, and EM techniques) in the study of carbonate solids involved in diagenetic processes.

4) Multidisciplinary analytical and numerical models require development. These models should involve considerations of equilibrium and irreversible thermodynamics and kinetics of carbonate mineral-organic matter-water interactions within a sound hydrodynamic and basin evolution framework.

5) Solution chemistry models at elevated P and T in saline and brine solutions require further development.

6) There is a need for further investigations of carbonate reactions and diagenesis and fluid migration in accretionary prisms associated with subduction zones.

7) More deep bore holes should be drilled in carbonate sequences not directly associated with hydrocarbon accumulations to obtain diagenetic data and information on carbonate composition as a function of geologic age.

Carbonate-Related Geochemical Cycles

1) There is a strong need both from a scientific and practical viewpoint for a more quantitative understanding of the inorganic and organic carbon cycles and their interactions immediately prior to important human intervention in these cycles.

2) Processes involving interactions between the carbon cycle and nutrient cycles require further quantitative analysis, and models of these interactive cycles should be developed.

3) Flux estimates related to the modern carbon cycle require refinement, particularly those involving coastal margin and other shoal-water area inputs and outputs.

4) Detailed chemical stratigraphic analyses are needed to define short-term, hydrosphere-atmosphere events during Earth's history.

5) Continued ice-core analyses are required to understand more fully the glacial-interglacial record of atmosphere-hydrosphere change. These records should be integrated more fully with the continental and deep-sea record of climate change.

6) Precambrian carbonates deserve more attention in terms of their composition, petrology, stratigraphy, sedimentology, and basin history. There is a dearth of stratigraphically closely-spaced data on carbonates of this Eon.

7) There is a need for further development of models that quantitatively predict past conditions of Earth's surface environment from data derived from carbonate rocks and other lithologies.

REFERENCES

Acker J.G., Byrne R.H., Ben-Yaakov S., Feely R.A. and Betzer P.R. (1987) The effect of pressure on aragonite dissolution rates in seawater. *Geochim. Cosmochim. Acta* **51**, 2171-2175.

Adamson A.W. (1982) *Physical Chemistry of Surfaces. 4th Ed.* John Wiley and Sons, New York, 664 pp.

Adelseck C.G. Jr. (1978) Dissolution of deep-sea carbonate: preliminary calibration of preservational and morphological aspects. *Deep-Sea Res.* **25**, 1167-1185.

Adelseck C.G., Jr. and Berger W.H. (1975) On the dissolution of planktonic foraminifera and associated microfossils during settling and on the sea floor. In *Dissolution of Deep-Sea Carbonates* (eds. A. Be and W. Berger), pp. 70-81. Cushman Foundation for Foraminiferal Research, Special Publication 13, W. Sliter.

Agegian C.R. (1985) The Biogeochemical Ecology of *Porolithon gardineri* (Fosllie). Ph.D. dissertation, Univ. of Hawaii.

Agegian C.R. and Mackenzie F.T. (1989) Calcareous organisms and sediment mineralogy on a mid-depth bank in the Hawaiian Archipelago. *Pacific Sci.* **43**, 56-66.

Agegian C.R., Mackenzie F.T., Tribble J.S. and Sabine C. (1988) Carbonate production and flux from a mid-depth bank ecosystem, Penguin Bank, Hawaii. *Natl. Undersea Res. Prog. Res. Report 88-1*, Rockville, MD, 5-32.

Aharon P., Socki R.A. and Chan L. (1987) Dolomitization of atolls by seawater convection flow: Test of a hypothesis at Niue, South Pacific. *J. Geol* **95**, 187-203.

Al-Shaieb Z. and Shelton J.W. (1981) Migration of hydrocarbons and secondary porosity in sandstones. *AAPG Bull.* **65**, 2433-2436.

Albright J.N. (1971) Vaterite stability. *Amer. Mineralogist* **56**, 620-624.

Alekin O.A. (1978) Water erosion of land surface. In *U.S.S.R. Committee for the Internatl. Hydrol. Decade World Water Balance and Water Resources of the Earth, Studies and Reports in Hydrology, 25*, UNESCO Press, Paris, 663 pp.

Alekin O.A. and Brazhnikova L.V. (1960) A contribution on runoff of dissolved substances on the world's continental surface. *Gidrochim. Mat.* **32**, 12-34 (in Russian).

Alekin O.A. and Brazhnikova L.V. (1962) The correlation between ionic transport and suspended sediment. *Dokl. Acad. Nauk. SSSR* **146**, 203-206, (in Russian).

Alekin O.A. and Brazhnikova L.V. (1968) Dissolved matter discharge and mechanical and chemical erosion. *Assoc. Int. Hydrol. Sci. Pub.* **78**, 35-41.

Alexandersson E.T. and Milliman J.D. (1981) Intragranular Mg-calcite cement in *Halimeda* plates from the Brazilian continental shelf. *J. Sediment. Petrol.* **51**, 1309-1314.

Allan J.R. and Matthews R.K. (1977) Carbon and oxygen isotopes as diagenetic and stratigraphic tools: data from surface and subsurface of Barbados, West Indies. *Geology* **5**, 16-20.

Allan J.R. and Matthews R.K. (1982) Isotope signatures associated with early meteoric diagenesis. *Sedimentology* **29**, 797-817.

Allen M.R. (1985) The origin of dolomites in the phosphatic sediments of the Miocene Pungo River Formation: North Carolina. M.S. thesis, Duke Univ., Durham, North Carolina.

Aller R.C. (1980) Diagenetic processes near the sediment-water interface of Long Island Sound. I. Decomposition and nutrient element geochemistry (S, N, P). *Adv. in Geophys.* **22**, 237-350.

Aller R.C. (1982) Carbonate dissolution in nearshore terrigenous muds: the role of physical and biological reworking. *J. Geol.* **90**, 79-95.

Aller R.C. and Rude P.D. (1988) Complete oxidation of solid phase sulfides by manganese and bacteria in anoxic marine sediments. *Geochim. Cosmochim. Acta* **52**, 751-765.

Aller R.C., Mackin J.E. and Cox R.T. Jr. (1986) Diagenesis of Fe and S in Amazon inner shelf muds: apparent dominance of Fe reduction and implications for the genesis of ironstones. *Cont. Shelf Res.* **6**, 263-289.

Amankonah J.F. and Somasundran P. (1985) Effects of dissolved mineral species on the electrokinetic behavior of calcite and apatite. *Colloids and Surfaces* **15**, 335-353.

Ameil A., Friedman G.M. and Miller D. S. (1973) Distribution and nature of incorporation of trace metals in argonite corals. *Sedimentology* **20**, 47-64.

Am. J. Sci. 1978, articles therein: **278**.

Andersen N.R. and Malahoff A. (1977) *The Fate of Fossil Fuel CO_2 in the Oceans.* Plenum Press, New York, 749 pp.

Anderson D.L. (1989) *Theory of the Earth.* Blackwell Scientific Publications, Boston, 366 pp.

Angus J.G., Rayner B.J. and Robson M. (1979) Reliability of experimental partition coefficients in carbonate systems: evidence for inhomogeneous distribution of impurity cations. *Chem. Geol.* **27**, 181-205.

Anthony S., Peterson F.L., Mackenzie F.T. and Hamlin S.N. (1989) Hydrogeochemical controls on the accusence and flow of groundwater within the Laura freshwater lens, Majuro Atoll, Marchall Islands. *Geol. Soc. Amer. Bull.* **101**, 1066-1075.

Archer D., Emerson S. and Reimers C. (1989) Dissolution of calcite in deep-sea sediments: pH and O_2 microelectrode results. *Geochim. Cosmochim. Acta* (in press).

Arnseth R.W. (1982) Carbonate and clay mineral reactions in a modern mixing zone environment, Salt River Estuary, St. Croix, U.S. Virgin Islands. Ph.D. dissertation, Northwestern Univ.

Arrhenius G. (1988) Rate of production, dissolution and accumulation of biogenic solids in the ocean. *Paleogeogr. Paleoclimatol. Paleoecol.* **67**, 119-146.

Arrhenius S. (1896) On the influence of carbonic acid in the air upon the temperature of the ground. *Philos. Mag.* **41**, 237-275.

Arthur M.A., Anderson T.F., Kaplan I.R., Veizer J. and Land L.S. (1983) *Stable Isotopes in Sedimentary Geology , SEPM Short Course No. 10,* Dallas, TX.

Arthur M.A., Dean W.E. and Schlanger S.O. (1985) Variations in the global carbon cycle during the Cretaceous related to climate, volcanism, and changes in atmospheric CO_2: Natural Variations Archean to Present. In *The Carbon Cycle and Atmospheric CO_2* (eds. E.T. Sundquist and W.E. Broecker), pp. 504-529. Amer. Geophys. Union Geophys. Monograph **32**, Washington, D.C.

Arthur M.A., Dean W.E. and Pratt L.M. (1988) Geochemical and climatic effects of increased marine organic carbon burial at the Cenomanian/Turonian boundary. *Nature* **335**, 714-717.

Atkinson M.J. and Smith S.V. (1983) C:N:P ratios of benthic marine plants. *Limnol. Oceanogr.* **28**, 568-575.

Atreya S.K., Pollack J.D. and Matthews M.S. (1989) *Origin and Evolution of Planetary and Satellite Atmospheres.* Univ. of Arizona Press, Tucson, 881 pp.

Bacastrow R.B. and Keeling C.D. (1979) Models to predict future atmospheric CO_2 concentrations. In *Workshop on the Global Effects of Carbon Dioxide from Fossil Fuels* U.S. Dept. Energy Conf. -770385, NTIS, Springfield, VA, 72-90.

Back W. and Hanshaw B.B. (1970) Comparison of chemical hydrology of the carbonate peninsulas of Florida and Yucatan. *J. Hydrol.* **10**, 330-368.

Back W., Hanshaw B.B., Herman J.S., and Van Driel J.N. (1986) Differential dissolution of a Pleistocene reef in the ground-water mixing zone of coastal Yucatan, Mexico. *Geology* **14**, 137-140.

Bacon M.P. (1984) Glacial to interglacial changes in carbonate and clay sedimentation in the Atlantic Ocean estimated from ^{230}Th measurements. *Isotope Geosci.* **2**, 97-111.

Badiozamani K. (1973) The Dorag dolomitization model--application to the Middle Ordovician of Wisconsin. *J. Sediment. Petrol.* **43**, 965-984.

Badiozamani K., Mackenzie F.T. and Thorstenson D.C. (1977) Experimental carbonate cementation: Salinity, temperature, and vadose-phreatic effects. *J. Sediment. Petrol.* **47**, 529-542.

Baertschi P. (1957) Messung und Deutung relativer Häufigkeitsvariationen von O^{18} and C^{13} in Karbonatgesteinen und Mineralien. *Schweiz. Mineral. Petrog. Mitt.* **37**, 73-152.

Baertschi P. (1976) Absolute ^{18}O content of standard mean ocean water. *Earth Planet. Sci. Lett.* **31**, 341.

Baes C.F., Jr., Goeller H.E., Olson J.S. and Rotty R.M. (1976) *The Global Carbon Dioxide Problem.* NTIS, U.S. Dept. Comm., Springfield, VA, 72 pp.

Baker P.A. (1981) The diagenesis of marine carbonate sediments: Experimental and natural geochemical observations. Ph.D. dissertation, Univ. California, San Diego.

Baker P.A. and Bloomer S.H. (1988) The origin of celestite in deep-sea carbonate sediments. Geochim. Cosmochim. Acta **52**, 335-340.

Baker P.A. and Burns S.J. (1985) Occurrence and formation of dolomite in organic-rich continental margin sediments. *AAPG* **69**, 1917-1930.

Baker P.A. and Kastner M. (1981) Constraints on the formation of sedimentary dolomite. *Science* **213**, 214-216.

Baker P.A., Kastner M., Byerlee J.D. and Lockner D.A. (1980) Pressure solution and hydrothermal recrystallization of carbonate sediments--an experimental study. *Mar. Geol.* **38**, 185-203.

Baker P.A., Gieskes J.M. and Elderfield J. (1982) Diagenesis of carbonates in deep-sea sediments-evidence from Sr/Ca ratios and interstitial dissolved Sr^{2+} data. *J. Sediment. Petrol.* **52**, 71-82.

Barnaby R.J. and Rimstidt J.D. (1989) Redox conditions of calcite cementation interpreted from Mn and Fe contents of authigenic calcites. *Geolog. Soc. Amer. Bull.* **101**, 795-804.

Barnola J.M., Raynaud D., Korotkevich Y.S. and Lorius C. (1987) Vostok ice core provides 160,000-year record of atmospheric CO_2. *Nature* **329,** 408-414.

Baronnet A. (1982) Ostwald ripening: The case of calcite and mica. *Estudios Geologie* **38**, 185-198.

Barth T.F.W. (1962) *Theoretical Petrology, 2nd Edition.* John Wiley and Sons, Inc., New York, 387 pp.

Bates N.R. and Brand U. (1990) Secular variation of calcium carbonate mineralogy: An evaluation of oöid and micrite chemistries. *Geologishe Rundschau* (in press).

Bates R.G. (1973) *Determination of pH.* Wiley Interscience, New York, 479 pp.

Bathurst R.C.G. (1958) Diagenetic fabrics in some British Dinantian limestones. *Liverpool Manchester Geol. J.* **2**, 11-36.

Bathurst R.G.C. (1967) Oölitic films on low energy carbonate sand grains, Bimini Lagoon, Bahamas. *Mar. Geol.* **5**, 89-109.

Bathurst R.G.C. (1974) Marine diagenesis of shallow water calcium carbonate sediments. *Ann. Rev. Earth and Planet. Sci.* **2**, 257-274.

Bathurst R.G.C. (1975) *Carbonate Sediments and Their Diagenesis. Developments in Sedimentology 12*. (Second Edition), Elsevier Publishing. Co., Amsterdam,   658 pp.

Bathurst R.G.C. (1987) Diagenetically enhanced bedding in argillaceous platform limestones: stratified cementation and selective compaction. *Sedimentology* **34**, 749-778.

Baumann J., Buhmann D., Dreybrodt W. and Schultz H.D. (1985) Calcite dissolution kinetics in porous media. *Chem. Geolog.* **53**, 219-228.

Beach D.K. (1982) Depositional and diagenetic history of Pleiocene-Pleistocene carbonates of northwestern Great Bahama bank: Evolution of a carbonate platform. Ph.D. dissertation, Univ. of Miami, Coral Gables, Florida.

Ben-Yaakov S. (1973) pH buffering of pore water of recent anoxic marine sediments. *Limnol. Oceanogr.* **18**, 86-94

Ben-Yaakov S. and Kaplan I.R. (1971) Deep-sea in situ calcium carbonate saturometry. *J. Geophys. Res.* **76**, 722-731.

Bender M.L. and Heggie D.T. (1984) Fate of organic carbon reaching the deep sea floor: a status report. *Geochim. Cosmochim. Acta* **48**, 977-986.

Berger W.H. (1967) Foraminiferal ooze: Solutions at depths. *Science* **156**, 83-85.

Berger W.H. (1968) Planktonic Foraminifera: Selective solution and paleo-climatic interpretation. *Deep-Sea Res.* **15**, 31-43.

Berger W.H. (1970) Planktonic Foraminifera: Selective solution and the lysocline. *Mar. Geol.* **8**, 111-138.

Berger W.H. (1976) Biogenous Deep Sea Sediments: Production, Preservation and Interpretation. In *Chemical Oceanography, Vol. 5, 2nd Edition*. (eds. J.P. Riley and R. Chester), pp. 266-388. Academic Press, London.

Berger W.H. (1977) Deep-sea carbonate and the deglaciation preservation spike in pteropods and foraminifera. *Nature* **269**, 301-304.

Berger W.H. (1978) Deep-sea carbonate: Pteropod distribution and the aragonite compensation depth. *Deep-Sea Res.* **25**, 447-452.

Berger W.H. and Piper D.J.W. (1972) Planktonic foraminifera: Differential settling, dissolution, and redeposition. *Limnol. Oceanogr.* **17**, 275-287.

Berger W.H. and Soutar A. (1970) Preservation of plankton shells in an anaerobic basin off California. *Geol. Soc. Amer. Bull.* **81**, 275-282.

Berger W.H. and Vincent E. (1986) Deep-sea carbonates: Reading the carbon-isotope signal. *Geologische Rundschau* **75**, 249-269.

Berger W.H., Adelseck C.G., Jr. and Mayer L.A. (1976) Distribution of carbonate in surface sediments of the Pacific Ocean. *J. Geophys. Res.* **81**, 2617-2627.

Berner E.K. and Berner R.A. (1987) *The Global Water Cycle: Geochemistry and Environment.* Prentice Hall, Englewood Cliffs, N.J., 398 pp.

Berner R.A. (1966) Chemical diagenesis of some modern carbonate sediments. *Amer. J. Sci.* **264**, 1-36.

Berner R.A. (1969) Chemical changes affecting dissolved calcium during the bacterial composition of fish and clams in seawater. *Mar. Geol.* **7**, 253-2

Berner R.A. (1971) *Principles of Chemical Sedimentology.* McGraw-Hill, New York, 240 pp.

Berner R.A. (1974) Sedimentary pyrite formation: an update. *Geochim. Cosmochim. Acta* **48**, 605-615.

Berner R.A. (1975) The role of magnesium in the crystal growth of calcite and aragonite from seawater. *Geochim. Cosmochim. Acta* **39**, 489-504.

Berner R.A. (1976) The solubility of calcite and aragonite at one atmosphere and 34.5 parts per thousand. *Amer. J. Sci.* **276**, 713-730.

Berner R.A. (1977) Sedimentation and dissolution of pteropods in the Ocean. In *The Fate of Fossil Fuel CO_2 in the Oceans* (eds. N.R. Anderson and A. Malahoff), pp. 243-260. Plenum Press, New York.

Berner R.A. (1978) Equilibrium kinetics, and the precipitation of magnesian calcite from seawater. *Amer. J. Sci.* **278**, 1435-1477.

Berner R.A. (1980) *Early Diagenesis, A Theoretical Approach..* Princeton Univ. Press, Princeton, NJ, 241 pp.

Berner R.A. (1986) Kinetic approach to chemical diagenesis. In *Studies in Diagenesis* (ed. F.A. Mumpton), pp. 13-20.U.S. Geol. Surv. 1598, Washington.

Berner R.A. (1987) Models for carbon and sulfur cycles and atmospheric oxygen: Application to Paleozoic geologic history. *Amer. J. Sci.* **287**, 177-196.

Berner R.A. (1989) Biogeochemical cycles of carbon and sulfur and their effect on atmospheric oxygen over Phanerozoic time. *Paleogeogr. Paleoclimatol. Paleoecol.* **75**, 97-122.

Berner R.A. and Canfield D.E. (1989) A new model for atmospheric oxygen over Phanerozoic time. *Amer. J. Sci.* **289**, 333-361.

Berner R.A. and Honjo S. (1981) Pelagic sedimentation of aragonite: Its geochemical significance. *Science* **211**, 940-942.

Berner R.A and Lasaga A.C. (1989) Modeling the geochemical carbon cycle. *Scientific Amer.* **260**, 74-81.

Berner R.A. and Morse J.W. (1974) Dissolution kinetics of calcium carbonate in seawater. IV: Theory of calcite dissolution. *Amer. J. Sci.* **274**, 108-135.

Berner R.A. and Raiswell R. (1983) Burial of organic carbon and pyrite sulfur in sediments over Phanerozoic time: A new theory. *Geochim. Cosmochim. Acta* **47**, 855-862.

Berner R.A. and Westrich J.T. (1985) Bioturbation and the early diagenesis of carbon and sulfur. *Amer. J. Sci.* **285**, 195-206.

Berner R.A., Scott M.R. and Thomlinson C. (1970) Carbonate alkalinity in the pore waters of anoxic sediments. *Limnol. Oceanogr.* **15**, 544-549.

Berner R.A., Berner E.K. and Keir R. (1976) Aragonite dissolution on the Bermuda Pedastal: Its depth and geochemical significance. *Earth Planet. Sci. Lett.* **30**, 169-178.

Berner R.A., Westrich J.T., Graber R., Smith J. and Martens C.S. (1978) Inhibition of aragonite precipitation from supersaturated seawater: A laboratory and field study. *Amer. J. Sci.* **278**, 816-837.

Berner R.A., Lasaga A.C. and Garrels R.M. (1983) The carbonate-silicate geochemical cycle and its effect on atmospheric carbon dioxide over the past 100 million years. *Amer. J. Sci.* **283**, 641-683.

Bernstein L.D. and Morse J.W. (1985) The steady-state calcium carbonate ion activity product of recent shallow water carbonate sediments in seawater. *Mar. Chem.* **15**, 311-326.

Bertram M. (1989) Temperature effects on magnesian calcite solubility and reactivity: Application to natural systems. M.S. thesis, Univ. Hawaii.

Betzer P.R., Byrne R.H., Acker J.G., Lewis C.S., Jolley R.R. and Feely R.A. (1984) The Oceanic Carbonate System: A Reassessment of Biogenic Controls. *Science* **226**, 1074-1077.

Betzer P.R., Byrne R.H., Acker J.G., Lewis C.S., Jolley R.R. and Feely R.A. (1986) Biogenic input to the oceanic carbonate system: Mass fluxes of pteropods and foraminifera in tropical, temperate, and sub-arctic regions of the western North Pacific, (manuscript).

Bilinski H. and Schindler P. (1982) Solubility and equilibrium constants of lead carbonate solutions (25ºC, I=0.3 mol dm-3). *Geochim. Cosmochim. Acta* **46**, 921-928.

Birnbaum S.J. and Wireman J.W. (1984) Bacterial sulfate reduction and pH: Implications for early diagenesis. *Chem. Geol.* **43**, 143-149.

Biscaye P.E., Kolla V. and Turekian K.K. (1976) Distribution of calcium carbonate in surface sediments of the Atlantic Ocean. *J. Geophys. Res.* **81**, 2595-2603.

Bischoff J.L. (1969) Temperature controls on aragonite-calcite transformation in aqueous solution. *Amer. Mineralogist* **54**, 149-155.

Bischoff J.L. and Dickson F.W. (1975) Seawater-basalt interaction at 200°C and 500 bars: Implications for origin of seafloor heavy metal deposits and regulation of seawater chemistry. *Earth Planet. Sci. Lett.* **25**, 385-397.

Bischoff J.L. and Fyfe W.S. (1968) Catalysis, inhibition, and the calcite-aragonite problem. 1. The aragonite-calcite transformation. *Amer. J. Sci.* **266**, 65-79.

Bischoff J.L. and Seyfried W.E. (1978) Hydrothermal chemistry of seawater from 25°C to 350°C. *Amer. J. Sci.* **278**, 838-860.

Bischoff W.D. (1985) Magnesian calcites: Physical and chemical properties and stabilities in aqueous solution of synthetic and biogenic phases. Ph.D. dissertation, Northwestern Univ..

Bischoff W.D. and Burke C.D. (1989) Phanerozoic carbonate skeletal mineralogy and atmospheric CO_2. Proceed of the Chapman Confer. on the Gaia Hypothesis, Amer. Geophys. Union, Washington, D.C., (in press).

Bischoff W.D., Bishop F.C. and Mackenzie F.T. (1983) Biogenically produced magnesian calcite: Inhomogeneities in chemical and physical properties; comparison with synthetic phases. *Amer. Mineral.* **68**, 1183-1188.

Bischoff W.D., Sharma S.K., and Mackenzie F.T. (1985) Carbonate ion disorder in synthetic and biogenic magnesian calcites: A Raman spectral study. *Amer. Mineral.* **71**, 581-589.

Bischoff W.D., Mackenzie F.T. and Bishop F.C. (1987) Stabilities of synthetic magnesian calcites in aqueous solution: Comparison with biogenicmaterials. *Geochim. Cosmochim. Acta* **51**, 1413-1423.

Blake D.R. and Rowland F.S. (1988) Continuing worldwide increase in tropospheric methane, 1978-1987. *Science* **239**, 1129-1131.

Blatt H., Middleton G. and Murray R. (1980) *Origin of Sedimentary Rocks.* Prentice-Hall, Inc., Englewood Cliffs, N.J., 782 pp.

Bockris J.O'M. and Reddy A.K.N. (1977) *Modern Electrochemistry.* (2 volumes), 3rd Edition. Plenum/Rosetta, New York.

Bodine M.W., Holland H.D. and Borcsik M. (1965) Coprecipitation of manganese and strontium with calcite. Symposium: Problems of Postmagmatic Ore Deposition, *Prague* **2**, 401-406.

Böggild O.B. (1930) The shell structure of the mollusks. *Danske Didensk. Selsk, Sr.* **9**, 235-326.

Bolin B. (1960) On the exchange of carbon dioxide between the atmosphere and the sea. *Tellus* **12**, 274-281.

Borowitzska M.A. (1981) Photosynthesis and calcification in the articulated coralline red algae *Amphiroa anceps* and *A. foliacea. Mar. Biol.* **62**, 17-23.

Bowers T.S., Jackson K.J., and Helgeson H.C. (1984) *Equilibrium Activity Diagrams (for Coexisting Minerals and Aqueous Solutions at Pressures and Temperatures of 5 KB and 600°C)*. Springer-Verlag, Berlin, 397 pp.

Boyle E.A. (1981) Cadmium, zinc, copper, and barium in foraminifera tests. *Earth Planet. Sci. Lett.* **53**, 11-35.

Boyle E.A. (1983) Manganese carbonate overgrowths on foraminifera tests. *Geochim. Cosmochim. Acta* **47**, 1815-1819.

Boyle E.A. (1986) Paired carbon isotope and cadmium data from benthic foraminifera: Implications for changes in oceanic phosphorous, oceanic circulation, and atmospheric carbon dioxide. *Geochim. Cosmochim. Acta* **50**, 265-276.

Boyle E.A. (1988a) The role of vertical chemical fractionation in controlling late Quaternary atmospheric carbon dioxide. *J. Geophys. Res.* **93**, 15701-15714.

Boyle E.A. (1988b) Vertical oceanic nutrient fractionation and glacial/interglacial CO_2 cycles. *Nature* **331**, 55-56.

Boyle E.A. and Keigwin L.D. (1985) Comparison of Atlantic and Pacific paleochemical records for the last 25,000 years: Changes in deep ocean circulation and chemical inventories. *Earth and Planet. Sci. Lett.* **76**, 135-150.

Bradshaw A.L. and Brewer P.G. (1988a) High precision measurements of alkalinity and total carbon dioxide in seawater by potentiometric titration - 1. Presence of unknown protolyte(s)? *Mar. Chem.* **23**, 69-86.

Bradshaw A.L. and Brewer P.G. (1988b) High precision measurements of alkalinity and total carbon dioxide in seawater by potentiometric titration - 2. Measurements on standard solutions. *Mar. Chem.* **24**, 155-162.

Bradshaw A.L., Brewer P.G., Shafer D.K. and Williams R.T. (1981) Measurements of total carbon dioxide and alkalinity by potentiometric titration in the GEOSECS program. *Earth Planet. Sci. Lett.* **55**, 99-115.

Bramlette M.N. (1958) Significance of coccolithophorids in calcium carbonate deposition. *Geol. Soc. Amer. Bull.* **69**, 121-126.

Bramlette M.N. (1961) Pelagic sediments. In *Oceanography* (ed. M. Sears), pp. 345-366. Amer. Assoc. Adv. Sci. Publ. **67**.

Brand U. (1982) The ^{18}O and ^{13}C contents of Carboniferous fossil components: Seawater effects. *Sedimentology* **29**, 139-147.

Brand U. and Veizer J. (1980) Chemical diagenesis of a multicomponent carbonate system 1: Trace elements. *J. Sediment. Petrol.* **50**, 1219-1236.

Brand U. and Veizer J. (1981) Chemical diagenesis of a multicomponent carbonate system--2, Stable isotopes. *J. Sediment. Petrol.* **51**, 987-997.

Bretz J.H. (1960) Bermuda: a partially drowned late mature Pleistocene karst. *Geol. Soc. Amer. Bull.* **71**, 1729-1754.

Brewer P.G., Wong G.T.F., Bacon M.P. and Spencer D.W. (1975) An oceanic calcium problem? *Earth Planet. Sci. Lett.* **26**, 81-87.

Brewer P.G., Nozaki Y., Spencer D.W. and Fleer A.P. (1980) Sediment trap experiments in the deep North Atlantic: Isotopic and elemental fluxes. *J. Mar. Res.* **38**, 703-728.

Bricker O.P. (1971) *Carbonate Cements.* John Hopkins Univ. Studies in Geology, **19**, 376 pp.

Bricker O.P. and Mackenzie F.T. (1970) Limestones and red soils of Bermuda: Discussion. *Geol. Soc. Amer. Bull.* **81**, 2523-2524.

Broecker W.S. (1970) A boundary condition on the evolution of atmospheric oxygen. *J. Geophys. Res.* **75**, 3553-3557.

Broecker W.S. (1982) Ocean chemistry during glacial time. *Geochim. Cosmochim. Acta* **46**, 1689-1705.

Broecker W.S. (1983) The ocean. *Scientific Amer.* **249**, 146-160.

Broecker W.S. and Denton G.H. (1989) The role of ocean-atmosphere reorganizations in glacial cycles. *Geochim. Cosmochim. Acta* **53**, 2465-2501.

Broecker W.S. and Denton G.H. (1990) What drives glacial cycles? *Scientific Amer.* **262**, 48-56.

Broecker W.S. and Peng T.-H. (1974) Gas exchange rates between air and sea. *Tellus* **26**, 21-35.

Broecker W.S. and Peng T.-H. (1982) *Tracers in the Sea.* Eldigio Press, Palisades, NY, 690 pp.

Broecker W.S. and Peng T-H. (1987) The role of $CaCO_3$ compensation in the glacial to interglacial atmospheric CO_2 change. *Global Biogeochemical Cycles* **1**, 15-30.

Broecker W.S. and Peng T.-H. (1989) The cause of the glacial to interglacial atmospheric CO_2 change: A polar alkalinity hypothesis. *Global Biogeochemical Cycles* **3**, 215-239.

Broecker W.S. and Takahashi T. (1966) Calcium carbonate precipitation on the Bahama Banks. *J. Geophys. Res.* **71**, 1575-1602.

Broecker W.S. and Takahashi T. (1977) Neutralization of fossil fuel CO_2 by marine calcium carbonate. In *The Fate of Fossil Fuel CO_2 in the Oceans* (eds. N.R. Anderson and A. Malahoff), pp. 213-241. Plenum Press, New York.

Broecker W.S., Takahashi T. and Peng V. (1985) *Reconstruction of Past Atmospheric CO_2 Contents from the Chemistry of the Contemporary Ocean: An Evaluation.* DOE/OR-857(TR020), Washington, D.C., 79 pp.

Broecker W.S., Ledwell J.R., Takahashi T., Weiss R., Merlivat L, Memery L., Peng T.-H., Jahne B. and Munnich K.O. (1986) Isotopic versus micrometeorologic ocean CO_2 fluxes: A serious conflict. *J. Geophys. Res.* **91**, 10517-10527.

Brooks J.M., Kennicutt II, M.C., Fischer C.R., Macko S.A., Cole K., Childress J.J., Bidigare R.R. and Vetter R.D. (1987) Deep-sea hydro-carbon seep communities: Evidence for energy in nutritional carbon sources. *Science* **238**, 1138-1142.

Brown W.H. Fyfe W.S. and Turner F.J. (1962) Aragonite in California glaucophane schists. *J. Petrol.* **3**, 566-587.

Buat-Menard P. (1983) Particle geochemistry in the atmosphere and oceans. In *Air-Sea Exchange of Gases and Particles* (eds. P.S. Liss and G.N. Shinn), pp. 455-532. D. Reidel Publish. Co., Dordrecht.

Buat-Menard P. and Chesselet R. (1979) Variable influence of the atmospheric flux on the trace metal chemistry of oceanic suspended matter. *Earth and Planet. Sci. Lett.*, **42**, 399-411.

Buchbinder L.G. and Friedman G.M. (1980) Vadose, phreatic and marine diagenesis of Pleistocene-Holocene carbonates in a borehole-Mediterranean coast of Isreal. *J. Sediment. Petrol.* **50**, 395-409.

Budd D.A. (1984) Freshwater diagenesis of Holocene oöid sands, Schooner Cays, Bahamas. Ph.D. dissertation, Univ. of Texas at Austin.

Budd D.A. (1988) Aragonite-to-calcite transformation during fresh-water diagenesis of carbonates: Insights from pore-water chemistry. *Geol. Soc. Amer. Bull.* **100**, 1260-1270.

Buddemeier R.W. and Smith S.V. (1988) Coral reef growth in an era of rapidly rising sea level: Predictions and suggestions for long-term research. *Coral Reefs* **7**, 51-56.

Buddemeier R.W., Schneider R.D. and Smith S.V. (1981) The alkaline earth chemistry of corals. *Proc. Fourth International Coral Reef Symp., Manila* **2**, 81-85.

Budyko M.I., Ronov A.B. and Yanshin A.L. (1985) *History of the Earth's Atmosphere*. Springer Verlag, New York, 139 pp.

Buhmann D. and Dreybrodt W. (1985a) The kinetics of calcite dissolution and precipitation in geologically relevant situations of Karst areas. 1. Open System. *Chem. Geol.* **48**, 189-211.

Buhmann D. and Dreybrodt W. (1985b) The kinetics of calcite dissolution and precipitation in geologically relevant situations of Karst areas. 2. Closed System. *Chem. Geol.* **53**, 109-124.

Buhmann D. and Dreybrodt W. (1987) Calcite dissolution kinetics in the system $H_2O-CO_2-CaCO_3$ with participation of foreign ions. *Chem. Geol.* **64**, 89-102.

Burke W.H., Denison R.E., Hetherington E.A., Koepnick R.B., Nelson H.F. and Otto J.B. (1982) Variation of seawater strontium 87/strontium 86 throughout Phanerozoic time. *Geology* **10**, 516-519.

Burns S.J. and Baker P.A. (1987) A geochemical study of dolomite in the Monterey Formation, California. *J. Sediment. Petrol.* **57**, 126-139.

Burton E.A. and Walter L.M. (1987) Relative precipitation rates of aragonite and Mg calcite from seawater: Temperature or carbonate ion control? *Geology* **15**, 111-114.

Burton W.K., Carbrera N. and Frank F.C. (1951) The growth of crystals and the equilibrium structure of their surfaces. *Phil Trans. Roy. Soc.*, A243: 299-358.

Busenberg E. and Plummer L.N. (1985) Kinetic and thermodynamic factors controlling the distribution of SO_4^{2-} and Na^+ in calcites and selected aragonites. *Geochim. Cosmochim. Acta* **49**, 713.

Busenberg E. and Plummer L.N. (1986a) The solubility of $BaCO_3(cr)$ (witherite) in CO_2-H_2O solutions between 0 and 90°C, evaluation of the association constants of $BaHCO_3^+(aq)$ and $BaCO_3^0(aq)$ between 5 and 80°C, and a preliminary evaluation of the thermodynamic properties of $Ba^{2+}(aq)$. *Geochim. Cosmochim. Acta* **50**, 2225-2233.

Busenberg E. and Plummer L.N. (1986b) A comparative study of the dissolution and crystal growth kinetics of calcite and aragonite. In *Studies in Diagenesis* (ed. F.A. Mumpton), pp. 139-168. U.S. Geol. Surv. Bull. **1578**.

Busenberg E. and Plummer L.N. (1989) Thermodynamics of magnesian calcite solid-solutions at 25°C and 1 atm pressure. *Geochim. Cosmochim. Acta* **53**, 1189-1208.

Busenberg E., Plummer L.N. and Parker V.B. (1984) The solubility of strontianite ($SrCO_3$) in CO_2-H_2O solutions between 2 and 91°C, the association constants of $SrHCO_3^+(aq)$ and $SrCO_3^0(aq)$ between 5 and 80°C, and an evaluation of the thermochemical properties of $Sr^{2+}(aq)$ and $SrCO_3(cr)$ at 25°C and 1 atm total pressure. *Geochim. Cosmochim. Acta* **48**, 2021-2035.

Butler J.N. (1982) *Carbon Dioxide Equilibria and Their Applications*. Addison-Wesley Pub. Co., Cambridge, MA., 259 pp.

Byrne R.H., Acker J.G., Betzer P.R., Feely R.A. and Cates M.H. (1984) Water column dissolution of aragonite in the Pacific Ocean. *Nature* **312**, 321-326.

Carbollo J.D., Land L.S. and Miser D.E. (1987) Holocene dolomitization of supratidal sediments by active tidal pumping, Sugarloaf Key, Florida. *J. Sediment. Petrol.* **57**, 153-165.

Carlson W.D. (1980) The calcite-aragonite equilibrium: effects of Sr substitution and anion orientational disorder. *Amer. Mineral.* **65**, 1252-1262.

Carlson W.D. (1983) The polymorphs of $CaCO_3$ and the aragonite-calcite transformation. In *Carbonates: Mineralogy and Chemistry* (ed. R.R. Reeder), pp. 191-225. Min. Soc. Amer., Bookcrafters Inc., Chelsea, MI.

Carpenter A.B. (1976) Dorag dolomitization model by K. Badiozamani-A discussion. *J. Sediment. Petrol.* **46**, 258-261.

Carpenter A.B. (1980) The chemistry of dolomite formation 1. The stability of dolomite. In *Concepts and Models of Dolomitization.* (eds. D.H. Zenger, J.B. Dunham and R.L Ethington), pp. 111-121. Soc. Econ. Paleontologist and Mineralogists, Spec. Pub. **28**. Tulsa, OK.

Carothers W.W. and Kharaka Y.K. (1978) Aliphatic acid anions in oil-field waters--implications for origin of natural gas. *AAPG Bull.* **62**, 2441-2453.

Carter P.W. (1978) Adsorption of amino-acid containing organic matter by calcite and quartz. *Geochim. Cosmochim. Acta* **42**, 1239-1242.

Carter P.W. and Mitterer R.M. (1978) Amino acid composition of organic matter associated with carbonate and non-carbonate sediments. *Geochim. Cosmochim. Acta* **42**, 1231-1238.

Cavaliere A.R., Barnes R.D. and Cook C.B. (1987) *Field Guide to the Conspicuous Flora and Fauna of Bermuda.* Bermuda Biological Station for Research Special Publication No. 28, Saint Georges, Bermuda, 82 pp.

Chafetz H.S. (1986) Marine peloids: A product of bacterially induced precipitation of calcite. *J. Sediment Petrol.* **56**, 812-817.

Chafetz H.S., Wilkinson B.H., and Love K.M. (1985) Morphology and composition of non-marine carbonate cements in near-surface settings. In *Carbonate Cements* (eds. N. Schneidermann and P.M. Harris), pp. 337-347. SEPM Spec. Pub. No. 36, Tulsa, OK.

Chafetz H.S., McIntosh A.G. and Rush P.F. (1988) Freshwater phreatic diagenesis in the marine realm of recent Arabian Gulf carbonates. *J. Sediment. Petrol.* **58**, 433-440.

Chamberlain T.C. (1898) An attempt to frame a working hypothesis of the cause of glacial periods on an atmospheric basis. *J. Geol.* **7**, 545.

Chave K.E. (1952) A solid solution between calcite and dolomite. *J. Geol.* **60**, 190-192.

Chave K.E. (1954a) Aspects of the biogeochemistry of magnesium 1. Calcareous marine organisms. *J. Geol.* **62**, 266-283.

Chave K.E. (1954b) Aspects of the biogeochemistry of magnesium 2. Calcareous sediments and rocks. *J. Geol.* **62**, 587-599.

Chave K.E. (1960) Carbonate skeletons to limestones: Problems. *Trans. N.Y. Acad. Sci.* **23**, 14-24.

Chave K.E. (1962a) Factors influencing the mineralogy of carbonate sediments. *Limnol. Oceanogr.* **7**, 218-223.

Chave K.E. (1962b) Processes of carbonate sedimentation. *Narragansett Marine Lab., Occ. Pub.* **1**, 77-85.

Chave K.E. (1964) Skeletal durability and preservation. In *Approaches to Paleoecology* (eds. J. Imbre and N. Newell), pp. 377-387. John Wiley and Sons, Inc., New York.

Chave K.E. (1967) Recent carbonate sediments-An unconventional view. *J. Geological Education* **15**, 200-204.

Chave K.E. (1981) Summary of background information on magnesium calcites. In *Some Aspects of the Role of the Shallow Ocean in Global Carbon Dioxide Uptake* (eds. R.M. Garrels and F.T. Mackenzie), pp. A4-A9. U.S.D.O.E. Conf. 8003115, Natl. Tech. Inform. Service, Springfield, VA.

Chave K.E. (1984) Physics and chemistry of biomineralization. *Ann. Rev. Earth Planet. Sci.* **12**, 293-305.

Chave K.E. and Schmalz R.F. (1966) Carbonate-seawater interaction. *Geochim. Cosmochim. Acta* **30**, 1037-1048.

Chave K.E. and Suess E. (1967) Suspended minerals in seawater. *N.Y. Acad. Sci. Trans.* **29**, 991-1000.

Chave K.E., Deffeyes K.S., Weyl P.K., Garrels R.M. and Thompson M.E. (1962) Observations on the solubility of skeletal carbonates in aqueous solutions. *Science* **137**, 33-34.

Chen C.-T. A., Pytkowicz R.M. and Olson E.J. (1982) Evaluation of the calcium problem in the South Pacific. *Geochem. J.* **16**, 1-10.

Chilingar G.V. (1956) Relationship between Ca/Mg ratio and geological age. *AAPG Bull.* **40,** 2256-2266.

Chilingar G.V. (1962) Dependence on temperature of Ca/Mg ratio of skeletal structures of organisms and direct chemical precipitates out of seawater. *Southern California Acad. Sci. Bull.* **61**, 45-60.

Choquette P.W. and James N.P. (1987) Diagenesis #12. Diagenesis in limestones-3. The deep burial environment. *Geoscience Canada* **14**, 3-35.

Choquette P.W. and Pray L.C. (1970) Geologic nomenclature and classification of porosity in sedimentary carbonates. *AAPG Bull.* **54**, 207-250.

Chou L., Garrels R.M. and Wollast R. (1989) Comparative study of the dissolution kinetics and mechanisms of carbonates in aqueous solutions. *Chem. Geol.* **78**, 269-282.

Christ C.L. and Hostetler P.B. (1970) Studies in the system $MgO-SiO_2-CO_2-H_2O$ (II): The activity-product constant of magnesite. *Amer. J. Sci.* **268**, 439-453.

Chyba C.F. (1990) Impact delivery and erosion of planetary oceans in the early inner Solar System. *Nature* **343,** 129-133.

Cicerone R.J. and Oremland R.S. (1988) Biogeochemical aspects of atmospheric methane. *Global Biogeochemical Cycles* **2,** 299-327.

Clark S.P.Jr. (1966) High-pressure phase equilibria. In *Handbook of Physical Constants* (ed. S.P. Clark, Jr.), pp. 345-370. Geol. Soc. America Memoir **97.**

Clarke F.W. (1924) *Data on Geochemistry.* U.S. Geol. Sur. Bull., **770,** 841 pp.

Clarke F.W. and Wheeler W.C. (1917) The inorganic constituents of marine invertebrates. *U.S. Geol. Surv. Profess. Paper* 124, 56 pp.

Claypool G.E., Holser W.T., Kaplan I.R., Sakai H. and Zak I. (1972) Sulfur and oxygen isotope geochemistry of evaporite sulfates. *Abstr. Geol. Soc. Amer.* **4,** 473.

Claypool G.E., Holser W.T., Kaplan I.R., Sakai H. and Zak I. (1980) The age curves of sulfur and oxygen isotopes in marine sulfate and their mutual interpretation. *Chem. Geol.* **28,** 199-260.

Clayton R.N., Friedman I., Graf D.L., Mayeda T.K., Meents W.F. and Shimp N.F. (1966) The origin of saline formation waters, I. Isotopic composition. *J. Geophys. Res.* **71,** 3869-3882.

Clayton R.N., Goldsmith J.R., Karel K.J., Mayeda T.K. and Newton R.C. (1975) Limits on the effect of pressure in isotopic fractionation. *Geochim. Cosmochim. Acta* **39,** 1197.

Cloud P.E., Jr. (1962a) Environment of calcium carbonate deposition west of Andros Island, Bahamas. *U.S. Geol. Surv. Profess. Papers* **350,** 1-138.

Cloud P.E., Jr. (1962b) Behaviour of calcium carbonate in sea water. *Geochim. Cosmochim. Acta* **26,** 867-884.

Cobler R. and Dymond J. (1980) Sediment trap experiment on the Galapagos spreading center. Equatorial Pacific. *Science* **209,** 801-803.

Compton J.S. (1988) Degree of supersaturation and precipitation of organogenic dolomite. *Geology* **16,** 318-321.

Compton J.S. and Siever R. (1984) Stratigraphy and dolostone occurrence in the Miocene Monterey Formation, Santa Maria Basin area, California. In *Dolomites of the Monterey Formation and Other Organic-Rich Units* (eds. R.E. Garrison, M. Kastner and D.H. Zenger), pp. 141-153. Society Economic Paleontologists and Mineralogists Pacific Section **41.**

Compton J.S. and Siever R. (1986) Diffusion and mass balance of magnesium during early dolomite formation, Monterey Formation. *Geochim. Cosmochim. Acta* **50,** 125-135.

Crocket J.H. and Winchester J.W. (1966) Coprecipitation of zinc with calcium carbonate. *Geochim. Cosmochim. Acta* **30,** 1093-1109.

Crowley T.J. (1985) Late quaternary carbonate changes in the north Atlantic and Atlantic/Pacific comparisons. In *The Carbon Cycle and Atmospheric CO2: Natural Variations Archean to Present* (eds. E.T. Sundquist and W.S. Broecker), pp. 271-284. Amer. Geophys. Union, Washington, D.C.

Curry W.B. and Lohmann G.P. (1985) Carbon deposition rates and deep water residence time in the equatorial Atlantic Ocean throughout the last 160,000 years. In *The Carbon Cycle and Atmospheric CO2: Natural Variations Archean to Present* (eds. E.T. Sundquist and W.S. Broecker), pp. 285-301. Amer. Geophys. Union, Washington, D.C.

Curry W.B. and Lohmann G.P. (1986) Late quaternary carbonate sedimentation at the Sierra Leone Rise (eastern equatorial Atlantic Ocean). *Mar. Geol.* **70**, 223-250.

Czerniakowski L.A., Lohmann K.C. and Wilson J.L. (1984) Closed system marine burial diagenesis: Isotopic data from the Austin Chalk and its components. *Sedimentology* **31**, 863-877.

Daly R.A. (1909) First calcareous fossils and evolution of limestones. *Geol. Soc. Amer. Bull.* **20**, 153-170.

Dardenne M. (1967) Etude experimentale de la distribution du zinc dans les carbonates de calcium. *Bull. Bur. Rech. Geol. Min.* **5**, 75-110.

Davies P.J., Bubela B. and Ferguson J. (1978) The formation of oöids. *Sedimentology* **25**, 703-730.

Davis J.A., Fuller C.C. and Cook A.D. (1987) A model for trace metal sorption processes at the calcite surface: Adsorption of Cd^{2+} and subsequent solid solution formation. *Geochim. Cosmochim Acta* **51**, 1477-1490.

Davis J.A. and Hayes K.F. (1986) Geochemical Processes at Mineral Surfaces: An Overview. In *Geochemical Processes at Mineral Surfaces* (eds. J.A. Davis and K.F. Hayes), pp. 2-19.

de Boer R.B. (1977) On the thermodynamics of pressure solution--interaction between chemical and mechanical forces. *Geochim. Cosmochim. Acta* **41**, 249-256.

Deevey E.S., Jr. (1973) Sulfur, nitrogen, and carbon in the biosphere. In *Carbon and the Biosphere* (eds. G.M. Woodwell and E.B. Peacan), pp. 182-190. WSAEC, Washington, D.C.

Degens E.T. (1976) Molecular mechanisms on carbonate, phosphate, and silica deposition in the living cell. *Topics in Current Chemistry* **64**, 1-111.

deKanel J. and Morse J.W. (1978) The chemistry of orthophosphate uptake from seawater onto calcite and aragonite. *Geochim. Cosmochim. Acta* **42**, 1335-1340.

Delaney A.C., Parkin D.W., Griffin J.J., Goldberg E.D. and Reiman B.E.F. (1967) Airborne dust collected at Barbados. *Geochim. Cosmochim. Acta* **31**, 885-909.

Delaney M.L., Bé A.W.H. and Boyle V. (1985) Li, Sr, Mg, and Na in foraminiferal calcite shells from laboratory culture, sediment traps, and sediment cores. *Geochim. Cosmochim. Acta* **49**, 1327-1341.

Delaney A.C., Parkin D.W., Griffin J.J., Goldberg E.D. and Reiman B.E.F. (1967) Airborne dust collected at Barbados. *Geochim. Cosmochim. Acta* **31**, 885-909.

Delwiche C.C. and Likens G.E. (1977) Biological response to fossil fuel combustion products. In *Global Chemical Cycles and Their Alterations by Man* (ed. W. Stumm), pp. 73-88. Dahlen Konferenzen, Berlin.

Desitter L.U. (1947) Diagenesis of oil-field brines. *AAPG Bull.* **31**, 2030-2040.

Des Marais D.J. (1985) Carbon exchange between the mantle and the crust, and its effect upon the atmosphere: Today compared to Archean time. In *The Carbon Cycle and Atmospheric CO_2: Natural Variations Archean to Present* (eds. E.T. Sundquist and W.E. Broecker), pp. 602-611. Am. Geophys. Union Geophys. Monograph **32**, Washington, D.C.

Detwiler R.P. and Hall C.A.S. (1988) Tropical forests and the global carbon cycle. *Science* **239**, 42-46.

Devery D.M. and Ehlmann A.J. (1981) Morphological changes in a series of synthetic Mg-calcites. *Amer. Mineral.* **66**, 592-595.

Dickson J.A.D. (1985) Diagenesis of shallow-marine carbonates. In *Sedimentology: Recent Developments and Applied Aspects.* (eds. P.J. Brenchley and B.P.J. Williams), pp.173-188. Blackwell Scientific Publications, Boston, MA.

Dodd J.R. (1967) Magnesium and strontium in calcareous skeletons: A review. *J. Paleontol.* **41**, 1313-1329.

Doerner H.A. and Hoskins W.M. (1925) Coprecipitation of radium and barium sulfates. *J. Amer. Chem. Soc.* **47**, 662-675.

Dorobek S.L. (1987) Petrology, geochemistry and origin of burial diagenetic facies, Siluro-Devonian Helderberg Group (carbonate rocks), central Appalachians. *AAPG Bull.* **71**, 492-514.

Dorobek S.L. and Read J.F. (1986) Sedimentology and basin evolution of the Siluro-Devonian Helderberg Group, central Appalachians. *J. Sediment. Petrol.* **56**, 601-613.

Dravis J. (1979) Rapid and widespread generation of recent oölitic hardgrounds on a high energy Bahamian platform, Eleuthera Bank, Bahamas. *J. Sediment. Petrol.* **49**, 195-208.

Drever J.L. (1974) The magnesium problem. In *The Sea. Vol. 5* (ed. E.D. Goldberg), pp. 337-358. Wiley Interscience, New York.

Drever J.L., Li Y.-H. and Maynard J.B. (1987) Geochemical cycles: The continental crust and the oceans. In *Chemical Cycles in the Evolution of the Earth* (eds. C.B. Gregor, R.M. Garrels F.T. Mackenzie and J.B. Maynard), pp. 17-54. John Wiley, New York.

Dromgoole E.L. and Walter L.M. (1987) Iron manganese incorporation into calcite cements: Implications for diagenetic studies. Geological Society of America Annual Meeting and Exposition, Phoenix, Arizona, 627.

Droxler A.W., Morse J. W. and Kornicker W. A. (1988a) Controls on carbonate mineral accumulation in Bahamian Basins and adjacent Atlantic Ocean sediments. *J. Sediment. Petrol.* **58**, 120-130.

Droxler A.W., Morse J.W. and Baker V. (1988b) Good agreement between carbonate mineralogical depth variations of surficial periplatform ooze and carbonate saturation levels of the overlying intermediate waters. New data from the Nicaragua Rise. *EOS* **91**, 1345.

Dunham R.J. (1962) Classification of carbonate rocks according to depositional texture. In *American Association Petroleum Geologists Memoir No. 1* (ed. W.E. Ham), pp. 108-121, Tulsa, OK.

Dunham R.J. (1969) Early vadose silt in Townsend mound (reef), New Mexico. In *Depositional Environments in Carbonate Rocks: A Symposium* (ed. G.M. Friedman), pp. 139-181. Soc. Econ. Paleontologists Mineralogists Spec. Publ. 14, Tulsa, OK.

Dunham R.J. (1971) Meniscus cement. In *Carbonate Cements* (ed. O.P. Bricker) pp. 297-300. AAPG Stud. in Geol. No. 19. John Hopkins Univ. Press, Baltimore, MD.

Dunlop J.S.R., Muir M.O., Milne V.A. and Groves D.I. (1978) A new microfossil assemblage from the Archean of Western Australia. *Nature* **274,** 676-678.

Durney D.W. (1976) Pressure-solution and crystallization deformation. *Phil. Trans. R. Soc. Lond. A* **283**, 229-240.

Ebhardt G. (1968) Experimental compaction of carbonate sediments. In *Recent Developments in Carbonate Sedimentology of Central Europe* (eds. G. Muller and G. Friedman), pp. 58-65. Springer-Verlag, Berlin.

Edmond J.M. and Von Damm K. (1983) Hot springs on the ocean floor. *Scientific American* **238**, 78-92.

Edmond J.M., Measures C., McDuff R.E., Chan L.H., Collier R., Grant B., Gordon L.I. and Corliss J.B. (1979) Ridge crest hydrothermal activity and the balances of the major and minor elements in the ocean: The Galapagos data. *Earth Planet. Sci. Lett.* **46**, 1-18.

Edmunds W.M., Cook J.M., Darling W.G., Kinniburgh D.G., Miles D.L., Bath A.H., Morgan-Jones M., and Andrews J.N. (1987) Baseline geochemical conditions in the Chalk aquifer, Berkshire, U.K.: A basis for groundwater quality management. *Applied Geochemistry* **2**, 251-274.

Ehhalt D.H. (1979) Der atmospharische krieslauf von methan. *Naturwissenschaften* **66**, 307-311.

Einsele G. (1982) Mechanism of sill intrusion into soft sediment and expulsion of pore water. *Initial Reports Deep Sea Drilling Project* **64**, 1169-1176.

Ekdale A.A. and Bromley R.G. (1988) Diagenetic microlamination in chalk. *J. Sediment. Petrol.* **58**, 857-861.

Elderfield H., Gieskes J.M., Baker P.A., Oldfield R.K., Hawkesworth C.J. and Miller R. (1982) $^{87}Sr/^{86}Sr$ and $^{18}O/^{16}O$ ratios, interstitial water chemistry and diagenesis in deep-sea carbonate sediments of the Ontong Java Plateau. *Geochim. Cosmochim. Acta* **46**, 2259-2268.

Emerson S.R. and Bender M.L. (1981) Carbon fluxes at the sediment-water inter in the deep-sea: Calcium carbonate preservation. *J. Mar. Res.* **39**, 139-162.

Emerson S., Jahnke R., Bender M., Froelich P., Klinkhammer G., Bowser C. and Setlock G. (1980) Early diagenesis in sediments from the eastern equatorial Pacific. I. Pore water nutrient and carbonate results. *Earth Planet. Sci. Lett.* **49**, 57-80.

Emerson S.R., Grundmanis V. and Graham V. (1982) Carbonate chemistry in marine pore waters: MANOP sites C and S. *Earth Planet. Sci. Lett.* **61**, 220-232.

Emerson S.R., Fischer K., Reimers C. and Heggie D. (1985) Organic carbon dynamics and preservation in deep-sea sediments. *Deep-Sea Res.* **32**, 1-21.

Emrich K., Ehhalt D.H. and Vogel J.C. (1970) Carbon isotope fractionation during the precipitation of calcium carbonate. *Earth Planet. Sci. Lett.* **8**, 363-371.

Engel A.E.J. and Engel C.G. (1970) Continental accretion and the evolution of North America. In *Adventures in Earth History* (ed. P. Cloud), pp. 293-312. W.H. Freeman and Co., San Francisco.

Enos P. and Sawatsky L.H. (1981) Pore networks in Holocene carbonate sediments. *J. Sediment. Petrol.* **51**, 961-985.

Fabricius F.H. (1977) *Origin of Marine Oöids. Contributions to Sedimentology*, **7**, E. Schweizerbartshe Verlagsbuchhandlung, Stuttgart. 113 pp.

Fanning K.A., Byrne R.H., Breland II, J.A. and Betzer P.R. (1981) Geothermal springs of the west Florida continental shelf: Evidence for dolomitization and radionuclide enrichment. *Earth and Planet. Sci. Lett.* **52**, 345-354.

Feely R.A. and Chen C.-T. A. (1982) The effect of excess CO_2 on the calculated calcite and aragonite saturation horizons in the northeast Pacific. *Geophys. Res. Lett.* **9**, 1294-1297.

Ferguson J., Bubela B. and Davies P.J. (1978) Synthesis and possible mechanism of formation of radial carbonate oöids. *Chem. Geol.* **22**, 285-308.

Fischer A.G. (1984) The two Phanerozoic super cycles. In *Catastrophies in Earth History* (eds. W.A. Berggren and J.A. Vancouvering), pp.129-148. Princeton Univ. Press, Princeton, NJ.

Fischer A.G. and Arthur M.A. (1977) Secular variations in the pelagic realm. *Special Publication Soc. Econ. Paleontologists and Mineralogists* **25**, 19-50.

Fischer A.T., Honjo S. and Garrison R.W. (1967) *Electron Micrographs of Limestones and Their Nannofossils.* Princeton Univ. Press, Princeton, New Jersey, 137 pp.

Folk R.L. (1959) Practical petrographic classification of limestones. *AAPG Bull.* **43**, 1-38.

Folk R.L. (1962) Spectral subdivision of limestone types. In *Classification of Carbonate Rocks* (ed. W.E. Ham) pp. 62-84. AAPG Mem. **1**, Tulsa, OK.

Folk R.L. (1973) Carbonate petrography in the Post-Sorbian age. In *Evolving Concepts in Sedimentology* (ed. R.N. Ginsburg), pp. 118-158. John Hopkins Univ. Press, Baltimore, MD.

Folk R.L. (1974) The natural history of crystalline calcium carbonate: Effect of magnesium content and salinity. *J. Sediment. Petrol.* **44**, 40-53.

Folk R.L. (1978) A chemical model for calcite crystal growth and morphology control. *J. Sediment. Petrol.* **48**, 345-347.

Folk R.L. and Land L.S. (1975) Mg/Ca ratio and salinity: Two controls over crystallization of dolomite. *AAPG Bull.* **59**, 60-68.

Foxall R., Peterson G.C., Rendall H.M. and Smith A.L. (1979) Charge determination at calcium salt/aqueous solution interface. *Faraday I* **75**, 1034-1039.

Frank J.R., Carpenter R.B. and Oglesby T.W. (1982) Cathodoluminescence and composition of calcite cement in the Taum Sauk limestone (Upper Cambrian), Southeast Missouri. *J. Sedimet. Petrol.* **52**, 631-638.

Franklin M.L. and Morse J.W. (1983) The interaction of manganese(II) with the surface of calcite in dilute solutions and seawater. *Mar. Chem.* **12**, 241-254.

Franks S.G. and Forester R.W. (1984) Relationships among secondary porosity, pore-fluid chemistry and carbon dioxide, Texas Gulf Coast. In *Clastic Diagenesis* (eds. D.A. McDonald and R.C. Surdam), pp. 63-79. AAPG, Tulsa, OK.

Frantz J.D. (1971) Bermudian ground and cave waters: Bermuda Biological Station for Research. Spec. Pub. **7**, 47-55.

Friedman G.M. (1969) *Depositional Environments in Carbonate Rocks. Soc. Econ. Paleontologists and Mineralogists, Spec. Publ.* **14**, Tulsa, OK., 209 pp.

Friedman G.M. (1980) Dolomite is an evaporite mineral: Evidence from the rock record and from sea-marginal ponds of the Red Sea. *Soc. Econ. Paleontologists and Mineralogists* Spec. Pub. 28. Tulsa, OK. 69-80.

Friedman G.M. (1988) Methane-derived organogenic carbonates formed by subduction-induced pore-water expulsion along the Oregon/Washington margin: Discussion and reply. *Geol. Soc. Amer. Bull.* **100**, 622-623.

Friedman G.M. and Sanders J.E. (1967) Origin and occurrence of dolostones. In *Carbonate Rocks: Developments in Sedimentology* (eds. G.V. Chilingar, H.J. Bissell and R.W. Fairbridge) pp. 267-348 9A: Elsevier Publishing Co., Amsterdam.

Frishman S.A. and Behrens E.W. (1969) Geochemistry of oölites, Baffin Bay, Texas. Geol. Soc. America Abs. with Programs, pt. **7**, 71.

Füchtbauer H. and Hardie L.A. (1976) Experimentally determined homogeneous distribution coefficients for precipitated magnesian calcites: Application to marine carbonate cements. *Ann. Mtg. Geol. Soc. Amer.* **8**, 877.

Fyfe W.S. and Bischoff J.L. (1965) The calcite-aragonite problem. *Soc. Econ. Paleontologists and Mineralogists, Spec. Pub.* **13**, 3-13.

Gaffey Susan J. (1988) Water in skeletal carbonates. *J. Sediment. Petrol.* **58**, 397-414.

Gaillard J.-F., Jeandel C., Michard G., Nicolas E. and Renard D. (1986) Interstitial water chemistry of Villefranche Bay sediments: Trace metal diagenesis. *Mar. Chem.* **18**, 233-247.

Gaines A.M. (1974) Protodolomite synthesis at 100°C and atmospheric pressure. *Science* **183**, 518-520.

Gaines A.M. (1977) Protodolomite redefined. *J. Sediment. Petrol.* **47**, 543-546.

Gaines A.M. (1980) Dolomitization kinetics: Recent experimental studies. In *Concepts and Models of Dolomitization* (eds. D.H. Zenger, J.B. Dunham, and R.L. Ethington), pp. 81-86. *Soc. of Econ. Paleontologists and Mineralogist*, Spec. Pub. **28**. Tulsa, OK.

Gammon R.H., Sundquist E.T. and Fraser P.J. (1985) History of carbon dioxide in the atmosphere. In *Atmospheric Carbon Dioxide and the Global Carbon Cycle* (ed. J.R. Trabalka), pp. 25-62. U.S.D.O.E. ER-0239, Washington, D.C.

Garlick G.D. (1969) The stable isotopes of oxygen: In *Handbook of Geochemistry*, V. II-1 (ed. K.H.Wedepohl), Springer-Verlag, Chapter 8B.

Garrels R.M. (1965) Silica: Role in the buffering of natural waters. *Science* **148**, 69.

Garrels R.M. (1976) A survey of low-temperature water-mineral reactions. In *Interpretation of Environmental Isotope and Hydrochemical Data in Ground-Water Hydrology*, pp. 65-84. Vienna International Atomic Energy Agency.

Garrels R.M. (1988) Sediment cycling during Earth history. In *Physical and Chemical Weathering in Geochemical Cycles* (eds. A. Lerman and M. Meybeck), pp. 341-356. Kluwer Academic Publishers, Boston.

Garrels R.M. and Berner R.A. (1983) The global carbonate-silicate sedimentary system--some feedback relations. In *Biomineralization and Biological Metal Accumulation* (eds. P. Westbroek and E.W. DeJong), pp. 73-87. D. Reidel Publishing Co., Dordrecht, Holland.

Garrels R.M. and Christ C.L. (1965) *Solutions, Minerals, and Equilibria*. Harper, New York, 450 pp.

Garrels R.M. and Lerman A. (1981) Phanerozoic cycles of sedimentary carbon and sulfur. *National Academy of Sciences Proceedings* **78**, 4652-4656.

Garrels R.M. and Lerman A. (1984) Coupling of the sedimentary sulfur and carbon cycles-an improved model. *Amer. J. Sci.* **284**, 989-1007.

Garrels R.M. and Thompson M.E. (1962) A chemical model for seawater at 25°C and one atmosphere total pressure. *Amer. J. Sci.* **260**, 57-66.

Garrels R.M. and Mackenzie F.T. (1971a) *Evolution of Sedimentary Rocks*. W.W. Norton and Co., New York, 397 pp.

Garrels R.M. and Mackenzie F.T. (1971b) Gregor's denudation of the continents. *Nature* **231**, 382-383.

Garrels R.M. and Mackenzie F.T. (1972) A quantitative model for the sedimentary rock cycle. *Mar. Chem.* **1**, 22-41.

Garrels R.M. and Mackenzie F.T. (1974) Chemical history of the oceans deduced from postdepositional changes in sedimentary rocks. In *Studies in Paleo-Oceanography Special Publication Society Economic Paleontologists and Mineralogists* **20** (ed. W. W. Hay), pp. 193-204. Tulsa, OK.

Garrels R.M. and Perry E.A., Jr. (1974) Cycling of carbon, sulfur and oxygen through geologic time. In *The Sea, Volume 5* (ed. E.D. Goldberg) pp. 303-336. Wiley-Interscience, New York.

Garrels R.M. and Wollast R. (1978) Discussion of: Equilibrium criteria for two-component solids reacting with fixed composition in an aqueous phase--example: The magnesian calcites. *Amer J. Sci.* **278**, 1469-1474.

Garrels R.M., Lerman A. and Mackenzie F.T. (1976) Controls of atmospheric O_2 and CO_2 - -past, present and future. *Amer. J. Sci.* **64,** 306-315.

Garrels R.M., Mackenzie F.T. and Hunt C. (1975) *Chemical Cycles and the Global Environment*. W. Kaufmann, Inc., Los Angeles, CA, 206 pp.

Garrels R.M., Thompson M.E. and Siever R. (1960) Stability of some carbonates at 25°C and one atmosphere total pressure. *Amer. J. Sci.* **258**, 402-418.

Garrison R.E. (1981) Diagenesis of oceanic carbonate sediments: A review of the DSDP perspective. *Soc. Econ. Paleontologists and Mineralogists Spec. Pub.* **32**, 181-207.

Garrison R.E., Kastner M. and Zenger D.H. (1984) Dolomites of the Monterey Formation and other organic-rich units. *Soc. Econ. Paleontologists and Mineralogists Pacific Section Special Publication* **41**, 215 pp.

Gastil R.G. (1960) The distribution of mineral dates in time and space. *Am. J. Sci.* **258**, 1-35.

Gavish E. and Friedman G.M. (1969) Progressive diagenesis in Quaternary to Late Tertiary carbonate sediments: sequence and time scale. *J. Sediment. Petrol.* **39**, 980-1006.

Genthon C., Barnola J.M., Raynaud D., Lorius C., Jouzel J., Barkov N.I., Korotkevich Y.S. and Kotlyakov V.M. (1987) Vostok ice core: Climate response to CO_2 and orbital forcing changes over the last climatic cycle. *Nature* **329**, 414-418.

Gieskes J.M. (1983) The chemistry of interstitial waters of deep sea sediments: Interpretation of deep sea drilling data. In *Chemical Oceanography*, Vol. 8 (eds. J.P. Riley and R. Chester), pp. 222-271. Academic Press, New York.

Ginsburg R.N. (1956) Environmental relationships of grain size and constituent particles in some South Florida carbonate sediments. *AAPG Bull.* **40**, 2384-2427.

Ginsburg R.N. and Schroeder J. (1973) Growth and submarine fossilization of algae cup reefs, Bermuda. *Sedimentology* **20**, 575-614.

Given R.K. and Wilkinson B.H. (1985a) Kinetic control of morphology composition and mineralogy of abiotic sedimentary carbonates. *J. Sediment. Petrol.* **55**, 109-119.

Given R.K. and Wilkinson B.H. (1985b) Kinetic control of morphology, composition, and mineralogy of abiotic sedimentary carbonates--Reply. *J. Sediment. Petrol.* **55**, 921-926.

Given R.K. and Wilkinson B.H. (1987) Dolomite abundance and stratigraphic age: Constraints on rates and mechanisms of Phanerozoic dolostone formation. *J. Sediment Petrol.* **57**, 457-469.

Glover E.D. and Sippel R.F. (1967) Synthesis of magnesium calcites. *Geochim. Cosmochim. Acta* **31**, 603-614.

Glover E.D. and Pray L.C. (1971) High-magnesian calcite and aragonite cementation within modern subtidal carbonate sediment grains. In *Carbonate Cements* (ed. O.P. Bricker), pp. 80-87. Johns Hopkins, Baltimore, MD.

Goff S.K. (1979) Early carbonate cementation and diagenesis as related to the present surficial water regime, Engebi Island, Enemetak Island. M.S. thesis, Louisiana State Univ., Baton Rouge, LA.

Goldman J.C. and Dennett M.R. (1983) Carbon dioxide exchange between air and seawater: No evidence for rate catalysis. *Science* **220**, 199-201.

Goldsmith J.R. (1953) A "simplexity principle" and its relation to "ease" of crystallization. *J. Geol.* **61**, 439-451.

Goldsmith J.R. (1983) Phase relations of rhomboheral carbonates. In *Reviews in Mineralogy: Carbonates - Mineralogy and Chemistry* (ed. R. J. Reeder), pp. 49-76. Mineralogical Society of America. Bookcrafters, Inc., Chelse, MI.

Goldsmith J.R., Graf D.L., Witters J., Northrop D. (1962) Studies in the system $CaCO_3$-$MgCO_3$-$FeCO_3$: 1. Phase relations; 2. A method for major element spectrochemical analysis; 3. Compositions of some foreign dolomites. *J. Geol.* **70**, 659-688.

Goldsmith J.R. and Heard H.C. (1961) Subsolidus phase relations in the system $CaCO_3$-$MgCO_3$. *J. Geol.* **69**, 45-74.

Goldstein R.H. (1988) Cement stratigraphy of Pennsylvanian Holder Formation, Sacramento Mountains, New Mexico. *Am. Assoc. Petrol. Geol. Bull.* **72**, 425-438.

Goodell H.G. and Garman R.K. (1969) Carbonate geochemistry of Superior deep test well, Andros Island, Bahamas. *Am. Assoc. Petrol. Geol. Bull.* **53**, 513-536.

Goreau T.F. (1963) Calcium carbonate deposition by coralline algae and corals in relation to their roles as reef-builders. *Ann. N.Y. Acad. Sci.* **109**, 127-167.

Gornitz V., Lebedeff S. and Hansen J. (1982) Global sea level trend in the past century. *Science* **215**, 1611.

Graham D.W., Bender M.L., Williams D.F. and Keigwin Jr. L.D. (1982) Strontium-calcium ratios in Cenozoic planktonic foraminifera. *Geochim. Cosmochim. Acta* **46**, 1281-1292.

Gran G. (1952) Determination of equivalence point in potentiometric titration: Part II. *Analyst* **77**, 661-671.

Gregg J.M. (1985) Regional epigenetic dolomitization in the Bonneterre dolomite (Cambrian), southeastern Missouri. *Geology* **13**, 503-506.

Gregor C.B. (1970) Denudation of the continents. *Nature* **226**, 273-275.

Gregor C.B. (1985) The mass-age distribution of Phanerozoic sediments. In *Geochronology and the Geologic Record, Mem. No. 10* (ed. N.J. Snelling), pp. 284-289. Geol. Soc. London.

Gregor C.B. (1988) Prologue: Cyclic processes in geology, a historical sketch. In *Chemical Cycles in the Evolution of the Earth* (eds. C.B. Gregor, R.M. Garrels, F.T. Mackenzie and J.B. Maynard), pp. 5-16. John Wiley & Sons, New York.

Gregor C.B., Garrels R.M., Mackenzie F.T. and Maynard J.B. (1988) *Chemical Cycles in the Evolution of the Earth.* John Wiley and Sons, New York, 276 pp.

Gross M.G. (1964) Variation in the $^{18}O/^{16}O$ and $^{13}C/^{12}C$ ratios of diagenetically altered limestones in the Bermuda Islands. *J. Geol.* **72**, 170-194.

Grossman E.L. (1984) Carbon isotopic fractionation in live benthic foraminifera - comparison with inorganic precipitate studies. *Geochim. Cosmochim. Acta* **48**, 1505-1512.

Grossman E.L. and Ku T.L. (1981) Aragonite-water isotopic paleotemperature scale based on the benthic foraminifera Hoeglundina elegans. *Geol. Soc. America Abstracts with Programs* **13**, 464.

Grover G., Jr. and Read J.F. (1983) Paleoaquifer and deep burial related cements defined by regional cathodoluminescent patterns, Middle Ordovician carbonates, Virginia. *AAPG Bull.* **67**, 1275-1303.

Gunatilaka A. (1987) The dolomite problem in the light of recent studies. *Modern Geology* **11**, 311-324.

Gunatilaka A., Saleh A. and Al-Temeemi A. (1985) Sulfate reduction and dolomitization in a Holocene lagoon in Kuwait, Northern Arabian Gulf. SEPM Mid-Year Meeting. Golden, Colorado, Abstracts. **38**.

Hafele W. (1979) A global and long-range picture of energy development. *Science* **209**, 174.

Hallam A. (1984) Pre-quaternary sea-level changes. In *Annual Reviews of Earth and Planetary Sciences* **12** (eds. G.W. Weatherill, A.L. Albee and F.G. Stehli), pp. 205-243. Annual Reviews Inc., Palo Alto, California.

Halley R.B. and Evans C.C. (1983) The Miami Limestone: A guide to selected outcrops and their interpretation. Miami Geological Society, 67 pp.

Halley R.B. and Harris P.N. (1979) Fresh-water cementation of a one thousand year-old oolite. *J. Sediment. Petrol.* **49**, 969-988.

Halley R.B. and Schmoker J.W. (1983) High porosity Cenozoic carbonate rocks of south Florida: Progressive loss of porosity with depth. *AAPG Bull.* **67**, 191-200.

Ham W.E. (1962) Classification of Carbonate Rocks, a Symposium. American Association Petroleum Geologists Memoir **1**, Tulsa, OK. 1279 pp.

Hancock J.N. and Scholle P.A. (1975) Chalk of the North Sea. In *Petroleum and the Continental Shelf of North-west Europe, Volume I. Geology* (ed. A.W. Woodland), pp. 413-425. Elsevier, New York.

Hanor J.S. (1978) Precipitation of beachrock cements: Mixing of marine and meteoric waters vs. CO_2 degassing. *J. Sediment. Petrol.* **48**, 489-501.

Hanor J.S. (1983) Fifty years of development of thought on the origin and evolution of subsurface sedimentary brines. In *Revolution in the Earth Sciences: Advances in the Past Half-Century* (ed. S.J. Boardman), pp. 99-111. Kendall/Hunt, Dubuque, IA.

Hansen J. and Lebedeff S. (1987) Global trends of measured surface air temeprature. *J. Geophys. Res.* **92**, 13345-13372.

Hanshaw B.B., Back W. and Deike R.G. (1971) A geochemical hypothesis for dolomitization by groundwater. *Econ. Geol.* **66**, 710-724.

Hansson I. (1972) An analytical approach to the carbonate system in seawater. Ph.D. dissertation, Univ. of Göteberg, Sweden.

Hansson I. (1973) A new set of pH scales and standard buffers for seawater. *Deep-Sea Res.*, **20**, 479-491.

Harbison G.R. and Gilmer R.W. (1986) Effects of animal behavior on sediment trap collections: Implications for the calculation of aragonite fluxes. *Deep-Sea Res.* **33**, 1017-1024.

Hardie L.A. (1987) Perspectives on dolomitization: A critical view of some current views. *J. Sediment. Petrol.* **57**, 166-183.

Harker R.I. and Tuttle O.F. (1955) Studies in the system $CaO-MgO-CO_2$, Part 1. *Am. J. Sci.* **253**, 209-224.

Harmon R.S., Schwarcz H.P. and Ford D.C. (1978) Late Pleistocene sea level history of Bermuda. *Quat. Res.* **9**, 205-218.

Harmon R.S., Land L.S., Mitterer R.M., Garrett P., Schwarcz H.P. and Larson G.J. (1981) Bermuda sea level during the last interglacial. *Nature* **289**, 481-483.

Harmon R.S., Mitterer R.M., Kriausakul N., Land L.S., Schwarcz H.P., Garrett P., and Larson G.J., Vacher H.L. and Rowe M. (1983) U-series and amino-acid racemization geochronology of Bermuda: Implications for eustatic sealevel fluctuation over the past 250,000 years. *Palaeogeogr., Palaeoclimatol., Palaeoecol* . **44**, 41-70.

Harmon R.S., White W.B., Drake J.J., and Hess J.W. (1975) Regional hydrochemistry of North American caronate terrains. *Water Resources Research* **11**, 963-967.

Harris P.J., Kendall G.ST.C. and Lerche I. (1985) Carbonate cementation--A brief review. In *Carbonate Cements* (eds. N. Schneidermann, N.and P.M. Harris), pp. 79-95. Society of Economic Paleontologists and Mineralogist, Tulsa, OK.

Harris, W.H. (1971) Groundwater-carbonate rock interactions, Barbados, W.I. Ph.D. dissertation, Brown Univ., Providence, R.I.

Harris W.H. and Matthews R.K. (1968) Subaerial diagenesis of carbonate sediments: Efficiency of the solution-reprecipitation process. *Science* **160**, 77-79.

Hart R. (1973) A model for chemical exchange in the basalt-seawater system of oceanic Layer II. *Canad. J. Earth Sci.* **10**, 799-816.

Harvie G.E. and Weare J.H. (1980) The prediction of mineral solubilities in natural waters: The Na-K-Mg-Ca-Cl-SO4-H2O system from zero to high concentration at 25°C. *Geochim. Cosmochim. Acta* **44**, 981-997.

Hay W.W. (1985) Potential errors in estimates of carbonate rock accumulating through geologic time. In *The Carbon Cycle and Atmospheric CO2: Natural Variations Archean to Present* (eds. E.T. Sundquist and W.S. Broecker), pp. 573-584. Geophysical Monograph **32**, American Geophysical Union, Washington, D.C.

Hay W.W. and Southam J.R. (1977) Modulation of marine sedimentation by the continental shelves. In *The Fate of Fossil Fuel CO2 in the Oceans* (eds. N.R. Anderson and A. Malahoff), pp. 569-604. Plenum Press, New York.

Hays J.D. and Pitman W.C. (1973) Lithospheric plate motion, sea level changes and climatic and ecological consequences. *Nature* **246**, 18-22.

Heath G.R. (1974) Dissolved silica and deep-sea sediments. In *Studies in Paleo-Oceanography* (ed. W.W. Hay), pp. 77-93. S.E.P.M. Spec. Publication **20**, Tulsa, OK.

Heath G.R. and Culberson C. (1970) Calcite: Degree of saturation, rate of dissolution, and the compensation depth in the deep oceans. *Geol. Soc. Amer. Bull.* **81**, 3157-3160.

Heath K.C. and Mullins H.T. (1984) Open-ocean, off-bank transport of fine-grained carbonate sediment in the northern Bahamas. In *Fine-Grained Sediments: Deep Water Processes and Facies* (eds. D.A.B. Stow and D.J.W. Piper), pp. 199-208. Blackwell Scientific Publication, London.

Heckel P.H. (1983) Diageneitic model for carbonate rocks in mid-continent Pennsylvanian eustatic cyclothems. *J. Sediment. Petrol.* **53**, 733-760.

Hecht F. (1933) Der Verbleib der organische Substanz der Tiere bei meerischer Einbettung. *Senckenbergiana* **15**, 165-219.

Heezen B.C. and Tharp M. (1965) Tectonic fabric of the Atlantic and Indian Oceans and continental drift. *Phil. Trans. Roy. Co. London, Series A* **258**, 90-106.

Helgeson H.C., Garrels R.M. and Mackenzie F.T. (1969) Evaluation of irreversible reactions in geochemical processes involving minerals and aqueous solutions: 2. Applications. *Geochim. Cosmochim. Acta* **33**, 455-481.

Heinrichs T.K. and Reimer T.O. (1977) A sedimentary barite deposit from the Archean Fig Tree Group of the Barberton Mountain Land (South Africa). *Econ. Geol.* **72**, 1426-1441.

Hemming N.G., Meyers W.J. and Grams J.C. (1989) Cathodoluminescence in diagenetic calcites: The roles of Fe and Mn as deduced from electron probe and spectrophotometric measurements. *J. Sediment. Petrol.* **59**, 404-411.

Henderson L.M. and Kracek F.C. (1927) The fractional precipitation of barium and radium chromates. *J. Am. Chem. Soc.* **49**, 739.

Hester K. and Boyle E. (1982) Water chemistry control of cadmium content in recent benthic foraminifera. *Nature* **298**, 260-262.

Heydari E. and Moore C.H. (1989) Burial diagenesis and thermochemical sulfate reduction, Smackover Formation, southeastern Mississippi salt basin. *Geology* **17**, 1080-1084.

Hiltabrand R.R., Ferrell R.E. and Billings G.K. (1973) Experimental diagenesis of Gulf Coast argillaceous sediment. *AAPG Bull.* **57**, 338-348.

Hinga K.R., Sieburth J. McN and Heath G. R. (1979) The supply and use of organic material at the deep-sea floor. *J. Mar. Res.* **37**, 557-579.

Hird K. and Tucker M.E. (1988) Contrasting diagenesis of two Carboniferous oölites from South Wales: a tale of climatic influence. *Sedimentology* **35**, 587-602.

Hoefs J. (1987) *Stable Isotope Geochemistry (3rd edition).* Springer-Verlag, New York. 241 pp.

Hoffman J.S., Keyes D. and Titus J.G. (1983) *Projecting Future Sea Level Rise.* U.S. EPA 230-09-007, Washington, D.C., 121 pp.

Holland H.D. (1966) *The coprecipitation of metallic ions with calcium carbonate. Final Report, U.S. AEC Contract No. AT (30-1)-2266,* Dept. Geology, Princeton Univ., NJ.

Holland H.D. (1973) Systematics of the isotopic composition of sulfur in the oceans during the Phanerozoic and its implications for atmospheric oxygen. *Geochim. Cosmochim. Acta* **37**, 2605-2616.

Holland H.D. (1978) *The Chemistry of the Atmosphere and Oceans.* Wiley Interscience, New York, 351 pp.

Holland H.D. (1984) *The Chemical Evolution of the Atmosphere and Oceans*. Princeton Univ. Press, Princeton, NJ, 582 pp.

Holland H.D., Borcsik M., Munoz J.L. and Oxburgh U.M. (1963) The coprecipitation of Sr^{2+} with aragonite and of Ca^{2+} with strontianite between 90 and 100ºC. *Geochim. Cosmochim. Acta* **27**, 957-977.

Holland H.D., Holland H.J. and Munoz J.L. (1964) The coprecipitation of cations with $CaCO_3$. II. The coprecipitation of Sr^{2+} with calcite between 90 and 100ºC. *Geochim. Cosmochim. Acta* **28**, 1287-1302.

Holland H.D., Lazar B. and McCaffrey M. (1986) Evolution of the atmosphere and oceans. *Nature* **320,** 27-33.

Holser W.T. (1979) Mineralogy of evaporites. *Mineral Soc. Amer. Rev. Mineral.*, **6**, 211-294.

Holser W.T. (1984) Gradual and abrupt shifts in ocean chemistry during Phanerozoic time. In *Patterns of Change in Earth Evolution* (eds. H.D. Holland and A.F. Trendall), pp. 123-144. Dahlem Workshop, Springer-Verlag, Berlin.

Holser W.T. and Kaplan I.R. (1966) Isotope geochemistry of sedimentary sulfates. *Chem. Geol.* **1,** 93-135.

Holser W.T., Schidlowski M., Mackenzie F.T. and Maynard J.B. (1988) Geochemical cycles of carbon and sulfur. In *Chemical Cycles in the Evolution of the Earth* (eds. C.B. Gregor, R.M. Garrels, F.T.Mackenzie and J.B. Maynard), pp. 105-174. John Wiley and Sons, New York.

Holser W.T., Schonlaub H-P., Attrep M., Jr., Boeckelmann K., Klein P., Magaritz M., Orth C.J., Fenninger A., Jenny C., Kralik M., Mauritsch H., Pak E., Schramm J-M., Stattegger K. and Schmöller R. (1989) A unique geochemical record at the Permian/Triassic boundary. *Nature* **337,** 39-44.

Honjo S. (1975) Dissolution of suspended coccoliths in the deep-sea water column and sedimentation of coccolith ooze. In *Dissolution of Deep-Sea Carbonates* (eds. W. Sliter, A.W.H. Be and W.H. Berger) pp. 115-128. Cushman Found. Foraminiferal Res., Spec. Publ. No. 13.

Honjo S. (1976) Coccoliths: Production, transportation and sedimentation. *Marine Micropaleontology* **1,** 65-79.

Honjo S. (1980) Material fluxes and modes of sedimentation in the mesopelagic and bathypelagic zones. *J. Mar. Res.* **38,** 53-97.

Horibe Y., Oba T. and Niitsuma N. (1969) Relationship between water temperature and oxygen isotopic ratio [in Japanese]. *Fossils, Special Issue,* **1,** 15-20.

Horibe Y., Endo K. and Tsubota H. (1974) Calcium in the South Pacific and its correlation with carbonate alkalinity. *Earth Planet. Sci. Lett.* **23,** 136-140.

Houghton R.A., Hobbie J.E., Melillo J.M., Moore B., Peterson B.J., Shaver G.R. and Woodwell G.M. (1983) Changes in the carbon content of terrestrial biota and soils between 1860-1980: A net release of CO_2 to the atmosphere. *Ecological Monographs* **53**, 235-262.

Houghton R.A., Boone R.D., Fruci J.R., Hobbie J.E., Melillo J.M., Palm C.A., Peterson B.J., Shaver G.R. and Woodwell G.M. (1987) The flux of carbon from terrestrial ecosystems to the atmosphere in 1980 due to changes in land use: Geographical distribution of the global flux. *Tellus* **39**, 122-139.

House W.A., Howard J.R. and Skirrow G. (1984) Kinetics of carbon dioxide transfer across the air/water interface. Faraday Discuss. *Chem. Soc.* **77**, 33-46.

Howland M., Talbot M.R., Qvale M.R., Olaussen S. and Aasverg L. (1987) Methane-related carbonate cements in pockmarks of the North Sea. *J. Sediment. Petrol.* **57**, 881-892.

Hubbert M.K. (1970) Energy resources for power production, Proc. IAEA Sym. Environmental Aspects of Nuclear Power Stations, IAEA-SM-146/1.

Hull H.S. and Turnbull A.G. (1973) A thermochemical study of monohydrocalcite. *Geochim. Cosmochim. Acta* **37**, 685-694.

Idso S.B. (1982) *Carbon Dioxide: Friend or Foe.* Institute for Carbon Dioxide Research Press, Tempe, AZ, 92 pp.

Ingle S.E. (1975) Solubility of calcite in the ocean. *Mar. Chem.* **3**, 301-319.

Irwin H. (1980) Early diagenetic carbonate precipitation and pore fluid migration in the Kimmeridge Clay of Dorset, England. *Sedimentology* **27**, 577-591.

Isaacs C.M. (1984) Disseminated dolomite in the Monterey Formation, Santa Maria and Santa Barbara areas, California. In *Dolomites of the Monterey Formation and Other Organic-Rich Units* (eds. R.E. Garrison, M. Kastner and D.H. Zenger), pp. 155-170. Society of Economic Paleontologists and Mineralogists Pacific Section **41**, Tulsa, OK.

Ishikawa M. and Ichikuni M. (1984) Uptake of sodium and potassium by calcite. *Chem. Geol.* **42**, 137-146.

Izuka S.K. (1988) The variation of magnesium concentration in the tests of recent and fossil benthic foraminifera. Ph.d. dissertation, Univ. of Hawaii.

Jackson T.A. and Bischoff J.L. (1971) The influence of amino acids on the kinetics of the recrystallization of aragonite to calcite. *J. Geol.* **79**, 493-497.

Jacobsen R.L. and Langmuir D. (1974) Dissociation constants of calcite and $CaHCO_3^+$ from 0 to 50°C. *Geochim. Cosmochim. Acta* **38**, 301-318.

Jacobson R.L. and Usdowski H.E. (1976) Partitioning of strontium between calcite, dolomite, and liquids. *Contrib. Mineral. Petrology* **59**, 171-185.

Jahnke R.A., Emerson S.R., Cochran J.K. and Hirschberg D.J. (1986) Fine scale distribution of porosity and particulate excess 210Pb, organic carbon and CaCO3 in surface sediments of the deep equatorial Pacific. *Earth Planet. Sci. Lett.* **77**, 59-69.

James N.P. (1979) Reefs. In *Facies Models* (ed. R.G. Walker), pp. 121-133. *Geoscience Canada Rept. Ser.* **1**.

James, N.P. and Bone Y. (1989) Petrogenesis of Cenazoic temperate water calcarenites, South Australia: A model for meteoric/shallow burial diagenesis of shallow water calcite sediments. *J. Sediment. Petrol.* **59**, 191-203.

James N.P. and Choquette P.W. (1983) Diagenesis 5: Limestones--The seafloor diagenetic environment. *Geoscience Canada* **10**, 162-179.

James N.P. and Choquette P.E. (1984) Diagenesis 9 - Limestones - The meteoric diagenetic environment. *Geoscience Canada* **11**, 161-194.

James N.P. and Mountjoy E.W. (1983) Shelf-slope break in fossil carbonate platforms: An overview. In *Shelfbreak: Critical Interface on Continental Margins* (eds. D.J. Stanley and G.T. Moore), pp. 189-206. SEPM Spec. Pub. No. 33, Tulsa, OK.

Javoy M., Pineau F. and Allegre C.-J. (1982) Carbon geodynamic cycle. *Nature* **300**, 171-173.

Javoy M., Pineau F. and Allegre C.-J. (1983) Carbon geodynamic cycle--Reply. *Nature* **303**, 731.

Johnson K.S. (1982) Solubility of rhodochrosite (MnCO3) in water and seawater. *Geochim. Cosmochim. Acta* **46**, 1805-1809.

Jones M.E., Bedford J. and Clayton C. (1984) Unnatural deformation mechanisms in the Chalk. *J. Geol. Soc. Lon.* **141**, 675-683.

Jouzel J., Lorius C., Petit J.R., Genthon C, Barkov N.I., Kotlyakov V.M. and Petrov V.M. (1987) Vostok ice core: A continuous isotope temperature record over the last climatic cycle (160,000 years). *Nature* **329,** 403-408.

Judson S. (1968) Erosion of the land. *Amer. Scientist* **56,** 356-374.

Kanwisher J. (1963) On the exchange of gases between the atmosphere and the sea. *Deep-Sea Res.* **10**, 195-207.

Kasting J.F. (1987) Theoretical constraints on oxygen and carbon dioxide concentrations in the Precambrian atmosphere. *Precambrian Research* **34,** 205-229.

Kasting J.R. (1989) Long-term stability of the Earth's climate. *Paleogeogr. Paleoclimatol. Paleoecol.* **75,** 83-95.

Kasting J.F., Toon O.V. and Pollack J.B. (1988) How climate evolved on the terrestrial planets. *Scientific Amer.* **258**, 90-97.

Katz A. (1973) The interaction of magnesium with calcite during crystal growth at 25-90ºC and one atmosphere. *Geochim. Cosmochim. Acta* **37**, 1563-1586.

Katz A. and Matthews A. (1977) The dolomitization of $CaCO_3$: An experimental study at 252-295ºC. *Geochim. Cosmochim. Acta* **41**, 297-308.

Katz A., Sass E., Starinsky A. and Holland H.D. (1972) Strontium behavior in the aragonite--calcite transformation: An experimental study at 40-98ºC. *Geochim. Cosmochim. Acta* **36**, 481-496.

Kaufman J., Cander H.S., Daniels L.D. and Meyers W.J. (1988) Calcite cement stratigraphy and cementation history of the Burlington-Keokuk Formation (Mississippian), Illinois and Missouri. *J. Sediment. Petrol.* **58**, 312-326.

Kazmierczak T.F., Tomson M.B. and Nancollas G.H (1982) Crystal growth of calcium carbonate. A controlled composition study. *J. Physical Chem.* **86**, 103-107.

Keeling C.D., Mook W.G. and Tans P.P. (1979) $^{13}C/^{12}C$ ratio of atmospheric carbon dioxide. *Nature* **217**, 121-123.

Keeling C.D., Bacastrow R.D. and Tans P.P. (1980) Predicted shift in the $^{13}C/^{12}C$ ratio of atmospheric carbon dioxide. *Geophys. Res. Lett.* **7**, 505-508.

Keeling C.D., Mook W.G. and Tans P.P. (1979) $^{13}C/^{12}C$ ratio of atmospheric carbon dioxide. *Nature* **217**, 121-123.

Kelts K. and McKenzie J. (1984) A comparison of anoxic dolomite from deep-sea sediments: Quaternary Gulf of California and Messinian Tripoli Formation of Sicily. In *Dolomites of the Monterey Formation and Other Organic-Rich Units* (eds. R.E. Garrison, M. Kastner, D.H. Zenger), pp. 19-28. Society Economic Paleontologists and Mineralogists, Pacific Section **41**, Tulsa, OK.

Keir R.S. (1980) The dissolution kinetics of biogenic calcium carbonates in seawater. *Geochim. Cosmochim. Acta* **44**, 241-252.

Keir R.S.(1984) Recent increase in Pacific $CaCO_3$ dissolution: A mechanism for generating old ^{14}C ages. *Mar. Geol.* **59**, 227-250.

Keir R.S. and Hurd D.C. (1983) The effect of encapsulated fine grained sediment and test morphology on the resistance of planktonic foraminifera to dissolution. *Mar. Micropaleontol.* **8**, 183-214.

Kellogg W.W. (1987) Mankind's impact on climate: The evolution of awareness. *Climatic Change* **10**, 113-136.

Kelts K. and McKenzie J. (1984) A comparison of anoxic dolomite from deep-sea sediments: Quaternary Gulf of California and Messinian Tripoli Formation of Sicily. In *Dolomites of the Monterey Formation and Other Organic-Rich Units* (eds. R.E. Garrison, M. Kastner, D.H. Zenger), pp. 19-28. Society Economic Paleontologists and Mineralogists, Pacific Section **41**, Tulsa, OK.

Kinsman D.J.J. (1969) Interpretation of Sr^{2+} concentrations in carbonate minerals and rocks. *J. Sediment. Petrol.* **49**, 937-944.

Kinsman J.J. and Holland H. D. (1969) The coprecipitation of cations with $CaCO_3$-IV. The coprecipitation of Sr^{2+} with aragonite between 16° and 96°C. *Geochim. Cosmochim. Acta* **33**, 1-17.

Kitano Y. (1962) The behavior of various inorganic ions in the separation of calcium carbonate from a bicarbonate solution. *Bull. Chem. Soc. Japan* **35**, 1973-1980.

Kitano Y. and Hood J.D.W. (1965) The influence of organic material on the polymorphic crystallization of calcium carbonate. *Geochim. Cosmochim. Acta* **29**, 29-41.

Kitano Y. and Kanamori N. (1966) Synthesis of magnesium calcite at low temperatures and pressures. *J. Geochem.* **1**, 1-11.

Klass M.J., Kersey D.G., Berg R.R. and Tieh T.T. (1981) Diagenesis and secondary porosity in Vicksburg Sandstones, McAllen Ranch Field, Hidalgo County, Texas. *Trans.-Gulf Coast Assoc. Geol. Socs.* **31**, 115-123.

Klein C., and Gole M.J. (1981) Mineralogy and petrology of parts of the Marra Mamba Iron-formation, Hamersley Basin, Western Australia. *Amer. Mineral.* **66**, 507-525.

Klotz I.M. (1950) *Chemical Thermodynamics.* Prentice-Hall, Englewood Cliffs, NJ, 369 pp.

Knauth L.P. and Lowe D.R. (1978) Oxygen isotope geochemistry of cherts from the Onverwacht Group (3.4 billion years), Transvaal, South Africa, with implications for secular variations in the isotopic composition of cherts. *Earth and Planet. Sci. Lett.* **41**, 209-222.

Koch B. (1989) Les Calcaires en Milieu Marin. D.S. dissertation, Universite de Liege, Liege, Belgium.

Kohlmaier G.H., Brohl H., Sire E.P., Plochl M. and Revelle R. (1987) Modeling stimulation of plants and ecosystem response of present levels of excess atmospheric CO_2. *Tellus* **39**, 155-170.

Kohout F.A. (1965) A hypothesis concerning the cyclic flow of salt water related to geothermal heating in the Floridian Aquifer. *Transactions New York Academy of Science Series 2,* **28**, 249-271.

Kohout F.A. (1967) Groundwater flow and the geothermal regime of the Floridian Plateau. Transactions Gulf Coast Association Geological Society **17**, 339-354.

Kohout F.A., Henry H.R. and Banks J.E. (1977) Hydrogeology related to geothermal conditions of the Floridian Plateau. In *The Geothermal Nature of the Floridian Plateau* (eds. K.L. Smith and G.M. Griffin). Florida Department of Natural Resources Bureau Geology Special Publication **21**, Blackwell Oxford, New York.

Kolla V., Bé A.W.H. and Biscaye P.E. (1976) Calcium carbonate distribution in the surface sediments of the Indian Ocean. *J. Geophys. Res.* **81**, 2605-2616.

Kretz R. (1982) A model for the distribution of trace elements between calcite and dolomite. *Geochim. Cosmochim. Acta* **46**, 1979-1981.

Kroopnick P.M. (1985) The distribution of ^{13}C of ΣCO_2 in the world oceans. *Deep-Sea Res.* **32**, 57-84.

Krumgalz B.S. and Millero F.J. (1982) Physico-chemical study of the Dead Sea water body. I. Activity coefficients of major ions in Dead Sea water. *Mar. Chem.* **11**, 209-222.

Kump L.R. (1989) Chemical stability of the atmosphere and ocean. *Paleogeogr. Paleoclimatol. Paleoecol.* **75,** 123-136.

Kump L.R. and Garrels R.M. (1986) Modeling atmospheric O_2 in the global sedimentary redox cycle. *Amer. J. Sci.* **286,** 337-360.

Lafon G.M. and Mackenzie F.T. (1974) Early evolution of the oceans-a weathering model. In *Studies in Paleo-Oceanography* (ed. W.W. Hay), pp. 205-218. Society Economic Paleontologists and Mineralogists, Special Publication No. 20, Tulsa, OK.

Lafon G.M. and Vacher H.L. (1975) Diagenetic reactions as stochastic processes: Application to the Bermudian eolianites. *Geol. Soc. Amer. Memoir* **142**, 187-204.

Lahann R.W. (1978) A chemical model for calcite crystal growth and morphology control. *J. Sediment. Petrol.* **48**, 337-344.

Lahann R.W. and Siebert R.M. (1982) A kinetic model for distribution coefficients and application to Mg-calcites. *Geochim. Cosmochim. Acta* **46**, 2229-2237.

Land L.S. (1966) Diagenesis of metastable skeletal carbonates. Ph.D. Dissertation, Lehigh Univ., Bethlehem, PA.

Land L.S. (1967) Diagenesis of skeletal carbonates. *J. Sediment. Petrol.* **37**, 914-930.

Land L.S. (1970) Phreatic vs vadose meteoric diagenesis of limestones: Evidence from a fossil water table. *Sedimentology* **14**, 175-185.

Land L.S. (1973) Holocene meteoric dolomitization of Pleistocene limestones , North Jamaica. *Sedimentology* **20**, 411-424.

Land L.S. (1979a) The fate of reef-derived sediment on the North Jamaica island slope. *Mar. Geol.* **29**, 55-71.

Land L.S. (1979b) Chert-chalk diagenesis: The Miocene island slope of Jamaica. *J. Sediment. Petrol.* **49**, 223-232.

Land L.S. (1980) The isotopic and trace element geochemistry of dolomite: The state of the art. *Soc. Econ. Paleontologist and Mineralogists,* Spec. Pub. **28**, 87-110.

Land L.S. (1985) The origin of massive dolomite. *J. Geol.* **33**, 112-125.

Land L.S. (1986) Limestone diagenesis--Some geochemical considerations. In *Studies in Diagenesis.* (ed. F.A. Mumpton), pp. 129-137. *U.S. Geol. Surv. Bull.*, **1578**, Wahington, D.C.

Land L.S. (1987) The major ion chemistry of saline brines in sedimentary basins. *Amer. Inst. Phys. Conf. Proc.* **154**, 160-179.

Land L.S. and Epstein S. (1970) Late pleistocene diagenesis and dolomitization, North Jamaica. *Sedimentology* **14**, 187-200.

Land L.S. and Goreau T.F. (1970) Submarine lithification of Jamaican reefs. *J. Sediment. Petrol.* **40**, 457-462.

Land L.S. and Hoops G.K. (1973) Sodium in carbonate sediments and rocks: A possible index to salinity of diagenetic solutions. *J. Sediment. Petrol.* **43**, 614-617.

Land L.S. and Mackenzie F.T. (1970) *Field Guide to Bermuda Geology.* Bermuda Biological Station for Research Spec. Publ. **4**, St. Georges, Bermuda, 14 pp.

Land L.S. and Prezbindowski D.R. (1981) The origin and evolution of saline formation water, Lower Cretaceous carbonates, south-central Texas. *J. Hydrology* **54**, 51-74.

Land L.S. and Prezbindowski D.R. (1985) Chemical constraints and origins of four groups of Gulf Coast reservoir fluids: Discussion. *AAPG Bull.* **69**, 119-126.

Land L.S., Mackenzie F.T. and Gould S.J. (1967) Pleistocene history of Bermuda. *Geol. Soc. Amer. Bull.* **78**, 993-1006.

Land L.S., Behrens E.W. and Frishman S.A. (1979) The oöids of Baffin Bay, Texas. *J. Sediment. Petrol.* **49**, 1269-1278.

Langmuir D. (1973) Stability of Carbonates in the system $MgO-CO_2-H_2O$. *J. Geol.* **73**, 730-754.

Lasaga A.C., Berner R.A. and Garrels R.M. (1985) An improved geochemical model of atmospheric CO_2 fluctuations over the past 100 million years. In *The Carbon Cycle and Atmospheric CO₂: Natural Variations Archean to Present* (eds. E.T. Sundquist and W.S. Broecker), pp. 397-411. American Geophysical Union, Geophysical Monograph **32**, Washington, D.C.

Last W.M. (1982) Holocene carbonate sedimentation in Lake Manitoba, Canada. *Sedimentology* **29**, 691-704.

Legasov V.A. and Kuz'min I.I. (1981) The problem of energy production. *Prioroda* **2**, 8.

Lerman A.L.D. (1979) *Geochemical Processes Water and Sediment Environments.* Wiley Interscience, NewYork, 481 pp.

Lerman A. L. D. and Dacey M.F. (1974) Stoke's settling and chemical reactivity of suspended particles in natural waters. In *Suspended Solids in Water* (ed. R.J. Gibbs), pp. 17-47. Plenum Press, New York.

Leutloff A.H. (1979) Regional distribution of microdolomite inclusions in Mississippian echinoderms from southwestern New Mexico. M.S. thesis, State Univ. of New York, Stony Brook.

Levy H. (1971) Natural atmosphere: Large radical and formaldehyde concentrations predicted. *Science* **173**, 141-143.

Li Y-H. (1972) Geochemical mass balance among lithosphere, hydrosphere and atmosphere. *Amer. J. Sci.* **272,** 119-137.

Li Y.-H., Takahashi T. and Broecker W.S. (1969) Degree of saturation of $CaCO_3$ in the oceans. *J. Geophys. Res.* **74**, 5507-5525.

Liebezeit G., Böhm L., Dawson R. and Wefer G. (1983) Estimation of algal carbonate input to marine aragonitic sediments on the basis of xylose content. *Mar. Geolog.* **54**, 249-262.

Lighty R.G. (1985) Preservation of internal reef porosity and diagenetic sealing of submerged ealy holocene Barrier Reef, southeast Florida Shelf. In *Carbonate Cements* (ed. O.P. Bricker), pp. 123-151. John Hopkins Press, Baltimore, MD.

Link G. (1903) Die Bildung der Oölithe und Rogensteine. *Neues Jahrb. Mineral., Geol., Paläontel.* **16**, 495-513.

Likens G.E., Mackenzie F.T., Richey J., Sedwell J.R. and Turekian K.K. (1981) Flux of organic carbon by rivers to the sea. U.S. DOE Conference Report 8009140, Washington, D.C., 397 pp.

Lindh T.B. (1983) Temporal variations in ^{13}C, ^{34}S and global sedimentation during the Phanerozoic. M.S. thesis, Univ. of Miami, Miami, FL.

Lindh T.B., Salzman E.S., Sloan J.L. II, Mattes B.W. and Holser W.T. (1981) A revised δC^{13}-age curve. *Geolog. Soc. Amer Abstract Programs* **13**, 498.

Lindquist S.J. (1977) Secondary porosity development and subsequent reduction, overpressured Frio Formation Sandstone (Oligocene), South Texas. *Trans.-Gulf Coast Assoc. Geol. Socs.* **27**, 99-107.

Lippmann F. (1960) Versuch zur Aufklarung der Bildungsbedeingrengen von Kalzit under Aragonit. *Fortschr. Mineral.* **38**, 156-161.

Lippmann F. (1973) *Sedimentary Carbonate Minerals.* Springer-Verlag, New York, 228pp.

Liss P.S.(1973) Processes of gas exchange across an airwater interface. *Deep-sea Res.* **20**, 221-238.

Livingston H.D. and Thompson G. (1971) Trace element concentrations in some modern corals. *Limnol. Oceanogr.* **16**, 786-796.

Livingstone D.A. (1963) Chemical composition of rivers and lakes. *U.S. Geol. Surv., Prof. pap.* **440-g**, 64 pp.

Lohmann K.C. (1982) Inverted J carbon and oxygen isotopic trends -- criteria for shallow meteoric phreatic diagenesis. (abstract). Ann. Meeting Geol. Soc. America, Abstract with Program, 548.

Lohmann K.C. (1988) Geochemical patterns of meteoric diagenetic systems and their application to paleokarst. In *Paleokarst* (eds. P.W. Choquette and N.P. James), pp. 58-80. Springer Verlag, New York.

Lohmann K.C. and Walker J.C.G. (1989) The $\delta^{18}O$ record of Phanerozoic abiotic marine calcite cements. *Geophys. Res. Lett.* **16**, 319-322.

Longman M.W. (1980) Carbonate diagenetic textures from near surface diagenetic environments. *Amer. Assoc. Petrol. Geol. Bull.* **64**, 461-487.

Lopatin N.V. (1971) Temperature and geologic time as factors in coalification. *Akad. Nauk. SSR Izv. Ser. Geol.* **3**, 95-106.

Loreau J.P. (1982) *Sediments aragonitiques et leur genese.* Mem. Mus. Nat.d'Histoire Naturelle, Serie C, Tome **XLVII**, 300 pp.

Lorens R.B. (1978) A study of biological and physical controls of the trace metal content of calcite and aragonite. Ph.D. Dissertation, Univ. of Rhode Island.

Lorens R.B. (1981) Sr, Cd, Mn and Co distribution coefficients in calcite as a function of calcite precipitation rate. *Geochim. Cosmochim. Acta* **45**, 553-561.

Lorens R.B. and Bender M.L. (1980) The impact of solution chemistry on Mytilus edulis calcite and aragonite. *Geochim. Cosmochim. Acta* **44**, 1265-1278.

Loughnan F.C. (1969) *Chemical Weathering of the Silicate Minerals.* Elsevier, New York, 154 pp.

Lowenstam H.A. (1954a) Factors affecting the aragonite: Calcite ratios in carbonate-secreting marine organisms. *J. Geol.* **62**, 284-322.

Lowenstam H.A. (1954b) Environmental relations of modifications compositions of certain carbonate-secreting marine invertebrates. *Natl. Acad. Sci. Proc.*, 39-48.

Lowenstam H.A. (1961) Mineralogy, $^{18}O/^{16}O$ ratios, and strontium and magnesium contents of recent and fossil brachiopods and their bearing on the history of the oceans. *J. Geol.* **69**, 241-260.

Lowenstam H.A. (1963) Sr/Ca ratio of skeletal aragonites from the recent marine biota at Palau and from fossil gastropods. In *Isotopic and Cosmic Chemistry* (ed. H. Craig et. al.), pp. 114-132. North Holland Publishing Co., Amsterdam.

Lowenstam H.A. and Epstein S. (1957) On the origin of sedimentary aragonite needles of the Great Bahama Bank. *J. Geol.* **65**, 364-375.

Lowenstam H.A. and Weiner S. (1989) *Biomineralization.* Oxford Univ. Press, New York, 324 pp.

Lumsden D.N. (1985) Secular variations in dolomite abundance in deep marine sediments. *Geology* **13**, 766-769.

Lumsden D.N. (1988) Characteristics of deep-marine dolomite. *J. Sediment. Petrol.* **58**, 1023-1031.

Lundegard P.D. and Land L.S. (1989) Carbonate equilibria and pH buffering by organic acids--response to changes in P_{CO_2}. *Chem. Geol.* **74**, 277-287.

Lundegard P.D., Land L.S. and Galloway W.E. (1984) Problem of secondary porosity: Frio Formation (Oligocene), Texas Gulf Coast. *Geology* **12**, 399-402.

Lyman J. (1957) Buffer mechanism of seawater. Ph.D. dissertation, Univ. California.

Machel H.G. (1985) Cathodoluminescence in calcite and dolomite and its chemical interpretation. *Geosci. Canada* **12**, 139-147.

Machel H.G. (1987) Saddle dolomite as a by-product of chemical compaction and thermochemical sulfate reduction. *Geology* **15**, 936-940.

Machel H.G. (1989) Relationships between sulfate reduction and oxidation of organic compounds to carbonate diagenesis, hydrocarbon accumulations, salt domes, and metal sulfide deposits. *Carbonates and Evaporites* **4**, 137-151.

Machel H.G. and Mountjoy E.W. (1986) Chemistry and environments of dolomitization - A reappraisal. *Earth-Science Reviews* **23**, 175-222.

Machta L., Hanson K. and Keeling C.D. (1977) Atmospheric carbon dioxide and some interpretations. In *The Fate of Fossil Fuel CO_2 in the Oceans* (eds. N.R. Anderson and A. Malahoff), pp. 131-144. Plenum Press, New York.

Mackenzie F.T. (1964a) Bermuda Pleistocene eolianites and paleowinds: *Sedimentology* **3**, 51-64.

Mackenzie F.T. (1964b) Geometry of Bermuda calcareous dune cross-bedding. *Science* **144**, 1449-1450.

Mackenzie F.T. (1975) Sedimentary cycling and the evolution of seawater. In *Chemical Oceanography, 2nd edition, Volume 1* (eds. J.P. Riley and G. Skirrow), pp. 309-364. Academic Press, London.

Mackenzie F.T. (1981) Global carbon cycle: Some minor sinks for CO_2. In *Flux of Organic Carbon from the Major Rivers of the World to the Ocean* (ed. G. Likens), pp. 360-384. U.S. DOE Rept. CONF-800940, Washington, D.C.

Mackenzie F.T. and Agegian C. (1986) Biomineralization, atmospheric CO_2 and the history of ocean chemistry. Proc. 5th Internatl. Conf. Biomineral., Dept. Geology, Univ. Texas, Arlington, 2.

Mackenzie F.T. and Agegian C. (1989) Biomineralization and tentative links to plate tectonics. In *Origin, Evolution and Modern Aspects of Biomineralization in Plants and Animals* (ed. R.E. Crick), pp. 11-28. Plenum Press, New York.

Mackenzie F.T. and Bricker O.P. (1971) Cementation of sediments by carbonate minerals. In *Carbonate Cements* (ed. O.P. Bricker), pp. 239-251. Johns Hopkins Univ., Studies in Geology No. 19, Baltimore, MD.

Mackenzie F.T. and Garrels R.M. (1966a) Chemical mass balance between rivers and oceans. *Amer. J. Sci.* **264**, 507-525.

Mackenzie F.T. and Garrels R.M. (1966b) Silica-bicarbonate balance in the ocean and early diagenesis. *J. Sediment. Petrol.* **36**, 1075-1084.

Mackenzie F.T. and Pigott J.P. (1981) Tectonic controls of Phanerozoic sedimentary rock cycling. *J. Geol. Soc. London* **138**, 183-196.

Mackenzie F.T. and Wollast R. (1977) Thermodynamic and kinetic controls of global chemical cycles of the elements. In *Global Chemical Cycles and their Alterations by Man* (ed. W. Stumm), pp. 45-60. Dahlem Konferenzed, Berlin.

Mackenzie F.T., Ristvet B.L., Thorstenson D.C., Lerman A. and Leeper, R.H. (1981) Reverse weathering and chemical mass balance in a coastal environment. In *River Inputs to Ocean Systems* (eds. J.M. Martin, J.D. Burton and D. Eisma). pp. 152-187. UNEP-UNESCO, Switzerland.

Mackenzie F.T., Bischoff W.B., Bishop F.C., Loijens M., Schoonmaker J. and Wollast R. (1983) Magnesian calcites: Low-temperature occurrence, solubility and solid-solution behavior, In *Carbonates: Mineralogy and Chemistry* (ed. R.J. Reeder), pp. 97-144. Mineral. Soc. Amer., Chelsea, Mich.

Malone M.J., Baker P.A. and Burns S.J. (1989) Geochemistry of periplatform carbonate sediments ODP Site 716: Maldives Archipelago, Indian Ocean (in press).

Manabe S. and Wetherald (1980) On the distribution of climate change resulting from an increase in CO_2 content of the atmosphere. *J. Atmos. Sci.* **37**, 99-118.

Manghnani M.H., Schlanger S.O. and Milholland P.D. (1980) Elastic properties related to depth of burial, strontium content and age, and diagenetic stage in pelagic carbonate sediments. In *Bottom Interacting Ocean Acoustics* (eds. W.A. Kupferman and F.D. Jensen), pp.?. Plenum Press, New York.

Manheim F.T. (1976) Interstitial waters of marine sediments. In *Chemical Oceanography, Vol 6* (eds. J.P. Riley and R. Schester), pp. 115-186. Academic Press, New York.

Mapstone N.B. (1975) Diagenetic history of a North Sea chalk. *Sedimentology* **22**, 601-614.

Marijns A. (1978) Composition des Carbonates en Milieu Marin. thesis, Univ. Libre de Bruxelles, Brussells, Belgium.

Martin J.M. and Meybeck M. (1978) The content of major elements in the dissolved and particulate load of rivers. In *Biochemistry of Estuarine Sediments*, pp. 95-110. UNESCO Press.

Martin J.M. and Meybeck M. (1979) Elemental mass-balance of material carried by major world rivers. *Mar. Chem.* **7**, 173-206.

Martin G.D., Wilkinson B.H. and Lohmann K.C. (1986) The role of skeletal porosity in aragonite neomorphism -- Strombus and Montastrea from the Pleistocene Key Largo Limestone, Florida. *J. Sediment. Petrol.* **56**, 194-203.

Matter A.R., Douglas R.G. and Perch-Nielsen K. (1975) Fossil preservation, geochemistry and diagenesis of pelagic carbonates from Shatsky Rise, Northwest Pacific. *Initial Reports of the Deep Sea Drilling Project* **32**, Washington, D.C., U.S. Government Printing Office, 891-919.

Mattes B.W. and Mountjoy E.W. (1980) Burial dolomitization of the Upper Devonian Miette Buildup, Jasper national Park, Alberta. In *Concepts and Models of Dolomitization* (eds. D.H. Zenger, J.B. Dunham and R.L. Ethington), pp. 259-297. Special Publication Soc. Econ. Paleontologists and Mineralogists **28**, Tulsa, OK.

Matthews R.K. (1968) Carbonate diagenesis: Equilibration of sedimentary mineralogy to the subaerial environment; coral cap of Barbados, West Indies. *J. Sediment. Petrol.* **38**, 1110-1119.

Matthews R.K. (1974) A process approach to diagenesis of reefs and reef associated limestones. In *Reefs in Time and Space* (ed. L.F. Laporte), pp. 234-256. SEPM Spec. Pub. No. 18, Tulsa, OK.

May H.M., Kinniburgh D.G., Helmke P.A. and Jackson M.L. (1986) Aqueous dissolution, solubilities and thermodynamic stabilities of common aluminosilicate clay minerals: Kaolinite and smectites. *Geochim. Cosmochim Acta* **50**, 1667-1677.

Maynard J.B. (1976) The long-term buffering of the oceans. *Geochim. Cosmochim. Acta* **40**, 1523-1532.

Maynard J.B. (1981) Carbon isotopes as indicators of disposal patterns in Devonian-Mississippian shales of the Appalachian Basin. *Geology* **9**, 262-265.

McBride M.B. (1979) Chemisorption and precipitation of Mn^{2+} at $CaCO_3$ surfaces. *J. Soil Sci. Soc. Amer.* **43**, 693-698.

McCauley J.W. and Roy R. (1974) Controlled nucleation and crystal growth of various $CaCO_3$ phases by the silica gel technique. *Amer. Mineral.* **59**, 947-963.

McConnaughey, T. (1989a) ^{13}C and ^{18}O isotopic disequilibrium in biological carbonates: I. Patterns. *Geochim. Cosmochim. Acta* **53**, 151-162.

McConnaughey, T. (1989b) ^{13}C and ^{18}O isotopic disequilibrium in biological carbonates: II. *In vitro* simulation of kinetic isotope effects. *Geochim. Cosmochim. Acta* **53**, 163-171.

McConnell J.C., McElroy M.B. and Wofsy S.C. (1971) Natural sources of atmospheric CO. *Nature* **223,** 187-188.

McCrea J.M. (1950) On the isotopic chemistry of carbonates and a paleotemperature scale. *J. Chem. Phys.* **18**, 849-847.

McHargue T.R. and Price R.C. (1982) Dolomite from clay in argillaceous or shale-associated marine carbonates. *J. Sediment. Petrol.* **52**, 873-886.

Mehrbach C., Culberson C.H., Hawley J.E. and Pytkowicz R.M. (1973) Measurements of the apparent dissociation constants of carbonic acid in seawater at atmospheric pressure. *Limnol. Oceanogr.* **18**, 897-907.

Meigen W. (1903) Beitrage zur Kenntnis des kohlensauren Kalk. *Naturwiss. Gesell. Freiburg, Berlin* **13**, 1-55.

Memery L. and Merlivat L. (1985) Influence of gas transfer on the CO_2 uptake by the ocean. *J. Geophys. Res.* **90**, 7361-7366.

Meshri I.D. (1986) On the reactivity of carbonic and organic acids and the generation of secondary porosity. In *Roles of Organic Matter in Sediment Diagenesis* (ed. D.L. Gautier), pp. 123-128. Soc. Economic Paleontologists and Mineralogists Spec, Pub. **38**, Tulsa, OK.

Metzger W.J. and Bernard W.M. (1968) Transformation of aragonite to calcite under hydrothermal conditions. *Amer. Mineral.* **53**, 295-299.

Meybeck M. (1976) Total mineral dissolved transport by world major rivers. *Hydrol. Sci. Bull.* **21**, 265-284.

Meybeck M. (1977) Dissolved and suspended matter carried by rivers: Composition, time and space variations and world balance. In *Interactions between Sediments and Fresh Water* (ed. H.L. Golterman), pp. 25-32. Junk and Pudoc, Amsterdam.

Meybeck M. (1979a) Concentrations des eaux fluivales en elements majeurs et apports en solution aux oceans. *Révue de Géologie Dynamique et de Géographie Physique* **21**, 215-246.

Meybeck M. (1979b) Pathways of major elements from land to ocean through rivers. In *Review and Workshop on River Inputs to Ocean Systems*, pp. 26-10. March, 1979, FAO, Rome.

Meybeck M. (1982) Carbon, nitrogen and phosphorus transport by world's rivers. *Am. J. Sci.* **282**, 401-450.

Meybeck M. (1987) Global chemical weathering of surficial rocks estimated from river dissolved loads. *Amer. J. Sci.* **287**, 401-428.

Meyers J.H. (1987) Marine vadose beachrock cementation by cryptocrystalline magnesian calcite--Maui, Hawaii. *J. Sediment. Petrol.* **57**, 558-570.

Meyers W.J. (1974) Carbonate cement stratigraphy of the Lake Valley Formation (Mississippian), Sacramento Mountains, New Mexico. *J. Sediment. Petrol.* **44**, 827-861.

Meyers W.J. (1978) Carbonate cements: Their regional distribution and interpretation in Mississippian limestones of southwestern New Mexico. *Sedimentology* **25**, 371-401.

Meyers W.J. (1980) Compaction in Mississippian skeletal limestones, southwestern New Mexico. *J. Sediment. Petrol.* **50**, 457-474.

Meyers W.J. and Lohmann K.C. (1985) Isotope geochemistry of regionally extensive calcite cement zones and marine components in Mississippian limestones, New Mexico. In *Carbonate Cements* (eds. N. Schneidermann and P.M Harris) pp. 223-239. SEPM Spec. Pub. No. **36**, Tulsa, OK.

Michard G. (1968) Coprecipitation de l'ion manganeaux avec le carbonate de calcium. *Comptes Rendus Acad. Sci. Paris Series D* **267**, 1685-1688.

Millero F.J. (1979) The thermodynamics of the carbonate system in seawater. *Geochim. Cosmochim. Acta* **43**, 1651-1661.

Millero F.J. (1982a) Use of models to determine ionic interactions in natural waters. *Thallasia Jugoslavica* **18**, 253-291.

Millero F.J. (1982b) The effect of pressure on the solubility of minerals in water and seawater. *Geochim. Cosmochim. Acta* **46**, 11-22.

Millero F.J. (1983) Influence of pressure on chemical processes in the sea. In *Chemical Oceanography, Vol. 8* (eds. J.P. Riley and G. Skirrow), pp.1-88, Academic Press, London.

Millero F.J. and Schreiber D.R. (1982) Use of ion pairing model to estimate activity coefficients of the ionic components of natural waters. *Amer. J. Sci.* **282**, 1508-1540.

Millero F.J., Milne P.J. and Thurmond V.L. (1984) The solubility of calcite, strontianite and witherite in NaCl solutions at 25°C. *Geochim. Cosmochim. Acta* **48**, 1141-1143.

Millero F. J., Morse J.W. and Chen C.-T. (1979) The carbonate system in the western Mediterranean Sea. *Deep-Sea Res.* **26A**, 1395-1404.

Milliken K.L. and Pigott J.D. (1977) Variation of oceanic Mg/Ca ratio through time - implications for the calcite sea (abst). Geol. Soc. America South-Cent. Mtg., 64-65.

Milliman J.D. (1974) *Recent Sedimentary Carbonates 1, Marine Carbonates*. Springer-Verlag, Berlin, 375 pp.

Milliman J. D. and Müller J. (1973) Precipitation and lithification of magnesian calcite in the deep-sea sediments of the eastern Mediterranean Sea. *Sedimentology* **20**, 29-45.

Milliman J. D. and Müller J. (1977) Characteristics and genesis of shallow-water and deep-water limestones. In *The Fate of Fossil Fuel CO2 in the Oceans* (eds. N.R. Anderson and A. Malahoff), pp. 655-672. Plenum Press, New York.

Mimran Y. (1975) Fabric deformation induced in Cretaceous chalks by tectonic stresses. *Tectonophysics* **26**, 309-316.

Mimran Y. (1977) Chalk deformation and large-scale migration of calcium carbonate. *Sedimentology* **24**, 333-360.

Mitchell J.T., Land L.S. and Miser D.E. (1987) Modern marine dolomite cement in a north Jamaican fringing reef. *Geology* **15**, 557-560.

Mitterer R.M. (1968) Amino acid composition of organic matrix in calcareous oölites. *Science* **167**, 1498-1499.

Mitterer R.M. (1971) Influences of natural organic matter on $CaCO_3$ precipitation. In *Carbonate Cements* (ed. O.P. Bricker), pp. 254-258. The Johns Hopkins Press, Baltimore.

Mitterer R.M. (1972) Biogeochemistry of aragonite mud and oölites. *Geochim. Cosmochim. Acta* **36**, 1407-1422.

Mitterer R.M. and Cunningham R., Jr. (1985) The interaction of natural organic matter with grains surfaces: Implications for calcium carbonate precipitation. In *Carbonate Cements* (eds. N. Schneidermann and P.M. Harris), pp. 17-31. Soc. Econom. Paleontoligists and Mineralogists, Tulsa, OK.

Mix A.C. (1989) Influence of productivity variations on long-term atmospheric CO_2. *Nature* **337**, 541-544.

Moberly R. (1968) Composition of magnesian calcites of algae and paleosopods by electron microprobe analysis. *Sedimentology* **11**, 61-82.

Möller P. and Sastri C.S. (1974) Estimation of the number of surface layers of calcite involved in Ca-^{45}Ca isotopic exchange with solution. *Zeitschrift Phys. Chem. Neue Folge* **89**, 80-87.

Möller P. and Rajagopalan G. (1976) A geochemical model for dolomitization based on material balance: Part 1. *Geol. Jb.* **20**, 41-56.

Monnin C. and Schott J. (1984) Determination of the solubility products of sodium carbonate minerals and an application to trona deposition in Lake Magadi (Kenya). *Geochim. Cosmochim. Acta* **48**, 571-581.

Moore C.H., Jr. (1973) Intertidal carbonate cementation, Grand Cayman, West Indies. *J. Sediment. Petrol.* **43**, 591-602.

Moore C.H. (1985) Upper Jurassic subsurface cements: A case history. In *Carbonate Cements* (eds. N. Schneiderman and P.M. Harris), pp. 291-308. Society Economic Paleontologists and Mineralogists Special Publication **36**, Tulsa, OK.

Moore C.H., Jr. (1989) *Carbonate Diagenesis and Porosity. Developments in Sedimentology 46*, Elsevier, Amsterdam, 338 pp.

Moore C.H. and Brock F.C. (1982) Discussion of paper by Wagner and Matthews. *J. Sediment. Petrol.* **52**, 19-23.

Moore C.H. and Druckman Y. (1981) Burial diagenesis and porosity evolution, Upper Jurassic Smackover, Arkansas and Louisiana. *Am. Assoc. Petrol. Geol. Bull.* **65**, 597-628.

Morrow D.W. (1982a) Diagenesis 1. Dolomite--Part 1. The chemistry of dolomitization and dolomite precipitation. *Geoscience Canada* **9**, 5-13

Morrow D.W. (1982b) Diagenesis 2. Dolomite--Part 2. Dolomitization models and ancient dolostones. *Geoscience Canada* **9**, 95-107.

Morse J.W. (1973) The dissolution kinetics of calcite: A kinetic origin for the lysocline. Ph.D. dissertation, Yale Univ.

Morse J.W. (1974) Dissolution kinetics of calcium carbonate in seawater. V: Effects of natural inhibitors and the position of the chemical lysocline. *Amer. J. Sci.* **274**, 638-647.

Morse J.W. (1978) Dissolution kinetics of calcium carbonate in seawater. VI: The near-equilibrium dissolution kinetics of calcium carbonate-rich deep sea sediments. *Amer. J. Sci.* **278**, 344-355.

Morse J.W. (1983) The kinetics of calcium carbonate dissolution and precipitation. In *Reviews in Mineralogy: Carbonates - Mineralogy and Chemistry* (ed. R. J. Reeder), pp. 227-264. Mineralogical Society of America. Bookcrafters, Inc., Chelse, MI.

Morse J.W. (1985) Discussion: Kinetic control of morphology, composition, and mineralogy of abiotic sedimentary carbonates. *J. Sediment. Petrol.* **55**, 919-934.

Morse J.W. (1986) The surface chemistry of carbonate minerals in natural waters: An overview. *Mar. Chem.* **20**, 91-112.

Morse J.W. and Bender M.L. (1989) Partition coefficients in calcite: an examination of factors influencing the validity of experimental results and their application to natural systems. *Chem. Geol.*, (in press).

Morse J.W. and Berner R.A. (1972) Dissolution of calcium carbonate in seawater II. A kinetic origin for the lysocline. *Amer. J. Sci.* **272**, 840-851.

Morse J.W. and Berner R.A. (1979) The chemistry of calcium carbonate in the deep oceans. In *Chemical Modelling--Speciation, Sorption, Solubility, and Kinetics in Aqueous Systems* (ed. E. Jenne), pp. 499-535. Amer. Chem. Soc., Washington, D.C.

Morse J.W. and Casey W.H. (1988) Ostwald processes and mineral paragenesis in sediments. *Amer. J. Sci.* **288**, 537-560.

Morse J.W. and Cook N. (1978) The distribution of phosphorous in North Atlantic deep sea and continental slope sediments. *Limnol. Oceanogr.* **23**, 825-830.

Morse J.W. and Mucci A. (1984) Composition of carbonate overgrowths produced on calcite crystals in Bahamian pore waters. *Sediment. Geol.* **40**, 287-291.

Morse J.W. and Casey W.H. (1988) Ostwald processes and mineral paragenesis in sediments. *Amer. J. Sci.* **288**, 537-560.

Morse J.W., deKanel J. and Harris J. (1979) Dissolution kinetics of calcium carbonate in seawater. VII: The dissolution kinetics of synthetic aragonite and pteropod tests. *Amer. J. Sci.* **279**, 482-502.

Morse J.W., Mucci A. and Millero F.J. (1980) The solubility of calcite and aragonite in seawater of 35‰ salinity and atmospheric pressure. *Geochim. Cosmochim. Acta* **44**, 85-94.

Morse J.W., Millero F.J., Thurmond V., Brown E. and Ostlund H.G. (1984) The carbonate chemistry of Grand Bahama Bank waters: after 38 years, another look. *J. Geophys. Res.* **89**, 3604-3614.

Morse J.W., Zullig J.J., Bernstein L.D., Millero F.J., Milne P., Mucci A. and Choppin G.R. (1985) Chemistry of calcium carbonate-rich shallow water sediments in the Bahamas. *Amer. J. Sci.* **285**, 147-185.

Morse J.W., Zullig J.J., Iverson R.L., Choppin G.R., Mucci A. and Millero F.J. (1987) The influence of seagrass beds on carbonate sediments in the Bahamas. *Mar. Chem.* **22**, 71-83.

Mottl M.J. (1976) Chemical exchange between seawater and basalt during hydrothermal alteration of the oceanic crust. Ph.D. dissertation, Harvard Univ., Cambridge.

Mottl J.D. and Holland H.D. (1978) Chemical exchange during hydrothermal alteration of basalt by seawater I. Experimental results for major and minor components of seawater. *Geochim. Cosmochim. Acta* **42**, 1103-1115.

Mottl M.J., Lawrence J.R. and Keigwin L.D. (1983) Elemental and stable isotope composition of porewaters and carbonate sediments from Deep Sea Drilling Project Sites 501/504 and 505. *Initial Reports DSDP* **69**, 461-473.

Moulin E, Jordens A. and Wollast R. (1985) Influence of the aerobic bacterial respiration on the early dissolution of carbonates in coastal sediments. *Proc. Progress in Belgium Oceanographic Research*, Brussels. 196-208.

Mount J. (1984) Mixing of silicaclastic and carbonate sediments in shallow shelf environments. *Geology* **12**, 432-435.

Mount J. (1985) Mixed silicaclastic and carbonate sediments: A proposed first-order textural and compositional classification. *Sedimentology* **32**, 435-442.

Mucci A. (1983) The solubility of calcite and aragonite in seawater at various salinities, temperatures and 1 atmosphere total pressure. *Amer. J. Sci.* **238**, 780-799.

Mucci A. (1986) Growth kinetics and composition of magnesian calcite overgrowths precipitated from seawater: Quantitative influence of orthophosphate ions. *Geochim. Cosmochim. Acta* **50**, 2255-2265.

Mucci A. (1987) Influence of temperature on the composition of magnesian calcite overgrowths precipitated from seawater. *Geochim. Cosmochim. Acta* **51**, 1977-1984.

Mucci A. (1988) Manganese uptake during calcite precipitation from seawater: Conditions leading to the formation of a pseudokutnahorite. *Geochim. Cosmochim. Acta* **52**, 1859-1868.

Mucci A. and Morse J.W. (1983) The incorporation of Mg^{2+} and Sr^{2+} into calcite overgrowths: influences of growth rate and solution composition. *Geochim. Cosmochim. Acta* **47**, 217-233.

Mucci A. and Morse J.W. (1984) The solubility of calcite in seawater of various magnesium concentrations, $I_t = 0.697$ m at 25^oC and one atmosphere total pressure. *Geochim. Cosmochim. Acta* **48**, 815-822.

Mucci A. and Morse J.W. (1985) Auger spectroscopy determination of the surface-most adsorbed layer composition on aragonite, calcite, dolomite, and magnesite in synthetic seawater. *Amer. J. Sci.* **285**, 306-317.

Mucci A. and Morse J.W. (1989) Chemistry of low-temperature abiotic calcite. *CRC Critical Reviews in Aquatic Chemistry*, (in press).

Mucci A., Morse J.W. and Kaminsky M.S. (1985) Auger spectroscopy analysis of magnesian calcite overgrowths precipitated from seawater and solutions of similar composition. *Amer. J. Sci.* **285**, 289-305.

Mucci A., Canuel R. and Zhong S. (1989) The solubility of calcite and aragonite in sulfate-free seawater and the seeded growth kinetics and composition of precipitates at 25^oC. *Chem. Geol.* **74**, 309-329.

Mullins H.T., Wise S.W. Jr., Gardulski A.F., Hinchey E.J., Master P.M. and Siegel D.I. (1985) Shallow subsurface diagenesis of Pleistocene periplatform ooze: northern Bahamas. *Sedimentology* **32**, 473-494.

Murray J. and Renard A.F. (1981) *Report on deep-sea deposits, from the Report on the Scientific Results of the Voyage H.M.S. Challenger*. Great Britain, Challenger office.

Murray J.W., Emerson S. and Jahnke R. (1980) Carbonate saturation and the effect of pressure on the alkalinity of interstitial waters from the Guatemala Basin. *Geochim. Cosmochim. Acta* **44**, 963-972.

Nagy K.L. (1988) The solubility of calcite in NaCl and Na-Ca-Cl brines. Ph.D. dissertation, Texas A&M Univ.

Nagy K.L. and Morse J.W. (1990) Temperature dependence of calcite solubility of NaCl solutions from 25° to 89°C at 1 atm. *Geochim. Cosmochim. Acta* (in press).

Nakayama F.S. (1968) Calcium activity, complex and ion pair in saturated $CaCO_3$ solutions. *Soil Sci.* **106**, 429-434.

Nancollas G.H. and Reddy M.M. (1971) The crystallization of calcium carbonate II. Calcite growth mechanism. *J. Coll. Interface Sci.* **37**, 824-830.

National Research Council (1983) *Changing Climate*. National Academic Press, Washington, DC, 496. pp.

Neev D. and Emery K.O. (1967) The Dead Sea: Depositional processes and environment of evaporites. *Geol. Survey Bull.* **41**, Ministry of Development, Jerusalem.

Neugebauer J. (1973) The diagenetic problem of chalk: The role of pressure solution and pore fluid. *Jahrb. Geol. Palaeontol. Abhandlungen* **143**, 223-245.

Neugebauer J. (1974) Some aspects of cementation of chalk. In *Pelagic Sediments: On Land and Under the Sea* (eds. K.J. Hsu and H.C. Jenkyns), pp. 149-176. International Association of Sedimentologists Special Publication No. 1, Blackwell Scientific Publishers, Oxford.

Neumann A.C. (1965) Processes of recent carbonate sedimentation in Harrington Sound Bermuda. *Bull. Mar. Sci.* **15**, 987-1035.

Neumann A.C. and Land L.A. (1975) Lime mud deposition and calcareous algae in the Bight of Abaco, Bahamas: A budget. *J.. Sediment. Petrol.* **45**, 763-786.

Nier A.O. (1950) A redetermination of the relative abundances of the isotopes of carbon, nitrogen, oxygen, argon, and potassium. *Phys. Rev.* **77**, 789.

Nielsen H. (1965) Schwefelisotope in marinen Kreislauf und das $\delta^{34}S$ des fruheren Meese. *Geologish Rundschau* **55**, 160-172.

Nielsen H. (1978) Sulfur isotopes in nature. In *Handbook of Geochemistry* (ed. K.H. Wedepohl), Part 16B. Springer Verlag, Berlin.

Niemann J.C., and Read J.F. (1988) Regional cementation from unconformity-recharged aquifer and burial fluids, Mississippian Newman Limestone, Kentucky. *J. Sediment. Petrol.* **58**, 688-705.

Nordstrom D.K. and Munoz J. L. (1986) *Geochemical Thermodynamics*. Blackwell Scientific Publications, Palo Alto, CA, 477 pp.

Northrop D.A. and Clayton R.N. (1966) Oxygen isotope fractionation in systems containing dolomite. *J. Geol.* **74**, 174.

Ohara M. and Reid R.C. (1973) *Modeling Crystal Growth Rates from Solution.* Prentice-Hall, Inc., Englewood Cliffs, NJ, 272 pp.

Ohde S. (1987) Geochemistry of sedimentary protodolomite. *J. Earth Sciences* **35**, 167-180.

Ohde S. and Kitano Y. (1978) Synthesis of protodolomite from aqueous solution at normal temperature and pressure. *Geochem. J.* **12**, 115-119.

Ohde S. and Kitano Y. (1984) Coprecipitation of strontium with marine Ca-Mg carbonates. *Geochem. J.* **18**, 143-146.

Okumura M., Kitano Y. and Idogaki M. (1983) Removal of anions by carbonate sedimentation from seawater. *Geochem. J.* **17**, 105-110.

Olsen, S. (1982) Computing climate. *Science*, **82**, 54.

O'Neil J.R. and Epstein S. (1966) Oxygen isotope fractionation in the system dolomite-calcite-carbon dioxide. *Science* **152**, 198-201.

O'Neil, T.J. (1974) Chemical interaction due to mixing of meteoric and marine waters in a Pleistocene reef complex, Rio Bueno, Jamaica. M.S. thesis, Louisiana State Univ., Baton Rouge, LA.

Onuma N., Masuda F., Hirano M. and Woda K. (1979) Crystal structure control on trace element partition in molluscan shell formation. *Geochemical Jour.* **13**, 187-189.

Oomori T., Kaneshima K. and Maezato Y. (1987) Distribution coefficient of Mg^{2+} ions between calcite and solution at 10-50°C. *Mar. Chem.* **20**, 327-336.

Oomori T., Kaneshima K. and Kitano Y. (1988) Solubilities of calcite, aragonite and protodolomite in supratidal brines of Minamidaito-Jima Island, Okinawa, Japan. *Mar. Chem.* **25**, 57-74.

Oppenheimer C.H. (1961) Note on the formation of spherical aragonatic bodies in the presence of bacteria from the Bahama Bank. *Geochim. Cosmochim. Acta* **23**, 295-299.

Palciauskas V.V. and Domenico P.A. (1976) Solution chemistry, mass transfer, and the approach to chemical equilibrium in porous carbonate rocks and sediments. *Geol. Soc. Amer. Bull.* **87**, 207-214.

Palmer M.R. and Elderfield H. (1986) Rare earth elements and neodymium isotopes in ferromanganese oxide coatings of Cenozoic foraminifera from the Atlantic Ocean. *Geochim. Cosmochim. Acta* **50**, 409-417.

Parker C.A. (1974) Geopressures and secondary porosity in the deep Jurassic of Mississippi. *Trans.-Gulf Coast Assoc. Geol. Socs.* **24**, 69-80.

Parker F.L. and Berger W.H. (1971) Faunal and solution patterns of planktonic Foraminifera in surface sediments of the South Pacific. *Deep-Sea Res.* **18**, 73-107.

Paull C.K., Hecker B., Commeau R., Freeman-Lynde R.P., Neumann A.C., Corso W.B., Golubic S., Hook J.E., Sikes E., and Curray J. (1984) Biological communities at the Florida Escarpment resemble hydrothermal vent taxa. *Science* **226**, 965-967.

Pedersen T.F. and Price N.B. (1982) The geochemistry of manganese carbonate in Panama Basin sediments. *Geochim. Cosmochim. Acta* **46**, 59-68.

Peng T.-H. (1986) Uptake of anthropogenic CO_2 by lateral transport models of the ocean based on the distribution of bomb-produced $14C$. *Radiocarbon* **28**, 363-375.

Perry E.C., Jr., Ahmad S.N. and Swulius T.M. (1978) The oxygen isotopic composition of 3800 m.y. old metamorphosed chert and iron formation from Isukasia, West Greenland. *J. Geol.* **86**, 223-239.

Peterson B.J. and Melillo J.M. (1985) The potential storage of carbon caused by eutrophication of the biosphere. *Tellus* **37**, 117-127.

Peterson L.C. and Prell W.H. (1985a) Carbonate dissolution in recent sediments of the eastern equatorial Indian Ocean: Preservation patterns and carbonate loss above the lysocline. *Mar. Geol.* **64**, 259-290.

Peterson L.C. and Prell W.H. (1985b) Carbonate preservation and rates of climatic change: An 800 kyr record from the Indian Ocean. In *The Carbon Cycle and Atmospheric CO_2: Natural Variations Archean to Present* (eds. E.T. Sundquist and W.S. Broecker), pp. 251-269.

Peterson M.N. (1966) Calcite: rates of dissolution in a vertical profile in the Central Pacific. *Science* **154**, 1542-1544.

Pettijohn F.J. (1957) *Sedimentary Rocks*, 2nd Ed., Harper & Brothers, New York. 718 pp.

Pestana H.R. (1977) *Guide for the Identification of Carbonate Grains and Carbonate-Producing Organisms.* Special Publications No. 13, Bermuda Biological Station for Research, Saint Georges, Bermuda, 53 pp.

Pigott J.D. and Land L.S. (1986) Interstitial water chemistry of Jamaican Reef sediment: sulfate reduction and submarine cementation. *Mar. Chem.* **19**, 355-378.

Pingitore N.E. (1976) Vadose and phreatic diagenesis: Processes, products and their recognition in corals. *J. Sediment. Petrol.* **46**, 985-1006.

Pingitore N.E. (1978) The behavior of Zn^{2+} and Mn^{2+} during carbonate diagenesis: Theory and applications. *J. Sediment. Petrol.* **48**, 799-814.

Pingitore N.E. and Eastman M.P. (1986) The coprecipitation of Sr^{2+} with calcite at 25°C and 1 atm. *Geochim. Cosmochim. Acta* **50**, 2195-2203.

Pingitore N.E., Eastman M.P., Sandidige K.O. and Freiha B. (1988) The coprecipitation of manganese(II) with calcite: An experimental study. *Mar. Chem.* **25**, 107-120.

Pinter N. and Gardner T.W. (1989) Construction of a polynomial model of glacio-eustatic fluctuation: Estimating paleo-sealevels continuously through time. *Geology* **17**, 295-298.

Pisciotto K.A. and Mahoney J.J. (1981) Isotopic survey of diagenetic carbonates, Deep Sea Drilling Project Leg 63. In *Initial Reports of the Deep Sea Drilling Project 63* (eds. Yeats et al.), pp. 595-609. La Jolla, CA.

Pitman W.C., Jr. (1978) Relationship between eustasy and stratigraphic sequences of passive margins. *Geol. Soc. Amer.Bull.* **89,** 1389-1403.

Pitzer K.E. (1979) Theory: Ion interaction approach. In *Activity Coefficients in Electrolyte Solutions, Vol. I* (ed. R.M. Pytkowicz), pp.157-208, CRC Press, Boca Raton, FL.

Plummer L.N. (1970) Soils and the diagenesis of Bermuda carbonate sands. Bermuda Biological Station for Research Spec. Pub. 7, St. Georges, Bermuda, 56-98.

Plummer L.N. (1975) Mixing of sea water with calcium carbonate ground water. *Geol. Soc. Amer. Memoir* **142**, 219-236.

Plummer L.N. (1977) Defining reactions and mass transfer in part of the Floridan Aquifer. *Water Resources Res.* **13**, 801-812.

Plummer L.N. and Busenberg E. (1982) The solubilities of calcite, aragonite and vaterite in CO_2-H_2O solutions between 0 and 90°C and an evaluation of the aqueous model for the system $CaCO_3$-CO_2-H_2O. *Geochim. Cosmochim. Acta* **46**, 1011-1040.

Plummer L.N. and Busenberg E. (1987) Thermodynamics of aragonite-strontianite solid solutions: Results from stoichiometric solubility at 25 and 76°C. *Geochim. Cosmochim. Acta* **51,**1393-1411.

Plummer L.N. and Mackenzie F.T. (1974) Predicting mineral solubility from rate data: Application to the dissolution of magnesian calcites. *Amer. J. Sci.* **274,** 61-83.

Plummer L.N., Wigley T.M.L. and Parkhurst D.L. (1978) The kinetics of calcite dissolution in CO_2-water systems at 5° to 60°C and 0.0 to 1.0 atm CO_2. *Amer. J. Sci.* **278**, 179-216.

Plummer L.N., Parkhurst D.L. and Wigley T.M.L. (1979) Critical review of the kinetics of calcite dissolution and precipitation. In *Chemical Modelling-- Speciation, Sorption, Solubility and Kinetics in Aqueous Systems* (ed. E. Jenne), pp. 537-573. American Chemical Society, Washington, D.C.

Plummer L.N., Parkhurst D.L. and Thorstenson D.C. (1983) Development of reaction models for ground-water systems. *Geochim. Cosmochim. Acta* **47**, 665-686.

Plummer L.N., Vacher H.L., Mackenzie F.T., Bricker O.P., and Land L.S. (1976) Hydrogeochemistry of Bermuda: A case history of ground-water diagenesis of biocalcarenites. *Geol. Soc. Amer. Bull.* **87**, 1301-1316.

Popp B.N., Anderson F.T. and Sandberg P.A. (1986) Brachiopods as indicators of original isotopic compositions in Paleozoic limestones. *Geol. Soc. Amer. Bull.* **97**, 1262-1269.

Prentice M.L. and Matthews R.K. (1988) Cenozoic ice-volume history: Development of a composite oxygen isotope record. *Geology* **16**, 963-966.

Prezbindowski D.R. (1985) Burial cementation--is it important? A case study, Steward City trend, south central Texas. In *Carbonate Cements* (eds. N. Schneidermann and P.M. Harris), pp. 241-264. Society Economic Paleontologists and Mineralogists Special Publication **36**, Tulsa, OK.

Pytkowicz R.M. (1965) Rates of inorganic calcium carbonate nucleation. *J. Geol.* **73**, 196-199.

Pytkowicz R.M. (1970) On the carbonate compensation depth in the Pacific Ocean. *Geochim. Cosmochim. Acta* **34**, 836-839.

Pytkowicz R.M. (1979) *Activity Coefficients in Electrloyte Solutions*, Vols. I and II. CRC Press, Boca Raton, FL.

Pytkowicz R.M. (1971) Sand-seawater interactions in Bermuda beaches. *Geochim. Cosmochim. Acta* **35**, 509-515.

Pytkowicz R.M. (1973) Calcium carbonate retention in supersaturated seawater. *Amer. J. Sci.* **273**, 515-522.

Pytkowicz M. and Cole M. R. (1980) Equilibrium and kinetic problems in mixed electrolyte solutions. In *Thermodynamics of Aqueous Systems with Industrial Applications* (ed. S.A. Newman), ACS Symp. Series, **133**, Washington, D.C., 644-652.

Raiswell R. (1988) Chemical model for the origin of minor limestone-shale cycles by anaerobic methane oxidation. *Geology* **16**, 641-644.

Raiswell R. and Brimblecombe P. (1977) The partition of manganese into aragonite between 30 and 60°C. *Chem. Geol.* **19**, 145-151.

Ramanathan V. (1988) The greenhouse theory of climate change: A test by an inadvertent global experiment, *Science* **240**, 293-299.

Read J.F. (1980) Carbonate ramp-to-basin transitions and foreland basin evolution, Middle Ordovician, Virginia Appalachians. *AAPG Bull.* **64**, 1575-1612.

Read J.F., Grotzinger J.P., Bova J.A. and Koerschner W.F. (1986) Models for generation of carbonate cycles. *Geology* **14**, 107-110.

Reading H.G. (1978) *Sedimentary Environments and Facies*. Elsevier, New York. 557 pp.

Reaves C.M. (1986) Organic matter metabolizability and calcium carbonate dissolution in nearshore marine muds. *J. Sediment. Petrol.* **56**, 486-494.

Reddy M.M. and Gaillard W.D. (1981) Kinetics of calcium carbonate (calacite)-seeded crystallization: Influence of solid/solution ratio on the reaction rate constant. *J. Colloid Interface Sci.* **80**, 171-178.

Reddy M.M., Plummer L.N. and Busenberg E. (1981) Crystal growth of calcite from calcium bicarbonate solutions at constant P_{CO_2} and 25°C: A test of the calcite dissolution model. *Geochim. Cosmochim. Acta* **45**, 1281-1291.

Redfield A.C., Ketchum B.H. and Richard F.A. (1963) The influence of organisms on the composition of seawater. In *The Sea* (ed. M.N. Hill), pp. 27-77. Wiley and Sons, New York.

Reeckmann S.A. (1988) Diagenetic alterations in temperate shelf carbonates. *Sediment. Geology* **60**, 209-219.

Reeder R.J. and Grams J.C. (1987) Sector zoning in calcite cement crystals: Implications for trace element distributions in carbonates. *Geochim. Cosmochim. Acta* **51**, 187-194.

Reeder R.J. and Prosky J.L. (1986) Compositional sector zoning in dolomite. *J. Sediment. Petrol.* **56**, 237-247.

Rees C.E. (1970) The sulfur isotope balance of the ocean: An improved model. *Earth. Planet. Sci. Lett.* **7**, 366-370.

Reiterer F., Johannes W. and Gamsjäger H. (1981) Semimicro determination of solubility constants: Copper(II) carbonate and iron(II) carbonate. *Mikrochimica Acta* **I**, 63-72.

Renard M. (1986) Pelagic carbonate chemostratigraphy (Sr, Mg, ^{18}O, ^{13}C). *Mar. Micropaleontol.* **10**, 117-164.

Revelle R. and Fairbridge R. (1957) Carbonate and carbon dioxide. *Geol. Soc. Amer. Mem.* **67**, 239 pp.

Revelle R. and Munk W. (1977) The global carbon dioxide cycle and the biosphere. In *Energy and Climate*, pp. 243-280. National Academy of Sciences Press, Washington, D.C.

Revelle R. and Suess H. (1957) Carbon dioxide exchange between atmosphere and ocean, and the question of an increased atmospheric CO_2 during the past decades. *Tellus* **9**, 18-27.

Ristvet B.L. (1971) The progressive diagenetic history of Bermuda. In *Bermuda Biol. Station for Research Spec. Pub.* **9**, 118-157.

Ristvet B. (1978) Reverse weathering reactions within recent nearshore marine sediments, Kaneohe Bay, Oahu. Test Directorate Field Command, Kirtland AFB, New Mexico, 314 pp.

Ritger S., Carson B. and Suess E. (1987) Methane-derived authigenic carbonates formed by subduction-induced pore-water expulsion along the Oregon/Washington margin. *Geol. Soc. Amer. Bull.* **98**, 147-156.

Robie R.A., Hemingway B.S. and Fisher J.R. (1979) *Thermodynamic Properties of Minerals and Related Substances at 298.15K and 1 Bar (10^2 Pascals) Pressure and at Higher Temperatures*. Geol. Survey Bull. 1452. U.S. Gov. Printing Office, Washington, D.C., 456 pp.

Robin P.-Y. F. (1978) Pressure solution at grain-to-grain contacts. *Geochim. Cosmochim. Acta* **42**, 1383-1389.

Roehl P.O. and Choquette P.W. (1985) Introduction. In *Carbonate Petroleum Reservoirs* (eds. P.O. Roehl and P.W. Choquette), pp. 1-18. Springer Verlag, New York.

Ronov A.B. (1964) Common tendencies in the chemical evolution of the Earth's crust, ocean and atmosphere. *Geochem. Intl.* **1**, 713-737.

Ronov A.B. (1968) Probable changes in the composition of seawater during the course of geological time. Invited lecture VII, International Sedimentological Congress, Edinburgh, Scotland.

Ronov A.B. (1980) *Sedimentary Cover of the Earth*. Nauka, Moscow, 78 pp. (in Russian).

Ronov A.B. (1982) The Earth's sedimentary shell (quantitative patterns of its structure, composition and evolution). The 20th V.I. Vernadskiy Lecture, March 12, 1978. In *The Earth's Sedimentary Shell (Quantitative Patterns of its Strcuture, Composition and Evolution)* (ed. Yaroshevskiy), pp. 1-80. Nauka, Moscow. (International Geological Review English Translations **24**, Nos. 11-12, 1313-1363 and 1365-1388, 1982).

Ronov A.B. and Migdisov A.A. (1970) Evolution of the chemical composition of the rocks in the shields and sediment cover of the Russian and North American Platform. *Geochem. Intl.* **7**, 294-324.

Ronov A.B. and Migdisov A.A. (1971) Evolution of the chemical composition of the rocks in the shields and sediment cover of the Russian and North American Platforms. *Sedimentology* **16**, 137-185.

Ronov A.B., Khain V.E., Valukhovsky A.N. and Seslavinsky K.B. (1980) Quantitative analysis of Phanerozoic sedimentation. *Sediment. Geol.* **25**, 311-325.

Rosenfeld J.K. (1979) Interstitial water and sediment chemistry of two cores from Florida Bay. *J. Sediment. Petrol.* **49**, 989-994.

Roth P.H., Mullin M.M. and Berger W.H. (1975) Coccolith sedimentation by fecal pellets: Laboratory experiment and field observations. *Geol. Soc. Amer. Bull.* **86**, 1079-1084.

Rothpletz A. (1892) Über die Bildung der Oölithe. *Bot. Centralbl.* **35**, 1-4.

Rotty R.M. (1979) Present and future production of CO_2 from fossil fuels - A global appraisal. In *Workshop on the Global Effects of Carbon Dioxide from Fossil Fuels*, pp. 36-43. USDOE Conf.-770385, NTIS, Springfield, VA.

Rowe G.T. and Gardner W.D. (1979) Sedimentation rates in the slope water of the northern Atlantic Ocean measured directly with sediment traps. *J. Mar. Res.* **37**, 581-600.

Rubinson M. and Clayton R.N. (1969) Carbon-13 fractionation between aragonite and calcite. *Geochim. Cosmochim. Acta* **33**, 997-1002.

Runnels D.D. (1969) Diagenesis, chemical sediments, and the mixing of natural waters. *J. Sediment. Petrol.* **39**, 1188-1201.

Rutter E.H. (1983) Pressure solution in nature, theory and experiment. *J. Geol. Soc. London* **140**, 725-740.

Saltzman E.S., Lindh T.B. and Holser W.T. (1982) Secular changes in $\delta^{13}C$ and $\delta^{34}S$, global sedimentation, P_{O2} and P_{CO2} during the Phanerozoic. *Geol. Soc. Amer. Abstr.* **14**, 607.

Sandberg P.A. (1975) New interpretation of great salt lake oöids and of ancient non-skeletal carbonate mineralogy. *Sedimentology* **22,** 497-538.

Sandberg P.A. (1983) An oscillating trend in Phanerozoic non-skeletal carbonate mineralogy. *Nature* **305**, 19-22.

Sandberg P.A. (1985) Nonskeletal aragonite and pCO_2 in the Phanerozoic and Proterozoic. In *The Carbon Cycle and Atmospheric CO2: Natural Variations Archean to Present* (eds. E.T. Sundquist and W.S. Broecker), pp.585-594. Geophys. Monograph Series **32**, Amer. Geophys. Union, Washington, D.C.

Sansone F.J. (1985) Methane in the reef flat porewaters of Davies Reef, Great Barrier Reef (Australia). *Proceedings of the Fifth International Coral Reef Congress, Tahiti* **3**, 415-420.

Sansone F.J., Andrews C.G., Buddemeier R.W. and Tribble G.W. (1988a) Well point sampling of reef interstitial water. *Coral Reefs* **7**, 19-22.

Sansone F.J., Tribble G.W., Buddemeier R.W. and Andrews C.C. (1988b) Time and space scales of anaerobic diagenesis within a coral reef framework. *Proceedings of the Sixth International Coral Reef Congress, Australia* **3**, 367-372.

Sansone F.J., Tribble G.W., Buddemeier R.W. and Andrews C.C. (1990) Microbially mediated diagenesis within the lithified coral reef frameworks. *Coral Reefs* (submitted).

Sass E. and Katz A. (1982) The origin of platform dolomites: New evidence. *Amer. J. Sci.* **282**, 1184-1213.

Sass E. and Bein A. (1989) Dolomites and salinity: A comparative geochemical study. In *Sedimentology and Geochemistry of Dolostones, SEPM Spec. Pub.,* Tulsa, OK, (in press).

Sass E., Morse J.W. and Millero F.J. (1983) Dependence of values of calcite and aragonite thermodynamic solubility products on ionic models. *Amer. J. Sci.* **283,** 218-229.

Savin S.M., Douglas R.G. and Stehli F.G. (1975) Tertiary marine paleotemperatures. *Geol. Soc. Amer Bull.* **86,** 1499-1510.

Sayles F.L. (1979) The composition and diagenesis of interstitial solutions; I. Fluxes across the seawater-sediment interface in the Atlantic Ocean. *Geochim. Cosmochim. Acta* **43,** 527-546.

Sayles F.L. (1980) The solubility of $CaCO_3$ in seawater at $2^{o}C$ based upon in-situ sampled pore water composition. *Mar. Chem.* **9,** 223-235.

Sayles F. L. (1985) $CaCO_3$ solubility in marine sediments: Evidence for equilibrium and non-equilibrium behavior. *Geochim. Cosmochim. Acta* **49,** 877-888.

Sayles R.W. (1931) Bermuda during the Ice Age. *Amer. Acad. Arts Sci.* **66,** 381-468.

Schidlowski M. (1979) Antiquity and evolutionary status of bacterial sulfate reduction: Sulfur isotope evidence. *Origins of Life* **9,** 299-331.

Schidlowski M. (1982) Content and isotopic composition of reduced carbon in sediments. In *Mineral Deposits and the Evolution of the Biosphere* (eds. H.D. Holland and M. Schidlowski), pp. 103-122. Springer Verlag, Berlin.

Schidlowski M. (1987) Application of stable carbon isotopes to early biochemical evolution on Earth. *Annual Review Earth Planet. Sci.* **15,** 47-72.

Schidlowski M. (1988) A 3800-million-year isotopic record of life from carbon in sedimentary rocks. *Nature* **333,** 313-318.

Schifano G. and Censi P. (1986) Oxygen and carbon isostope composition, magnesium and strontium contents of calcite from a subtidal *patella coerulea* shell. *Chem. Geol.* **58,** 325-331.

Schink D.R. and Guinasso N.L., Jr. (1977) Modelling the influence of bioturbation and other processes on calcium carbonate at the sea floor. In *The Fate of Fossil Fuel CO_2 in the Oceans* (eds. N.R. Anderson and A. Malahoff), pp. 375-400. Plenum Press, New York.

Schlager W. and James N.P. (1978) Low-magnesian calcite limestones forming at the deep-seafloor, Tongue of the Ocean, Bahamas. *Sedimentology* **25,** 675-702.

Schlanger S.O. and Douglas R.G. (1974) The pelagic ooze-chalk-limestone transition and its implications for marine stratigraphy. *Special Publication of the Intl. Assoc. Sedimentol.* **1,** 117-148.

Schmalz R.F. (1965) Brucite in carbonate secreted by the Red Alga Goniolithon sp. *Science* **149**, 993-996.

Schmalz R.F. (1967) Kinetics and diagenesis of carbonate sediments. *J. Sediment. Petrol.* **37**, 60-67.

Schmalz R.F. (1971) Beachrock formation at Eniwetok Atoll, In *Carbonate Cements* (ed. O.P. Bricker), pp. 17-24. John Hopkins Univ. Press, Baltimore, MD.

Schmitt G.W. (1973) Interstitial water composition and geochemistry of deep Gulf Coast shales and sandstones. *AAPG Bull.* **57**, 321-337.

Schmitt V. and McDonald D.A. (1979) The role of secondary porosity in the course of sandstone diagenesis. *Soc. of Econ. Paleontologists and Mineralologists Spec. Publ.* **26**, 175-208.

Schmoker J.W. (1984) Empirical relationship between carbonate porosity and thermal maturity: An approach to regional porosity prediction. *AAPG Bull.* **68**, 1697-1703.

Schmoker J.W., Krystink K.B. and Halley R.B. (1985) Selected characteristics of limestone and dolomite reservoirs in the United States. *AAPG Bull.* **69**, 733-741.

Schneidermann N. (1973) Deposition of coccoliths in the compensation zone of the Atlantic Ocean. *Proceedings of Symposium on Calcareous Nannofossils* (eds. L. A. Smith and J. Hardenbol), pp. 140-151. Gulf Coast Section SEPM, Houston, TX.

Schneidermann N. (1977) Chapter 12. Selective Dissolution of Recent Coccoliths in the Atlantic Ocean. In *Oceanic Micropalaeontology, Volume 2* (ed. A.T.S. Ramsay), 1009-1046. Academic Press Inc., New York.

Schneidermann N. and Harris P.M. (1985) *Carbonate Cements*, Society of Economic Paleontologists and Mineralogists, Tulsa, OK. 379 pp.

Schoffin, T.P. (1987) *An introduction to carbonate sediments and rocks*. Chapman and Hall, N.Y. 274 pp.

Scholle P.A. (1977) Chalk diagenesis and its relation to petroleum exploration: Oil from chalk, a modern miracle? *AAPG Bull.* **61**, 982-1009.

Scholle P.A. and Halley R.B. (1985) Burial diagenesis: Out of sight, out of mind. In *Carbonate Cements* (eds. N. Schneidermann and P.M. Harris), pp. 309-334. Society Economic Paleontologists and Mineralogists Special Publication **36**, Tulsa, OK.

Scholle P.A., Bebout D.G. and Moore C.H. (1983) *Carbonate Depositional Environments*. American Association of Petroleum Geologists, Tulsa, OK., 708 pp.

Schoonmaker J.E. (1981) Magnesian calcite-seawater reactions solubility and recrystallization behavior. Ph.D. dissertation, Northwestern Univ., Evanston, IL.

Schoonmaker J., Mackenzie F.T. and Garrels R.M. (1982) Behavior of calcite in seawater: experimental evidence for magnesian calcite overgrowths (abstract). *Geol. Soc. Amer. Ann. Mtg.* **14**, 612.

Searl A. (1988) The limitations of "cement stratigraphy" as revealed in some Lower Carboniferous oölites from south Wales. *Se. Geol.* **57**, 171-183.

Seiler W. (1974) The cycle of atmospheric CO. *Tellus* **27**, 116-135.

Senum G.L. and Gaffney J.S. (1985) A reexamination of the tropospheric methane cycle: Geophysical implications. In *The Carbon Cycle and Atmospheric CO_2: Natural variations Archean to Present* (eds. E.T. Sundquist and W.S. Broecker), pp. 61-69. AGU Geophys. Monograph **32**, Washington, D.C.

Seyfried W.E., Jr. (1976) Seawater-basalt interaction from 25-300 and 1-500 bars: Implications for the origin of submarine metal-bearing hydrothermal solutions and regulation of ocean chemistry. Ph.D. dissertation, Univ. Southern California.

Shackleton N.J. (1985) Oceanic carbon isotope constraints on oxygen and carbon dioxide in the Cenozoic atmosphere. In *The Carbon Cycle and Atmospheric CO_2: Natural Variation Archean to Present* (eds. E.T. Sundquist and W.S. Broecker), pp. 412-417. AGU Geophys. Monograph **32**, Washington, D.C.

Shackleton N.J. and Pisias N.G. (1985) Atmospheric carbon dioxide, orbital forcing and climate. In *The Carbon Cycle and Atmospheric CO_2: Natural Variation Archean to Present* (eds. E.T. Sundquist and W.S. Broecker) pp. 303-317. Geophysical Monograph **32**, American Geophysical Union, Washington, D.C.

Shannon R.D. (1976) Revised effective ionic radii and systematic studies of interatomic distances in halides and chalcogenides. *Acta Crystallog.* **A32**, 751-767.

Shaver G.R. and Woodwell G.M. (1983) Changes in the carbon content of terrestrial biota and soils between 1860 and 1980: A net release of CO_2 to the atmosphere. *Ecological Monograph* **53**, 235-262.

Shaw D.B. and Weaver C.E. (1965) The mineralogical composition of shales. *J. Sediment. Petrol.* **35**, 213-222.

Shen G.T. and Boyle E.A. (1988) Determination of lead, cadmium and other trace metals in annually-banded corals. *Chem Geol.* **67**, 47-62.

Shiller A.M. and Gieskes J.M. (1980) Processes affecting the oceanic distributions of dissolved calcium and alkalinity. *J. Geophys. Res.* **85**, 2719-2727.

Shinn E.A. (1969) Submarine lithification of Holocene carbonate sediments in the Persian Gulf. *Sedimentology* **12**, 109-144.

Shinn E.A., Halley R.B., Hudson J.H. and Lidz B.H. (1977) Limestone compaction: An enigma. *Geology* **5**, 21-24.

Shinn E.A., Steinen R.P., Lidz B.H. and Swart P.K. (1989) Whitings, a sedimentologic dilemma. *J. Sediment. Petrol.* **59**, 147-161.

Sholkovitz E. (1973) Interstitial water chemistry of the Santa Barbara basin sediment. *Geochim. Cosmochim. Acta* **37**, 2043-2073.

Short F.T., Davis M.W., Gibson R.A. and Zimmerman C.F. (1985) Evidence for phosphorous limitation in carbonate sediments of the seagrass *Syringodium filiforme*. *Estuarine Coastal Shelf Sci.* **20**, 419-430.

Sibley D.F., Dedoes R.E. and Bartlett T.R. (1987) Kinetics of dolomitization. *Geology* **15**, 1112-1114.

Siegenthaler U. and Oeschger H. (1987) Geospheric CO_2 emissions during the past 200 years, reconstructed by deconvolution of ice core data. *Tellus* **39**, 140-154.

Siever R. (1979) Plate-tectonic controls on diagenesis. *J Geol.* **87**, 127-155.

Silliman B. (1846) On the chemical composition of the calcareous corals. *Amer. J. Sci.* **1**, 189-199.

Simmons J.A.K. (1983) The biogeochemistry of the Devonshire Lens, Bermuda. M.S. thesis, Univ. of New Hampshire.

Simmons J.A.K. (1987) Major and minor ion geochemistry of groundwaters from Bermuda. Ph.D. dissertation, Univ. of New Hampshire.

Simone L. (1981) Oöids: a review. *Earth-Science Reviews* **16**, 319-355.

Sjoberg E.L. (1978) Kinetics and mechanism of calcite dissolution in aqueous solutions at low temperatures. Acta Univ. Stockholmiensis, Stockholm Contr. *Geology* **332**, 1-92.

Sleep N.H. (1976) Platform subsidence mechanisms and eustatic sea-level change. *Tectonophysics* **36**, 45-56.

Sloss L.L. (1963) Sequences in the cratonic interior of North America. *Bull. Geol. Soc. Amer.* **74**, 93-114.

Sloss L.L. (1976) Areas and volumes of cratonic sediments, Western North America and Eastern Europe. *Geology* **4**, 272-276.

Sloss L.L. (1988a) Forty years of sequence stratigraphy. *Geol. Soc. Amer. Bull.* **100**, 1661-1665.

Sloss L.L. (1988b) Tectonic evolution of the craton in Phanerozoic time. In *Sedimentary Cover of the Craton: U.S.* (ed. L.L. Sloss), pp. 25-52. Geol. Soc. America, Geology of North America D-2.

Sloss L.L. and Speed R.C. (1974) Relationships of cratonic and continental margin tectonic episodes. In *Special Publication Soc. Econ. Paleontologists and Mineralogists* **22**, 98-119.

Smalley P.C., Raiheim A., Dickson J. A.D. and Emery D. (1988) [87]Sr/[86]Sr in waters from the Lincolnshire Limestone aquifer, England, and the potential of natural strontium isotopes as a tracer for a secondary recovery seawater injection process in oilfields. *Applied Geochem.* **3**, 591-600.

Smart, P.I., Dawans, J.M. and Whitaker, F. (1988) Carbonate dissolution in a modern mixing zone. *Nature* **335**, 811-813.

Smith A.D. and Roth A.A. (1979) Affect of carbon dioxide concentration on calcification in the red coralline alga <u>Bossiella orbigniana</u>. *Mar. Biol.* **52**, 217-225.

Smith R.M. and Martel A.E. (1976) *Critical Stability Constants, Vol. 4.* Plenum, 37 pp.

Smith S.D. and Jones E.P.(1985) Evidence for wind-pumping of the air-sea gas exchange based on direct measurements of CO_2 fluxes. *J. Geophys. Res.* **90**, 869-875.

Smith S.D. and Jones E.P.(1986) Isotopoc and micrometeorological ocean CO_2 fluxes: different time and space scales. *J. Geophys. Res.* **91**, 10,529-10,532.

Smith S.V. (1978) Coral reef area and contributions of reefs to processes and resources of the world's oceans. *Nature* **273**, 225-226.

Smith S.V. (1985) Physical, chemical and biological characteristics of CO_2 gas flux across the air-water interface. *Plant. Cell Environ.* **8**, 387-398.

Smith S.V. and Mackenzie F.T. (1987) The ocean as a net heterotrophic system: Implications from the carbon biogeochemical cycle. *Global Biogeochem. Cycles* **1**, 187-198.

Smith S.V., Buddemeier R.W., Redaije R.D. and Houck J.E. (1979) Strontium-calcium thermometry in coral skeletons. *Science* **204**, 404-407.

Smith S.V., Dygas J.A. and Chave K.E. (1968) Distribution of calcium carbonate in pelagic sediments. *Mar. Geol.* **6**, 391-400.

Solomon A.M., Trabalka J.R., Reichle D.E. and Voorhees L.D. (1985) The global cycle of carbon. In *Atmospheric Carbon Dioxide and the Global Carbon Cycle* (ed. J.R. Trabalka), pp. 3-24. U.S. Dept. Energy, DOE/ER-0239, Washington D.C.

Somasundaran P. and Agar G.E. (1967) The zero point of charge of calcite. *J. Colloid Interface Sci.* **24**, 433-440.

Sorby H.C. (1863) The Bakerian lecture. On the direct correlation of mechanical and chemical forces. *Roy. Soc. Lond. Proc.* **12**, 538-550.

Sorby H.C. (1879) The structure and origin of limestones. *Proc. Geolog. Soc. London* **35**, 56-95.

Southam J.R. and Hay W.W. (1981) Global sedimentary mass balance and sea level changes. In *The Oceanic Lithoshpere, The Sea, Volume 7* (ed. C. Emiliani), pp. 1617-1684. John Wiley, New York.

Speer J.A. (1983) Crystal chemistry and phase relations of orthorhombic carbonates. In *Reviews in Mineralogy: Carbonates - Mineralogy and Chemistry* (ed. R. J. Reeder), pp. 145-189. Mineralogical Society of America. Bookcrafters, Inc., Chelse, MI.

Sperber C.M., Wilkinson B.H. and Peacor D.R. (1984) Rock composition, dolomite stoichiometry, and rock/water reactions in dolomitic carbonate rocks. *J. Geol.* **92**, 609-622.

Spotts J.H. and Silverman S.R. (1966) Organic dolomite from Point Fermin, California. *Amer. Mineralogist* **51**, 1144-1155.

Stallard R.F. (1980) Major elements chemistry of the Amazon river system. Ph.D. dissertation, MIR-WHOI, Woods Hole Inst. Oceanography.

Stehli F.G. and Hower J. (1961) Mineralogy and early diagenesis of carbonate sediments. *J. Sediment. Petrol.* **31**, 358-371.

Steinen R.P. (1974) Phreatic and vadose diagenetic modification of Pleistocene limestone: Petrographic observations from subsurface of Barbados, West Indies. *Amer. Assoc. Petrol. Geol. Bull.* **58**, 1008-1024.

Steinen R.P. and Matthews R.K. (1973) Phreatic vs vadose diagenesis: stratigraphy and mineralogy of a cored bore hole on Barbados, W.I. *J. Sediment. Petrol.* **43**, 1012-1020.

Stoessel R.K. (1987) Chemistry and environments of dolomitization - A reappraisal (discussion). *Earth-Science Reviews* **24**, 211-215.

Stoessell R.K. and Moore C.H. (1983) Chemical constraints and origins of four groups of Gulf Coast Reservoir fluids. *AAPG Bull.* **67**, 896-906.

Stoessell R.K., Ward W.C., Ford B.H. and Schuffert J.D. (1989) Water chemistry and $CaCO_3$ dissolution in the saline part of the open-flow mixing zone, coastal Yucatan Peninsula, Mexico. *Geol. Soc. Amer. Bull.* **101**, 159-169.

Stommel H. (1957) A survey of ocean current theory. *Deep-Sea Res.* **4**, 149-184.

Stumm W. (1987) *Aquatic Surface Chemistry*. John Wiley and Sons, New York, NY., 520 pp.

Stumm W. and Leckie J.O. (1970) Phosphate exchange with sediments: Its role in the productivity of surface waters. In *Advances in Water Pollution Research, Vol. 2.* pp. 26/1-26/16. Pergamon Press, Oxford.

Suess E. and Fütterer D. (1972) Aragonitic oöids: experimental precipitation from seawater in the presence of humic acid. *Sedimentology* **19**, 129-139.

Suess E. and von Huene R. (1988) Ocean drilling program Leg 112, Peru continental margin: Part II, Sedimentary history and diagenesis in a coastal upwelling environment. *Geology* **16**, 939-943.

Suess E. and Whiticar M.J. (1989) Methane-derived CO_2 in pore fluids expelled from the Oregon subduction zone. *Paleogeog., Paleoclimatol. Paleoecol.* **71**, (in press).

Suess E. Balzer W., Heese K.-F., Müller P.J., Ungerer C.A. and Wefer G. (1982) Calcium carbonate hexahydrate from organic-rich sediments of the Antarctic shelf: Precursursors to glendonites. *Science* **216**, 1128-1130.

Sundquist E.T. (1985) Geological perspectives on carbon dioxide and the carbon cycle. In *The Carbon Cycle and Atmospheric CO_2: Natural Variations Archean to Present* (eds. E.T. Sundquist and W.S. Broecker), pp. 5-59. AGU Geophys. Monograph **32**, Washington, D.C.

Sundquist E.T. and Broecker W.S. (1985) *The Carbon Cycle and Atmospheric CO_2: Natural Variations Archean to Present*. Amer. Geophys. Union, Washington, D.C., 627 pp.

Sundquist E.T., Plummer L.N. and Wigley T.M.L. (1979) Carbon dioxide in the surface ocean: The homogeneous buffer factor. *Science* **204**, 1203-1204.

Surdam R.C. and Crossey L.J. (1985) Organic-rich reactions during progressive burial: Key to porosity/permeability enchancement and/or preservation. Phil. Trans. Roy. Soc. London, Series A **315**, 135-156.

Surdam R.C., Boese S.W. and Crossey L.J. (1984) The chemistry of secondary porosity. In *Clastic Diagenesis* (eds. D.A. McDonald and R.C. Surdam), pp. 127-149. AAPG, Tulsa, OK.

Sverdrup H.U., Johnson M.W. and Fleming R.H. (1942) *The Oceans*. Prentice-Hall, Inc., Englewood Cliffs, NJ 1087 pp.

Sverjensky D.A. (1984) Prediction of Gibbs free energies of calcite-type carbonates and the equilibrium distribution of trace elements between carbonates and aqueous solutions. *Geochim. Cosmochim. Acta* **48**, 1127-1134.

Swart P.K. (1983) Carbon and oxygen isotope fractionation in scleractinian corals: A review. *Earth-Sci. Rev.* **19**, 51-80.

Swart P.K. and Hubbard J.A.E.B. (1982) Uranium in scleractinian coral skeletons. *Coral Reefs* **1**, 13-19.

Symes J.L. and Kester D.R. (1984) Thermodynamic solubility studies of the basic copper carbonate mineral, malachite. *Geochim. Cosmochim. Acta* **48**, 2219-2229.

Tada R. and Siever R. (1986) Experimental knife-edge pressure solution of halite. *Geochim. Cosmochim. Acta*, **50**, 29-36.

Taft W.H. (1967) Physical chemistry of formation of carbonates. In *Carbonate Rocks, pt. B* (eds. G.V. Chilingar, H.J. Bissell and R.W. Fairbridge), pp. 151-168. Elsevier Publishing Co., Amsterdam.

Taft W.H., Arrington F., Haimovitz A., MacDonald C. and Woolheater C. (1968) Lithification of modern carbontae sediments at Yellow Bank, Bahamas. *Mar. Sci. Gulf Caribbean Bull.* **18**, 762-828.

Takahashi K. and Be A.W.H. (1984) Planktonic foraminifera: factors controlling sinking speeds. *Deep-Sea Res.* **31**, 1477-1500.

Takahashi T. and Broecker W.S. (1977) Mechanisms for calcite dissolution on the sea floor. In *The Fate of Fossil Fuel CO2 in the Oceans* (eds. N.R. Anderson and A. Malahoff), pp. 455-477. Plenum Press, New York.

Takahashi T., Broecker W.S., Werner S.R. and Bainbridge A.E. (1980a) Carbonate chemistry of the surface waters of the world oceans. In *Isotope Marine Chemistry* (eds. E.D. Goldberg, Y. Horibe and K. Saruhashi). Uchida Rokakuho Publishing Company.

Takahashi T., Broecker W.S., Bainbridge A.E. and Weiss R.F. (1980b) Carbonate Chemistry of the Atlantic, Pacific and Indian Oceans: The Results of the GEOSECS Expeditions, 1972-1978. Technical Report No. 1, CV-1-80, Lamont-Doherty Geological Observatory, Palisades, New York, 15 pp.

Takahashi T., Chipman D. and Volk T. (1983) Geographical, seasonal and secular variations of the partial pressure of CO2 in surface waters of the North Atlantic Ocean: The results of the TTO Program. In *Carbon Dioxide Science and Consensus (CONF-820970)*, II.123-II.146, U.S. Dept. of Energy, Washington , D.C. (available from NTIS, Sprinfield, VA.).

Tarutani T., Clayton R.N. and Mayeda T.K. (1969) The effects of polymorphism and magnesium substitution on oxygen isotope fractionation between calcium carbonate and water. *Geochim. Cosmochim. Acta* **33**, 987-996.

Terjesen S.G., Erga O., Thorsen G. and Ve A. (1961) II. Phase boundary processes as rate determining steps in reactions between solids and liquids. *Chem. Engn. Sci.* **74**, 277-288.

Thode H.G., Shima M., Rees C.E. and Krishnamurty K.V. (1965) Carbon-13 isotope effects in systems containing carbon dioxide, bicarbonate, carbonate, and metal ions. *Can. J. Chem.*. **43**, 582-595.

Thompson G. (1983a) Basalt-seawater interaction. In *Hydrothermal Processes at Seafloor Spreading Centers* (eds. P.A. Rona, K. Bostrom, L. Laubjer and K.L. Smith), pp. 225-278. Plenum Press, New York.

Thompson G. (1983b) Hydrothermal fluxes in the ocean. In *Chemical Oceanography* (eds. P. Riley and R. Chester), pp. 272-338. Academic Press, New York.

Thompson T.G. and Chow T.J. (1955) The strontium calcium atom ratio on carbonate-secreting marine organisms. In *Papers in Marine Biology and Oceanography, Deep-Sea Res. Suppl.*, **3**, 20-39.

Thomson J. (1862) On crystallization and liquefaction, as influenced by stresses tending to change of form in crystals. *Phil. Mag.* **24**, 395-401.

Thortenson D.C. and Mackenzie F.T. (1974) Time variability of pore water chemistry in recent carbonate sediments, Devil's Hole, Harrington Sound. Bermuda. *Geochim. Cosmochim. Acta* **38**, 1-19.

Thorstenson D.C. and Plummer L.N. (1977) Equilibrium criteria for two component solids reacting with fixed composition in an aqueous phase -- example: The magnesian calcites. *Amer. J. Sci.* **277**, 1203-1223.

Thorstenson D.C., Mackenzie F.T. and Ristvet B.L. (1972) Experimental vadose and phreatic cementation of skeletal carbonate sand. *J. Sediment. Petrol.* **42**, 162-167.

Thrailkill J. and Robl T.L. (1981) Carbonate geochemistry of vadose water recharging limestone aquifers. *J. Hydrology* **54**, 195-208.

Thunell R.C. (1982) Carbonate dissolution and abyssal hydrography in the Atlantic Ocean. *Mar. Geol.* **47**, 165-180.

Thunell R.C. and Honjo S. (1981) Planktonic foraminiferal flux to the deep oceanic sediment trap results from the tropical Atlantic and the central Pacific. *Mar. Geol.* **40**, 237-253.

Trabalka J.R. (1985) *Atmospheric Carbon Dioxide and the Global Carbon Cycle.* USDOE Rept. ER-0239, Office Sci. Tech. Inform., Oak Ridge, TN, 314 pp.

Tribble G.W. (1990) Early diagenesis in a coral reef framework. PhD. dissertation. Univ. Hawaii at Manoa, Honolulu, HI.

Tribble G.W., Sansone F.J. and Smith, S.V. (1988) Material fluxes from a reef framework. *Proc. 6th Internatl. Coral Reef Cong.* **2**, 577-582.

Tribble G.W., Sansone F.J. and Smith, S.V. (1990) Stoichiometric modeling of carbon diagenesis within a coral reef framework. *Geochim. Cosmochim. Acta*, (submitted).

Trichet J. (1968) Étude de la composition de la fraction organique des oolites. Comparaison aree celle des membranes des bactéries et des cyanophycé es. *Compt. Rend.* **267**, 1492-1494.

Tsuesue A. and Holland H.D. (1966) The coprecipitation of cations with $CaCO_3$-- III. The coprecipitation of zinc with calcite between 50 and 250°C. *Geochim. Cosmochim. Acta* **30**, 439-453.

Tsunogai S. and Watanabe Y. (1981) Calcium in the North Pacific water and the effect of organic matter on the calcium-alkalinity relation. *Geochem. Jour.* **15**, 95-107.

Tucker M.E. (1984) Calcitic, aragonitic and mixed calcitic-aragonitic oöids from the mid-Proterozoic Belt Supergroup, Montana. *Sedimentology* **31**, 627-644.

Turcott D.L. and Burke K. (1978) Global sea-level changes and the thermal structure of the Earth. *Earth Planet. Sci. Lett.* **41**, 341-356.

Turekian K.K. (1964) The geochemistry of the Atlantic Ocean basin. *Trans. of the New York Acad. Sci.* **26**, 312-330.

Turnbull A.G. (1973) A thermochemical study of vaterite. *Geochim. Cosmochim. Acta* **37**, 1593-1601.

Turner J.V. (1982) Kinetic fractionation of carbon-13 during calcium carbonate precipitation. *Geochim. Cosmochim. Acta* **46**, 1183-1191.

Urmos J., Sharma S.K. and Mackenzie F.T. (1986) Characterization of biomineral with Raman spectroscopy. EOS **67**,1721.

US/USSR Workshop (1981) Report of US/USSR workshop on the Climatic Effects of Increased Atmospheric Carbon Dioxide. Project 02.08-11, US/USSR Agreement on Protection of the Environment, Leningrad, 99 pp.

Vacher H.L. (1971) Late Pleistocene sea-level history: Bermuda evidence. Ph.D thesis, Northwestern Univ., Evanston, IL.

Vacher H.L. (1973) Coastal dunes of younger Bermuda. In *Coastal Geomorphology* (ed. D.R. Coates), pp.335-391. State Univ. of New York, Binghamtonk, N.Y.

Vacher H.L. (1974) *Groundwater Hydrology of Bermuda..* Government of Bermuda, Public Works Dept., Hamilton, Bermuda, 87 pp.

Vacher H.L. (1978) Hydrogeology of Bermuda -- Significance of an across-the island variation in permeability. *J. Hydrol.* **39**, 207-226.

Vacher, H.L. and Ayers, J.F. (1980) Hydrology of small oceanic islands -- utility of an estimate of recharge inferred from the chloride concentration of the freshwater lenses. *Jour. Hydrol.* **45**, 21-37.

Vacher H.L. and Harmon R.S. (1987) *Penrose Conference Field Guide to Bermuda Geology*, Bermuda Biological Station for Research, St. Georges, Bermuda, 50 pp.

Vacher H.L. and Hearty P. (1989) History of stage-5 sea level in Bermuda: Review with new evidence of a brief rise to present sea level during substage 5A. *Quaternary Sci. Rev.* **8**, 159-168.

Vacher H.L., Bengtsson T.O. and Plummer L.N. (1989) Hydrology of meteoric diagenesis - Ground-water residence times and aragonite-calcite stabilization rates in island fresh-water lenses. *Geol. Soc. Amer. Bull.* (in press).

Vacher H.L., Rowe M. and Garrett P. (1989) Geologic Map of Bermuda. Public Works Dept. Bermuda, Hamilton, Bermuda.

Vail P.R., Mitchum R.W. and Thompson S. (1977) Seismic stratigraphy and global changes of sea level. 4, Global cycles of relative chambes of sea level. *AAPG Mem.* **26**, 83-97.

van Andel T.J.H. (1975) Mesozoic/Cenozoic calcite compensation depth and the global distribution of calcareous sediments. *Earth Planet. Sci. Lett.* **26**, 187-194.

Van Houten F.B. and Bhattacharyya (1982) Phanerozoic oölitic ironstones-geologic record and facies. *Ann. Rev. Earth Plan. Sci.* **10,** 441-458.

Veizer J. (1973) Sedimentation in geologic history: recycling vs. evolution or recycling with evolution. *contrib. Mineral. Petrol.* **38,** 261-278.

Veizer J. (1978) Secular variations in the composition of sedimentary carbonate rocks, II. Fe, Mn, Ca, Mg, Si, and minor constituents. *Precambrian Research* **6,** 381-413.

Veizer J. (1983) Trace elements and isotopes in sedimentary carbonates. In *Carbonates: Mineralogy and Chemistry, Reviews in Mineralogy, 11* (ed. R.J. Reeder), pp. 265-300. Mineral. Soc. America, Book Crafters, Incorporated, Chelsea, MI.

Veizer J. (1985) Carbonates and ancient oceans: Isotopic and chemical record on time scales of 10^7 - 10^9 years. In *The Carbon Cycle and Atmospheric CO_2; Natural Variations, Archean to Present* (eds. E.T. Sundquist and W.S. Broecker), pp. 595-601. Geophys. Monographs, Washington, D.C.

Veizer J. (1988) The evolving exogenic cycle. In *Chemical Cycles in the Evolution of the Earth* (eds. C.B. Gregor, R.M. Garrels, F.T. Mackenzie and J.B. Maynard), pp. 175-261. John Wiley and Sons, New York.

Veizer J. (1989) Strontium isotopes in seawater through time. In *Annual Reviews Earth and Planet. Sci.* **17** (ed. G.W. Weatherill), pp. 141-167. Annual Reviews Inc., Palo Alto, California.

Veizer J. and Garrett D.E. (1978) Secular variations in the composition of sedimentary rocks, I. Alkali metals. *Precambrian Research* **6,** 367-380.

Veizer J. and Hoefs J. (1976) The nature of $^{18}O/^{16}O$ and $^{13}C/^{12}C$ secular trends in sedimentary carbonate rocks. *Geochim. Cosmochim. Acta* **40,** 1387-1395.

Veizer J. and Jansen S.L. (1979) Basement and sedimentary recyling and continental evolution. *J. Geol.* **87,** 341-370.

Veizer J. and Jansen S.L. (1985) Basement and sedimentary recycling-2: Time dimension to global tectonics. *J. Geol.* **93,** 625-643.

Veizer J., Holser W.T. and Wilgus C.K. (1980) Correlation of $^{13}C/^{12}C$ and $^{34}S/^{32}S$ secular variations. *Geochim. Cosmochim. Acta* **44,** 579-587.

Veizer J., Fritz P. and Jones B. (1986) Geochemistry of brachiopods: Oxygen and carbon isotopic records of Paleozoic oceans. *Geochem. Cosmochim. Acta* **50,** 1679-1696.

Verrill A.E. (1907) The Bermuda Islands, Part IV, Geology and paleontology, and Part V, an account of the coral reefs. *Connecticut Acad. Arts Aci. Trans.* **12,** 45-348.

Videtich P.E. (1985) Electron microprobe study of Mg distribution in recent Mg calcites and recrystallized equivalents from the Pleistocene and Tertiary. *J. Sediment. Petrol.* **55,** 421-429.

Vinogradov A.P. (1959) *The Geochemistry of Rare and Dispersed Chemical Elements in Soils.* Consult Bur., New York, 209 pp.

Vinogradov A.P. and Ronov A.B (1956a) Composition of the sedimentary rocks of the Russian platform in relation to the history of its tectonic movements. *Geochemistry* **6,** 533-559.

Vinogradov A.P. and Ronov A.B. (1956b) Evolution of the chemical composition of clays of the Russian platform. *Geochemistry* **2,** 123-129.

Volat J.L., Pastouret L. and Vergnaud-Grazzini C. (1980) Dissolution and carbonate fluctuations in Pleistocene deep-sea cores: A review. *Mar. Geol.* **34,** 1-28.

Volk T. (1989a) Rise of angiosperms as a factor in long-term climatic cooling. *Geology* **17,** 107-110.

Volk T. (1989b) Sensitivity of climate and atmosheric CO_2 to deep-ocean and shallow-ocean carbonate burial. *Nature* **337,** 637-640.

von Bubnoff S. (1954) *Fundamentals of Geology.* 3rd ed., Oliver and Lloyd Ltd., Edinburgh, 287 pp.

Von Damm K.L. (1983) Chemistry of submarine hydrothermal solutions at 21°N, East Pacific Rise and Guaymas Basin, Gulf of California. Ph.D. dissertation, Massachusetts Institute of Technology, Cambridge, MA.

Vonnegut K., Jr. (1967) *Slaughterhouse-Five or the Childrens Crusade.* Delta Dell Publishing Co., Inc., New York, 186 pp.

Wagner P.D. and Matthews R.K. (1982) Porosity preservation in the upper Smackover (Jurassic) carbonate grainstone, Walker Creek field, Arkansas: Response of Paleophreatic lenses to burial processes. *J. Sediment. Petrol.* **52,** 3-18.

Walls R.A., Ragland P.C. and Crisp. E.L. (1977) Experimental and natural early diagenetic mobility of Sr in Mg in biogenic carbonates. *Geochim. Cosmochim. Acta* **41,** 1731-1737.

Walker J.C.G. (1977) *Evolution of the Atmosphere.* MacMillan Publishing Co., New York, 318 pp.

Walter L.M. (1983) The dissolution kinetics of shallow water carbonate grain types: Effects of mineralogy, microstructure and dolution chemistry. Ph.D. dissertation, Univ. Miami, Miami, FL.

Walter L.M. (1985) Relative reactivity of skeletal carbonates during dissolution: implications for diagenesis. In *Carbonate Cements* (eds. N. Schneidermann and P.M. Harris), pp. 3-16. Soc. Econ. Paleontologist and Mineralogists, Spec. Pub. **2.** Tulsa, OK.

Walter L.M. (1986) Relative efficiency of carbonate dissolution and precipitation during diagenesis: A progress report on the role of solution chemistry. *J. Sediment. Petrol.* **56,** 1-11.

Walter L.M. and Burton E.A. (1986) The effect of orthophosphate on carbonate mineral dissolution rates in seawater. *Chem. Geol.* **56**, 313-323.

Walter L.M. and Burton E.A. (1987) Active dissolution in modern shallow marine carbonate sediments: Global implications. Geological Society of America Annual Meeting and Exposition, Phoenix Arizona, 880.

Walter L.M. and Hanor J.S. (1979) Effect of orthophosphate on the dissolution kinetics of biogenic magnesian calcites. *Geochim. Cosmochim. Acta* **43**, 1377-1385.

Walter L.M. and Morse J.W. (1984a). Magnesian calcite solubilities: A reevaluation. *Geochim. Cosmochim. Acta* **48**, 1059-1069.

Walter L.M. and Morse J.W. (1984b) Reactive surface area of skeletal carbonates during dissolution: Effect of grain size. *J. Sediment. Petrol.* **54**, 1081-1090.

Walter L.M. and Morse J.W. (1985) The dissolution kinetics of shallow marine carbonates in seawater: A laboratory study. *Geochim. Cosmochim. Acta* **49**, 1503-1513.

Wangersky P.J. and Joensuu O. (1964) Strontium, magnesium and manganese in fossil foraminiferal carbonates. *J. Geol.* **72**, 477-483.

Wanless H.R. (1979) Limestone response to stress: Pressure solution and dolomitization. *J. Sediment. Petrol.* **49**, 437-462.

Waples D.W. (1980) Time and temperature in petroleum formation: Application of Lopatin's method to petroleum exploration. *AAPG Bull.* **64**, 916-926.

Ward G.K. and Millero F.J. (1974) The effect of pressure on the ionization of boric acid in aqueous solutions from molal volume data. *J. Sol. Chem.* **3**, 417-430.

Wardlaw N., Oldershaw A. and Stout M. (1978) Transformation of aragonite to calcite in a marine gastropod. *Can. J. Earth Sci.* **15**, 1861-1866.

Watabe N. (1981) Crystal growth of calcium carbonate in the invertebrates. *Prog. Crystal Growth Charact.* **4**, 99-147.

Weast R.C. (ed.) (1985) *The Handbook of Chemisty and Physics 1984-1985.* 65th edition. CRC Press Inc., Boca Raton, FL.

Weaver C.E. (1967) Potassium, Illite and the Ocean. *Geochim. Cosmochim. Acta* **31**, 2181-2196.

Weber J.N. (1973) Incorporation of strontium into reef coral skeletal carbonates. *Geochim. Cosmochim. Acta* **37**, 2173-2190.

Weber J.N. and Kaufman J.W. (1965) Brucite in the calcareous algae Gonolithon. *Science* **149**, 996-997.

Weil S.M., Buddemeir R.W., Smith S.V. and Kroopnick P.M. (1981) The stable isotopic composition of coral skeletons: Control by environmental variables. *Geochim. Cosmochim. Acta* **45**, 1147-1153.

Weiss R.F. (1974) Carbon dioxide in water and seawater: The solubility of a non-ideal gas. *Mar. Chem.* **2**, 203-215.

Wells A.J. and Illing L.V. (1964) Present-day precipitation of calcium carbonate in the Persian Gulf. In *Delataic and Shallow Water Marine Deposits* (ed. van Straaten) pp. 429-435. Elsevier Developments in Sedimentology Series, Elsevier, New York.

Wethered E. (1890) On the occurence of the genus Girranella in oölitic rocks and remarks on oolitic structure. *Quart. J. Geol. Soc. London* **46**, 270-281.

Wetzel A. (1989) Influence of heat flow on ooze/chalk cementation: Quantification from consolidation parameters in DSDP sites 504 and 505 sediments. *J. Sediment. Petrol.* **59**, 539-547.

Weyl P.K. (1959) Pressure solution and the force of crystallization--a phenomenological theory. *J. Geophys. Res.* **64**, 2001-2025.

Weyl P.K. (1965) The solution behavior of carbonate materials in seawater. EPR Publ., 428, Shell Development Co., 1-59.

Weyl P.K. (1967) The solution behavior of carbonate materials in sea water. Int'l. Conf. Tropical Oceanography, Univ. Miami, Florida., 178-228.

White A.F. (1977) Sodium and potassium coprecipitation in aragonite. *Geochim. Cosmochim. Acta* **41**, 613-625.

White A.F. (1978) Sodium coprecipitation in calcite and dolomite. *Chem. Geol.* **23**, 65-72.

White D.E. (1965) Saline waters of sedimentary rocks. In *Fluids in Subsurface Environments* (eds. A. Young and J.E. Galley), pp. 342-366. AAPG Memoir **4**.

White D.E., Hem J.D. and Waring G.A. (1963) Chemical composition of subsurface waters: Data of Geochemistry, 6th ed., U. S. Geol. Surv. Profess. Paper, 440-F, Washington, D.C.

White W.B. (1974) The carbonate minerals. In *The Infra-red Spectra of Minerals* (ed. V.C. Farmer), pp. 227-284, Mineralogical Society of America Monograph **4**, Mineralogical Society, London.

Whitfield M. (1975) Seawater as an electrolyte solution. In *Chemical Oceanography*, Vol. I (eds. P. Riley and G. Skirrow), pp. 44-172, Academic Press, London.

Whitfield M. (1979) Activity coefficients in natural waters. In *Activity Coefficients in Electrolyte Solutions,* Vol. II (ed. R.M. Pytkowixz), pp.153-300, CRC Press, Boca Raton, FL.

Wildeman T.R. (1969) The distribution of Mn^{2+} in some carbonates by electron paramagnetic resonance. *Chem. Geol.* **5**, 167-177.

Wilkinson B.H. (1979) Biomineralozation, paleoceanography, and the evolution of calcareous marine organisms. *Geology* **7**, 524-527.

Wilkinson B.H. and Algeo, T.J. (1989) Sedimentary carbonate record of calcium-magnesium cycling. *Amer. J. Sci.* **289,** 1158-1194.

Wilkinson B.H. and Given R.K. (1986) Secular variation in abiotic marine carbonates: constraints on Phanerozoic atmospheric carbon dioxide contents and oceanic Mg/Ca ratios. *J. Geol.* **94,** 321-334.

Wilkinson B.H. and Walker J.C.G. (1989) Phanerozoic cycling of sedimentary carbonate. *Amer. J. Sci.* **289,** 525-548.

Wilkinson B.H., Bucznski C. and Owen R.M. (1984) Chemical control of carbonate phases: implications from upper Pennsylvanian calcite-aragonite oöids of southeastern Kansas. *J. Sediment. Petrol.* **54,** 932-947.

Wilkinson B.H., Owen R.M. and Carroll A.R. (1985) Submarine hydrothermal weathering, global eustasy and carbonate polymorphism in Phanerozoic marine oolites. *J. Sediment. Petrol.* **55,** 171-183.

Williams J. (ed.), (1978) *Carbon Dioxide, Climate and Society.* Pergamon Press, New York, 312 pp.

Wilson J.L. (1975) *Carbonate Facies in Geologic History.* Springer-Verlag, New York, 471 pp.

Wilson R.C.L. (1967) Particle nomenclature in carbonate sediments. *N. Jb. Geol. Paläont. Mh.,* 498-510.

Windom H.L. (1976) Lithogenous material in marine sediments. In *Chemical Oceanography Vol. 5,* 2nd ed (eds. J.P. Riley and R. Chester) , pp. 103-136. Academic Press, New York.

Winland H.D. (1969) Stability of calcium carbonate polymorphs in warm, shallow seawater. *J. Sediment. Petrol.* **39,** 1579-1587.

Wise D.U. (1974) Continental margins, freeboard and the volumes of continents and oceans through time. In *The Geology of Continental Margins* (eds.K. Burke and C.L. Drake) pp. 45-58. Springer, Berlin/New York.

Wolery T.J. (1978) Some chemical aspects of hydrothermal processes at midoceanic ridges - A theoretical study, I. Basalt-seawater reaction and chemical cycling between the oceanic crust and the oceans, II. Calculation of chemical equilibrium between aqueous solutions and minerals. Ph.D. dissertation, Northwestern Univ. Evanston, IL.

Wolery T.J. and Sleep N.H. (1976) Hydrothermal circulation and geochemical flux at mid-ocean ridges. *J. Geol.* **84,** 249-275.

Wolery T.J. and Sleep N.H. (1988) Interactions of geochemical cycles with the mantle. In *Chemical Cycles in the Evolution of the Earth* (eds. C.B. Gregor, R.M. Garrels, F.T. Mackenzie and J. B. Maynard), pp. 77-104. John Wiley, New York.

Wollast R. (1971) Kinetic aspects of the nucleation and growth of calcite from aqueous solutions. In *Carbonate Cements* (ed. O.P. Bricker), pp. 264-273. . John Hopkins Univ. Press, Baltimore, MD.

Wollast R (1974) The silica problem. In *The Sea Vol 5*: (ed. E.D. Goldberg) pp. 359-392. Wiley Interscience, New York.

Wollast R. (1983) Interactions in estuaries and coastal waters. In *The Major Biogeochemical Cycles and Their Interactions* (eds. B. Bolin and R.E. Cook), pp. 385-410. J. Wiley and Sons, New York.

Wollast R. and Mackenzie F.T. (1983) Global cycle of silica. In *Silicon Geochemistry and Biogeochemistry* (ed. S.R. Aston), pp. 39-76. Academic Press, New York.

Wollast R. and Mackenzie F.T. (1989) Global biogeochemical cycles and climate. In *Climate and Geosciences* (ed. A. Berger), pp. 453-473. NATO ASI Series, D. Reidel Publishing Co., Dordrecht, Holland.

Wollast R. and Marijns A. (1980) Mecanisme de dissolution des calcites magnesiennes naturelles. Cong. "Cristallisation, Deformation, Dissolution des Carbonates." Bordeaux, Nov. 1980, 477-486.

Wollast R. and Reinhard-Derie D. (1977) Equilibrium and mechanism of dissolution of magnesium calcites. In *The Fate of Fossil Fuel CO$_2$ in the Oceans* (eds. N.R. Andersen and A. Malahoff), pp. 479-492. Plenum Press. New York.

Wollast R., Garrels R.M. and Mackenzie F.T. (1980) Calcite-seawater reactions in ocean waters. *Amer. J. Sci.* **280**, 831-848.

Wood J.R. (1987) A model for dolomitization by pore fluid flow. In *Physics and Chemistry of Porous Media II* (eds. J.R. Banavar, J. Koplik and K.W. Winkler), pp. 17-36. Amer. Inst. Physics Proceedings **154**, Amer. Inst. Physics, New York.

Woods T.L. (1988) Calculated solution solid relations in the low temperature system CaO-MgO-FeO-CO$_2$-H$_2$O. Ph.D. dissertation, Univ. South Florida.

Woodwell G.M. (1983) Biotic effects on atmospheric carbon dioxde: A review and projection in changing climate. National Academy of Sciences Press, Washington, D.C., 216-241.

Woronick R.E. and Land L.S. (1985) Late burial diagenesis lower Cretaceous Pearsall and lower Glen Rose formations, south Texas. In *Carbonate Cements* (eds. N. Schneidermann and P.M. Harris), pp. 265-275. Society Economic Paleontologists and Mineralogists Special Publication **36**, Tulsa, OK.

Worsley T.R., Nance R.D. and Moody J. B. (1984) Global tectonics and eustasy for the past two billion years. *Mar. Geol.* **58**, 373-400.

Zachos J.C., Arthur M.A. and Dean W.E. (1989) Geochemical evidence for suppression of pelagic marine productivity at the Cretaceous/Tertiary boundary. *Nature* **337,** 61-64.

Zenger D.H. (1989) Dolomite abundance and stratigraphic age: Constraints on rates and mechanisms of Phanerozoic dolostone formation. *J. Sediment. Petrol.* **59,** 162-164.

Zenger D.H. and Mazzullo S.J. (1982) *Dolomitization,* Benchmark Papers in Geology, Vol. 65, Hutchinson Ross Publishing. Stroudsburg, PA, 426 pp.

Zenger D.H., Dunham J.B. and Ethington R.L. (1980) Concepts and Models of Dolomitization. *Soc. Econ. Paleontologists and Mineralogists,* Spec. Pub. **28,** Tulsa, OK. 320 pp.

Zhong S. and Mucci A. (1989) Calcite and aragonite precipitation from seawater solutions of various salinities: Precipitation rates and overgrowth compositions. *Chem. Geol.* (in press).

Zimmerman H.B., Shackleton N.J., Backman J., Kent D.V., Baldauf J.G., Kaltenback A.J. and Morton A.C. (1981) History of plio-pleistocene climate in the northeastern Atlantic, deep sea drilling project hole 552A. DSDP **81,** 861-875.

Zullig J.J. and Morse J.W. (1988) Interaction of organic acids with carbonate mineral surfaces in seawater and related solutions I. Fatty acid adsorption. *Geochim. Cosmochim. Acta* **52,** 1667-1678.

INDEX

A single bar indicates a subentry and double bars a sub-subentry. Alphabetization is letter by letter of first significant word, inclusive of chemical terminology.

Printed and bound by CPI Group (UK) Ltd, Croydon, CR0 4YY

03/10/2024

01040331-0015